Lecture Notes in Artificial Intelligence 9924

Subseries of Lecture Notes in Computer Science

LNAI Series Editors

Randy Goebel
University of Alberta, Edmonton, Canada
Yuzuru Tanaka
Hokkaido University, Sapporo, Japan
Wolfgang Wahlster
DFKI and Saarland University, Saarbrücken, Germany

LNAI Founding Series Editor

Joerg Siekmann
DFKI and Saarland University, Saarbrücken, Germany

More information about this series at http://www.springer.com/series/1244

Petr Sojka · Aleš Horák
Ivan Kopeček · Karel Pala (Eds.)

Text, Speech, and Dialogue

19th International Conference, TSD 2016
Brno, Czech Republic, September 12–16, 2016
Proceedings

 Springer

Editors
Petr Sojka
Faculty of Informatics
Masaryk University
Brno
Czech Republic

Ivan Kopeček
Faculty of Informatics
Masaryk University
Brno
Czech Republic

Aleš Horák
Faculty of Informatics
Masaryk University
Brno
Czech Republic

Karel Pala
Faculty of Informatics
Masaryk University
Brno
Czech Republic

ISSN 0302-9743   ISSN 1611-3349 (electronic)
Lecture Notes in Artificial Intelligence
ISBN 978-3-319-45509-9   ISBN 978-3-319-45510-5   (eBook)
DOI 10.1007/978-3-319-45510-5

Library of Congress Control Number: 2016949127

LNCS Sublibrary: SL7 – Artificial Intelligence

Printed on acid-free paper

This Springer imprint is published by Springer Nature
The registered company is Springer International Publishing AG Switzerland

Preface

The annual Text, Speech, and Dialogue Conference (TSD), which originated in 1998, is approaching the end of its second decade. In the course of this time thousands of authors from all over the world have contributed to the proceedings. TSD constitutes a recognized platform for the presentation and discussion of state-of-the-art technology and recent achievements in the field of natural language processing. It has become an interdisciplinary forum, interweaving the themes of speech technology and language processing. The conference attracts researchers not only from Central and Eastern Europe but also from other parts of the world. Indeed, one of its goals has always been to bring together NLP researchers with different interests from different parts of the world and to promote their mutual cooperation.

One of the declared goals of the conference has always been, as its title says, twofold: not only to deal with language processing and dialogue systems as such, but also to stimulate dialogue between researchers in the two areas of NLP, i.e., between text and speech people. In our view, the TSD conference was successful in this respect in 2016 again. We had the pleasure to welcome three prominent invited speakers this year: Hinrich Schütze presented a keynote talk about a current hot topic, deep learning of word representation, under the title *Embeddings! For Which Objects? For Which Objectives?*; Ido Dagan's talk dealt with *Natural Language Knowledge Graphs*; and Elmar Nöth reported on *Remote Monitoring of Neurodegeneration through Speech*. Invited talk abstracts are attached below.

This volume contains the proceedings of the 19th TSD conference, held in Brno, Czech Republic, in September 2016. During the review process, 62 papers were accepted out of 127 submitted, an acceptance rate of 49 %.

We would like to thank all the authors for the efforts they put into their submissions and the members of the Program Committee and reviewers who did a wonderful job selecting the best papers. We are also grateful to the invited speakers for their contributions. Their talks provide insight into important current issues, applications, and techniques related to the conference topics.

Special thanks are due to the members of Local Organizing Committee for their tireless effort in organizing the conference.

We hope that the readers will benefit from the results of this event and disseminate the ideas of the TSD conference all over the world. Enjoy the proceedings!

July 2016

Aleš Horák
Ivan Kopeček
Karel Pala
Petr Sojka

Organization

TSD 2016 was organized by the Faculty of Informatics, Masaryk University, in cooperation with the Faculty of Applied Sciences, University of West Bohemia in Plzeň. The conference webpage is located at http://www.tsdconference.org/tsd2016/

Program Committee

Nöth, Elmar (General Chair) (Germany)
Agirre, Eneko (Spain)
Baudoin, Geneviève (France)
Benko, Vladimir (Slovakia)
Cook, Paul (Australia)
Černocký, Jan (Czech Republic)
Dobrisek, Simon (Slovenia)
Ekstein, Kamil (Czech Republic)
Evgrafova, Karina (Russia)
Fiser, Darja (Slovenia)
Galiotou, Eleni (Greece)
Garabík, Radovan (Slovakia)
Gelbukh, Alexander (Mexico)
Guthrie, Louise (UK)
Haderlein, Tino (Germany)
Hajič, Jan (Czech Republic)
Hajičová, Eva (Czech Republic)
Haralambous, Yannis (France)
Hermansky, Hynek (USA)
Hlaváčová, Jaroslava (Czech Republic)
Horák, Aleš (Czech Republic)
Hovy, Eduard (USA)
Khokhlova, Maria (Russia)
Kocharov, Daniil (Russia)
Konopík, Miloslav (Czech Republic)
Kopeček, Ivan (Czech Republic)
Kordoni, Valia (Germany)
Král, Pavel (Czech Republic)
Kunzmann, Siegfried (Germany)
Loukachevitch, Natalija (Russia)

Magnini, Bernardo (Italy)
Matoušek, Václav (Czech Republic)
Mihelić, France (Slovenia)
Mouček, Roman (Czech Republic)
Mykowiecka, Agnieszka (Poland)
Ney, Hermann (Germany)
Oliva, Karel (Czech Republic)
Pala, Karel (Czech Republic)
Pavesić, Nikola (Slovenia)
Piasecki, Maciej (Poland)
Psutka, Josef (Czech Republic)
Pustejovsky, James (USA)
Rigau, German (Spain)
Rothkrantz, Leon (The Netherlands)
Rumshinsky, Anna (USA)
Rusko, Milan (Slovakia)
Sazhok, Mykola (Ukraine)
Skrelin, Pavel (Russia)
Smrž, Pavel (Czech Republic)
Sojka, Petr (Czech Republic)
Steidl, Stefan (Germany)
Stemmer, Georg (Germany)
Tadić Marko (Croatia)
Varadi, Tamas (Hungary)
Vetulani, Zygmunt (Poland)
Wiggers, Pascal (The Netherlands)
Wilks, Yorick (UK)
Woliński, Marcin (Poland)
Zakharov, Victor (Russia)

Additional Referees

Benyeda, Ivett
Beňuš, Štefan
Brychcin, Tomáš
Doetsch, Patrick
Feltracco, Anna
Fonseca, Erick
Geröcs, Mátyás
Goikoetxea, Josu
Golik, Pavel
Guta, Andreas

Hercig, Tomáš
Karčová, Agáta
Laparra, Egoitz
Lenc, Ladislav
Magnolini, Simone
Makrai, Márton
Simon, Eszter
Uhliarik, Ivor
Wawer, Aleksander

Organizing Committee

Aleš Horák (Co-chair), Ivan Kopeček, Karel Pala (Co-chair), Adam Rambousek (Web System), Pavel Rychlý, Petr Sojka (Proceedings)

Sponsors and Support

The TSD conference is regularly supported by the International Speech Communication Association (ISCA). We would like to express our thanks to Lexical Computing Ltd. and IBM Česká republika, spol. s r. o. for their kind sponsoring contribution to TSD 2016.

Abstract Papers

Embeddings! For Which Objects?
For Which Objectives?

Hinrich Schütze

Chair of Computational Linguistics, University of Munich (LMU)
Oettingenstr 67, 80538 Muenchen, Germany
hinrich@hotmail.com

Natural language input in deep learning is commonly represented as embeddings. While embeddings are widely used, fundamental questions about the nature and purpose of embeddings remain. Drawing on traditional computational linguistics as well as parallels between language and vision, I will address two of these questions in this talk. (1) Which linguistic units should be represented as embeddings? (2) What are we trying to achieve using embeddings and how do we measure success?

Natural Language Knowledge Graphs

Ido Dagan

Natural Language Processing Lab
Department of Computer Science, Bar Ilan University, Ramat Gan, 52900, Israel
dagan@cs.biu.ac.il

How can we capture the information expressed in large amounts of text? And how can we allow people, as well as computer applications, to easily explore it? When comparing textual knowledge to formal knowledge representation (KR) paradigms, two prominent differences arise. First, typical KR paradigms rely on pre-specified vocabularies, which are limited in their scope, while natural language is inherently open. Second, in a formal knowledge base each fact is encoded in a single canonical manner, while in multiple texts a fact may be repeated with some redundant, complementary or even contradictory information.

In this talk I will outline a new research direction, which we term Natural Language Knowledge Graphs (NLKG), that aims to represent textual information in a consolidated manner, based on the available natural language vocabulary and structure. I will first suggest some plausible requirements that such graphs should satisfy, that would allow effective communication of the encoded knowledge. Then, I will describe our current specification for NLKG structure, motivated by a use case of representing multiple tweets describing an event. Our structure merges individual proposition extractions, created in an Open-IE flavor, into a representation of consolidated entities and propositions, adapting the spirit of formal knowledge graphs. Different mentions of entities and propositions are organized into entailment graphs, which allow tracing the inference relationships between these mentions. Finally, I will review some concrete research components, including a proposition extraction tool and lexical inference methods, and will illustrate the potential application of NLKGs for text exploration.

Remote Monitoring of Neurodegeneration through Speech

Elmar Nöth

Pattern Recognition Lab, Friedrich-Alexander-Universität
Erlangen-Nürnberg (FAU)
Erlangen, Germany
elmar.noeth@fau.de

Abstract. In this talk we will report on the results of the workshop on "Remote Monitoring of Neurodegeneration through Speech", which was part of the "Third Frederick Jelinek Memorial Summer Workshop"[1] and took place at Johns Hopkins University in Baltimore, USA from June 13th to August 5th, 2016.

Keywords: Neurodegeneration, Pathologic speech, Telemonitoring

Alzheimer's disease (AD) is the most common neurodegenerative disorder. It generally deteriorates memory function, then language, then executive function to the point where simple activities of daily living (ADLs) become difficult (e.g., taking medicine or turning off a stove). Parkinson's disease (PD) is the second most common neurodegenerative disease, also primarily affecting individuals of advanced age. Its cardinal symptoms include akinesia, tremor, rigidity, and postural imbalance. Together, AD and PD afflict approximately 55 million people, and there is no cure. Currently, professional or informal caregivers look after these individuals, either at home or in long-term care facilities. Caregiving is already a great, expensive burden on the system, but things will soon become far worse. Populations of many nations are aging rapidly and, with over 12 % of people above the age of 65 having either AD or PD, incidence rates are set to triple over the next few decades.

Monitoring and assessment are vital, but current models are unsustainable. Patients need to be monitored regularly (e.g., to check if medication needs to be updated), which is expensive, time-consuming, and especially difficult when travelling to the closest neurologist is unrealistic. Monitoring patients using non-intrusive sensors to collect data during ADLs from speech, gait, and handwriting, can help to reduce the burden.

Our goal is to design and evaluate a system for remotely monitoring neurodegenerative disorders, e.g., over the phone or internet, as illustrated in Fig. 1. The doctor or an automatic system can contact the patient either on a regular basis or upon detecting a significant change in the patient's behavior. During this interaction, the doctor or the automatic system has access to all the sensor-based evaluations of the patient's ADLs and his/her biometric data (already stored on the server).

---

[1] http://www.clsp.jhu.edu/workshops/16-workshop/

The doctor/system can remind these individuals to take their medicine, can initiate dedicated speech-based tests, and can recommend medication changes or face-to-face meetings in the clinic.

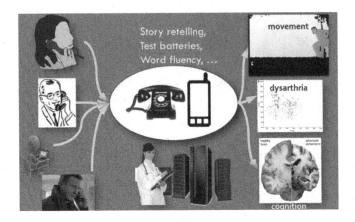

Fig. 1. General Methodology

The analysis is based on the fact that these disorders cause characteristic lexical, syntactic, and spectro-temporal acoustic deviations from non-pathological speech, and in a similar way, deviations from non-pathologic gait and handwriting.

In order to detect a significant change in the patients behavior, we envision an unobtrusive monitoring of the patient's speech during his ADLs, mainly his/her phone conversations. The advantage is, that the patient is not aware of it and doesn't feel the pressure of constantly being tested. To achieve that, we are working on a smartphone app, that will record the patient's part of any phone conversation. After the conversation the speech data are analyzed on the smartphone, the results are used to accumulate statistics, the updated statistics are transmitted to a server, and the speech signal is deleted. The statistics concern the communication behavior in general as well as that of a phone call. Typical parameters are:

– Communication behavior

  • How often does (s)he use the phone?
  • How often does (s)he call people from the phone book?
  • How often does (s)he initiate the call?
  • At what time of the day does (s)he communicate?

– Call behavior

  • How long does (s)he call?
  • Percentage of speaking time
  • Average duration of turns
  • Variation in fundamental frequency/energy ($\rightarrow$ emotional/interest level)

In order to achieve this, we need to systematically release restrictions on existing data collections of AD and PD data. Typically, for data collections, as many circumstances are kept constant as possible, e.g.,

– Exclusion of non-natives
– Exclusion of persons with other known neurological disease
– Identical task
– Identical recording conditions
– Accompanying expert

Then, the speech of the patients is compared to an age/sex matched control group.

We will report on experiments that will allow to monitor the patient during regular phone conversations where there is no control over the recording condition and the topic/vocabulary of the conversation.

Contents

Speech

Text

Grammatical Annotation of Historical Portuguese: Generating a Corpus-Based Diachronic Dictionary

Eckhard Bick[1] and Marcos Zampieri[2,3(✉)]

[1] University of Southern Denmark, Odense, Denmark
eckhard.bick@mail.dk
[2] Saarland University, Saarbrücken, Germany
marcos.zampieri@dfki.de
[3] German Research Center for Artificial Intelligence (DFKI),
Saarbrücken, Germany

Abstract. In this paper, we present an automatic system for the morphosyntactic annotation and lexicographical evaluation of historical Portuguese corpora. Using rule-based orthographical normalization, we were able to apply a standard parser (PALAVRAS) to historical data (Colonia corpus) and to achieve accurate annotation for both POS and syntax. By aligning original and standardized word forms, our method allows to create tailor-made standardization dictionaries for historical Portuguese with optional period or author frequencies.

Keywords: Historical corpus · Corpus annotation · Dictionary

1 Introduction

Historical texts are notoriously difficult to treat with language technology tools. Problems include document handling (hand-written manuscripts, scanning, OCR), conservation of meta-data, and orthographical and standardization issues. This paper is concerned with the latter, and we will show how a modified parser for standard Portuguese can be used to annotate historical texts and to generate an on-the-fly dictionary of diachronic variation in Portuguese for a specific corpus, mapping spelling variation in a particular period, author or text collection. The target and evaluation data for our experiments come from the Colonia Corpus [16] whereas for the annotation pipeline we use the PALAVRAS parser [1].

Several large projects handling historical Portuguese are worth mentioning, among them the syntactically oriented Tycho Brahe Corpus [3,5] and the lexicographical HDBP project [8] aiming at the construction of a historical dictionary of Brazilian Portuguese. A third one is the online 45M word Corpus do Português [4] which provides a diachronic cross section of both European and Brazilian Portuguese. Spelling variation was an important issue in both the Tycho Brahe

© Springer International Publishing Switzerland 2016
P. Sojka et al. (Eds.): TSD 2016, LNAI 9924, pp. 3–11, 2016.
DOI: 10.1007/978-3-319-45510-5_1

and the HDBP projects. Though the Tycho Brahe project originally used tagger lexicon extensions, both projects ended up basing their variation handling on a rule-based regular expression methodology suggested by Hirohashi [7]. The HDBP version, called Siaconf, lumps variants around a common 'base form', but not necessarily the modern form, favoring precision (almost 100 %) over recall [8]. Hendrickx and Marquilhas [6] adapted a statistical spelling normalizer to Portuguese, recovering 61 % of variations, 97 % of which were normalized to the correct standard form. They also showed that spelling normalization improved subsequent POS tagging, raising accuracy about 2/3 of the distance between unmodified and manual gold standard input. Rocio et al. [12], assigned neural-network-learned and post-edited POS tags *before* morphological analysis, after hand-annotating 10,000 words per text *without* normalisation, then adding partial syntactic parses for the output, using 250 definite clause grammar rules developed for partial parsing of contemporary Portuguese. In our own approach, like HDBP, we adopt a rule-based normalization approach [2], but aiming at exclusively modern forms, both for lexicographical reasons, and to support tagging and parsing with standard tools *without* the need of hand-annotated data.

2   Motivation

A historical dictionary can take different forms, spanning from the purely philological aspect to automatically extracted corpus data and frequency lists. Thus, Silvestre and Villalva [15] aim at producing a historical root dictionary for Portuguese, based on lexical analysis, etymology and using other dictionaries, rather than corpora, as their source. By contrast, the HDBP dictionary is based on 10M words of corpus data, providing definitions and quotations for historical usage [9]. Spelling variation is not the primary focus of either, and the published HDBP lumps variants under modern-spelled entries (10,500). However, the HDBP group also provides an automatically extracted glossary of 76,000 spelling variants for 31,000 'common' forms, as well as a manually compiled list of 20,800 token fusions (junctions). While this glossary constitutes an extensive and valuable resource, there are number of gaps filled by our project:

1. The HDBP glossary uses only Brazilian sources, while Colonia is a cross-variant depository with a potentially broader focus.
2. Unlike our parser-based resource, the glossary does not resolve POS ambiguity, nor does it offer inflectional analysis.
3. At least in its current form, the glossary does not differentiate periods or authors, something our proposed live system is able to generate on the fly.
4. Modern and historical entries are mixed, and it is not possible to tell one from the other. Thus, consonant gemination is mostly regarded as a variant, listing *villa* under *vila*, but modern *tão* and *chamam*, for instance, are listed under the entry of *tam* and *xamam*.
5. Contributing to the last problem, the glossary strips acute and circumflex accents in its entries, creating ambiguity even in the modern, standardized form, e.g. *continua* ADJ (*contínua*) vs. V (*continùa/continûa/continúa*). And

though ã and õ are maintained, a grapheme like -ão is not disambiguated with regard to -am, which it historically often denoted. Thus, the entry *matarão* may really mean *mataram*, which becomes clear from the entry *vierão* which is not ambiguous like *matarão*, but can only mean *vieram*.

6. The glossary contains fusions, not marked as such (e.g. *foime*), and does apparently not make use of the separate junction lexicon.

While some of these problems (4–6) could be addressed by reorganizing the data and aligning it with a modern lexicon, we believe that a live, automated system with a flexible source management, parser support, contextual disambiguation, and a clear variant2standardized entry structure can still contribute something to the field. We evaluate our method and results using the Colonia corpus, but our approach can easily be adapted for new or different data sets.

3   Corpus and Annotation

The Colonia corpus[1] is considered to be the largest historical Portuguese corpus to date. It contains Portuguese manuscripts, some of them available in other corpora (e.g. Tycho Brahe [5] and the GMHP corpus[2]), published from 1500 to 1936 divided into 5 sub-corpora per century. Texts are balanced in terms of variety, 48 European Portuguese texts and 52 Brazilian Portuguese texts (Table 1).

Table 1. Corpus size by century

Century	Texts	Tokens
16th	13	399,245
17th	18	709,646
18th	14	425,624
19th	38	2,490,771
20th	17	1,132,696
Total	100	5,157,982

Colonia has been used for various research purposes including temporal text classification [11,17], diachronic morphology [10], and lexical semantics [13].

Grammatical annotation adds linguistic value to a corpus, complementing existing philological mark-up (source, date, author, comments) and allowing quantitative or qualitative linguistic research not easily undertaken on raw-text corpora. Since it is time consuming to annotate a corpus by hand, automatic annotation is often chosen as a quick means to allow statistical access to corpus data. Obviously, a historical corpus will present special difficulties in this respect, since the performance of a parser built for modern text may be impaired by non-standard spelling and unknown words. In addition, historical Portuguese

---

[1] (1) Original version: http://corporavm.uni-koeln.de/colonia; (2) With our annotation and normalized lemmas: http://corp.hum.sdu.dk/cqp.pt.html.

[2] http://www.usp.br/gmhp/CorpI.html.

is difficult to tokenize, because word fusion may follow prosodic rules and occur for many function and even content words not eligible for fusion in modern Portuguese. For our work, we tackled these issues by adding pre-processing modules and a lexical extension to the PALAVRAS parser [2]. Our annotation method involves the following steps (Fig. 1):

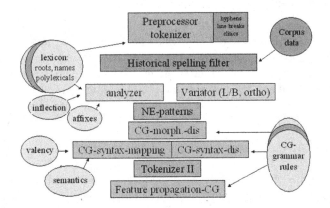

Fig. 1. System components and data flow

1. A pre-processor handling tokenization issues such as Spanish-style clitics (fused without hyphen), preposition fusion and apostrophe fusion at vowel/ vowel word borders.
2. A voting-based language filter blocking non-Portuguese segments from getting false-positive Portuguese analyses in the face of the orthographical 'relaxation' necessary for historical text.
3. A historical spelling filter, recognizing historical letter combinations and inflexion paradigms, and replacing words with their modern form where possible. Using a 2-level annotation, the original form is stored, while the standardised form is passed on to the parser. This module existed, but was extended with hundreds of patterns.
4. A fullform lexicon of modern word forms, built by generating all possible inflexion forms from the lemma base forms in the parser lexicon (1). This word list is used to validate candidates from (3) and for accent normalization.
5. An external dictionary and morphological analyzer, supplementing the parser's own morphological module. The module adds (historical and Tupi-Brazilian) readings to the (heuristic) ones for unknown words, allowing contextual Constraint Grammar (CG) rules to decide in cases of POS-ambiguity.

PALAVRAS' annotation scheme uses the following fields: (1) Word - (2) lemma [...] - (3) secondary tags (sub class, valency or semantics) - (4) part of speech (PoS), (5) inflexion, (6) syntactic function (@...) and (7) numbered dependency relations. For orthographically standardized historical words, (1) is the original word form, while the lemma (2) will indicate the modern lexeme. A special <OALT:...> tag in field (3) is used for normalized versions of the word form (1).

```
Esta [este] <dem> DET F S @>N #1->2
povoaçam [povoação] <OALT:povoação> <Lciv> N F S @SUBJ> #2->3
he [ser] <OALT:é> V PR 3S IND VFIN @FS-STA #3->0
uma [um] <arti> DET F S @>N #4->5
Villa [vila] <OALT:Vila> <Lciv> N F S @<SC #5->3
mui [muito] <OALT:muito> <quant> ADV @>A #6->7
fermosa [fermoso] <ORTO:formoso> ADJ F S @N< #7->5
```

Because the added historical-orthographical information is contained in angle-bracketed tags, this annotation scheme is fully compatible with all PALAVRAS post-processing tools, allowing easy conversion into constituent tree format, MALT xml, TIGER xml, CoNLL format, PENN and UD treebank formats etc. However, in order to handle ambiguity and avoid false positives, normalisation patterns should only be applied for out-of-lexicon words, and multiple filtering options must be constrained by a modern lexicon. For this purpose we used a list of about half a million distinct word forms inflexionally constructed from PALAVRAS' lemma lexicon, as well a modern spell-checking list. Accent-carrying forms were checked both with and without accents, to allow for the fact that historical Portuguese was often more explicit in marking a vowel as closed or open, respectively (*cêdo, gastára, afóra*). The fullform list was also used to handle word fusion (junctions). For this task, unknown forms were systematically stripped of 'particle candidates' (prepositions, adverbs, pronouns/clitics), checking the remaining word stem against the modern word list. The following are the orthographical topics that were treated by a pattern-matching pre-processor:

```
geminated and triple consonants: (attenção, accumula, soffra, affligir)
word fusion: heide, hade -> hei de, há de
"Greek" spelling: ph->f, th->t, y -> i (mathematica, authores, systema)
nasals: em[dt]->en (bemdito), om[df]->on (comforme), aon->ão (christaons)
    chaotic -ão/am and -ões: áo, ào, âo, aõ, aò, àm, ao,ões, óes, oens
extra hiatus-h: sahiu, incoherente, comprehender
z/s-dimorphism: isa -> iza, [aeu]z -> s, [óú]s$ -> z
s/c-ellision: sci -> ci, cqu -> qu: descifrada, sciência
lack of tonic accents: aniversario, malicia, razoavel, providencia, fariamos
superfluous accets: dóe, pessôa, enfrê
fluctuating accents: nòs, serà, judaïsmo
```

To evaluate the effectiveness of orthographical filtering on the performance of the PALAVRAS parser, we did a small, inspection-based evaluation of 1 random sample per century, comparing the modified PALAVRAS with the original Treetagger [14] annotation as a baseline (provided on the Colonia website and produced without additional modules). Since the older texts require more orthographical intervention, the 20th century figures can also be used as a kind of baseline for PALAVRAS itself. Percentages in the table are for non-punctuation tokens (words), and the unknown/heuristic lemma count is without proper nouns.[3]

---

[3] TreeTagger does not distinguish between common and proper nouns, but for the 'unknown' count, names were removed by inspection.

The table data indicates that the modified PALAVRAS outperforms the baseline for all centuries, and that the expected performance decrease for older texts is buffered by orthographical filtering. For both parsers a correlation between lexical coverage and tagging accuracy can be observed. A notable exception of the age-accuracy correlation is the 17th century text, which on inspection proved very modern in its orthography, probably due to the fact of it being a newspaper text (Manuel de Galhegos: 'Gazeta de Lisboa'), and as such subject to conscious standardization and proof-reading.[4]

Syntactic function assignment profited from orthographical filtering only indirectly, and historical syntactic variation (e.g. VS, VOS and OVS word order) were not addressed directly, leading to a moderate decrease in performance compared to modern text (Table 2).

Table 2. System performance per century

Century	Words	Treetagger unknown (- PROP)	PALAVRAS heuristic lemma (- PROP)	Treetagger accuracy (POS)	PALAVRAS accuracy modified (POS)	PALAVRAS accuracy modified (synt. function)
16th	473	15.2	0.4	80.1	96.6	91.1
17th	432	0.7	0.0	96.5	98.8	94.4
18th	477	21.8	0.6	81.1	97.7	91.6
19th	372	1.3	0.8	95.2	98.1	93.3
20th	446	0.2	0.0	97.3	99.6	96.0

4   Generating a Diachronic Dictionary

From the automatically annotated Colonia corpus we extracted all wordforms that had undergone orthographical normalization (Table 3).

Table 3. Frequency of non-standard forms across centuries

Century	Words	Orthographically non-standard	Fused	Fused (relative)
16th	528K	4.11 %	0.93 %	22.63 %
17th	577K	2.09 %	0.25 %	12.31 %
18th	456K	2.88 %	0.25 %	8.68 %
19th	2,459K	0.30 %	0.08 %	28.23 %
20th	857K	0.17 %	0.04 %	23.52 %

---

[4] At the time of writing it was not clear if this text had been subject to philological editing in its current form, which might explain its fairly modern orthography.

This happened either by in-toto filtering or by inflexion-based lookup in the add-on lexicon. The frequency of such forms decreased, as one would expect, over the centuries. Token fusion followed this trend, but was lowest in the 18th century in relative terms (i.e. out of all orthographical changes).[5] Another trend across time is the decreasing use of Latin and Spanish. Our language identification module identified foreign chunks and excluded them from the analysis (Table 4). As can be seen, Latin and Spanish had a certain presence in Portuguese writing in the first 3 centuries of the Colonia period, enough to disturb lexicographical work if no language-filtering was carried out.[6]

Table 4. Distribution of non-Portuguese text across centuries

Century	All Foreign	Latin	Spanish	Italian	French
16th	0.78 %	0.49 %	0.26 %	0.01 %	-
17th	0.78 %	0.51 %	0.24 %	-	0.01 %
18th	0.23 %	0.17 %	0.05 %	-	-
19th	0.03 %	0.02 %	-	-	0.01 %
20th	0.03 %	0.01 %	0.01 %	-	0.01 %

Together, the orthographically non-standard words constitute a dictionary of historical Portuguese spelling with 10,400 wordform types, representing 52,000 corpus tokens. The dictionary contains 862 non-standard word fusion types (e.g. *ess'outra, fui-lh'eu, estabeleceremse*), representing around 5,000 tokens.

```
capitaens <OALT:capitães> (14; - 17th:10 18th:4 - -)
capitaes <OALT:capitães> (1; 16th:1 - - - -)
capitaina <ORTO:capitânia (5; 16th:5 - - - -)
capitam <OALT:capitão> (5; 16th:4 - 18th:1 - -)
capitan <OALT:capitão> (1; 16th:1 - - - -)
capitanîas <OALT:capitanias> (1; - 17th:1 - - -)
capitaõ <OALT:capitão> (3; - 17th:1 18th:2 - -)
```

5 Conclusions and Outlook

In this paper, we have shown that with an orthographical standardization module, a tagger/parser for modern Portuguese (PALAVRAS) can achieve reasonable performance across a wide range of historical texts, outperforming an unaltered statistical tagging baseline (Treetagger) by a large margin. Standardization was

---

[5] Parts of fused tokens were counted individually in the statistics, the token count is therefore higher than it would be counting the original text tokens as-is.

[6] Note that the figures constitute a lower bound. In order to achieve a precision close to 100 %, only chunks with at least 4 (clear Latin 3) non-name words were treated, so individual loan words or mini-quotes are not included.

most important for the 16th–18th century, although some individual 17th century texts in our corpus already showed signs of standardization. Syntactically motivated grammar adaptations were not part of the current project, but are likely to further enhance performance, so future work should focus on this area.

An important result from the new annotation of the Colonia corpus is a method for automatically producing tailor-made spelling dictionaries of historical Portuguese. The resulting dictionary for Colonia itself contains almost 10,000 entries with century frequency information. We hope that both the method and the resource will be useful not only for linguistic-lexicographical purposes, but also as a language-technology resource, making it possible to reduce the out-of-vocabulary problem encountered by statistical taggers when used on historical text. A problematic aspect of the fullform substitution strategy for unknown words are false negatives, where a word matches an existing modern form, but still should have been changed (e.g. *noticia* V? vs. *notícia* N), and ambiguous cases like *estillo*, where the substitution *estilho* V? (Spanish ll/lh) was allowed to preclude the correct *estilo* N (gemination variant). Frequency ranking might help, but only to a certain degree, and an alternative strategy - as yet untried - would be to pass both readings on to the CG grammar module for contextual resolution, based on the differences in POS or inflection.

References

1. Bick, E.: PALAVRAS, a constraint grammar-based parsing system for Portuguese. In: Working with Portuguese Corpora, pp. 279–302 (2014)
2. Bick, E., Módolo, M.: Letters and editorials: a grammatically annotated corpus of 19th century Brazilian Portuguese. In: Proceedings of the 2nd Freiburg Workshop on Romance Corpus Linguistics, pp. 271–280 (2005)
3. Britto, H., Finger, M., Galves, C.: Computational and linguistic aspects of the Tycho Brahe parsed corpus of historical Portuguese. In: Romance Corpus Linguistics: Corpora and Spoken Language, pp. 137–146 (2002)
4. Davies, M.: Creating and using the corpus do Português and the frequency dictionary of Portuguese. In: Working with Portuguese Corpora, pp. 89–110 (2014)
5. Galves, C., Faria, P.: Tycho Brahe Parsed Corpus of Historical Portuguese (2010). http://www.tycho.iel.unicamp.br/tycho/corpus/en/index.html
6. Hendrickx, I., Marquilhas, R.: From old texts to modern spellings: an experiment in automatic normalisation. JLCL **26**(2), 65–76 (2011)
7. Hirohashi, A.: Aprendizado de Regras de Substituição para Normatização de Textos Históricos (2005)
8. Junior, A.C., Aluísio, S.M.: Building a corpus-based historical Portuguese dictionary: challenges and opportunities. TAL **50**(2), 73–102 (2009)
9. Murakawa, C.D.A.A.: A Construção de um Dicionário Histórico: o Caso do Dicionário Histórico do Português do Brasil-séculos XVI, XVII e XVIII. Estudos de Lingüística Galega **6**, 199–216 (2014)
10. Nevins, A., Rodrigues, C., Tang, K.: The rise and fall of the L-shaped morpheme: diachronic and experimental studies. Probus **27**(1), 101–155 (2015)
11. Niculae, V., Zampieri, M., Dinu, L.P., Ciobanu, A.M.: Temporal text ranking and automatic dating of texts. In: Proceedings of EACL, pp. 17–21 (2014)

12. Rocio, V., Alves, M.A., Lopes, J.G., Xavier, M.F., Vicente, G.: Automated creation of a medieval Portuguese partial treebank. In: Abeillé, A. (ed.) Treebanks, pp. 211–227. Springer, Heidelberg (2003)
13. Santos, D., Mota, C.: A Admiração à Luz dos Corpos. Oslo Stud. Lang. **7**(1), 57–77 (2015)
14. Schmid, H.: Probabilistic part-of-speech tagging using decision trees. In: Proceedings of International Conference on New Methods in Language Processing, pp. 44–49 (1994)
15. Silvestre, J.P., Villalva, A.: A morphological historical root dictionary for Portuguese, pp. 967–971 (2014)
16. Zampieri, M., Becker, M.: Colonia: corpus of historical Portuguese. ZSM Studien, Special Volume on Non-standard Data Sources in Corpus-Based Research, pp. 77–84 (2013)
17. Zampieri, M., Malmasi, S., Dras, M.: Modeling language change in historical corpora: the case of Portuguese. In: Proceedings of LREC, pp. 4098–4104 (2016)

Generating of Events Dictionaries from Polish WordNet for the Recognition of Events in Polish Documents

Jan Kocoń[(✉)] and Michał Marcińczuk

Department of Computational Intelligence, Wroclaw University of Technology,
Wybrzeze Wyspianskiego 27, 50-370 Wroclaw, Poland
{jan.kocon,michal.marcinczuk}@pwr.edu.pl

Abstract. In this article we present the result of the recent research in the recognition of events in Polish. Event recognition plays a major role in many natural language processing applications such as question answering or automatic summarization. We adapted TimeML specification (the well known guideline for English) to Polish language. We annotated 540 documents in Polish Corpus of Wrocław University of Technology (KPWr) using our specification. Here we describe the results achieved by Liner2 (a machine learning toolkit) adapted to the recognition of events in Polish texts.

Keywords: Information extraction · Event recognition · Polish wordnet

1 Introduction

Event recognition is one of the information extraction tasks. In the general understanding an *event* is anything what takes place in time and space, and may involve agents (executor and participants). In the context of text processing, event recognition relies on identification of textual mentions, which indicate events and describe them. In the literature there are two main approaches to this task: *generic* and *specific*. The *generic* approach assumes a coarse-grained categorization of events and is focused mainly on recognition of event mentions (textual indicators of events). Such approach is exploited in the TimeML guideline [1]. In turn the *specific* approach is focused on a detailed recognition of some predefined events including all components which describe them. This approach assumes that there is a predefined set of event categories with a complete description of their attributes. For example the ACE English Annotation Guidelines for Events [2] defines a *transport* as an event which "occurs whenever an ARTIFACT (WEAPON or VEHICLE) or a PERSON is moved from one PLACE (GPE, FACILITY, LOCATION) to another". The *specific* approach is a domain- or task-oriented for dedicated applications. In our research we have focused on the *generic* approach as it can be utilized in any domain-specific task. According to our best knowledge this is the first research on automatic recognition of generic events for Polish.

© Springer International Publishing Switzerland 2016
P. Sojka et al. (Eds.): TSD 2016, LNAI 9924, pp. 12–19, 2016.
DOI: 10.1007/978-3-319-45510-5_2

2   Event Categories

In our research we have exploited the coarse-grained categorization of events defined in TimeML Annotation Guidelines Version 1.2.1 [1]. TimeML defines seven categories of events, i.e., *reporting, perception, aspectual, intentional action, intentional state, state* and *occurrence*. We use a modified version of the TimeML guidelines[1]. One of the most important changes is the extension of the *occurrence* category. According to TimeML *occurrence* refers only to specific temporally located events. Instead, we use an *action* category which include also generics — actions which refer to some general rules (for example, "Water boils in 100°C"). We argue that the distinction between specific and generic actions is much more complex task than the identification of action mentions and may require discourse analysis. Also the event generality applies to the other categories of events as well. Thus, it should be treated as an event's attribute rather than a category. The other important modification, comparing the original TimeML guidelines, is the introduction of the *light predicates* category. This category is used to annotate synsemantic verbs which occur with nominalizations. This type of mentions does not contain enough semantic information to categorize the event. They carry only a grammatical and very general but sufficient lexical meaning which can be useful in further processing. A similar category was introduced by [4] in their research on event recognition for Dutch. The remaining categories have the same definition as in the TimeML guidelines. The final set of event categories contains: *action, reporting, perception, aspectual, i_action, i_state, state* and *light predicate*.

3   Data Sets

3.1   Corpus

In the research we used 540 documents from the Corpus of Wrocław University of Technology [5] which were annotated with events by two linguists according to our guidelines (see Sect. 2). We prepared two divisions for the purpose of the evaluation, which are presented in Table 1.

3.2   Inter-annotator Agreement

The inter-annotator agreement was measured on randomly selected 200 documents from KPWr. We used the positive specific agreement [6] as it was measured for T3Platinum corpus [7]. Two linguists annotated the randomly selected subset. We calculated the value of the positive specific agreement (PSA) for each category. The results are presented in Table 2.

According to [7] the best quality of data was achieved for TempEval-3 platinum corpus (T3Platinum, which contains 6375 tokens) and it was annotated and reviewed by the organizers. Every file was annotated independently by

---

[1] The comprehensive description of the modified guidelines is presented in [3].

Table 1. Description of two divisions of 540 documents from KPWr annotated with events. The first division is used to establish a baseline (see Sect. 6.1) and the second division is used to evaluate the impact of the generated dictionary features added to the baseline feature set for the result of the events recognition (see Sect. 6.2).

Division	Data set	Documents	Part of whole [%]
1	train1	270	50
	test1	135	25
	tune1	135	25
2	train2_p1	216	40
	train2_p2	216	40
	test2	108	20

Table 2. The value of positive specific agreement (PSA) calculated on the subset of 200 documents from KPWr, annotated independently with events by two domain experts. *A and B* means all annotations in which annotators A and B agreed. *Only A* is the number of annotations made only by annotator A and *only B* – the number of annotations made only by annotator B.

Category	A and B	Only A	Only B	PSA [%]
action	5268	1042	771	85.32
aspectual	100	31	23	78.74
perception	53	58	15	59.22
reporting	64	34	27	67.72
i_action	86	243	18	39.72
i_state	409	123	112	77.68
state	681	281	335	68.86
light predicate	84	119	61	48.28
Σ	6745	1931	1362	80.38

atleast two expert annotators and a third was dedicated to adjudicating between annotations. The result of overall T3Platinum inter-annotator positive specific agreement (PSA) at the level of annotating of events only was 0.87 and for the agreed entity set (exact matches) it was 0.92. It means that annotators agreed at the type of annotation at 0.92 for the annotations, which extents were agreed at 0.87, which for the task of manual annotation of both boundaries and event category is approximately 0.80. In our case for 200 randomly selected documents the PSA value achieved for the task of manual annotation of both boundaries and event categories was also 0.80. Unfortunately, for the corpus presented in [7] we see only the overall result for all event categories and we cannot compare the results for each category separately.

4   Generating of Event Dictionaries

The underlying hypothesis of this approach is that generalisation of specific words (*event mentions* in our case) in a subset of documents from a corpus allows to locate synsets in a wordnet, for which we can reconstruct dictionaries, which describe the observed phenomenon and allows to distinguish between different semantic categories of words (in our case — *event categories*) observed in the same set of documents. The algorithm consists of the following steps:

Construction of the helper graph — for each synset w from wordnet synsets W we add a subset of child lemmas C_w, which contains all lexical units from the synset and lexical units of its all hyponyms.

Building the corpus category vector — for the subset S of documents from the corpus and for the number of observed categories T (in our case 8 categories of events + *0 category* for words which do not indicate any event) we build $|T|$ vectors V. For each vector V^t describing the category $t \in T$, the length $|V^t|$ is equal to the number of words from the subset S and the value on n-th position (which represents n-th word in S) equals 1 if word S_n belongs to category t, 0 otherwise.

Building the corpus synset vector — for each $(w, C_w) \in W$ we build a vector A_w. The length $|A_w|$ is equal to the number of words from subset S and the value on n-th position (which represents n-th word in S) equals 1 if word $S_n \in C_w$, 0 otherwise.

Calculating the Pearson's correlation — for each $w \in W$ and each $t \in T$ we calculate the value of a Pearson's correlation $P_w^t = pearson(V^t, A_w)$.

Selection of the best nodes in hyponym branches — for each $t \in T$ we selected only these synsets from W, for which the value of P_w^t was the highest and the lowest in each hyponym branch. B_+^t is the subset of synsets and their child lemmas with the highest positive Pearson's correlation values in each hyponym branch of WordNet, and B_-^t is the subset of synsets and their child lemmas with the lowest negative Pearson's correlation values in each hyponym branch of wordnet. The whole process can be also driven with a given threshold p, which means the minimum absolute value of calculated Pearson's correlation to add a synset to B_+^t or B_-^t. In our experiments we used $p = 0.001$.

Selection of the best B_+, B_- subsets — we built a method for each $t \in T$ to combine the best nodes in hyponym branches to construct a pair of subsets (L^t, H^t) where $L^t \in B_-^t$ and $H^t \in B_+^t$ of the best nodes for which the value of Pearson's correlation calculated between V^t and a modified corpus synset vector M^t built on a pair (L^t, H^t) would be the highest. A length of a modified vector $|M^t|$ is equal to a number of words from subset S and the value on n-th position (which represents n-th word in S) equals 1 if word $S_n \in H^t \vee S_n \notin L^t$ and 0 otherwise. Constructing of (L^t, H^t) is iterative and requires to construct only H^t first. To do that in each step we try to add $b \in B_+^t$ to H^t, recalculate M^t and check if $pearson(V^t, M^t)$ is higher. In each step we find $b \in B_+^t$ which gives the highest gain to the value of

$pearson(V^t, M^t)$ and we add b to H^t and we remove b from B^t_+. We do that as long as we have a positive gain. Then, having H^t, we do the same with B^t_- and L^t.

Generating of dictionaries — for each $t \in T$ we generate separate positive D^t_+ and negative D^t_- dictionaries for H^t and L^t:

$$\forall t \in T \forall (w, C_w) \in H^t, D^t_+ = D^t_+ \cup C_w$$
$$\forall t \in T \forall (w, C_w) \in L^t, D^t_- = D^t_- \cup C_w$$

5   Recognition

Many state of the art systems which recognize events use supervised sequence labeling methods, mostly Conditional Random Fields (CRFs) [8]. Recent workshops about the comparison of event recognition systems like TempEval-2 and TempEval-3 [9] show a shift in the state-of-the-art. Currently the recognition of events is done best by supervised sequential classifiers instead of rule-engineered systems [9]. The best machine learning system reported by UzZaman [9] — TIPSem [10] — utilizes CRFs in the recognition of events.

Our approach is based on the *Liner2* toolkit[2] [11], which uses CRF++[3] implementation of CRF. This toolkit was successfully used in other natural language engineering tasks, like recognition of Polish named entities [11,12] and temporal expressions [13].

6   Evaluation

6.1   Feature Selection and Baseline Features

In recognition, the values of features are obtained at the token level. As a *baseline* we used a result of the selection of features from the *default* set of features available in the Liner2 tool. It contains 4 types of features: morphosyntactic, ortographic, semantic and dictionary. We described the *default* set of 46 features in the article about the recognition of Polish temporal expressions [13].

The detailed description of the selection process is available in [13]. Table 3 presents the result of the feature selection from the *default* set of 46 features, based on the F_1-score of 10-fold cross-validation on *tune1* data set.

Table 4 presents the comparison of average F_1-score for all event categories achieved on *train1* and *test1* data sets and for two feature sets: *default* and *baseline*.

We analyzed the statistical significance of differences between two feature sets on two different data sets. To check the statistical significance of F_1-score difference we used paired-differences Student's t-test based on 10-fold cross-validation with a significance level $\alpha = 0.05$ [14]. Differences are not statistically significant for both data sets, but the reduction of a feature space is from 46 to only 6 features, which compose a *baseline* set of features for the further evaluation.

---

[2] http://nlp.pwr.wroc.pl/en/tools-and-resources/liner2.
[3] http://crfpp.sourceforge.net/.

Table 3. Result of the feature selection for Polish events recognition used in this work as a *baseline*. Used measure: average *exact match* F_1-score of 10-fold cross-validation on *tune1* set. Initial set of features: default 46 Liner2 features.

Iteration	Selected feature	F_1 [%]	Gain [pps]
1	class	62.13	62.13
2	hypernym-1	73.23	11.10
3	top4hyper-4	74.68	1.45
4	prefix-4	75.26	0.58
5	struct	75.85	0.59
6	synonym	76.30	0.45

Table 4. Comparison of results (F_1-score) achieved on two data sets (*train1* – 10-fold cross-validation on *train1* set; test1 – model is trained on *train1* set and evaluated on *test1* set) and two feature sets (*default* – 46 default features available in Liner2; *baseline* – result of the feature selection on *default* feature set and *tune1* data set).

Set	Default [%]	Baseline [%]
train1	77.47	77.53
test1	78.90	78.34

6.2 Baseline with Dictionary Features

We generated two sets of dictionaries for each part of *train2* set (these parts are fully separated). Dictionaries were created using the plWordNet [15] — the largest wordnet for Polish. We used dictionary features generated on *train2_p1* to evaluate the model on *train2_p2* data set, and then we used dictionary features generated on *train2_p2* to evaluate the model on *train2_p1* data set. The last two models (first trained on *train2_p1* and second trained on *train2_p2*) were evaluated using *test2* data set.

Table 5. Comparison of results (F_1-score) achieved on two **Part**s of *train2* data set: *p1* and *p2*. These data sets were also dictionary **Source**s for **B+dict** feature set, to compare results with **Baseline** feature set. We performed two types of **Eval**uation: *CV* (10-fold cross-validation on a part of *train2* set) and *test2* (the model is trained on a part of *train2* set and the evaluation is performed on a *test2* set).

Part	Eval.	Source	B [%]	B+dict [%]
p1	CV	p2	77.20	79.67
p2	CV	p1	76.92	78.92
p1	test2	p2	77.65	79.87
p2	test2	p1	77.81	79.82

We performed 4 tests to evaluate the impact of generated dictionary features added to the baseline feature set for the result of the events recognition. The result is presented in Table 5. We see that in each test we achieved better results with the set of features extended with dictionaries. We analyzed the statistical significance of differences between these results for each test. To check the statistical significance of F_1-score difference we used paired-differences Student's t-test based on 10-fold cross-validation with a significance level $\alpha = 0.05$ [14]. All differences are statistically significant.

7 Conclusions

In Table 6 we present the comparison of detailed results for each event category, achieved on both parts of *train2* data set (as a sum of True Positives, False Positives and False Negatives of 10-fold cross-validation on *train2_p1* and 10-fold cross-validation on *train2_p2*).

Table 6. Comparison of detailed results for each event **Category** achieved on both parts of *train2* data set (the result is the sum of TP, FP and FN of 10-fold cross-validation on *train2_p1* and *train2_p2*). **Baseline+dict** variant is a set of **Baseline** features extended with dictionary features. The last column shows the value of **PSA** (positive specific agreement), described in Sect. 3.2.

Category	Baseline			Baseline+dict			PSA
	P [%]	R [%]	F [%]	P [%]	R [%]	F [%]	[%]
action	80.05	83.14	81.57	82.49	83.87	83.18	85.32
aspectual	93.00	39.08	55.03	87.58	59.24	70.68	78.74
i_action	63.97	27.87	38.82	63.56	40.92	49.79	59.22
i_state	85.82	75.08	80.09	85.19	77.56	81.20	67.72
light_predicate	50.00	11.03	18.07	56.76	15.44	24.28	39.72
perception	96.55	23.14	37.33	85.90	55.37	67.34	77.68
reporting	73.91	44.24	55.35	71.13	51.30	59.61	68.86
state	70.81	50.19	58.74	68.10	62.17	65.00	48.28
$\sum$	79.54	74.75	77.07	80.88	77.82	79.32	80.38

We see that adding dictionary features statistically significantly increased the result of events recognition. Detailed analysis performed on separate event categories showed that the major improvement can be observed with categories which are underrepresented in corpus and for which the PSA value was smaller. For models with dictionary features the F-measure is more close to PSA values for all categories of events. Dictionary features increased the values of both precision and recall.

Acknowledgments. Work financed as part of the investment in the CLARIN-PL research infrastructure funded by the Polish Ministry of Science and Higher Education.

References

1. Saurí, R., Littman, J., Gaizauskas, R., Setzer, A., Pustejovsky, J.: TimeML Annotation Guidelines, Version 1.2.1 (2006)
2. LCD: ACE (Automatic Content Extraction) English Annotation Guidelines for Events (Version 5.4.3). Technical report, Linguistic Data Consortium (2005)
3. Marcińczuk, M., Oleksy, M., Bernaś, T., Kocoń, J., Wolski, M.: Towards an event annotated corpus of Polish. Cogn. Stud. Études Cogn. **15**, 253–267 (2015)
4. Schoen, A., van Son, C., van Erp, M., van der Vliet, H.: NewsReader document-level annotation guidelines - Dutch. NWR-2014-08. Technical report, VU University Amsterdam (2014)
5. Broda, B., Marcińczuk, M., Maziarz, M., Radziszewski, A., Wardyński, A.: WUTC: towards a free corpus of Polish. In: Proceedings of the Eighth Conference on International Language Resources and Evaluation (LREC 2012), Istanbul, Turkey, 23–25 May 2012 (2010)
6. Hripcsak, G., Rothschild, A.S.: Agreement, the F-measure, and reliability in information retrieval. J. Am. Med. Inform. Assoc. **12**, 296–298 (2005)
7. UzZaman, N., Llorens, H., Allen, J.F., Derczynski, L., Verhagen, M., Pustejovsky, J.: TempEval-3: evaluating events, time expressions, and temporal relations. CoRR abs/1206.5333 (2012)
8. Lafferty, J.D., McCallum, A., Pereira, F.C.N.: Conditional random fields: probabilistic models for segmenting and labeling sequence data. In: Proceedings of the Eighteenth International Conference on Machine Learning, ICML 2001, pp. 282–289. Morgan Kaufmann Publishers Inc., San Francisco (2001)
9. UzZaman, N., Llorens, H., Derczynski, L., Verhagen, M., Allen, J., Pustejovsky, J.: SemEval-2013 task 1: TEMPEVAL-3: evaluating time expressions, events, and temporal relations, Atlanta, Georgia, USA, p. 1 (2013)
10. Llorens, H., Saquete, E., Navarro, B.: TipSEM (English and Spanish): evaluating CRFs and semantic roles in TempEval-2. In: Association for Computational Linguistics, pp. 284–291 (2010)
11. Marcińczuk, M., Kocoń, J., Janicki, M.: Liner2 – a customizable framework for proper names recognition for Polish. In: Bembenik, R., Skonieczny, Ł., Rybiński, H., Kryszkiewicz, M., Niezgódka, M. (eds.) Intelligent Tools for Building a Scientific Information. SCI, vol. 467, pp. 231–254. Springer, Heidelberg (2013)
12. Marcińczuk, M., Kocoń, J.: Recognition of named entities boundaries in Polish texts. In: ACL Workshop Proceedings (BSNLP 2013) (2013)
13. Kocoń, J., Marcińczuk, M.: Recognition of Polish temporal expressions. In: Proceedings of Recent Advances in Natural Language Processing (RANLP 2015) (2015)
14. Dietterich, T.G.: Approximate statistical tests for comparing supervised classification learning algorithms. Neural Comput. **10**, 1895–1923 (1998)
15. Maziarz, M., Piasecki, M., Szpakowicz, S.: Approaching plWordNet 2.0. In: Proceedings of the 6th Global Wordnet Conference, Matsue, Japan (2012)

Building Corpora for Stylometric Research

Jan Švec[(✉)] and Jan Rygl

Natural Language Processing Centre, Faculty of Informatics,
Masaryk University, Botanická 68a, 602 00 Brno, Czech Republic
{svec,rygl}@fi.muni.cz

Abstract. Authorship recognition, machine translation detection, pedophile identification and other stylometry techniques are daily used in applications for the most widely used languages. On the other hand, under-represented languages lack data sources usable for stylometry research. In this paper, we propose novel algorithm to build corpora containing meta-information required for stylometry experiments (author information, publication time, document heading, document borders) and introduce our tool *Authorship Corpora Builder* (ACB). We modify data-cleaning techniques for purposes of stylometry field and add a heuristic layer to detect and extract valuable meta-information.

The system was evaluated on Czech and Slovak web domains. Collected data have been published and we are planning to build collections for other languages and gradually extend existing ones.

Keywords: Authorship corpora · Stylometry · Web structure detection · Corpora building

1 Introduction

Internet users are regularly confronted with hundreds of situations in which they could use stylometry techniques. These situations include spam and machine translation classification; age and gender recognition (pedophile detection [1]); and authorship detection (anonymous threats, false product reviews [2,3]).

For dominant languages, there are many valuable data sources which can be used for a stylometry research (e-mail corpus Enron [4], age and gender corpus of Koppel [5]). The existence of these data sources enables fast implementation and comparison of the best techniques and facilitates further research.

For under-represented languages, applicable data sources are limited. We can divide available data sources into following categories:

1. General web corpora: documents are crawled from whole Internet, filtered by language, deduplicated and boilerplate is removed (e.g. for Czech [6] and Slovak [7]). These corpora are unsuitable for several reasons:
   - meta-information is missing (we do not know genres and publication times of texts; authors' identities, ages and genders);
   - document borders are unclear;

© Springer International Publishing Switzerland 2016
P. Sojka et al. (Eds.): TSD 2016, LNAI 9924, pp. 20–27, 2016.
DOI: 10.1007/978-3-319-45510-5_3

- formatting is omitted (stylometry can use typography which cannot be witnessed in vertical corpus format)[1].
2. Classic corpora: they are usually limited to one data source (newspapers, books, e.g. Hungarian Szeged corpus [8]). They can be used for stylometric analysis on limited data domain. The main disadvantage of these corpora is presence of text correction, translation or co-authoring.
3. Specialized stylometric corpora: there is exactly one stylometric corpus for Czech language containing 1694 manually texts written by pupils at school [9].
· The process of building manually collected corpora is very slow and resource consuming process.

Development of stylometry tools for under-represented languages (such as Visegrád Four languages) is slow due to lack of quality data. Therefore, we should contribute to building stylometry data sources before conducting further stylometry experiments in minor languages. This work focus on authorship determination problem and building stylometric corpora usable for authorship recognition tasks. Other tasks as age and gender detection can use extracted names (names in Slavic languages distinguish gender; or names in general can be used to match user profile tables with the rest of information).

We propose a novel approach for building internet stylometry corpora and a modular system for collecting documents with meta-information, which are suitable for authorship research. Current systems for document crawling and text extraction are predominantly used for general web corpora building, which lacks useful meta-information for stylometry. Selecting the most suitable algorithms and adding a layer of heuristic enables fully automated data acquisition. The collected data are automatically annotated using information from the website. Documents without meta-information are omitted.

The system was successfully used for collecting Czech and Slovak data[2] and we are planning to build collections for other minor languages and publish them.

2   Information Extraction from Web

Building stylometry web corpora based on internet articles consists of downloading data from web (predominant crawlers can be used); detecting the structure of the web page (classic algorithms are optimized for boilerplate removal; we need to modify them); text extraction (we modify text processing to keep valuable information for stylometry); and novel heuristic evaluation of extracted data.

The relevant text data on web pages is usually surrounded by much other information, for example: menus, navigation parts, banners etc. These non-relevant parts must be properly identified and removed from the main content if we want to use the text for further analysis.

The work is focused on building web corpora which can be used for language research, but it is also a unique source of meta-information about the data.

---

[1] Word-per-line (WPL) text, as defined at the University of Stuttgart in the 1990s.
[2] Data are available at http://nlp.fi.muni.cz/projekty/acb/preview.

We always provide information about the author, so the corpora will be used for stylometric research. In addition, we provide information about article title and date of creation, along with URL address of the source. One of the many examples is using it for determining authorship of anonymous documents.

Stylometry corpora differ from classic web corpora by giving more emphasis on the relationship between the author and his documents. It focuses on documents, which are associated with certain author, so we can simply get various information from it. We can determine what vocabulary the author is using, how he is constructing sentences etc [10].

2.1 Site Style Tree Algorithm

The most important part of our work is to detect the structure of the web and extract parts which contain the main text of the article. There are various methods to detect structure (*Wrapper Induction: Efficiency and Expressiveness* [11], *Data Extraction based on Partial Tree Alignment* [12], *Information Extraction based on Pattern Discovery* [13], *Augmenting Automatic Information Extraction with Visual Perceptions* [14], *Site style tree* [15]). We have chosen *Site style tree* (SST), as the best method due the reasons described in Sects. 2.1 and 2.2.

SST is based on analysis of templates of HTML pages. It assumes that the important information on the page differs in content, size and shape opposite for non-important parts which have the same structure among many pages on the same domain.

The method uses a data structure called *Style Tree* (ST). Pages are first parsed into a DOM tree and then transferred to *Style Tree* that consists of two types of nodes, namely, *style nodes* and *element nodes*.

Style node represents a layout or presentation style and it consists of two components. First is a sequence of element nodes, second is the number of pages that has this particular style. *Element node* is similar to node in the DOM tree, but differs in his pointer to child nodes – which in element node is set to sequence of style nodes. Interconnection of *style node* and *element node* creates *Style tree*.

2.2 Determining Nodes with Relevant Information

In the SST we determine the important nodes, which have informational value like this: The more different child nodes *element node* has, the more important it is and, vice versa. We use weight for this attribute, which value can be between 0 and 1.

Greyed-out parts on picture no.: 1 can be labeled as non-important, because they have a high page count, which means they often repeat across the pages on the same domain. On the other hand, highlighted element node *Div* (by red color), is marked as important. Because it has many different children with a low page count. For example on 35 sites, it contains only element P, on 15 pages it contains elements P, *img*, A etc.

We used a metric, which measures the importance of the *element node* [16]:

$$NodeImp(E) = \begin{cases} -\sum_{i=1}^{l} p_i \cdot \log_m \cdot p_i & if \ m > 1 \\ 1 & if \ m = 1 \end{cases} \tag{1}$$

For particular element node E:

m is number of pages containing E,
l is number of child style nodes of E,
p is number of pages of particular child style node of E.

From the equation, we can see that when E contains little child nodes, the value of NodeImp(E) will be small, and vice versa. For example – we can count NodeImp for Table (from Fig. 1), like this:

$-0.35 \log_{100} 0.35 - 2 * (0.25 \log_{100} 0.25) - 0.15 \log_{100} 0.15 = 0.292 > 0.$

The computed value is afterwards compared with threshold. We can set the threshold in settings of our application. When the value is lower than threshold, *element node* is marked as non-informative, and vice versa.

2.3   Modified SST for Stylometry Research

For stylometry research, we introduce a novel approach how to build SST. We first build ST for first page, then for second page – these two ST are merged into SST. Adding ST of all other pages one by one (create ST for n pages, and merge them to SST *from 1st. to $n-1$th. page*), we build final SST. Built SST represents the structure of all pages on one specific domain. An example of the SST can be seen in Fig. 1.

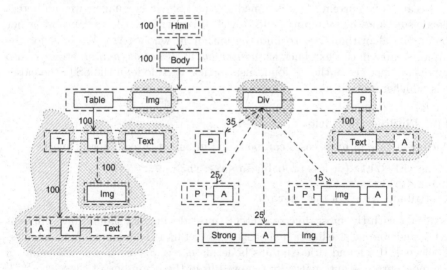

Fig. 1. Site style tree built from 100 pages (Colour figure online).

We also propose a novel modification of SST algorithm, in according way. If the input data contains a portal, which is unique between the pages, due to the *NodeImp* equation it receives a value of one. Because we need to distinguish pages based on the same template and remove the boilerplate, portal pages and various pages which are very different are not the subject of our interest. We decided to remove unique pages from our analysis. For our needs we marked these pages as non-important.

Non-important nodes are afterwards removed from the modified SST. We use this approach to remove boilerplate from each web page. Afterwards it contains the main text with meta data of the article - author, date and title.

2.4 Finding the Article Text

As an output from SST we obtain main text with meta-data. Text can contain boilerplate text – parts such as *a share button, a rate the article button, a link to comments*, etc. We can safely remove common text parts which are shorter than 20 characters[3] and doing so, boilerplate is removed.

Main text also contains meta-data (author, date and title); we must find it, store it[4] and delete from the main text. After we successfully clean the main text, it is stored in our corpus.

2.5 Finding the Author

Authorship Corpora Builder tries to find an expression "author(s)" in attributes of every tag in Slovak, Czech and English language (also tag `<author>` is checked). The score of each candidate tag is counted to determine the author (content must in most of cases abide name rules – regular expression preferring two words starting with a capital letter).

Because our current work is aimed at small Slavic languages, we can detect female surnames by searching suffix "ová". If tag with attribute "author" is not found, the algorithm tries to find the author in whole page. We look for the author's name in context, such as: *written by, published by, author*, etc. It is also important that the author is found near article text node in the SST, the closer it is – higher score it has.

2.6 Finding the Title

Document title in most cases can be found in one of three different places:

1. tag `<h1></h1>` (or less probably in `<h2>`, `<h3>`, . . . ;
2. tag `<title></title>`;
3. attribute "title" of tag `<meta/>`.

A variant with the most diverse content (most documents should have different titles) and abiding size limits (experimentally set min and max text length limits) is selected. If a found title contains boilerplate, e.g. *Server name: title name*, a common phrase is automatically removed from the beginning of the text.

---

[3] We have experimentally set the size limit to 20 after analysis of 5 different data sources.

[4] More detailed description is in next subsections.

2.7   Finding the Date

The algorithm searches the expression "date" in all examined languages. If we find this expression between attributes of an HTML tag, we check the content of this tag by regular expression, recognizing date format (we also check relative times such as *today, yesterday, ...*). Our regular expression works with an EU date format (starts with a day, then month and year), including months in full word. The regular expression can recognize three separators – dot, space, dash and slash.

If there are multiple dates in one article, algorithm prefers the tag, which contains the attribute "date" and its content corresponds to a regular expression. It also prefers when date is found near article text in SST. If the algorithm cannot find the expressions such as "date", it looks for all texts corresponding to date regular expression. If the matching text is found, its value is saved.

3   Results

The application was tested on 20 Czech and Slovak web sites, from each we extracted 500 articles. We created a corpus containing 10 000 documents and we plan to increase that number gradually. Collected data is published on http:// nlp.fi.muni.cz/projekty/acb/preview.

Four categories were observed: author of the article, title, date and main text. If a correct author was found on 450 pages from 500, authors' accuracy was 90 %; text of the article was found on 475 pages from 500, accuracy is 95 %, etc. We count accuracy for each category and we also compute averages for each category and for whole algorithm. The success rate of the algorithm can be seen in Figs. 2 and 3.

The average success rate of our algorithm is 90,68 % (if we omit missing values, it is almost 100 %). Results indicate that algorithm can be used to build

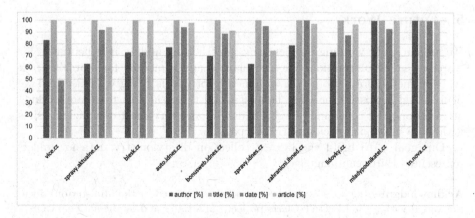

Fig. 2. Success rate of the first 10 sites

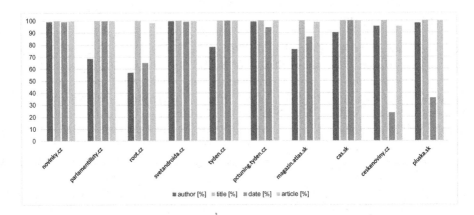

Fig. 3. Success rate of next 10 sites

the stylometry corpora. The most important meta-information is author of the article, therefore if the algorithm doesn't find the author, the article is discarded – the corpora will always contain relevant data.

4   Conclusions

Our application has shown that it can be efficiently used to build corpora for stylometry research in Czech and Slovak language. The corpus along with full text of the article and reference to the web source contains information about author, document title and creation date. The results are grouped by authors and web domains. The main strength of application ACB is that the process of building a corpus is fully automatic – user is only required to insert start URL, and optionally to adjust the settings.

5   Future Work

We are currently working on increasing precision by adding new heuristics for searching meta-information. We also plan to add a module for searching new types of meta-information. Our stylometry corpora is being extended and we plan to add heuristics for other languages. Last, but not least, we are working on full support of collecting of web discussions and very short documents – sites with more than one article per HTML page.

Our goal is to build the biggest collection of stylometry data for underrepresented European languages.

Acknowledgments. This work has been partly supported by the Ministry of Education of CR within the LINDAT-Clarin project LM2015071 and by the national COST-CZ project LD15066.

References

1. Peersman, C., Vaassen, F., Asch, V.V., Daelemans, W.: Conversation level constraints on pedophile detection in chat rooms. In: CLEF 2012 Evaluation Labs and Workshop, Online Working Notes, Rome, Italy, 17–20 September 2012
2. Chaski, C.E.: Who wrote it? Steps toward a science of authorship identification. Natl. Inst. Justice J. **233**, 15–22 (1997)
3. Joula, P.: Authorship attribution. Found. Trends Inf. Retr. **1**, 233–334 (2008)
4. Klimt, B., Yang, Y.: Introducing the Enron corpus. In: CEAS 2004 – First Conference on Email and Anti-Spam, Mountain View, California, USA, 30–31 July 2004
5. Koppel, M., Schler, J., Argamon, S., Pennebaker, J.: Effects of age and gender on blogging. In: AAAI 2006 Spring Symposium on Computational Approaches to Analysing Weblogs (2006)
6. Suchomel, V.: Recent Czech web corpora. In: Horák, A., Rychl, P. (eds.) Proceedings of 6th Workshop on Recent Advances in Slavonic Natural Language Processing, RASLAN 2012, Brno, Czech Republic, Tribun EU, pp. 77–83 (2012)
7. Medved', M., Jakubíček, M., Kovář, V., Němčík, V: Adaptation of Czech parsers for Slovak. In: Horák, A., Rychlý, P. (eds.) Proceedings of 6th Workshop on Recent Advances in Slavonic Natural Language Processing, Brno, Czech Republic, Tribun EU, pp. 23–30 (2012)
8. Csendes, D., Csirik, J.A., Gyimóthy, T.: The Szeged corpus: a POS tagged and syntactically annotated hungarian natural language corpus. In: Sojka, P., Kopeček, I., Pala, K. (eds.) TSD 2004. LNCS (LNAI), vol. 3206, pp. 41–47. Springer, Heidelberg (2004)
9. ÚČNK FF UK: SKRIPT2012: akviziční korpus psané češtiny– přepisy písemných prací žáků základních a středních škol v ČR (in English: acquisition corpus of Czech written language - transcripts of the written work of pupils in primary and secondary schools in the Czech Republic) (2013)
10. Koppel, M., Argamon, S., Shimoni, A.: Automatically categorizing written texts by author gender (2003)
11. Kushmerick, N.: Wrapper induction: efficiency and expressiveness. Artif. Intell. **118**(1–2), 15–68 (2000)
12. Zhai, Y., Liu, B.: Web data extraction based on partial tree alignment. In: Proceedings of the 14th International Conference on World Wide Web, WWW 2005, pp. 76–85. ACM, New York (2005)
13. Chang, C.H., Lui, S.C.: IEPAD: information extraction based on pattern discovery. In: Proceedings of the 10th International Conference on World Wide Web, WWW 2001, pp. 681–688. ACM, New York (2001)
14. Simon, K., Lausen, G.: ViPER: augmenting automatic information extraction with visual perceptions. In: Proceedings of the 14th ACM International Conference on Information and Knowledge Management, CIKM 2005, pp. 381–388. ACM, New York (2005)
15. Deepa, R., Nirmala, D.R.: Noisy elimination for web mining based on style tree approach. Int. J. Eng. Technol. Comput. Res. (IJETCR) **3**, 23–26 (2013)
16. Yi, L., Liu, B., Li, X.: Eliminating noisy information in web pages for data mining. In: Proceedings of the Ninth ACM SIGKDD International Conference on Knowledge Discovery and Data Mining, KDD 2003, pp. 296–305. ACM, New York (2003)

Constraint-Based Open-Domain Question Answering Using Knowledge Graph Search

Ahmad Aghaebrahimian[(✉)] and Filip Jurčíček

Faculty of Mathematics and Physics,
Institute of Formal and Applied Linguistics, Charles University in Prague,
Malostranské náměstí 25, 11800 Praha 1, Czech Republic
{Ebrahimian,Jurcicek}@ufal.mff.cuni.cz

Abstract. We introduce a highly scalable approach for open-domain question answering with no dependence on any logical form to surface form mapping data set or any linguistic analytic tool such as POS tagger or named entity recognizer. We define our approach under the Constrained Conditional Models framework which lets us scale to a full knowledge graph with no limitation on the size. On a standard benchmark, we obtained competitive results to state-of-the-art in open-domain question answering task.

Keywords: Question answering · Constrained conditional models · Knowledge graph · Vector representation

1 Introduction

We consider the task of simple open-domain question answering [4], where the answers can be obtained only by knowing one entity (i.e. popular things, people, or places) and one property (i.e. entities' attributes). The answer to such question is an entity or a set of entities. For instance, in the question *"What is the time zone in Dublin?"*, Dublin is an entity and time zone is a property. Hence, we have a pipeline which consists of two modules; property detection and entity recognition. For the first module, we train a classifier to estimate the probability of each property given each question. In the second module, given a question in natural language, we first estimate the distribution of each property using the classifier in the first module. Then, we retrieve entities constrained to some metadata about the properties. We use Freebase [3] as the knowledge graph to ground entities in our experiment. It contains about 58 million entities and more than 14 thousand properties. Hence, the entities which we obtain from knowledge graph are in many cases ambiguous. We extract metadata provided in the knowledge graph and integrate them into our system using Constrained Conditional Model framework (CCM) [13] to disambiguate the entities.

In WebQuestions [1], a data set of 5,810 questions which are compiled using the Google Suggest API, 86% of the questions are answerable by knowing only one entity [4]. It suggests that the majority of the questions which ordinary

P. Sojka et al. (Eds.): TSD 2016, LNAI 9924, pp. 28–36, 2016.
DOI: 10.1007/978-3-319-45510-5_4

people ask on the Internet are simple ones, and it emphasizes the importance of simple question answering systems. However, the best result of this task is 63.9 % path accuracy [4] which shows open-domain simple QA is still a challenging task in natural language processing (NLP).

Simple QA is not a simple task; flexible and unbound number of entities and their properties in open-domain questions makes entity recognition a real challenge. However, knowledge graphs can help a lot by providing a structural knowledge base on entities. The contributions of the current paper are a highly scalable QA model and a high-performance entity recognition model using knowledge graph search.

We organized the rest of this article in the following way. After a brief survey of freebase structure in Sect. 2, we describe our method in Sect. 3. We explain the settings of our experiment and the corresponding result in Sects. 4 and 5. In Sect. 6, we mention some related works and we discuss our approach in Sect. 7 before we conclude in Sect. 8.

2 The Knowledge Graph (Freebase)

Knowledge graphs contain significant amounts of factual information about entities (i.e., well-known places, people and things) and their attributes, such as place of living or profession. Large knowledge graphs cover numerous domains, and they may be a solution for scaling up domain-dependent QA systems to open-domain ones by expanding their boundary of entity and property recognition. Besides, knowledge graphs are instances of the linked data technologies. In other words, they can be connected easily to any other knowledge graph, and this increases their domain of recognition.

A knowledge graph is a graph structure in which source entities are linked to target ones through directed and labeled edges. In a typical knowledge graph, connecting a source entity to a target one using a directed edge forms the smallest structure in a KG, which is usually called an assertion. Large knowledge graphs such as Freebase [3] contain billions of such assertions. The entities in an assertion (i.e., source and target entity) are identified using a unique ID which is called machine MD or only *MID*. Entities and connecting edges are objects which mean they have some attributes called properties. These properties connect source entities to some other target entities in the graph. For the purpose of this paper, it is enough to know about *ID*, *MID*, *name*, *alias*, *type*, and *expected_type*.

Each entity in the graph has one unique *ID* and one unique *MID*. While *MID*'s value is always a code, *ID*'s value is sometimes a human readable string. In another word, in contrast to *MID* which has no meaningful association with the entity, *ID* sometimes has significant lexical similarity with the entity's *name* property (see Fig. 1). *name* is a surface form of the respective entity and it is usually a literal in a form of raw text, date or a numerical value. *alias* contains other aliases or surface forms for a given entity. A *type* defines an *IS A* relation with each entity. For instance, entity *Dublin* has types */common/topic*,

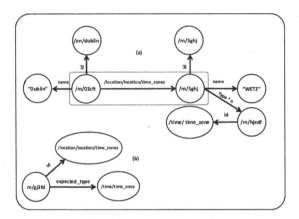

Fig. 1. Freebase structure; (a): type, id and name nodes, (b): expected type node

/book/book_subject[1] and many others, which says *Dublin* is not only a topic but also is the name of a book, ...

The *ID*, *MID*, *name*, *type* and *alias* properties are defined for all entities' (Fig. 1(a)). *expected_type* is defined only for edges and maintains what should you expect to get as a target entity when traversing the graph through this property. Each edge has at most one expected type. For instance */location/location/time_zones* (Fig. 1 (b)) has */time/time_zone* as expected type, and "*/film/film/produced_by*" has no expected type.

3   Our QA Learning Model

Constrained Conditional Model (CCM) [13] provides means of fine-tuning the result of a statistical model by enforcing declarative expressive constraints on them. For instance, if one asks 'What time is it in Dublin?', the answer should be directly related to a temporal notion, not something else. Knowing the fact that the question is about time, we can define a declarative constraint to state that the type of the answer should be the same as the expected type of the question's property. Constraints are essentially Boolean functions which can be generated using available metadata about entities and properties in Freebase.

As we illustrated in Fig. 1, Freebase assigns a set of *types* to each entity. It also assigns a unique *expected_type* to each property. Intuitively, the *type* of an answer to a question should be the same as the *expected_type* of the property for that question. A set of properties each of which with different probabilities and different *expected_types* is assigned to each test question. Some of the properties do not have *expected_type* and the types assigned to answers are not usually unique. Therefore, choosing the best property and entity for each question

---

[1] Logical symbols are given in an abbreviated form to save space. For instance, the full form of the second type above is http://rdf.freebase.com/book/book_subject.

requires searching in a huge space. Due to the enormous number of entities and their associated types in large knowledge graphs, translating a typical constraint like the one above into a feature set for training a statistical model is practically infeasible. Instead, using some constraints and Integer Linear Programming (ILP), we can make the search space much more efficient and manageable. In this way, we simply penalize the results of a statistical model which are not in line with our constraints.

Let's define P as the space of all properties assigned to a test question and E as the space of entities in it. For each question like *What is my time zone in Dublin?*, We intend to find the tuple $(p, e) \in P \times E$ for which the probability of p is maximal given some features and constraints. In our example question, we would like to get */location/location/time_zones* as the best property and *"/en/Dublin"* as the best-matching entity.

We decompose the learning model into two steps, namely; property detection and entity recognition. In property detection, we decide which property best describes the question. We model the assignment of properties given questions using the probability distribution in Eq. 1. We use logistic regression to train the model and use the model for N-best property assignment to each question at test time.

$$P(p|q) = \frac{\exp(\omega_p^T \phi(q))}{\Sigma_{p_i} \exp(\omega_{p_i}^T \phi(q))} \tag{1}$$

Given a question q, the aim is to find N-best properties which best describe the question and generate the correct answer when querying against the knowledge graph. ϕ in the model is a feature set representing the questions in vector space. ω, the features' weights, are optimized (Sect. 4) using gradient descent.

Training questions are accompanied with their knowledge graph assertions, each of which includes an entity, a property, and an answer. Entities and answers are in their *MID* forms. We also have access to the Freebase KG [8]. First, we chunk each question into its tokens and compute the features $\phi(q)$ by replacing each token with its vector representation. To train our classifier, we assign a unique index to each property in the training data set and use the indexes as labels for training questions.

The next step in the model is entity recognition which includes entity detection and entity disambiguation.

In entity detection, we distinguish the main entity from irrelevant ones. A typical question usually contains many spans that are available in the knowledge graph while only one of them is the main focus of the question. For instance, in the question *"what is the time zone in Dublin?"*, there are eleven valid entities available in the knowledge graph (*time, zone, time zone, ... and Dublin*) while the focus of the question is on *Dublin*.

In entity disambiguation, we disambiguate the detected entities in the last step. Given an entity like *'Dublin*, we need to know what Dublin (i.e., Dublin in Ireland, Dublin in Ohio, ...) is the focus of the question. To help the system with entity disambiguation, we use heuristics as a constraint to improve the chance of correct entities.

Entity recognition is done only at test time and on testing data. We use an Integer Linear Program model for assigning the best-matching entity to each test question (Eq. 2).

$$\text{best entity}(q) = \arg \max_e (\alpha^T s(p_q, e_q))$$
$$(p_q, e_q) \in P_q \times E_q \tag{2}$$

P_q are the N-best properties for a given question q and E_q are valid entities in q. $s(p_q, e_q)$ is a vector of p_q probabilities. α represents a vector of indicator variables which are optimized subject to some constraints. These constraints for each question are divided into two categories:

- Constraints in the first category enforce the type of answers of (p_q, e_q) to be equal to the expected type of p_q in each question (i.e., type constraints).
- Constraints in the second category dictate that the lexical similarity ratio between the values of *name* and *ID* properties connected to an entity should be maximal (i.e., similarity constraints).

Type constraints help in detecting the main focus of a given question among other valid entities. Despite the assigned properties, each question has $|E|$ valid entities from the KG. By valid, we mean entities which are available in the knowledge graph. After property detection, N-best properties are assigned to each question, each of which has no or at most one *expected_type*. The product between N-best properties and E valid entities gives us $N \times E$ tuples of *(entity, property)*. We query each tuple and obtain the respective answer from knowledge graph. Each of the answers has a set of *types*. If the *expected_type* of the property of each tuple was available in the set of its answer's *types*, type constraint for the tuple is satisfied.

Similarity constraints help in entity disambiguation. As we depicted it in Fig. 1, each entity has an *ID* and a *name* property. Ambiguous entities usually have the same *name* but different *ID*s. For instance, entities *"/m/02cft"* and *"/m/013jm1"* both have *Dublin* as their *name*, while the *ID* for the first one is */en/dublin* and for the second one is */en/dublin_ohio* (plus more than 40 other different entities for the same name). *name* determines a surface form for entities, and this is the property which we use for extracting valid entities in the first place. In our example, the similarity constraint for the entity */m/02cft* holds true because among all other entities, it has the maximal similarity ratio between its *name* and the *ID* property values. For some entities, the value of *ID* property for an entity is the same as its *MID*. In such cases, instead of *ID*, we use the *alias* property which contains a set of surface forms for entities.

3.1   Entity Detection

Instead of relying on external lexicons for mapping surface forms in the question to logical forms, we match surface forms and their *MID*s directly using the knowledge graph at test time. For entity detection, we extract spans of tokens in questions which correspond to surface forms of entities stored in the knowledge graph. We query a live and full version of Freebase using the Meta-Web Query Language (MQL). MQL is a template-base querying language which uses Google API service for querying Freebase in real time. We query the entity *MID* of each span. We have two alternatives to obtain initial entity *MID*s: greedy and full. In the greedy approach, only the longest valid entities are considered and their substrings, which may be still valid, are disregarded. In the full approach, however, all the entities are considered. For instance, in a simple span like "time zone", the greedy approach returns only *time zone* while the full approach returns *time*, *zone* and *time zone*.

3.2 Entity Disambiguation

Detected entities in the last step are ambiguous in many cases. Entities in massive knowledge graphs each have different meanings and interpretations. In a large knowledge graph, it is possible to find *Dublin* as the name of a city as well as the name of a book. Moreover, when it is the name of a city, that name is not still unique as we saw in the earlier section. We consider similarity constraints as true, if the lexical similarity ratio between *ID* and *name* properties connected to that entity is maximal among other ambiguous entities. It heuristically helps us to obtain an entity which has the highest similarity with the surface form in a given question.

4 Our Experiment

For testing our system, we used the SimpleQuestions data set [4]. SimpleQuestions contains 108,442 questions accompanied with their knowledge graph assertions. The questions in the SimpleQuestions data set are compiled manually based on the assertions of a version of a Freebase limited to 2 million entities (FB2M). Therefore, all answers of the questions can be found in FB2M. To make our result comparable to the results of SimpleQuestion authors, we used the official data set separation into 70 %, 10 %, and 20 % portions for train, validation, and test sets, respectively.

The inputs for the training step in our approach are the training and validation sets with their knowledge graph assertions. Using the Word2Vec toolkit [11], we replaced the tokens in the data sets with their vector representations to use them as features $\phi(q)$ in our model. We pruned questions with more than 20 tokens in length (only ten questions in the whole set). Using these features and the model described above, we trained a logistic regression classifier. We used the classifier at test time for detecting N-best properties for each test question.

We used path-level accuracy for evaluating the system. In path-level accuracy, a prediction is considered correct if the predicted entity and the property both are correct. Path-level accuracy is the same evaluation metric which is used by the data set authors. We obtained our best validation accuracy using the greedy approach for entity recognition and 128-dimensional vectors for property detection. Using the same configuration, we reported the accuracy of our system on test data.

5 Result

We used SimpleQuestions to train our system (Table 1). In this setting, with 99 % coverage, we obtained 61.2 % path accuracy, which is competitive to the results in [4] when training on the same data set (61.6 %). However, their results are reported on a limited version of Freebase (FB5M). Therefore, as we used the full knowledge graph, we hope that our system can answer every possible simple question whose answer is available in the full knowledge graph.

Table 1. Experimental results on test set of SimpleQuestions data set.

	Trained on	Knowledge graph	Path accuracy
Bordes et al.	SimpleQuestions	FB5M	61.6
Constraint-based: (Ours)	SimpleQuestions	FULL FB	61.2

6 Related Works

Domain specific QA has been studied well [7,10,14,17,18] for many domains. In a majority of these studies, a static lexicon is used for mapping surface forms of the entities to their logical forms. As opposed to KGs, Scaling up such lexicons which usually contain from hundreds to thousands of entities is neither easy nor efficient.

Knowledge graphs proved to be beneficial for different tasks in NLP including question answering. There are plenty of studies around using knowledge graphs for question answering either through information retrieval approach [4,15] or semantic parsing [1,2,5,9]. Unlike our work, they tend to use KGs only for validation and depend on predefined lexicons. Even in these studies, there is still a list of pre-defined lexicons for entity recognition (e.g., [1,5]). Essentially, they use knowledge graphs only for validating their generated logical forms and for entity recognition they still depend on initial lexicons. Dependence on pre-defined lexicons limits the scope of language understanding only to those predefined ones. In our approach, we do not use any data set or lexicon for entity recognition. Instead, we obtain valid entities by querying the knowledge graph at test time, and then we apply constraints on valid entities to get the correct entity for each question.

As regards CCM, it is first proposed by Roth and Yih [13] for reasoning over classifier results. It was used for different other problems in NLP [6,12]. [7] proposed a semantic parsing model using Question-answering paradigm on Geoquery [16] under CCM framework. Our work differs from them, first, by the size of the questions and the knowledge graph and second, by answering open-domain questions.

7   Discussion

The class of entities in open-domain applications is open and expanding. Training a statistical model for classifying millions of entities is practically infeasible due to the cost of training and lack of enough training data. Since entity decisions for one question have no effect on the next one, the entity model can be optimized for each question at test time and on the constraints particular to that question.

. In contrast, the class of properties is closed, and the property decisions usually are global ones. Therefore, a property model can be optimized once at training time and for all the questions. In this way, by making decisions on properties first, we decrease the decision space for entities by a large extent, and it makes our approach insensitive to the size of the knowledge graph. As we demonstrated in our experiment, optimizing entity recognition model on a single question lets the system scale easily to large knowledge graphs.

8   Conclusion

We introduced a question answering system with no dependence on external lexicons or any other tool. Using our system and on a full knowledge graph, we obtained competitive results compared to state-of-the-art systems with a limited knowledge graph. A 0.5 % decrease in the performance of the system in [4] when scaling from FB2M to FB5M with 3 million more entities suggests that QA in a full knowledge graph with more than 58 million entities is a much more difficult task. We showed that by means of enforcing expressive constraints on statistical models, our approach can easily scale up QA systems to a large knowledge graph irrespective of its size.

Acknowledgments. This research was partially funded by the Ministry of Education, Youth and Sports of the Czech Republic under the grant agreement LK11221, core research funding, SVV project number 260 333 and GAUK 207-10/250098 of Charles University in Prague. This work has been using language resources distributed by the LINDAT/CLARIN project of the Ministry of Education, and Sports of the Czech Republic (project LM2010013). The authors gratefully appreciate Ondřej Dušek for his helpful comments on the final draft.

References

1. Berant, J., Chou, A., Frostig, R., Liang, P.: Semantic parsing on freebase from question-answer pairs. In: Proceedings of EMNLP (2013)
2. Berant, J., Liang, P.: Semantic parsing via paraphrasing. In: Proceedings of ACL (2014)
3. Bollacker, K., Evans, C., Paritosh, P., Sturge, T., Taylor, J.: A collaboratively created graph database for structuring human knowledge. In: Proceedings of ACM (2008)
4. Bordes, A., Usunier, N., Chopra, S., Weston, J.: Large-scale Simple Question Answering with Memory Networks (2015). arXiv preprint arXiv:1506.02075

5. Cai, Q., Yates, A.: Large-scale semantic parsing via schema matching and lexicon extension. In: Proceedings of ACL (2013)
6. Chang, M., Ratinov, L., Roth, D.: Structured learning with constrained conditional models. Mach. Learn. **88**, 399–431 (2012)
7. Clarke, J., Goldwasser, D., Chang, M., Roth, D.: Driving semantic parsing from the world's response. In: Proceedings of the Conference on Computational Natural Language Learning (2010)
8. Google: Freebase Data Dumps (2013). https://developers.google.com/freebase/data
9. Kwiatkowski, T., Eunsol, C., Artzi, Y., Zettlemoyer, L.: Scaling semantic parsers with on-the-fly ontology matching. In: Proceedings of EMNLP (2013)
10. Kwiatkowski, T., Zettlemoyer, L., Goldwater, S., Steedman, M.: Inducing probabilistic CCG grammars from logical form with higher-order. In: Proceedings of EMNLP (2010)
11. Mikolov, T., Chen, K., Corrado, G., Dean, J.: Efficient estimation of word representations in vector space. In: ICLR Workshop (2013)
12. Punyakanok, V., Roth, D., Yih, W.: The importance of syntactic parsing and inference in semantic role labeling. Comput. Linguist. **34**, 257–287 (2008)
13. Roth, D., Yih, W.: Integer linear programing inference for conditional random fields. In: International Conference on Machine Learning (2005)
14. Wong, Y.-W., Mooney, R.: Learning synchronous grammars for semantic parsing with lambda calculus. In: Proceedings of ACL (2007)
15. Yao, X., Van Durme, B.: Information extraction over structured data: question answering with freebase. In: Proceedings of ACL (2014)
16. Zelle, J.M.: Using inductive logic programming to automate the construction of natural language parsers. Ph.D. thesis, Department of Computer Sciences, The University of Texas at Austin (1995)
17. Zelle, J.M., Mooney, R.J.: Learning to parse database queries using inductive logic programming. In: Proceedings of the National Conference on Artificial Intelligence (1996)
18. Zettlemoyer, L., Collins, M.: Learning to map sentences to logical form: structured classification with probabilistic categorial grammars. In: Proceedings of the Annual Conference in Uncertainty in Artificial Intelligence (UAI) (2005)

A Sentiment-Aware Topic Model for Extracting Failures from Product Reviews

Elena Tutubalina(✉)

Kazan (Volga Region) Federal University, Kazan, Russia
tutubalinaev@gmail.com

Abstract. This paper describes a probabilistic model that aims to extract different kinds of product difficulties conditioned on users' dissatisfaction through the use of sentiment information. The proposed model learns a distribution over words, associated with topics, sentiment and problem labels. The results were evaluated on reviews of products, randomly sampled from several domains (automobiles, home tools, electronics, and baby products), and user comments about mobile applications, in English and Russian. The model obtains a better performance than several state-of-the-art models in terms of the likelihood of a held-out test and outperforms these models in a classification task.

Keywords: Information extraction · Problem phrase extraction · Mining product defects · Topic modeling · LDA · Opinion mining

1 Introduction

During the last decades, consumer products have grown in complexity, and consumer dissatisfaction is increasingly caused by usability problems, in addition to problems with technical failures. Therefore, many products are being returned to their manufacturers even though they work correctly according to their manufacturers' internal controls. In this paper, we investigate an association between problem phrases and sentiment (negative, neutral, positive) phrases in the task of problem extraction of user reviews about consumer products.

Since users compare product performance to their expectations [6], individual problem situations can be followed by different sentiment expressions. Many users describe technical problems like malfunctions or failures without negative feedback since the use of a mechanical product entails some potential for failure and they are aware of warranties and refund policies (e.g., car repair). In contrast, non-technical difficulties, called *soft usability problems* [1], are described in conjunction with an opinion's sentiment since it falls short of expectations (e.g., "visual design of Windows 8 is terrible"). Moreover, people could be satisfied with overall product quality but be bothered with short cables or uncomfortable shapes in the electronic domain. Therefore, extraction of sentiment-specific complaints allows companies to focus closely on managing customer satisfaction. Such differences in user feedback explain the motivation for this paper.

© Springer International Publishing Switzerland 2016
P. Sojka et al. (Eds.): TSD 2016, LNAI 9924, pp. 37–45, 2016.
DOI: 10.1007/978-3-319-45510-5_5

Problem extraction and identification of defects have been studied in several papers [7,8,10,11]. However, recent works have been more focused on creating task-specific dictionaries [8,10], supervised classifiers [2,11], patterns to extract bug reports [3,7] and have not investigated how existing data about words' sentiment could influence the task.

Topic models have become the main method for aspect-based opinion mining due to unsupervised learning capability to identify reviews' latent topics with sentiments towards them. In this study, we focus on applying probabilistic modeling techniques to identify product problems. We propose the topic-sentiment-problem model (TSPM) with two variants: the first model has a distribution of (document, topic, sentiment) triples over problem labels, while in the second model, we assume that a document-specific distribution over problem labels could be summarized to extract overall product problems. This model allows us to find failures reflected by many reviews rather than a particular bug.

The research questions we aim to answer are the following: *What is the impact of sentiment as topic-specific information about user opinions? Do high-tech and mechanical problems reflect only negative opinions due to customer dissatisfaction?* Our contribution is to show that incorporating topics, sentiment, and problem information with the model's variables and asymmetric priors can improve problem phrase extraction from user reviews.

The paper is organized as follows. Section 2 provides a brief overview of related work. The models are presented in Sect. 3. Experimental results are reported in Sect. 4. Section 5 concludes the paper.

2   Related Work

Our work is related to studies on sentiment analysis and mining product failures.

Aspect-Based Sentiment Analysis. There have been many studies in the area of opinion mining; Liu gave a good overview of this field in [5]. A sentiment lexicon plays a central role in most methods. State-of-the-art methods explore the use of probabilistic topic models, such as latent Dirichlet allocation (LDA) and its extensions [4,12,13]. In general, a textual document like a review is represented by a mixture of topics, where an author discusses topics or aspects such as *design, comfort, appearance, reliability, safety* (e.g., for the automobile domain). These models assume that there is a document-specific distribution over sentiments since sentiment depends on document's topics. Models' priors are based on the lexicons. In a joint sentiment-topic model (JST), topics are dependent on sentiments, whereas Reverse-JST has an opposite topic-sentiment condition [4]. Jo and Oh proposed a model called ASUM [13], where all words in a sentence are generated from one topic with the same sentiment. In [12], a model called USTM incorporates user metadata with topics and sentiments. In this model, topics depend on the document's tags, words are conditioned on the latent topics, sentiments and tags. USTM outperformed JST and ASUM in sentiment prediction. However, authors did not analyze the minimum number of tags for high-performance training without possible overfitting.

Mining Failures. Recent studies have applied task-specific dictionaries or patterns to extract problem phrases [7,8,10,11]. Incorporating sentiment and problem phrases to extract a description of problems with products has been studied in [8,10,11]. In [11], a supervised classifier is used on a set of sentiment and syntactic features without complex feature selection choices to show a need for sentiment-based features. In [10], the authors created classification rules, based on negative words from a dictionary, to detect whether or not a sentence contains a problem. They did not use non-negative sentiment words. In [7], standard LDA was used to categorize reviews' words to show that bigrams and verb phrases are more informative than single words for mining defects. To our knowledge, the extensions of LDA have not been applied for jointly considering sentiment and complaints across reviews' topics rather than sentiment.

3   Model Description

In this section, we describe topic-sentiment-problem models (TSPMs); a general graphical representation is shown in Fig. 1. The primary goal of TSPMs is to discover relations between sentiments and product problems associated with those sentiments across different topics. Further, we discuss how TSPM(DP) for document problems and TSPM(GP) for global problems achieve this goal.

Let $D = \{d_1, d_2, \ldots, d_D\}$ be a collection of reviews; a review consists of N_d words. V is the vocabulary size; K, S, R are the total number of topics, sentiment labels, and problem labels, respectively. In this paper, the problem label is a binary variable (R equals 2), while sentiment could be negative, neutral, or positive (S equals 3). α, η, γ, β are Dirichlet smoothing parameters.

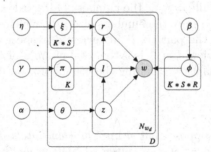

Fig. 1. Graphical representation of TSPM

3.1   TSPM(DP)

In TSPM(DP), we assume that *a problem label of a word depends on a local context, i.e. document's topics and sentiments*, as shown in Fig. 1. We describe a procedure of generating a word w in the review d using TSPM with the several steps. The model chooses a topic z from the document distribution over topics θ_d. There is the distribution over sentiments π_d^z for each topic z to sample a sentiment label l. Then a problem label r is chosen from the document-specific distribution $\xi_d^{(z,l)}$ for the pair (topic z, sentiment label l). Finally, words are chosen in condition with the combinations of topics, sentiment and problem labels.

TSPM(DP) assumes the following generative process: for each triple of topic z, sentiment l and problem label r, draw a word distribution $\phi_{z,l,r} \sim Dir(\beta_{lr})$ ($l \in \{neg, neu, pos\}$, $r \in \{pw, no\text{-}pw\}$). Then, for each review d:

- draw a topic distribution $\theta_d \sim Dir(\alpha)$
- for each topic z, draw a sentiment distribution $\pi_d^z \sim Dir(\gamma)$
- for each pair (z, l), draw a problem distribution $\xi_d^{z,l} \sim Dir(\eta)$
- For each of the N_d words in the review d:
  - draw a topic $z_{d,w_i} \sim Mult(\theta_d)$
  - for the chosen topic, draw a sentiment $l_{d,w_i} \sim Mult(\pi_d^z)$; for the (topic, sentiment) pair, draw a problem label $r_{d,w_i} \sim Mult(\xi_d^{z,l})$
  - for the chosen triple $(z_{d,pw_i}, l_{d,w_i}, r_{d,w_i})$, draw a word w_i from the distribution over words.

We use the Gibbs sampling for posterior inference. The assigned latent parameters for a word i in a review d can be sampled as follows:

$$P(\mathbf{z}_{d,i} = k, \mathbf{l}_{d,i} = l, \mathbf{r}_{d,i} = r | \mathbf{w}_{d,i} = w, \mathbf{z}_{-(d,i)}, \mathbf{l}_{-(d,i)}, \mathbf{r}_{-(d,i)}, \alpha, \beta, \gamma, \eta) \propto$$

$$\frac{n_{k,l,r,w}^{-(d,i)} + \beta_{l,r}^w}{n_{k,l,r}^{-(d,i)} + \sum_{j=1}^V \beta_{l,r}^j} \frac{n_{d,k,l,r}^{-(d,i)} + \eta}{n_{d,k,l}^{-(d,i)} + R*\eta} \frac{n_{d,k,l}^{-(d,i)} + \gamma}{n_{d,k}^{-(d,i)} + L*\gamma} \frac{n_{d,k}^{-(d,i)} + \alpha}{n_d^{-(d,i)} + K*\alpha} \qquad (1)$$

Here, $n_{k,l,r,w}$ is the number of times word w was assigned to topic k, sentiment l and problem label r, while $n_{k,l,r}$ is the number of all words assigned to the triple (k, l, r). $n_{d,k,l,r}$ and $n_{d,k,l}$ are the numbers of words, which were assigned to the triple (k, l, r) and to the pair (k, l) in review d, respectively. n_d is the total number of words in review d, $n_{d,k}$ is is the total number of n_d appeared in topic k. $-(d,i)$ denotes a quantity excluding the counts of a word i in review d.

3.2 TSPM(GP)

In **TSPM(GP)**, we suppose that *words' problem labels depend on a corpus-specific distribution of pairs (topic, sentiment) independently from a review*. This assumption refers to identifying common products' issues experienced by users rather than problems in a particular review. Problem labels are sampled from a document-independent distribution of pairs (topic, sentiment), which allows the combination of documents' distributions over problem labels. We describe a procedure of generating a word w in review d. The model chooses a topic z from θ_d, and then chooses a sentiment label l for each topic z from π_d^z. A problem label r is chosen from the corpus-based distribution $\xi^{(z,l)}$ over the pair (topic z, sentiment label l). Finally, words are chosen in condition with the combinations (z, l, r).

The generative process is similar to TSPM(DP). We sample the latent parameters for a word i in review d from Eq. 1, where items $n_{d,k,l,r}$ and $n_{d,k,l}$ in the second factor are summarized over all documents. Sentiment and problem information are incorporated via priors β_{lrw} as a sum of β_{lw} and β_{rw}, which are based on seed words from lexicons.

4   Experiments and Evaluation

To train models[1], we used reviews about several types of products from an Amazon data set in English[2] about electronics, cars, home tools, and baby products. We crawled Google Play to collect comments about financial applications in Russian. During a preprocessing step, we removed all stopwords, the punctuation, converted word tokens to lowercase. We applied stemming and lemmatization for all words in English and Russian, respectively, using the NLTK and Mystem libraries. If a word is preceded by a negation in English, we marked this word with the negation mark "neg_". We reduced datasets' words that occurred less than three times and adopted a dataset with labeled sentences from [8]. Statistics is presented in Table 1.

Table 1. Summary of statistics of review datasets.

Product domain	#reviews with overall rating r						#labeled examples (test)	
	$r = 1$	$r = 2$	$r = 3$	$r = 4$	$r = 5$	$|V|$	Problem	No-problem
Electronics	831	589	923	2196	5462	26513	603	747
Baby products	363	216	279	614	1647	6138	780	363
Home tools	860	508	983	1930	5720	16209	611	239
Cars	443	271	393	1267	4199	8674	827	171
Applications	2741	1308	1862	10597	10199	21375	1654	1216

We used state-of-the-art approaches as baselines, which are JST and Reverse-JST (further marked as RJST) [4], ASUM [13], and USTM [12]. Each dataset in English contained reviews from top-25 writing users' cities, which were used as metadata tags for USTM ($T = 25$). Following recent studies, posterior inference for all models was drawn using 1000 Gibbs iterations and set $K = 5$, $\alpha = \frac{50}{K}$, and priors $\gamma = \frac{0.01*AvgL}{S}$, $\eta = \frac{0.01*AvgL}{R}$, where $AvgL$ is the average review length.

4.1   Asymmetric Priors

We adopted MPQA[3] as a sentiment lexicon (SL). We used words and words' negations from a *Problem Words* dictionary as a problem lexicon (PL), described

---

[1] The code of TSPM is available at https://bitbucket.org/tutubalinaev/tspm/.

[2] https://snap.stanford.edu/data/web-Amazon.html.

[3] http://mpqa.cs.pitt.edu/lexicons/subj_lexicon/.

in [10]. To incorporate knowledge into β_{lrw}, we set the priors as follows: first, we first set the β priors for all words to $\beta_{*w} = 0.01$; for a positive word in SL, $\beta_{lw} = (1, 0.01, 0.001)$ (1 for pos, 0.01 for neu, and 0.001 for neg); for a negative word, $\beta_{lw} = (0.001, 0.01, 1)$; for a problem indicator from the PL, $\beta_{rw} = (1, 0.001)$ (1 for pw, 0.001 for $no\text{-}pw$); and for an indicator with the negation, $\beta_{rw} = (0.001, 1)$. Second, to avoid limitation of the fixed priors and dictionaries, we optimize priors β_{lrw} for all words, using the EM-algorithm with an update step after every 200 iterations of training [9]:

$$\beta_{lrw} = \frac{1}{\tau(i)} n_{l,r,w}, \tau(i) = n_w * \max(1, 200/i) \tag{2}$$

4.2   Quantitative Results

In order to evaluate the quality of topic models, we randomly used 10% of the reviews to compute perplexity, applying 90% reviews for training. We set K to 5 after experiments with a number of topics in TSPM. We used both lexicon separately to set baseline models' priors. Results are presented in Table 2. The proposed models achieved better results in terms of perplexity over ASUM, JST, and RJST, in which sentences and words are conditioned on latent topics and sentiments without observed tags. Therefore, the latent problem label is integrated into a topic-sentiment model without loss of quality.

Table 2. Perplexity of the topic models.

Models	Electronics	Cars	Tools	Baby	Applications
JST+SL	1549.45	951.16	1270.57	851.93	1575.30
RJST+SL	1662.26	1185.29	1406.71	966.98	1875.25
ASUM+SL	1854.15	1321.35	1528.28	1189.61	1576.41
USTM+SL	1764.29	547.784	880.77	420.29	no tags (n/a)
JST+PL	1594.23	994.86	1306.57	877.15	1438.59
RJST+PL	1705.90	1184.47	1479.44	995.51	1712.14
ASUM+PL	1811.04	1357.86	1551.76	1040.45	1436.14
USTM+PL	1840.84	**504.59**	**814.01**	**361.74**	no tags (n/a)
TSPM(DP)	1421.77	862.54	1136.04	810.30	**1078.66**
TSPM(GP)	**1318.69**	791.83	1065.84	846.35	1392.11

Second, the classification results are presented on the testing dataset in Table 3. A document d was classified to the problem class if its probability of problem label $P(r_{pw}|d)$ was higher than its probability of $P(r_{no\text{-}pw}|d)$ and vice versa. To calculate $P(r|d)$ we derive this distribution from sentiment-specific problem subtopics, ϕ, by marginalizing out topics, z, and sentiments, l:

$$P(r|d) \propto P(d|r) = \prod_{w \in \boldsymbol{w}_d} \sum_{z=1}^{K} \sum_{l=1}^{S} P(w|r, l, z) \tag{3}$$

Table 3. Accuracy and F1-measure of problem prediction using the topic models.

Models	Electronics		Cars		Tools		Baby		Applications	
	A	F1	A	F1	A	F1	A	F1	A	F1
JST+SL	0.516	0.218	0.277	0.276	0.361	0.290	0.402	0.341	0.651	0.421
RJST+SL	0.512	0.173	0.258	0.207	0.358	0.250	0.399	0.286	0.656	0.436
ASUM+SL	0.378	0.483	0.352	0.402	0.429	0.443	0.494	0.475	0.608	0.586
USTM+SL	0.501	0.502	0.462	0.242	0.535	0.344	0.499	0.303	no tags (n/a)	
JST	0.488	0.498	0.670	0.790	0.580	0.691	*0.611*	*0.711*	0.647	0.698
RJST	0.502	0.605	*0.679*	*0.797*	0.593	0.712	0.582	0.673	0.640	0.709
ASUM	*0.526*	*0.525*	0.571	0.697	0.511	0.605	0.587	0.665	*0.673*	*0.716*
USTM	0.470	0.619	0.616	0.748	*0.616*	*0.743*	0.546	0.643	no tags (n/a)	
TSPM(DP)	0.504	0.528	0.587	0.718	0.611	0.723	0.535	0.607	**0.675**	**0.727**
TSPM(GP)	0.472	**0.624**	**0.680**	**0.804**	**0.664**	**0.785**	**0.628**	**0.734**	0.613	0.645

TSPM(GP) performed better than TSPM(DP) in mechanical domains (2nd, 3rd) while TSPM(DP) are more accurate for high-tech products (1st, 5th) with complex functionality, where difficulties are more user-specific. TSPM(GP) and TSPM(DP) outperformed other models for English and Russian, respectively (Wilcoxon signed-rank test, $p < 0.001$). We also compared the results with the baselines, based on SL, using a negative label as a problem label. These models achieved the avg. decrease of 45 % of F1-measure than the same models, in which priors were based on PL. It confirms that there is no one-to-one reflection of a negative class to a problem class.

4.3 Qualitative Analysis

The aim of adding the sentiment variable is to summarize the opinions with different sentiment polarities to better understand a decline in customer satisfaction over products' problems. Table 4 contains examples of problem-specific topics (stemmed) over bigrams. Following the topics, a customer can like the design of a product and a product technically works, but a product's shape is uncomfortable or there is difficulty in finding a specific function ("imposs adjust", "tight fit", "not activ"). There is a difference with negative topics where a product does not work as expected and a customer is dissatisfied with product defects or discussions with a tech support ("zipper broke", "safeti issu", "wait hour", "replac part"). Therefore, positive reviews about product weakness could be treated as an additional way to investigate usability on products, while the companies could improve quality control in manufacturing processes considering products which classified as "defective" and cause users' frustration. TSPM(GP) extracts more verbs and adjectives, while TSPM(DP) extracts nouns aspects.

Table 4. Topics discovered about baby products (1), electronics (2), tools (3).

#	Type	Problem topics using TSPM(DP)	Problem topics using TSPM(GP)
1	NEU	car trip, tri mani, *seat difficult, fabric back*	love toy, *lock accid, break easier, not funct*
	POS	strap work, *imposs adjust*, pretti color	*difficult put*, son sensit, work realli, *tight fit*
	NEG	not work, return item, *zipper broke*	long trip, *safeti issu*, return day, strap little
2	NEU	screen protector, internet cabl, *not present*	*player freez, surround nois*, small instruct
	POS	time week, set option, good remot, *not activ*	navig button, function properli, *bit slow*
	NEG	noth happen, *tech support*, wait hour	*problem product, bit frustrat*, tight budget
3	NEU	product work, anyth els, pay ship, handi tool	stainless steel, water pressur, *cold water*
	POS	realli nice, bought set, batteri backup, *spare die*	stabl kid, *warm die*, free drop, *issu grit*
	NEG	send back, *replac part*, poor qualiti	home depot, hardwar store, *extens cord*

5   Conclusion

In this paper, we have proposed a probabilistic topic-sentiment-problem model (TSPM) that discovers information about products' problems in conjunction with reviews' topics and words' sentiments. We have presented two modifications of the TSPM: (i) TSPM(DP), in which a word's problem label is sampled from a distribution over topics and sentiments, and (ii) TSPM(GP), which summarize a document-specific distribution over problem labels to extract more global product problems. Qualitative analysis shows that users gain overall positive experience and emphasize functionality or uncomfortable mechanical structure in order to improve a product while the lack of technical quality or non-effective service recovery causes negative feedback and should be carried out by product services with the highest priority. In our future work, we plan to consider an impact of rating as an observed value.

Acknowledgements. This work was supported by the Russian Science Foundation grant no. 15-11-10019.

References

1. Kim, C., Christiaans, H., van Eijk, D.: Soft problems in using consumer electronic products. In: IASDR Conference (2007)
2. Gupta, N.K.: Extracting descriptions of problems with product and service from Twitter data. In: Proceedings of the 3rd Workshop on Social Web Search and Mining (SWSM 2011) in Conjunction with SIGIR 2011 (2011)
3. Iacob, C., Harrison, R., Faily, S.: Online reviews as first class artifacts in mobile app development. In: Memmi, G., Blanke, U. (eds.) MobiCASE 2013. LNICST, vol. 130, pp. 47–53. Springer, Heidelberg (2014)

4. Lin, C., He, Y., Everson, R., Rüger, S.: Weakly supervised joint sentiment-topic detection from text. IEEE Trans. Knowl. Data Eng. **24**(6), 1134–1145 (2012)
5. Liu, B.: Sentiment Analysis: Mining Opinions, Sentiments, and Emotions. Cambridge University Press, Cambridge (2015)
6. McCollough, M.A., Berry, L.L., Yadav, M.S.: An empirical investigation of customer satisfaction after service failure and recovery. J. Serv. Res. **3**(2), 121–137 (2000)
7. Moghaddam, S.: Beyond sentiment analysis: mining defects and improvements from customer feedback. In: Hanbury, A., Kazai, G., Rauber, A., Fuhr, N. (eds.) ECIR 2015. LNCS, vol. 9022, pp. 400–410. Springer, Heidelberg (2015). doi:10.1007/978-3-319-16354-3_44
8. Tutubalina, E.: Dependency-based problem phrase extraction from user reviews of products. In: Král, P., et al. (eds.) TSD 2015. LNCS, vol. 9302, pp. 199–206. Springer, Heidelberg (2015). doi:10.1007/978-3-319-24033-6_23
9. Tutubalina, E., Nikolenko, S.: Inferring sentiment-based priors in topic models. In: Pichardo Lagunas, O., et al. (eds.) MICAI 2015. LNCS, vol. 9414, pp. 92–104. Springer, Heidelberg (2015). doi:10.1007/978-3-319-27101-9_7
10. Solovyev, V., Ivanov, V.: Dictionary-based problem phrase extraction from user reviews. In: Sojka, P., Horák, A., Kopeček, I., Pala, K. (eds.) TSD 2014. LNCS, vol. 8655, pp. 225–232. Springer, Heidelberg (2014)
11. Walid, M., Hadeer, N.: Bug report, feature request, or simply praise? On automatically classifying app reviews. In: RE 2015 (2015)
12. Yang, Z., Kotov, A., Mohan, A., Lu, S.: Parametric and non-parametric user-aware sentiment topic models. In: Proceedings of the 38th International ACM SIGIR Conference on Research and Development in Information Retrieval, pp. 413–422. ACM (2015)
13. Yohan, J., Alice, O.: Aspect and sentiment unification model for online review analysis. In: Proceedings of the 4th ACM International Conference on Web Search and Data Mining, WSDM 2011, pp. 815–824 (2011)

Digging Language Model – Maximum Entropy Phrase Extraction

Jakub Kanis[(✉)]

Faculty of Applied Sciences, NTIS – New Technologies for the Information Society, University of West Bohemia, Univerzitní 8, 306 14 Pilsen, Czech Republic
jkanis@kky.zcu.cz

Abstract. This work introduces our maximum entropy phrase extraction method for the Czech – English translation task. Two different corpora and language models of the different sizes were used to explore a potential of the maximum entropy phrase extraction method and phrase table content optimization. Additionally, two different maximum entropy estimation criteria were compared with the state of the art phrase extraction method too. In the case of a domain oriented translation, maximum entropy phrase extraction significantly improves translation precision.

Keywords: Machine translation · Phrase extraction · Maximum entropy

1 Introduction

The phrase table and the language model are two important data components of the phrase based automatic translation system. Currently using state of the art method for the phrase extraction is based on the heuristic approach which extracts all possible phrases regardless of the other components of the translation system. However in the process of the decoding (searching for the best translation), only the translation hypotheses suggested by the phrase table are next evaluated. All the possible translations of the source translated text parts (translation options) are gathered through the analyzing of the source text by the phrase table containing the translation pairs. The translation options serve as a first step in the searching for the best (the most probable) translation performed by the following system components (language model, distortion model, etc.). The phrase table content, i.e. the translation pairs, is crucial for the best translation searching because only the hypotheses contained in the table can be considered. Thus, connect the phrase extraction to the whole translation chain and optimize phrase table content with regard to the other translation system components could bring additional improvements in a translation precision.

This work was supported by the project LO1506 of the Czech Ministry of Education, Youth and Sports. Access to computing and storage facilities owned by parties and projects contributing to the National Grid Infrastructure MetaCentrum, provided under the programme "Projects of Large Research, Development, and Innovations Infrastructures" (CESNET LM2015042), is greatly appreciated.

© Springer International Publishing Switzerland 2016
P. Sojka et al. (Eds.): TSD 2016, LNAI 9924, pp. 46–53, 2016.
DOI: 10.1007/978-3-319-45510-5_6

The main idea of this work is a broad comparison of a maximum entropy phrase extraction method originally introduced in our previous work [1] dealing with Sign language translation with the heuristic state of the art phrase extraction method in the task of the majority language translation. The original method was enhanced by adding word alignment features and changing of a weight optimization manner.

2 Phrase Extraction Based on Minimal Loss Principle

2.1 Minimal Loss Principle

Our task is to find for each source phrase $\bar{s}$ its translation, i.e. the corresponding target phrase $\bar{t}$. We suppose that we have a sentence-aligned bilingual corpus (pairs of the source and target sentences). We start with the source sentence $\mathbf{s} = w_1, ..., w_J$ and the target sentence $\mathbf{t} = w_1, ..., w_I$ and generate a bag β of all possible phrases up to the given length l:

$$\beta\{\mathbf{s}\} = \{\bar{s}_m\}_{m=1}^l : \{\bar{s}_m\} = \{w_n, ..., w_{n+m-1}\}_{n=1}^{J-m+1} \tag{1}$$

$$\beta\{\mathbf{t}\} = \{\bar{t}_m\}_{m=1}^l : \{\bar{t}_m\} = \{w_n, ..., w_{n+m-1}\}_{n=1}^{I-m+1} \tag{2}$$

The source phrases longer than one word are keeping for next processing only if they have been seen in the corpus at least as much as given threshold τ_a. All target phrases are keeping regardless of the number of their occurrence in the corpus. Each target phrase is considered to be a possible translation of each kept source phrase

$$\forall \bar{s} \in \beta\{\mathbf{s}\} : |\bar{s}| = 1 \vee N(\bar{s}) \geq \tau_a : \bar{s} \rightarrow \beta\{\mathbf{t}\}, \tag{3}$$

where $N(\bar{s})$ is number of occurrences of phrase $\bar{s}$ in the corpus. Now for each possible translation pair $(\bar{s}, \bar{t}) : \bar{t} \in T(\bar{s}), T(\bar{s}) = \{\bar{t}\} : \bar{s} \rightarrow \bar{t}$ we compute its corresponding score:

$$c(\bar{s}, \bar{t}) = \sum_{k=1}^K \lambda_k h_k(\bar{s}, \bar{t}), \tag{4}$$

where $h_k(\bar{s}, \bar{t}), k = 1, 2, ..., K$ is set of K features, which describe the relationship between the pair of phrases $(\bar{s}, \bar{t})$. The MERT training can be used for weights λ_k optimization. The resulting scores $\mathbf{c} = \{c\}$ are stored in a hash table, where the source phrase $\bar{s}$ is the key and data are all possible translations $\bar{t} \in T(\bar{s})$ with its score $c(\bar{s}, \bar{t})$. We process the whole training corpus and store the scores for all possible translation pairs.

The next step is choosing only "good" translations pairs $\{\bar{s} \rightarrow \bar{t}\}$ from all possible translations pairs for each sentence pair $\mathbf{s}, \mathbf{t}$. We generate the bag of all phrases up to the given length l for both sentences. Then for each $\bar{s} \in \beta(\mathbf{s})$ we compute a **translation loss** $\mathbf{L_T}$ for each $\bar{t} \in T(\bar{s}) = \beta(\mathbf{t})$. The translation loss L_T for the source phrase $\bar{s}$ and its possible translation $\bar{t}$ is defined as:

$$L_T(\bar{s}, \bar{t}) = \frac{\sum_{\tilde{s}_i \in \beta(\mathbf{s}), \tilde{s}_i \neq \bar{s}} c(\tilde{s}_i, \bar{t})}{c(\bar{s}, \bar{t})} \tag{5}$$

Algorithm of phrase extraction based on minimal loss principle
for all sentence pairs *(s,t)* **do**
    Make a bag of phrases for each side of translation
    for all phrase pairs $(\bar{s},\bar{t})$ **do**
        Count values of all features $h_k(\bar{s},\bar{t})$
        Count and save into hash table value of final score: $c(\bar{s},\bar{t}) = \sum_{k=1}^{K} \lambda_k h_k(\bar{s},\bar{t})$
    end for
end for
for all sentence pairs *(s,t)* **do**
    Make a bag of phrases for each side of translation
    for all phrases $\bar{s} \in \beta(\mathbf{s})$ **do**
        for all translations $\bar{t} \in \beta(\mathbf{t})$ **do**
            Count and save translation loss: $L_T(\bar{s},\bar{t}) = \dfrac{\sum_{\tilde{s}_i \in \beta(s), \tilde{s}_i \neq \bar{s}} c\ (\tilde{s}_i,\bar{t})}{c(\bar{s},\bar{t})}$
        end for
        Find best L_{Tmax} from all L_T and select all translation pairs with $L_T \geq L_{Tmax} - \tau$
    end for
end for

Fig. 1. Phrase extraction algorithm based on minimal loss principle.

We compute how much probability mass we lost for the rest of the source phrases from the bag $\beta(\mathbf{s})$ if we translate $\bar{s}$ as $\bar{t}$. For each $\bar{s}$ we store all translation losses $L_T(\bar{s},\bar{t})$ for all $\bar{t} \in \beta(\mathbf{t})$. Now we find the best L_T score L_{Tmax} (the one which brings the lowest loss in translation probability; for the log defined score it is the highest one) from all the stored scores L_T and keep only translation pairs $\bar{s} \rightarrow \bar{t}$, which L_T score is better then L_{Tmax} minus the given threshold τ. We process all sentence pairs and get a new phrase table. This table comprises only "good" translation pairs and the numbers of how many times a particular translation pair was determined as a "good" translation. These information can be then used for example for calculation of translation probabilities ϕ. The whole process is in Fig. 1.

2.2   Features

We used features based on the number of occurrences of the translation pairs and the particular phrases in the training corpus. We collected these numbers: the number of occurrences of each considered source phrase $N(\bar{s})$, the number of occurrences of each target phrase $N(\bar{t})$, the number of occurrences of each possible translation pair $N(\bar{s},\bar{t})$ and the number of how many times was the given source or the target phrase considered as the translation $N_T(\bar{s})$ and $N_T(\bar{t})$ respectively (it corresponds to the number of all phrases in all sentence pairs for which the given phrase was considered as a possible translation). These numbers are used to compute the following features: a translation probability ϕ, a probability p_T that the given phrase is a translation - all for both translation directions and a translation probability p_{MI} based on a mutual information. The translation probability ϕ is defined on the base of the relative frequencies as in [3]. And the probability p_T, that the given phrase is a translation, i.e. it appears together with the considered phrase as its translation, are defined as:

$$\phi(\bar{s}|\bar{t}) = \frac{N(\bar{s},\bar{t})}{N(\bar{t})}, \phi(\bar{t}|\bar{s}) = \frac{N(\bar{s},\bar{t})}{N(\bar{s})}; \quad p_T(\bar{s}|\bar{t}) = \frac{N(\bar{s},\bar{t})}{N_T(\bar{t})}, p_T(\bar{t}|\bar{s}) = \frac{N(\bar{s},\bar{t})}{N_T(\bar{s})} \quad (6)$$

The translation probability p_{MI} based on the mutual information is defined in [2] as:

$$p_{MI}(\bar{s}, \bar{t}) = \frac{MI(\bar{s}, \bar{t})}{\sum_{\bar{t} \in T(\bar{s})} MI(\bar{s}, \bar{t})} \qquad (7)$$

We can use both numbers N and N_T for the computing p_{MI} and p_{MI_T}:

$$MI(\bar{s}, \bar{t}) = p(\bar{s}, \bar{t}) \log \frac{p(\bar{s}, \bar{t})}{p(\bar{s}) \cdot p(\bar{t})}, p(\bar{s}, \bar{t}) = \frac{N(\bar{s}, \bar{t})}{N_S}, p(\bar{s}) = \frac{N(\bar{s})}{N_S}, p(\bar{t}) = \frac{N(\bar{t})}{N_S} \quad (8)$$

$$MI_T(\bar{s}, \bar{t}) = p_T(\bar{s}, \bar{t}) \log \frac{p_T(\bar{s}, \bar{t})}{p_T(\bar{s}) \cdot p_T(\bar{t})}, \qquad (9)$$

$$p_T(\bar{s}, \bar{t}) = \frac{N(\bar{s}, \bar{t})}{N_T}, p_T(\bar{s}) = \frac{N_T(\bar{s})}{N_T} p_T(\bar{t}) = \frac{N_T(\bar{t})}{N_T}, \qquad (10)$$

where N_S is the number of all sentence pairs in the corpus and N_T is the number of all possible considered translations, i.e. if the source sentence length is 5 and the target sentence length 9 then we add 45 to N_T.

Next used features are based on the word alignment model(s). We used three features: an extract word alignment probability p_{EWA}, a decode word alignment probability p_{DWA} and a within phrase pair consistency ratio $wppcr$. The extract probability calculation is based on the lexical translation probability $w(s_i|t_j)$ of word pair s_i, t_j (i.e. the probability that t_j is a translation of s_i) acquired from the established word alignment. It is defined as:

$$p_{EWA}(\bar{s}|\bar{t}) = \frac{\prod_{i=1}^{m} \sum_{j=1}^{n} w(s_i|t_j)}{n^m}, \qquad (11)$$

where $\bar{s} = s_1, \ldots, s_m$, $\bar{t} = t_1, \ldots, t_n$. The decode word alignment probability is a standard probability feature used by Moses decoder in its phrase table and originally introduced in [3]. The calculation is based on the lexical translation probability $w(s_i|t_j)$ and the alignment $A = \{a_k\}_{k=1}^{K} : s_i \rightarrow t_j$ of the words in the parallel sentence pair.

$$p_{DWA}(\bar{s}|\bar{a}, \bar{t}) = \prod_{i=1}^{m} \frac{1}{|\{j|(i,j) \in \bar{a}|} \sum_{\forall (i,j) \in \bar{a}} w(s_i|t_j), \qquad (12)$$

where $\bar{a}$ is word alignment between the word positions $i = 1, \ldots, m$ in $\bar{s}$ and $j = 1, \ldots, n$ in $\bar{t}$ including NULL word position. The within phrase pair consistency ratio $wppcr$ of the given phrase pair is defined in [4] as follow: the number of consistent word links associated with any words within the phrase pair divided by the number of all word links associated with any words within the phrase pair. The consistent word link connects only the words inside the phrase pair and vice versa the inconsistent word link connects a word within the phrase to a word outside the phrase pair. All features are counted for both translation directions.

Finally we have twelve features for the phrase extraction, six features based on the relative counts: $\phi(\bar{s}|\bar{t})$, $\phi(\bar{t}|\bar{s})$, $p_T(\bar{s}|\bar{t})$, $p_T(\bar{t}|\bar{s})$ and $p_{MI}(\bar{s}, \bar{t})$, $p_{MI_T}(\bar{s}, \bar{t})$ and six features based on the word alignment (in the case that we are using only one word alignment model since these feature can be computed for each word alignment model separately): $p_{EWA}(\bar{s}|\bar{t})$, $p_{EWA}(\bar{t}|\bar{s})$, $p_{DWA}(\bar{s}|\bar{a}, \bar{t})$, $p_{DWA}(\bar{t}|\bar{a}, \bar{s})$ and $wppcr(\bar{s}|\bar{t})$, $wppcr(\bar{t}|\bar{s})$.

Table 1. Training and development data statistics.

CzEng	Train 50k		Train 151k		dev (csen_eval)	
	cs	en	cs	en	cs	en
# sent	49 719		150 405		15 000	
# tokens	708 782	795 063	2 115 822	2 373 192	216 412	243 737
# words	80 736	45 421	155 122	84 516	37 099	21 946
dev OOV [%]	8.01	4.09	4.81	2.47	-	-
dev ppl	323.69	152.33	268.40	124.88	-	-
EuCsEn	Train 50k		Train 640k		dev (newstest2013)	
	cs	en	cs	en	cs	en
# sent	49 735		643 076		3 000	
# tokens	1 159 962	1 332 595	15 363 462	17 758 623	60 038	67 810
# words	60 361	25 353	192 492	71 148	16 284	9 679
dev OOV [%]	12.91	6.93	6.11	3.37	-	-
dev ppl	743.42	340.51	850.30	328.59	-	-

3   Experiments

3.1   Data

We mainly focus on the translation between Czech and English in our research. We used two freely available Czech - English parallel corpora to compare our phrase extraction method with the state-of-the-art method. The first corpus is CzEng 1.0[1] which contains 15 million parallel sentences. The second one is Czech - English part of the Europarl[2] corpus (referred as EuCsEn) which contains more than 640 thousands of parallel sentences. For training, we used only the same portion of the first 50k of sentences of both corpora to be able to compute and compare the experimental results. The detail data description is in Table 1. The CzEng development data (csen_eval) is a portion of the first 15 thousands of sentences from the DevTest section of the corpus. And analogously the CzEng test data (csen_heldout) is a portion of the first 15 thousands of sentences from the EvalTest section of the corpus.

3.2   Results

The results of the comparison of the state of the art (the label *moses* in the table) and our phrase extraction method (the label *me* in the table) are in Tables 2 and 3. The Moses system[3] was used as the state of the art phrase

---

[1] http://ufal.mff.cuni.cz/czeng/.

[2] http://www.statmt.org/europarl/.

[3] http://www.statmt.org/moses/.

Table 2. Development data results – BLEU scores: statistically better results are marked with [+], statistically worse with [-] (only bold columns), the best result is bold.

CzEng - csen_eval	me val1	me val2	me val3	me gen	me nm	moses	**me mert**	moses mert
map_o1_50klm	24.89	24.89	24.9	24.9	24.92	24.32	**25.61**[+]	24.67
mlp_o1_50klm	24.89	24.89	24.95	25.12	25.15		25.59[+]	
map_o5_50klm	24.04	24.05	24.07	24.12	24.16		24.74	
mlp_o5_50klm	23.81	23.93	23.97	23.97	24.1		24.53	
map_o1_151klm	26.19	26.19	26.19	26.19	26.19	25.05	26.91[+]	25.38
mlp_o1_151klm	25.97	26.02	26.09	26.25	26.38		**26.98**[+]	
map_o5_151klm	25.19	25.24	25.25	25.27	25.3		25.94[+]	
mlp_o5_151klm	24.79	24.99	25.01	25.01	25.06		25.56[+]	
EuCsEn - newstest2013								
map_o1_50klm	11.81	11.83	11.83	11.86	11.89	12.23	12.46	12.77
mlp_o1_50klm	12.07	12.15	12.22	12.34	12.47		12.63	
map_o5_50klm	11.71	12.05	12.1	12.12	12.16		12.57	
mlp_o5_50klm	11.93	11.94	12.32	12.43	12.49		**12.67**	
map_o1_640klm	12.97	12.97	13.0	13.0	13.04	13.1	**13.83**	13.59
mlp_o1_640klm	12.9	12.99	13.07	13.2	13.3		13.59	
map_o5_640klm	13.0	13.04	13.1	13.11	13.15		13.82	
mlp_o5_640klm	13.02	13.13	13.23	13.34	13.4		13.66	

Table 3. Test data results – BLEU scores: statistically better results are marked with [+], statistically worse with [-] (only bold columns), the best result is bold.

CzEng - csen_heldout	me 50klm	moses 50klm	**me mert 50klm**	moses mert 50klm	me 151klm	moses 151klm	**me mert 151klm**	moses mert 151klm
map_o1	25.88	25.06	**26.4**[+]	25.29	27.13	25.85	**27.59**[+]	26.0
mlp_o1	26.03		26.19[+]		27.18		27.58[+]	
map_o5	25.1		25.46		26.11		26.43[+]	
mlp_o5	24.85		25.1		25.8		25.93	
EuCsEn - newstest2012	me 50klm	moses 50klm	**me mert 50klm**	moses mert 50klm	me 640klm	moses 640klm	**me mert 640klm**	moses mert 640klm
map_o1	10.34	11.14	10.69[-]	11.25	11.68	11.74	**12.03**	11.88
mlp_o1	10.68		10.73[-]		11.69		11.73	
map_o5	10.73		**10.9**[-]		11.65		12.02	
mlp_o5	10.73		10.87[-]		11.62		11.7	

extractor (the default setting of the phrase extraction method was used) and decoder with standard feature set: translation probability ϕ, decode word alignment probability p_{DWA} (all for both directions), language model probability and the word and phrase penalty. All experiments were performed with the monotone reordering and maximal 20 translations for each source phrase. Two additional features: extract word alignment probability p_{EWA} (for both directions) and the additional phrase penalty (Euler number e for each phrase in the phrase table) were used in the case of decoding with me phrase table. Because consider each target phrase as the possible translation of each source phrase is computational challenging, all results are reported for the phrases of the maximal length three (usually there is a small performance difference between the phrases of the length three and longer phrases).

Two different estimation strategies were used for the maximum entropy based phrase extraction. The first one is the standard maximum a posteriori estimation, denoted as map prefix in the table. The second one is our minimal loss principle estimation (see Sect. 2.1), denoted as mlp prefix in the table. The prefix $o1$ then mark using of all source phrases longer than one word and the prefix $o5$ using of only longer phrases seeing minimal of five times. The string $50klm$ means using of language model trained strictly from the phrase table extraction data. Analogously $151klm$ ($640klm$) means using of a bigger language model trained from the additional parallel corpus training data. All used language models were standard trigram models with Kneser-Ney backoff prepared by KenLM[4] toolkit. A lexical probabilities provided by Moses system (the files $lex.0-0.*$) were used to compute word alignment based features.

The final me phrase table is the result of the intersection of both me phrase tables extracted for each translation direction (i.e. only the reciprocal translation pairs are selected). We need to optimize a relatively lot of variables to get the best result for the maximum entropy phrase extraction. There are seven weights for the features mentioned in the Sect. 2.2 because there is only one score (the geometric mean of the values) for each bidirectional feature. Next there are three default lexical translation probabilities w_{no_wa} (one for each feature: p_{EWA}, p_{DWA} and $wppcr$) which are used in the word alignment feature scores computation if appropriate $w(s|t)$ is null (p_{EWA}, p_{DWA}) or there are no word links between phrase words ($wppcr$). Finally, there is a threshold tau for the target phrase selection for each source phrase and a threshold tau_R for the final selection of the phrases included in the resulting phrase table. At last we have 23 variables (11 for each direction + tau_R) which are optimized. To be able to optimize this number of variables we use a combination of sequential optimization, grid search and Nelder-Mead method.

In the second column of Table 2 are the results for the w_{no_wa} values changing with the rest of variables fixed. All three variables have the same value in each step going from $1e-03$ to $1e-14$. In the third column are the results of the threshold tau changing with w_{no_wa} variables set to the best value from the previous optimization round and fixed values of the rest variables. And finally, in

---

[4] https://kheafield.com/code/kenlm/.

the fourth column are the results of the threshold tau_R changing with previously tuned variables set to the best value and fixed values of the rest variables. Then a set of all combinations of reasonable variable values based on the previous optimization rounds is generated and evaluated (results in the fifth column). This provisional best result is then used as a starting point for the MERT optimization by the Nelder-Mead algorithm. The final results of the phrase extraction variables optimization are in the sixth column. The baseline results of the Moses decoder are in the seventh column and the results after MERT decoder weights optimization are in the two last columns (labeled as *me mert* and *moses mert*). The analogous results for the test data are in Table 3.

4   Conclusion

Whereas in the case of the CzEng corpus maximum entropy phrase extraction improves the baseline by 1.11 % (*50klm*) and 1.59 % BLEU (*151klm*) respectively on the test data, in the case of the EuCsEn corpus the improvement, if any, is only modest and not statistically significant. The main reason is by our opinion a different character of both corpora or more accurately a rate of the divergence between the training part of the corpus and its development part from the language model point of the view. The perplexity values in Table 1 show higher text similarity (more than two times) between training and development data for the CzEng corpus than for the EuCsEn's ones. In addition, the higher translation precision for the lower perplexity (BLEU increase: 1.11 % → 1.59 % for ppl decrease: 152.33 → 124.88) supports our hypothesis[5]. It seems then, that optimization of the phrase table against the language model is beneficial especially in the case of a domain oriented translation. Both estimation strategies *map* and *mlp* perform with almost the same translation precision. On the other hand, the *mlp* criterion produce significantly smaller phrase tables (2 449 012 × 3 373 065 phrases for the best result: 27.58 % × 27.59 % BLEU) but still far away from the *moses* phrase table size (326 573 phrases).

References

1. Kanis, J., Müller, L.: Advances in Czech – signed speech translation. In: Matoušek, V., Mautner, P. (eds.) TSD 2009. LNCS, vol. 5729, pp. 48–55. Springer, Heidelberg (2009)
2. Lavecchia, C., Langlois, D., Smaïli, K.: Phrase-based machine translation based on simulated annealing. In: Proceedings of the LREC, Marrakech, Morocco. ELRA (2008)
3. Koehn, P., Och, F.J., Marcu, D.: Statistical phrase-based translation. In: HLT/NAACL (2003)
4. Venugopal, A., Vogel, S., Waibel, A.: Effective phrase translation extraction from alignment models. In: Proceedings of the ACL, pp. 319–326 (2003)

---

[5] There is a perplexity reduction for the EuCsEn corpus too, but is only 3.5 % relatively in opposite to the 18 % relative reduction for the CzEng corpus.

Vive la Petite Différence!
Exploiting Small Differences for Gender Attribution of Short Texts

Filip Graliński, Rafał Jaworski[✉], Łukasz Borchmann, and Piotr Wierzchoń

Adam Mickiewicz University in Poznań, Poznań, Poland
{filipg,rjawor,wierzch}@amu.edu.pl, borchmann@rainfox.org

Abstract. This article describes a series of experiments on gender attribution of Polish texts. The research was conducted on the publicly available corpus called "He Said She Said", consisting of a large number of short texts from the Polish version of Common Crawl. As opposed to other experiments on gender attribution, this research takes on a task of classifying relatively short texts, authored by many different people.

For the sake of this work, the original "He Said She Said" corpus was filtered in order to eliminate noise and apparent errors in the training data. In the next step, various machine learning algorithms were developed in order to achieve better classification accuracy.

Interestingly, the results of the experiments presented in this paper are fully reproducible, as all the source codes were deposited in the open platform *Gonito.net*. Gonito.net allows for defining machine learning tasks to be tackled by multiple researchers and provides the researchers with easy access to each other's results.

Keywords: Gender attribution · Text classification · Corpus · Common Crawl · Research reproducibility

1 Introduction

Gender classification of written language has been a subject of linguistic studies for decades now. Note, for instance, a ground-breaking book [8], describing characteristic features of women's language. In more recent years, linguists and socio-linguists researching this subject have been aided by statisticians, see for instance: [10,11]. Furthermore, development of social media opened a possibility of building large-scale text corpora, annotated with meta information regarding the author's gender. This resulted in numerous projects aimed at automatic gender classification based on training data acquired from the Web. The annotated text resources used to build the corpora were often taken from blogs, e.g. [10].

However, we believe that gender annotated corpora scraped from the selected Web sources are prone to some flaws. Firstly, they may suffer from thematic bias – women tend to write about different subjects than men, see [11]. Secondly, there might be significantly more text written by authors of either gender. This is due to the fact that if a corpus is a collection of gender annotated items such

© Springer International Publishing Switzerland 2016
P. Sojka et al. (Eds.): TSD 2016, LNAI 9924, pp. 54–61, 2016.
DOI: 10.1007/978-3-319-45510-5_7

as blog entries, these items might differ significantly in length. And lastly – the volume of the corpus may not be sufficient for reliable statistical analysis. All these flaws result in a situation, where gender classification is heavily biased by the subject women and men write about and not the language they are using.

In this paper we describe an experiment on gender classification of texts based on a custom built corpus which tries to avoid the flaws mentioned above. The corpus itself is publicly available. Gender classification mechanism is designed using statistical algorithms. It is also worth noting that the results of the described experiment are fully reproducible, as the research was conducted on the *Gonito.net* platform.

Section 2 of the paper describes similar experiments on automatic gender classification. The corpus used in our experiment is described in Sect. 3. Section 4 describes the classification mechanism itself, while Sect. 6 lists the conclusions.

2 Related Work

2.1 Argamon et al.

Work on gender classification has been described in [1]. Its authors carried out an experiment on data from the British National Corpus. The corpus contained documents labeled as being authored either by a woman or a man. Total volume of the corpus reached 604 documents, 302 for each gender, coming from different fiction and non-fiction genres (such as science, arts, commerce). Average number of words per document was 42,000 words, which brought the overall word count of the corpus to over 25 million words. Furthermore, the corpus was POS-tagged.

The authors used this data to train a custom version of an algorithm referred to as EG, described in [7]. For the sake of the training a feature set of 1000 features was prepared. It consisted of 467 function words and over 500 most frequent part-of-speech n-grams.

The experiment determined that female and male languages differ in terms of the use of pronouns and certain types of noun modifiers – females tend to use more pronouns, while males use more noun specifiers. Generally speaking, the authors describe female writing style as "involved" and male writing as "informative".

The work by Argamon et al. laid ground for research on automatic statistical gender classification. Some methods used by the authors, specifically the classification algorithm and feature selection procedure, could be and were improved in subsequent experiments.

2.2 Gender Classification of Blog Entries

In the first decade of the 21st century many solutions to the automatic gender classification problem exploited textual data from blog entries. Main advantages of blog corpora were considerable sizes and relatively reliable gender annotations (thanks to metadata about the authors of the blogs). Disadvantages, on the other hand, included significant thematic bias.

Nevertheless, a large number of classification algorithms was proposed. They relied on features such as content words, dictionary based content analysis or part-of-speech tags, see for instance [2,12] or [14].

The method proposed by [10] is based on two novel ideas. The first is a new feature group – variable length POS sequence patterns, while the second is a novel feature selection algorithm. The authors report the accuracy of their system to reach 88.56 %, while the previously mentioned systems by [2,12,14] only score 77.86 %, 79.63 % and 68.75 % respectively.

2.3   Other Interesting Solutions

The work presented in [11] explores different statistical approaches to gender classification. Interestingly, it is one of very few and possibly the first publication to acknowledge the problem of topic bias. The authors tried to remove the topic and genre bias completely and experiment with an evenly balanced corpus. Furthermore, they take the challenge of cross-topic and cross-genre gender classification. They try to find gender-specific features in language alone, with the help of stylometric techniques. After performing the experiment, the authors concluded that previous results in gender classification might have been overly optimistic due to the topic bias issues. Among various machine learning techniques used to perform classification on the bias-free corpus, simple character based models surprisingly proved the most robust.

Another interesting approach is presented in [3]. The authors managed to reproduce the results of [10] by using neural networks. Considering that [3] present merely preliminary research results, the deep learning approach may have a significant potential in the gender classification task.

3   Corpus Used in the Experiment

3.1   Original Corpus

The corpus used in our experiment is based on the "He Said She Said" Polish corpus (referred to as HSSS) described in detail in [5]. While the corpora used in research on gender-related differences have been based on *metadata* supplied manually, for example, statements of gender by text authors, or gender tags added manually by the corpus creators, the creators of the HSSS corpus propose a different approach: to look in the text itself for *gender-specific first-person expressions*. Not all languages have these (almost none of them to be found in English), but they are quite frequent in the Slavic languages and they also occur, to some extent, in the Romance languages. The procedure of preparing the HSSS corpus was to take Common Crawl-based Web corpus[1] of Polish [4] and grep for lines containing gender-specific first-person expressions to create a gender-specified subcorpus. The procedure was applied to Polish, a language in which the frequency of gender-specific first-person expressions is particularly

---

[1] http://data.statmt.org/ngrams/raw/.

high. The resulting HSSS corpus contains short fragments of texts, annotated as authored by a female or a male, accompanied by the URL pointing to the place where the fragment was published. Texts came from 92,834 different websites, by which we conclude that the HSSS corpus is likely to contain texts authored by 50–100 K different people.

The HSSS corpus differs from other gender annotated corpora as it contains short fragments of texts coming from a wide variety of thematic domains. Other researchers performed their experiments on narrow domain corpora (such as thematic blogs) which could make the classification task significantly easier.

3.2   Filtering the Corpus

Even though the HSSS corpus is an interesting linguistic dataset, it is not flawless. For example, it is not free from the topic bias problem. Most male texts come from websites presumably visited by men (such as technical, engineering or sport portals), while majority of female texts comes from more stereotypically women-like portals (e.g. sites about pregnancy). For the sake of our experiment, the original HSSS corpus underwent filtering procedures. Firstly, male/female balance was introduced by using the following principle – within fragments coming from one web domain (website) there should be an equal number of female and male texts. To ensure this, within each website, all the fragments that made one gender outnumber the other, were discarded. For example, if the corpus contained 3 male and 100 female texts from a parenting portal, only the 3 male and the first 3 female texts were taken into consideration. And if a site contained texts of only one gender, it was discarded completely. This procedure was performed in order to reduce the effects of the topic bias problem.

Another type of problems in the HSSS corpus are so called "leaks". We refer as leaks to gender-specific expressions which were not turned into male versions during the creation of the HSSS corpus. Examples of leaks we managed to filter out were some forms of adjectives and participles, which in Polish have genders (e.g. "green" in Polish is "zielona" in the feminine form but "zielony" in masculine).

Errors were also found in annotation of texts, which contained gender-specific first-person expressions, but the gender of the author could not be determined based on these expressions alone. For example, the name of the popular TV series "How I Met Your Mother" in Polish is "Jak poznałem waszą matkę" and the word "poznałem" is a male-specific first-person verb. Therefore, in the HSSS corpus we found reviews of the series written by females automatically annotated to be written by males. In order to overcome this problem, we filtered out all the texts containing the name of this and a few other titles causing the same problem.

Some noise was also observed in the HSSS corpus. It included texts which were clearly not written by Polish native speaker, but were a result of a very poor, probably automatic translation. In some of these texts we observed an oddly high frequency of the term "negacja logiczna" (logical negation), which we attributed to a machine translation mechanism, trying to translate the simple word "not" into Polish. As a result, we decided to discard all the texts containing the phrase "negacja logiczna".

3.3 Train, Dev and Test Sets

The training, development and test sets for the classification task were not selected from the corpus in a completely random manner, but instead the division respected the websites, i.e. one set of websites was denoted the training set, another – development set and the third – test set. Thus, the classification task became even more challenging, as the mechanism is to be trained on different thematic domains than it would be tested on. Importantly, all the gender-specific expressions identified during the creation of the corpus were turned to male versions, even in the female texts.

For these reasons, a classifier prepared for this task should in fact recognize gender-specific features in female and male texts and not base its judgements on simple topical differences.

4   Classification Problem Formulation and Solution

The modified version of the HSSS corpus was turned into a text classification challenge on Gonito.net (which is an open platform for research competition, cooperation and reproducibility). The training and developments sets as well as the input for the test set are publicly available there as a Git repository.[2] Accuracy was chosen as an evaluation metric.

The accuracy for the null model (always returning male or female authors) is of course 50 % {86dd91}.

(For each classification method described in this paper, the output files and all the source codes are available as submissions to the Gonito.net platform. Git commit SHA1 prefixes are always given here in curly brackets. In the electronic edition of this paper, the above commit number prefix is clickable as http:// gonito.net/q/86dd914ad99a4dd77ba1998bb9b6f77a6b076352 – alternatively, the ID 86dd91 could be entered manually at http://gonito.net/q or directly pasted into a link (http://gonito.net/q/86dd91) – and leads to a submission summary with an URL to a publicly accessible Git repository.)

Using a hand-crafted regular expression for known "leaked" feminine first-person expressions has an accuracy of only 51.03 % {f98f7d}.

4.1 Logistic Regression with Vowpal Wabbit

A logistic regression model was trained with the Vowpal Wabbit open-source learning system [9]. Lower-cased tokens (no other normalisation was done) were used as features. With this simple set-up, we obtained accuracy of 67.54 % {36ff5b}. Using a simple sigmoidal feedforward network with 6 hidden units yields a slightly better result (68.32 %, {d96cfc}), whereas adding prefixes {8f9557} or suffixes {8e0e25} as features does not improve much.

---

[2] git://gonito.net/petite-difference-challenge.git.

4.2   Language Models

We used KenLM language modelling toolkit [6] to create two separate 3-gram language models: one for male texts and one for female texts. During the classification, the class with the higher probability is simply chosen. Even though this method was substantially different from the Vowpal Wabbit classifier the results were quite similar (67.98 %, {85317f}). That's why we decided to combine both methods with another layer of neural network. This way, the best result so far was achieved: 71.06 % {12b6d8}.[3]

4.3   Morphosyntactic Tags Only

In order to completely avoid topical imbalance in vocabulary, we also trained classifiers using only morphosyntactic tags. TreeTagger [13] with a Polish model was used to tag the texts, i.e. each word was assigned its part of speech and other tags such as person, number, gender, tense.

A classifier was trained with Vowpal Wabbit with the following features:

- part-of-speech 3-grams,
- morphosyntactic tags for a given word treated as one feature,
- morphosyntactic tags for a given word treated as separate features.

The accuracy on the test set for this classifier was 59.17 % {b4e142}.

A similar, but slightly better result (60.58 %) was obtained with 6-gram language models trained on morphosyntactic tags {edfdfc}.

We find these results quite satisfactory as they were obtained on short tags using only morphosyntactic tags.

5   Back to the Corpus

The models described in Sect. 4 were analysed to find the most distinctive features. It turned out that a large number of them are actually "leaks" (expressions which should have been identified as gender-specific and normalised when HSSS was created but were not):

- gender-specific inflected forms of verbs absent from the lexicon of inflected forms,
- frequent inflected forms written with a spelling mistake (in particular, without a diacritic),
- verb *być* (*be*) with a longer adjective phrase (e.g. *jestem bardzo zadowolony/ zadowolona = I am very glad*),
- the word *sam/sama* (= *myself*, which has a different masculine and feminine form and which could mean *himself/herself*),

Some other problems were also identified in the balanced corpus: automatically generated spam not filtered out and gender-specific forms found in the film titles.

---

[3] The output files and source codes are available for inspection and reproduction at Git repository git://gonito.net/petite-difference-challenge, branch submission-00085.

6   Conclusions

The paper presented research on gender classification performed on a corpus created in a different way than in the work described so far. Gender annotations in the corpus were obtained by exploiting certain linguistic features of the Polish language, rather than by relying on meta-data. Furthermore, for the needs of the experiments the corpus was balanced by websites, in order to minimize the effect of gender and topic bias. Training data prepared in this manner is unique at least for the Polish language.

Developed classification algorithm achieved a maximum gender prediction accuracy of 71.06 %. The algorithm relied on language modelling (KenLM toolkit) and the Vowpal Wabbit machine learning system. The two methods were combined using a neural network. Classification results revealed some noise in the training data that can and should be filtered out. Nonetheless, the prediction accuracy of above 70 % can be viewed as a success, considering the competitiveness of the task.

Future work plans include further filtering of the corpus based on the information obtained during the gender classification task. Classification algorithms themselves will also be further optimized. This work will be facilitated by the Gonito.net platform.

Acknowledgements. Work supported by the **Polish Ministry of Science and Higher Education** under the **National Programme for Development of the Humanities**, grant 0286/NPRH4/H1a/83/2015: "50,000 słów. Indeks tematyczno-chronologizacyjny 1918–1939".

References

1. Argamon, S., Koppel, M., Fine, J., Shimoni, A.R.: Gender, genre, and writing style in formal written texts. TEXT **23**, 321–346 (2003)
2. Argamon, S., Koppel, M., Pennebaker, J.W., Schler, J.: Mining the blogosphere: age, gender and the varieties of self-expression. First Mon. **12**(9) (2007). http:// pear.accc.uic.edu/ojs/index.php/fm/article/view/2003
3. Bartle, A., Zheng, J.: Gender Classification with Deep Learning (2015)
4. Buck, C., Heafield, K., van Ooyen, B.: N-gram counts and language models from the common crawl. In: Proceedings of the Language Resources and Evaluation Conference, Reykjavk, Icelandik, Iceland, May 2014
5. Graliński, F., Borchmann, L., Wierzchoń, P.: "He said she said" – male/female corpus of polish. In: Proceedings of the Language Resources and Evaluation Conference LREC 2016 (2016)
6. Heafield, K.: KenLM: faster and smaller language model queries. In: Proceedings of the EMNLP 2011 Sixth Workshop on Statistical Machine Translation, Edinburgh, Scotland, UK pp. 187–197, July 2011. http://kheafield.com/professional/avenue/ kenlm.pdf
7. Kivinen, J., Warmuth, M.K.: Additive versus exponentiated gradient updates for linear prediction. In: Proceedings of the Twenty-seventh Annual ACM Symposium on Theory of Computing, STOC 1995, pp. 209–218. ACM, New York (1995). http://doi.acm.org/10.1145/225058.225121

8. Lakoff, R.: Language and woman's place. Harper colophon books, Harper & Row (1975). https://books.google.pl/books?id=0dFoAAAAIAAJ
9. Langford, J., Li, L., Zhang, T.: Sparse online learning via truncated gradient. In: Advances in Neural Information Processing Systems, NIPS 2008, vol. 21, pp. 905–912 (2009)
10. Mukherjee, A., Liu, B.: Improving gender classification of blog authors. In: Proceedings of the 2010 Conference on Empirical Methods in Natural Language Processing, EMNLP 2010, pp. 207–217. Association for Computational Linguistics, Stroudsburg (2010). http://dl.acm.org/citation.cfm?id=1870658.1870679
11. Sarawgi, R., Gajulapalli, K., Choi, Y.: Gender attribution: tracing stylometric evidence beyond topic and genre. In: Proceedings of the Fifteenth Conference on Computational Natural Language Learning, CoNLL 2011, pp. 78–86. ACL, Stroudsburg (2011). http://dl.acm.org/citation.cfm?id=2018936.2018946
12. Schler, J., Koppel, M., Argamon, S., Pennebaker, J.: Effects of age and gender on blogging. In: Proceedings of AAAI Spring Symposium on Computational Approaches for Analyzing Weblogs, March 2006
13. Schmid, H.: Probabilistic part-of-speech tagging using decision trees. In: Proceedings of the International Conference on New Methods in Language Processing, vol. 12, pp. 44–49 (1994)
14. Yan, X., Yan, L.: Gender classification of weblog authors. In: Proceedings of the AAAI Spring Symposia on Computational Approaches, pp. 27–29 (2006)

Towards It-CMC: A Fine-Grained POS Tagset for Italian Linguistic Analysis

Claudio Russo[✉]

Foreign Languages and Cultures Department, University of Turin,
Via Verdi 8, 10124 Turin, Italy
clrusso@unito.it
http://en.unito.it/

Abstract. The present work introduces It-CMC, a fine-grained POS tagset that aims at combining linguistic accuracy and computational sustainability. It-CMC is tailored on Italian data from Computer-Mediated Communication (CMC) and, across the sections of the paper, a sistematically comparison with the current tagset of the *La Repubblica* corpus is provided. After an early stage of performance monitoring carried out with Schmid's TreeTagger, the tagset is currently involved in a workflow that aims at creating an Italian parameter file for RFTagger.

Keywords: Morphological tagging · Syntactic tagging · Fine-grained POS tagset · Italian corpora · Linguistic analysis

1 Introduction

Recent decades have witnessed a very strong commitment towards POS-tagging tasks across a constantly increasing number of languages. Within this framework, Italian falls among the lucky well-documented languages: its very first POS tagset recommendation came from Monachini [11] within the EAGLES guidelines specification; over the years, some more tagsets have been proposed: Stein [18] drafted the tagset used in TreeTagger's first parameter files, a fine-grained tagset has been drafted by Barbera [3,4] for the *Corpus Taurinense*[1] and the first, inductively generated tagset has been designed by [6]; more recently, Petrov et al. [12] included Italian in their universal set of progressively updated POS mappings generated inductively.

Among such resources, probably the best known tagset – among linguists - has been illustrated in Baroni et al. [5]: conceived to tag the *La Repubblica* newspaper corpus [9], it is currently used as the official Italian tagset of the Sketch Engine platform as well as being included, sometimes with slight adaptations, in many other linguistic studies. Despite its deserved success, this tagset still seems to be too coarse-grained a resource to perform some particular linguistic analysis: on the one hand, some of its generalizations can be tackled by specific queries, where the lemma is explicitly specified; on the other hand, uniting several linguistic phenomena under the same tag makes searching the tagged corpus

---

[1] A POS-tagged corpus of ancient Italian.

© Springer International Publishing Switzerland 2016
P. Sojka et al. (Eds.): TSD 2016, LNAI 9924, pp. 62–73, 2016.
DOI: 10.1007/978-3-319-45510-5_8

a trickier task (although it is an effective solution against data sparseness): the most evident example of such generalization is the *CHE* tag, which includes any token of *che*, which, from a linguistic standpoint, falls under five different categories (namely interrogative pronoun, exclamative pronoun, relative pronoun, indefinite pronoun and subordinative conjunction) according to the grammar published by [17][2].

The present work introduces a finer-grained POS tagset that combines linguistic accuracy and computational sustainability. As a tagset tailored on Italian in Computer-Mediated Communication (CMC), some of its features may be perceived as too specific: to prevent the computational backfire generated by this increased complexity, an easy remapping to Petrov's universal tagset has been a condition pursued across the tagset's development processes. The tagset's hierarchy-defined features (HDF) will be presented and compared with the tagset derived from Baroni et al. [5] and currently used in the La Repubblica corpus; in a second stage, its morpho-syntactic features (MSF) will be listed. The concluding section contains the results of two parallel tagging processes, made with two POS-taggers made respectively by Schmid [15] and Schmid and Laws [16], namely TreeTagger and RFTagger.

2   The It-CMC Tagset: Introduction and General Comparison with *La Repubblica*

The It-CMC tagset counts 66 labels divided into 14 categories. For sake of brevity, the theoretical discussion about the terminological differences between It-CMC and the La Repubblica tagset has been skipped; a mapping table between the two tagsets is available in Appendix 6.

The two tagsets count 23 equivalent matchings on similar phenomena such as, for instance, qualifying adjectives, conjunctions and articulated prepositions. Some *La Repubblica* tags are missing in the It-CMC tagset, namely the negation tag *NEG*, the extralinguistic category *NOCAT*, the seven *VER:...:cli* tags and the tag *CHE*; the abbreviation phenomena were originally included in Baroni's tagset but have been deleted from the current *La Repubblica* tagset: in this work, they have been processed as MSFs, according to the workflow in which It-CMC is involved. On the other hand, the It-CMC tags *77* (Formula) and *69* (Emoticon) were added, clearly because they proved necessary to tag linguistic data from CMC; the tag *36* (Relative Pronoun-Determiner) partly covers the *CHE*-tagged phenomena: their treatment will be explained in a further, dedicated section of this work. Only two *La Repubblica* tags, *DET:wh* and *WH*, have been united into the coarser It-CMC tag 35 (Interrogative Pronoun-Determiners), but most of its inherent linguistic items usually fall under several categories. *La Repubblica*'s tags which underwent the heaviest modifications are the verbal tags *VER:fin*, *AUX:fin* and *VER2:fin*: each of them has been divided into 9 tags, one for each

---

[2] Chosen over the monumental work by Renzi et al. [13] thanks to its wider consensuality, following the recommendations by Leech [10].

mode-tense combination available in Italian. Each of the tags *ADV*, *ART*, *CON* and *PRE* have been split in two, to deal with general vs. connective adverbs, determinative vs. non-determinative articles, coordinating vs. subordinating conjunctions and simple vs. peudo-prepositions[3].

3   Analysis of It-CMC and La Repubblica's Tagset

3.1   One-to-One Correspondences and Partial Overlapping

As mentioned in the general presentation, both tagsets include dedicated tags for 23 phenomena, namely:

- Proper and common nouns;
- Qualifying adjectives;
- Indefinite and possessive pronouns;
- Articulated prepositions;
- Numeral adjectives/pronouns;
- Final and non-final punctuation;
- Loan words;
- Digits;
- Present and past participle, infinitive forms and gerundive forms of main, auxiliary and modal verbs.

While such tags correspond at a superficial reckoning, their usage is not always equivalent across the two corpora: an instance of known discrepancy in the annotation processes is represented by ordinal numerals (tagged as such with the It-CMC and as *ADJ* in *La Repubblica*).

3.2   Discrepancies

Pronouns and Determiners Across the Two Tagsets. The tagset presented in Baroni et al. [5], as well as its modified version currently used in the *La Repubblica* corpus, categorizes pronouns and determiners in separate sets. The It-CMC tagset follows a different path and keeps closer to the EAGLES guidelines, so that pronouns and determiners are merged into the single Pronoun-Determiner category (P-D) [2].

Strong and Weak Demonstrative P-Ds. The It-CMC tagset includes two contiguous tags for strong (30) and weak (31) demonstrative P-Ds. Such a distinction has been made because of the diastratic variation range of the corpus: sometimes the informal weak forms *'sto, 'sta, 'sti* and *'ste* are preferred to their strong counterparts *questo, questa, questi* and *queste*. As a newspaper corpus, *La Repubblica* contains no examples of this particular phenomenon and its two tags, *PRO:demo* and *DET:demo*, suitably reflects this sociolinguistic feature.

---

[3] [17] defines pseudo-preposition as "words that are mainly adverbs used in prepositional function [...]".

Personal P-Ds and Clitics. Four different tags for personal P-Ds and pronominal clitics have been included in the It-CMC tagset, namely *37* for P-Ds in nominative case, *38* for stressed P-Ds in oblique case (i.e. any case but nominative), *39* for unstressed P-Ds in oblique case. As a matter of fact, tags *38* and *39* allow users to distinguish accusative (stressed) pronouns from dative (unstressed) pronouns. According to this classification, the final user would also be able to distinguish extended dative constructions (1) from contracted ones (2) with one single query.

1. Questo è il mio cane. Lo affido a *te.*   -   38
   This is my dog. I entrust it to you.

2. Questo è il mio cane. *Te* lo affido.   -   39
   This is my dog. I entrust it to you.

Tag *37* is applied to P-Ds in nominative case, which do not suffer from homography with other pronouns; the pronouns in (1) and (2) are tagged with the broader tag *PRO:pers* in the *La Repubblica* corpus.

Ci, *vi* and *ne* are also potentially ambiguous because of their adverbial homographic counterparts (tagged in It-CMC with the label *46*, particle adverb). According to their context, these particles can either carry locative value as in (3) or pronominal personal value, as in (4)

3. *Ci*   sono   andat-o   ieri
   There   be-1SG.PRES   go.PFV.M.SG   yesterday
   I went there yesterday

4. *Ci*   h-anno   fa-tti   uscire
   1PL.DAT   have-3PL.PRES   make-PFV.M.PL   exit-INF
   They let us out

Such phenomena are correctly tagged as *CLI* according to the tagset of *La Repubblica*; nonetheless, such a tag simply supplies a distributional description, since clitics either appear before a verb or attached to it in final position. A combination of a phase of clitic identification combined with the two It-CMC tags (*46* and *39*) would empower the users to look for such distinct phenomena more effectively.

General and Connective Adverbs. Although [17] performed a deeper semantic analysis of the adverbial values (thus identifying seven categories), It-CMC presents only two continuous tags for general (*45*) and connective (*47*) adverbs: this choice was made mainly to prevent data sparseness, but the implementation of such information may be taken into account in the near future.

Within the training corpus, adverbs with connective value used in sentence-initial position have been tagged with tag *47*, as in (5):

5. *Beninteso,*   niente   di   male: ...
   Of course   nothing   of   evil: ...

Of course, there's nothing bad about it: ...

Any other adverb has been tagged as general (*45*). Within *La Repubblica*'s current tagset, adverbs are divided into adverbs with *-mente* endings and any other adverb. The It-CMC tagset does not provide such a distinction at the current stage, leaving the search of *-mente* adverbs to the user, for example by means of regular expression operators and/or advanced CQL queries.

Simple and Pseudo-Prepositions. Another slight modification to the tagset used in *La Repubblica* involves the treatment of simple and pseudo-prepositions. It-CMC tags simple prepositions with the tag *56* and pseudo-prepositions[4] with the tag *59*; within the *La Repubblica* corpus, instead, there is no such distinction and any non-articulated preposition is tagged as *PRE*. Tagging simple and pseudo-prepositions separately may prove useful, as far as linguistic inquiry is concerned: since pseudo-prepositions involve a much larger number of items than their counterpart, a dedicated tag can let the user search them selectively by simply inserting the POS in the query. On the other hand, this adverb/preposition overlapping is usually harder to detect by POS-tagging software and can require a remarkable number of distinctive items within the training corpus.

Definite vs. Indefinite Articles. The It-CMC tagset includes two tags for definite (*60*) and indefinite articles (*61*). This higher degree of specification does not burden the computational sustainability and, compared to *La Repubblica*'s single *ART* label, it allows users to perform quicker queries.

Finite Forms of Main, Auxiliary and Modal Verbs. The most prolific splitting operation has been performed on *La Repubblica*'s three broad finite verb tags. The labels for auxiliary, modal and main verbs' finite forms (namely tokens of the indicative, subjunctive, conditional and imperative mode) amount to 24 distinct tags: each of the three verbal categories counts 4 tags for indicative verbs, 2 for subjunctive, one for conditional, and one for imperative. On the one hand, such complexity proves to be problematic, at least in the earliest tagging stages, because of the formal correspondence of a number of noun and verb endings in Italian; on the other hand, such label precision is priceless in terms of linguistic analysis: a more detailed verbal tagging would enable sentence analysis in terms of hypothetical constructions (whose configuration can use six different verbal choices), temporal subordination and its narrative development (both

---

[4] [17] defines pseudo-preposition as "[...] words that are mainly adverbs but are used in prepositional function: [...]".

progressive and regressive), finer analysis of morphological errors in learners' corpora and many other potential topics.

As far as modal verbs are concerned, choosing between a present conditional verbal realization and a present indicative one carries solid social implications[5]: a single tag *VER2:fin*, unfortunately, prevents such queries from being carried out swiftly.

The CHE Tag and Its It-CMC Counterparts. From a linguistic perspective, the *CHE* tag in the *La Repubblica* corpus represents the most difficult label to handle when searching the corpus. The It-CMC proposal includes five distinct labels, based on [17]: *32* for *che* as indefinite P-D, *35* for interrogative P-D, *36* for relative P-D, *40* for exclamative P-D and *51* for subordinating conjunction; some simple examples and the corresponding It-CMC tags are listed in (6)–(10).

6. Ha   un   che   di familiare.
   have.3SG.PRES a   something of familiar.
   (He/She) is somewhat familiar.   - 32

7. Che   fai?
   What   do.2SG.PRES
   What do you do?   - 35

8. Ti   ringrazio   per   il   dono,   che
   2SG.DAT thank.1SG.PRES for the gift,   which

   è   stato   -   apprezzato.
   be.3SG.PRES be.PFV.M.SG appreciate.PFV.M.SG

   I thank you for the gift, which has been appreciated.   - 36

9. Che   noia!
   What   boredom!
   How boring!   - 40

10. Non ha   detto   che sia
    Not have.3SG.PRES said.PFV.M.SG that be.CONJ.PRES.3SG

    remunerativo.
    profitable.M.SG

    (He/She) didn't say that it would be profitable.   - 51

Such variety of labeling surely requires a much larger amount of training for any statistical POS-tagger, but also has remarkable potential in terms of linguistic queries and might improve the performance of Italian NLP tools mostly

---

[5] See Austin [1].

related to parsing and anaphora resolution. The most difficult disambiguation task lies in separating the tokens of *che* as a relative P-D from those with conjunctive function; nonetheless, such a distinction might be exploited by NLP tools to improve the verbs' referent identification rate, since the subject of the main proposition is (most likely) different from the subject of the subordinate one where *che* has subordinating value.

Excluded Tags. Eleven tags currently used in the *La Repubblica* corpus have not been included in It-CMC: ten of such tags are coveded by the set *VER:...:cli*, conceived to annotate verbs combined with clitic particles. As stated above, It-CMC is part of a workflow that involves a stage of clitic recognition and separation; after such splittings, clitics have been tagged with the weak personal P-D tag *39*. The remaining excluded tag is the residual tag *NOCAT*, originally used to tag non-linguistic elements such as percentage signs, interjections, hyphens and arithmetic signs.

Newly-Added Tags. It-CMC counts 2 newly added tags which are required by the nature of the linguistic data collected in the corpus: the first tag, *69*, is dedicated to emoticons, whose notable presence is attested throughout the corpus; the other one, *77*, is used to label formulae, such as arithmetic signs and bits of programming code. A tag for interjections (INT) had been included in Baroni [5] and suppressed in the current *La Repubblica* tagset: as it is useful in labeling CMC data, the interjection tag *68* has been included in the It-CMC tagset.

4   It-CMC Tagset: Morpho-Syntactic Features (MSF)

At the current stage, six categories of morphosyntactic information are represented in the It-CMC tagset. The tagset's morphosyntactic annotation is filled in six slots with fixed value, with the following structure:

entry POS.person.gender.number.degree.abbreviation.dialectalism

where values are separated by a full stop (.) and each position is filled with a suitable value from the set presented in Table 1.

During the tagging process, each tag's appropriate feature is filled in the corresponding slot; non-pertinent features are marked with 0. The last two slots (namely *abbreviation* and *dialectalism*) only admit values 1 or 0: only when a particular item appears in an abbreviated form or comes from a dialectal inventory, does the appropriate value shift to 1; this tagging strategy aims at covering the *ADJ:abbr*, *NOM:abbr*, *NPR:abbr* and *ADV:abbr* tags in Baroni [5][6], in order to prevent the tagset to cause further computational burden.

Sometimes, the mere morphology of some Italian words does not supply enough information to specify their gender or number[7]: in such instances, the

---

[6] Such tags have been suppressed in the current *La Repubblica* tagset.

[7] In such cases, gender/number information can be inferred from the word's context.

Table 1. MSF representation in the It-CMC tagset

MSF	Value	Feature	Position
Person	1	pers = 1	2
	2	pers = 2	
	3	pers = 3	
Gender	4	gend = masc	3
	5	gend = fem	
	4;5	gend = c	
Number	6	numb = sg	4
	7	numb = pl	
	6;7	numb = n	
Degree	8	degr = pos	5
	9	degr = comp	
	10	degr = sup	
Abbreviation	1	abbr = true	6
Dialectalism	1	dial = true	7

tags bear a value of *4;5* to represent common gender and the number slot is filled
with the invariant value *6;7*, before undergoing manual disambiguation during
the training sessions. As for the superlative degree of adjectives, the value *10*
does not cause any ambiguity, since it appears among MFS separators like any
other value.

An example of a fully-formed tag for a feminine, singular qualifying adjective
in its superlative degree would appear (in its explicit form) as

calmissima   26.NULL.feminine.singular.superlative.NULL.NULL

and in its numeric form as
calmissima   26.0.5.6.10.0.0

5   Tagging Sessions and Tagset Comparison

A TreeTagger parameter file modeled on the *La Repubblica* tagset has been
used to tag a sub-corpus of general and informal CMC data: in this annotation
process, the output correctly tagged 94.9 % of the tokens. Tagging the same sub-
corpus with It-CMC outcame in 93.9 % of correct guesses. Although 98.8 % of the
instances of *che* have been correctly identified, more tests on samples with wider
syntactic variation are still needed to reliably detect potential data sparseness.

For both taggers, the most tricky textual features are represented by multiple
orthographic realizations[8] and diatopically-marked lexical items.

---

[8] e.g. *é, e* and *è* vs. the standard form *è*.

6   Conclusions and Future Work

As regards the amount of dedicated resources and documentation, Italian can be considered a fairly well documented language. Although it is a balanced compromise between linguistic accuracy and computational burden, some of its features tend to be too coarse-grained for linguistic research: fitting examples of such features are the *CHE* tag and the *VER:fin* subset.

This work presents It-CMC, a fine-grained POS tagset that aims at supplying a precise linguistic description of Italian Computer-Mediated Communication data: the tagset contains 66 tags (compared to the 52 of the modified version of the *La Repubblica* tagset) divided into 14 subsets. In order to assess its computational sustainability, two parallel tagging paths[9] were tried in the early stages of its development, one using TreeTagger, the other with RFTagger [14]; in five loops of recursive training and tagging on data from different semantic domains: with a training corpus of 10,569 tokens, the RFTagger error percentage decreased by about 11 points, while TreeTagger's precision has increased by 2 % points only. Nonetheless, such amount of data can't be considered statistically significant for a proper performance assessment.

At present, RFTagger has been chosen over TreeTagger to perform the POS-tagging loops of the project, for two reasons: on the one hand, the constant improvement shown in the earliest tagging sessions: as the training corpus size increases and its semantic variation widens, RFTagger's performance is expected to improve further; should that not be the case, it will always be possible to remap the output extracting the POS labels so as to recreate a training corpus without MSF notation; on the other hand, an Italian parameter file for RFTagger is still lacking.

The project currently aims at compiling a training corpus of 250,000 tokens to create a solid statistical model of Italian in Computer-Mediated Communication: in terms of language resource development, such a model might prove useful also to annotate Italian learners' corpora, since the tricky presence of orthographic and typographic mistakes represents a common feature.

---

[9] every training loop involved a remapped version of De Mauro [8].

Appendix 1: Direct Comparison Between *La Repubblica* and CMC

Repubblica	Label	It-CMC	Label
NPR	Proper Noun	20	Proper Noun
NOUN	Common Noun	21	Common Noun
ADJ	Adjective	26	Qualifying Adjective
DET:demo	Demonstrative Determiner	30	Demonstrative P-D (strong)
PRO:demo	Demonstrative Pronoun	31	Demonstrative P-D (weak)
DET:indef	Indefinite Determiner	32	Indefinite P-D
PRO:indef	Indefinite Pronoun		
DET:poss	Possessive Determiner	33	Possessive P-D
PRO:poss	Possessive Pronoun		
DET:wh	Interrogative	35	Interrogative P-D
WH	Wh word		
		36	Relative P-D
		32	Indefinite P-D
CHE	"Che"	35	Interrogative P-D
		40	Exclamative P-D
PRO:pers	Personal Pronoun	37	Personal P-D (nominative, strong)
		38	Personal P-D (other case, strong)
CLI	Clitic	39	Personal P-D (other case, weak)
		46	Locative "ci", "vi", "ne"
not labeled		40	Exclamative P-D
ADV	Adverb	45	General Adverb
		47	Connective Adverb
ADV:mente	Adverb ending in "-mente"	45	General Adverb
CON	Conjunction	50	Coord. Conjunction
		51	Subord. Conjunction
ARTPRE	Articulated Prep.	58	Articulated Prep.
PRE	Preposition	56	Simple Preposition
		59	Pseudo-preposition
ART	Article	60	Definite Article
		61	Indefinite Article
DET:num	Numeral Determiner	64	Cardinal
PRO:num	Numeral Pronoun		numeral
ADJ	Ordinal numeral	65	Ordinal numeral
SENT	Final punctuation	70	Final punctuation
PUN	Non-final punctuation	71	Non-final punctuation
LOA	Loan Word	75	Foreign Word

		77	Formula
NOCAT		77	Formula
NUM	Digits	79	Digits

		111	Main verb (indicative, present)
		112	Main verb (ind., imperfect)
		113	Main verb (ind., simple past)
VER:fin	Verb (finite form)	114	Main verb (ind., future)
		115	Main verb (subjunctive, present)
		116	Main verb (subj., imperfect)
		117	Main verb (conditional, present)
		118	Main verb (imperative, present)

VER:infi	Verb (infinitive form)	121	Main verb (infinitive)
VER:ppre	Verb (finite form)	122	Main verb (participle, present)
VER:ppast	Verb (past participle)	123	Main verb (participle, past)
VER:geru	Verb (gerundive form)	124	Main verb (gerund, present)

		211	Auxiliary verb (indicative, present)
		212	Auxiliary verb (ind., imperfect)
		213	Auxiliary verb (ind., simple past)
AUX:fin	Auxiliary verb (finite form)	214	Auxiliary verb (ind., future)
		215	Auxiliary verb (subjunctive, present)
		216	Auxiliary verb (subj., imperfect)
		217	Auxiliary verb (conditional, present)
		218	Auxiliary verb (imperative, present)

AUX:infi	Auxiliary verb (infinitive)	221	Auxiliary verb (infinitive)
AUX:ppre	Auxiliary verb (participle, present)	222	Auxiliary verb (participle, present)
AUX:ppast	Auxiliary verb (participle, past)	223	Auxiliary verb (participle, past)
AUX:geru	Auxiliary verb (gerund, present)	224	Auxiliary verb (gerund, present)

		311	Modal verb (indicative, present)
		312	Modal verb (ind., imperfect)
		313	Modal verb (ind., simple past)
VER2:fin	Modal/causative verb (finite form)	314	Modal verb (ind., future)
		315	Modal verb (subjunctive, present)
		316	Modal verb (subj., imperfect)
		317	Modal verb (conditional, present)
		318	Modal verb (imperative, present)

VER2:infi	Modal/causative verb (infinitive form)	321	Modal verb (infinitive)
VER2:ppre	Modal/causative verb (finite form)	322	Modal verb (participle, present)
VER2:ppast	Modal/causative verb (past participle)	323	Modal verb (participle, past)
VER2:geru	Modal/causative verb (infinitive form)	324	Modal verb (gerund, present)

		68	Interjections
NOCAT	Non-linguistic element	69	Emoticons
		77	Formulae

1-2 NEG	"non"	45	General Adverb

References

1. Austin, J.L.: How To Do Things With Words. University Press, Oxford (1976)
2. Barbera, E.: Pronomi e determinanti nell'annotazione dell'italiano antico. La POS "PD" del Corpus Taurinense. Parallela IX. Atti del IX incontro italo-austriaco dei linguisti, pp. 35–51. Gottfried Egert Verlag, Wilhelmsfeld (2002)

3. Barbera, E.: Mapping dei tagset in bmanuel.org/corpora.unito.it. Tra guidelines e prolegomeni. In: Barbera, E., Corino, E., Onesti, C. (eds.) Corpora e linguistica in rete, pp. 373–388, Guerra Edizioni, Perugia (2007)
4. Barbera, E.: Un tagset per il Corpus Taurinense. Italiano antico e linguistica dei corpora. In: Barbera, E., Corino, E., Onesti, C. (eds.). Corpora e linguistica in rete, pp. 135–168, Guerra Edizioni, Perugia (2007)
5. Baroni, M., et al.: Introducing the La Repubblica Corpus: a large, annotated, TEI(XML)-compliant corpus of newspaper Italian. In: Proceedings of the Fourth International Conference on Language Resources and Evaluation (LREC 2004), pp. 1771–1774, ELRA (2004)
6. Bernardi, R., Bolognesi, A., Seidenari, C., Tamburini, F.: POS tagset design for Italian. In: Proceedings of the 5th International Conference on Language Resources and Evaluation, LREC 2004 (2004)
7. Bernardi, R., Bolognesi, A., Seidenari, C., Tamburini, F.: Automatic induction of a POS tagset for Italian. In: Proceedings of the Australasian Language Technology Workshop, pp. 176–183, Sidney, Australia (2005)
8. De Mauro, T.: GRADIT: Grande dizionario italiano dell'uso. UTET, Turin (2003)
9. La Repubblica Corpus. http://dev.sslmit.unibo.it/corpora/corpus.php?path=& name=Repubblica
10. Leech, G.: Introducing corpus annotation. In: Corpus Annotation: Linguistic Information from Computer Text Corpora, pp. 1–18. Longman, London (1997)
11. Monachini, M.: ELM-IT: EAGLES specification for Italian morphosintax Lexicon specification and classification guidelines. EAGLES Document EAG CLWG ELM IT/F
12. Petrov, S., Dipanjan, D., McDonald, R.: A universal part-of-speech tagset. In: Calzolari, N., Choukri, H., Declerck, T., Uur Doan, M., Maegaard, B., Mariani, J., Moreno, A., Odijk, J., Piperidis, S. (eds.) Proceedings of the Eight International Conference on Language Resources and Evaluation (LREC 2012), ELRA (2012)
13. Renzi, L., Salvi, G., Cardinaletti, A. (eds.): Grande Grammatica Italiana Di Consultazione. Il Mulino, Bologna (1991)
14. RFTagger. http://www.cis.uni-muenchen.de/~schmid/tools/RFTagger/
15. Schmid, H.: Probabilistic part-of-speech tagging using decision trees. In: Proceedings of International Conference on New Methods in Language Processing, Manchester, UK (1994)
16. Schmid, H., Laws, F.: Estimation of conditional probabilities with decision trees and an application to fine-grained POS tagging. In: Proceedings of the COLING 2008, pp. 777–784, Brighton, UK (2008)
17. Serianni, L.: Italiano: grammatica, sintassi, dubbi. Garzanti, Milano (2000)
18. TreeTagger. http://www.cis.uni-muenchen.de/~schmid/tools/TreeTagger/

FAQIR – A Frequently Asked Questions Retrieval Test Collection

Mladen Karan[✉] and Jan Šnajder

Faculty of Electrical Engineering and Computing,
Text Analysis and Knowledge Engineering Lab,
University of Zagreb, Unska 3, 10 000 Zagreb, Croatia
{mladen.karan,jan.snajder}@fer.hr

Abstract. Frequently asked question (FAQ) collections are commonly used across the web to provide information about a specific domain (e.g., services of a company). With respect to traditional information retrieval, FAQ retrieval introduces additional challenges, the main ones being (1) the brevity of FAQ texts and (2) the need for topic-specific knowledge. The primary contribution of our work is a new domain-specific FAQ collection, providing a large number of queries with manually annotated relevance judgments. On this collection, we test several unsupervised baseline models, including both count based and semantic embedding based models, as well as a combined model. We evaluate the performance across different setups and identify potential venues for improvement. The collection constitutes a solid basis for research in supervised machine-learning-based FAQ retrieval.

Keywords: Frequently asked questions · Information retrieval · Question answering

1 Introduction

Frequently asked question (FAQ) retrieval is the task of finding relevant FAQ question-answer pairs (FAQ-pairs) from a FAQ-collection. Unlike in standard, ad hoc retrieval, in FAQ retrieval the user query is typically formulated as a question, making FAQ retrieval more similar to question answering. However, unlike question answering, FAQ retrieval operates on a predefined collection of question-answer pairs. Furthermore, FAQ collections are often domain-specific – a case in point are the FAQ collections provided by customer support services, such as those of telecom companies, banks, or government institutions.

Effective semantic search for this type of collections would both reduce workload on customer support agents and improve the user experience. Furthermore, methodology used for FAQ retrieval may also be adapted for similar tasks, such as automated e-mail answering. However, answer retrieval in this setting is a very challenging task due to (1) the brevity of texts and (2) the requirement for domain specific knowledge. Both these aspects aggravate the *lexical gap* between the user's query and the FAQ-pairs.

© Springer International Publishing Switzerland 2016
P. Sojka et al. (Eds.): TSD 2016, LNAI 9924, pp. 74–81, 2016.
DOI: 10.1007/978-3-319-45510-5_9

One possible way to improve FAQ retrieval is to leverage the domain specificity of a FAQ-collection. Research in this direction [11] has demonstrated that, manually encoding domain knowledge addresses both above challenges effectively. However, this requires a lot of manual effort. A less labor intensive and more robust alternative is the data-driven approach that relies on supervised machine learning. This approach requires a FAQ-collection with queries and annotated relevance judgments. In many practical cases such a collection can be cheaply obtained by click log analysis. Unfortunately, for research purposes, existing FAQ-collections are either (1) not publicly available, (2) too small for training a supervised machine learning model, or (3) too general (not domain specific). In this paper, we present a FAQ retrieval test collection that addresses these deficiencies, and which is therefore a suitable test-bed for machine learning FAQ retrieval methods.

The contribution of this paper is twofold. First, we describe the creation of FAQIR – a new, domain-specific, multilevel relevance test collection for developing and evaluating FAQ retrieval models. Secondly, we perform an evaluation of different baseline models across many different settings, and explore the effects of several parameters on the retrieval performance. We make our collection freely available, in the hope that it will facilitate research in machine learning-based FAQ retrieval.

2 Related Work

One of the earliest research that focused specifically on FAQ collections is that of Burke et al. [2], who proposed a system based on various text similarity metrics. In [10], the FAQ retrieval task is tackled using manually constructed templates, which aim to alleviate lexical variation problems typical for this task.

More recently, Surdeanu et al. [12] proposed a supervised system for answer retrieval. They introduced various knowledge-based and distributional text similarity metrics for a supervised learning-to-rank model. The system was trained on a large domain collection of question-answer pairs crawled from the web. While quite large, this collection is arguably not ideal for FAQ retrieval research, as it maps questions to answers instead to FAQ-pairs, and, unlike typical FAQ collections, covers a very general domain.

FAQ IR test collections similar to the one we describe here have been developed for Asian languages [5,14]. A collection based on SMS messages is described in [6]. Collections most similar to ours can be found in [1,4], albeit focused on general-purpose FAQ retrieval. In contrast, our collection is domain-specific, and it also contains a larger number of queries than those presented in previous work, hence it is more suitable for supervised learning approaches to FAQ retrieval.

3 Retrieval Models

To reduce the annotation effort involved in obtaining the relevance judgments, we employ the commonly used pooling technique [13]. We use the following four information IR models to build the pool of FAQ-pairs for relevance judgments:

- BM25 [9] – the well-known bag-of-words ranking model that gives excellent performance; .
- VS – a classical vector space model using the tf-idf weighting computed on the FAQ-collection;
- W2V – a semantic compositionality model that represents each FAQ-pair/query by summing up semantic vectors of each content word. We use the freely available[1] semantic word vectors produced by word2vec [8]. Following [10], we apply information content based weighting;
- Combined – a joint model where the score of a document is the sum of ranks obtained using the above three models.

The FAQ-pairs and queries are preprocessed by lowercasing and the removal of all words shorter than three characters as well as words containing non alpha-numerical characters.

4   Test Collection

FAQ-Pairs. We obtain the FAQ-pairs from the freely-available[2] collection of English QA-pairs compiled by [12], originally derived from *Yahoo Answers*. The collection is topically very diverse. As we are interested in producing a domain-specific collection, we consider only FAQ-pairs belonging to a single category, namely the "maintenance & repairs" category. The resulting FAQ-pair collection contains 4313 FAQ-pairs.

Queries. The next step was to create queries that target the FAQ-pairs. To reduce annotation efforts, this was done in three phases, involving three annotators:

- One annotator (A1) built 50 *query templates*. A *query template* (QT) is a description of an information need, which facilitates creation of queries in the next step. An example QT is *"Useful information on how to keep insects out of a house/apartment/home."*;
- Annotator A1 proceeded to write about 8 different queries for each of the QTs. It was allowed to add some background context information that did not alter the basic information need specified by the QT. E.g., for a QT *"How to get stains out of carpet."*, the query might be *"My son spilled cocoa on my brand new carpet what can I do to fix it?"*. This makes devising different queries for the same QT much easier, and also makes the final retrieval task both more challenging and more realistic (context-rich FAQ questions are fairly frequent on *Yahoo Answers*);
- Annotators A2 and A3 were given the same task. Their fresh perspective on the QTs would allow them to think of new queries that annotator A1 did not think of. Consequently, they help to increase the variance of the possible

---

[1] https://code.google.com/p/word2vec/.
[2] https://webscope.sandbox.yahoo.com/.

queries for each QT. However, as they may drift too far and create queries that no longer match the information need defined by the QT, we provided each annotator with a sample of two queries from the previous step as reference points. After this step, we collected about 24 queries per QT. These were checked by annotator A1, to ensure no semantic drift occurred.

The obtained queries do not cover all FAQ-pairs in our collection. We decided to retain the FAQ-pairs that are not the target of any query. The motivation behind this decision is that in a realistic scenario we would not know the queries in advance, thus removing untargeted FAQ-pairs in advance would be impossible. Note that, while queries cover only a subdomain of our chosen "maintenance & repairs" category, this does not contradict the property of the collection being domain specific.

Relevance Judgments. The last step is to annotate the relevance judgments. Relevance is annotated on the level of QTs. Consequently, queries derived from the same QT share the same set of relevant FAQ-pairs. This scheme reduces the required amount of annotation considerably. To get a set of relevant FAQ candidates for each QT, we resort to pooling [13]: for each QT, we run BM25, VS, and W2V models on all queries from the given QT (on average 72 runs per QT). We pool the relevant candidates for a given QT as the union of top 10 results across all its runs.

To account for the different cases of relevance in FAQ retrieval, we depart from the usual binary relevance scheme, but instead devise a multilevel relevance scheme, as follows:

- *Relevant (R)* – a FAQ-pair is perfectly in line with the QT;
- *Useful (U)* – a FAQ-pair is not completely in line with the QT but may still give information that might be useful to the user. E.g., QT *"Information about removing bad odors from car"* and FAQ-pair containing *"Removing gasoline smell from garage"*;
- *Useless (X)* – a FAQ-pair is topically related to the QT but does not offer any useful information at all. E.g., QT – *"Information on getting mold out of a fridge."* and FAQ-pair matching the question very well but having *"With a chainsaw."* in the answer part;
- *Non-relevant (N)* – the FAQ-pair is completely unrelated to the QT.

Note that these labels represent the degree of relevance with respect to satisfying the information need specified in the QT. A similar scheme has already been used by [1].

Inter-annotator Agreement. A single annotator A1 annotated the entire data set (a time effort of about 40 h). In addition, for the purpose of estimating the inter-annotator agreement, a sample of five randomly chosen queries was annotated by another annotator A4 (a time effort of about 3 h). Upon inspecting the mismatches, we identified the following causes of systematic disagreements:

– Slips of attention – errors of this kind mostly occur when a *useful* FAQ-pair has its useful information subtly presented in a large text, which can be easily missed;
– Different understanding of QTs – in some cases annotators had slightly different interpretations of the QTs, causing them to disagree mostly between *useful* and *relevant* labels;
– Different relevance estimation – subjective differences in estimating the relevance. E.g., for a QT *"Information on removing bad odor from car"* and a FAQ-pair containing *"Making your kitchen smell good."*, both a "useful" and "useless" label might be justified.

Based on the inter-annotator analysis, we asked both annotators to calibrate their understanding of the QTs and revise their decision on all instances where they disagreed. Revising did not require changing annotations, but simply double checking them. Out of 101 annotations on which the A1 and A4 disagreed, annotator A1 revised 18 and annotator A4 revised 78 annotations.

While we report agreement on all four classes, we also consider two ways of interpreting the *"relevant"* and *"irrelevant"* classes:

– *RU-XN scheme* – joins R and U into a *relevant* class while the rest are *irrelevant*;
– *R-UXN scheme* – considers only R to be in the *relevant* class and all other labels in *irrelevant*.

The results of agreement are given in Table 1. We calculated agreement using Fleiss kappa [3]. Before revision, the agreement on the *RU-XN scheme* is much higher than on the *R-UXN scheme*. This indicates that many disagreements are between R and U labels. All scores become quite high after the revision step, which eliminated errors of the first two kinds. Thus, the disagreement still left after revision is due to the subjectivity of annotators. Because the rest of the data set was annotated by a single annotator, no further revision could be made. However, note that scores for the *RU-XN* were quite high even without revision. This implies that, with respect to this scheme, the annotations are quite consistent for the entire set.

Table 1. Inter-annotator agreement

	R-U-X-N	R-UXN	RU-XN
Initial	0.623	0.545	0.844
Revised	0.843	0.855	0.908

Test Collection Summary. The final FAQIR test collection contains 4313 FAQ-pairs and 1233 queries (derived from 50 QTs). The collection totals 201,349 words and 11,752 unique words. Additional statistics are given in Table 2.

Table 2. FAQIR test collection statistics

	Min	Max	Mean	Median
Query length	1	26	7.3	7
FAQ-question length	1	94	12.3	8
FAQ-answer length	1	376	33	22
Queries per QT (A1)	6	9	7.8	8
Queries per QT (A2)	6	8	7.5	8
Queries per QT (A3)	8	11	8.3	8
Pooled/group	72	358	202.8	200
Relevant/group	1	56	9.8	5
Useful/group	0	29	6.7	4

5   Evaluation

Here we present the IR performance of the baseline models described in Sect. 3 on the FAQIR collection. We evaluate the models by considering both the *R-UXN* and *RU-XN* scheme; the results are shown in Tables 3 and 4, respectively. In addition, we consider three possible ways of retrieving a FAQ-pair: considering (1) only the question (Q only), (2) only the answer (A only), and (3) both the question and answer (Q + A). We use classical IR measures of mean average precision (MAP) and mean reciprocal recall (MRR) [7]. Moreover, we use precision at rank 5 (P@5), which we consider particularly insightful since,

Table 3. Results for relevance defined by the *R-UXN* scheme

	Q only			A only			Q + A		
	MAP	MRR	P@5	MAP	MRR	P@5	MAP	MRR	P@5
BM25	0.204	0.518	0.253	0.121	0.362	0.173	**0.203**	**0.527**	0.268
VS	0.197	0.523	0.247	0.092	0.306	0.156	0.168	0.477	0.244
W2V	0.183	0.504	0.249	0.115	0.370	0.175	0.161	0.474	0.247
Combined	**0.208**	**0.548**	**0.285**	**0.119**	**0.387**	**0.194**	0.184	0.518	**0.282**

Table 4. Results for relevance defined by the *RU-XN* scheme

	Q only			A only			Q + A		
	MAP	MRR	P@5	MAP	MRR	P@5	MAP	MRR	P@5
BM25	**0.163**	0.613	0.355	**0.103**	0.481	0.251	**0.166**	0.638	0.375
VS	0.159	0.615	0.352	0.077	0.412	0.222	0.140	0.579	0.357
W2V	0.139	0.593	0.341	0.097	0.486	0.263	0.130	0.588	0.352
Combined	**0.163**	**0.645**	**0.396**	0.102	**0.515**	**0.284**	0.150	**0.642**	**0.410**

in a realistic scenario, a user would probably expect to find the relevant answers in the top five retrieved FAQ-pairs.

Some patterns are consistent over all evaluation settings.

Combined Model. The Combined model mostly performs best or very competitively – this indicates that the individual models may not make the same mistakes and that a combination is useful. A more sophisticated combination, such as a supervised learning to rank, may yield even better results.

Question vs. Answer. The performance in the "A only" scenario is consistently considerably worse than for the "Q only" and "Q + A". This indicates that matching a user query to the question part of a FAQ-pair is more useful than matching to the answer part. There are only slight performance differences between the "Q only" and "Q + A". Nonetheless, it may be useful to match to the question and answer part separately, combining the results. This we leave for future work.

Comparing Schemes. When considering MRR and P@5, results for the *RU-XN* scheme are consistently better than those for the *R-UXN* scheme. This is not surprising, as the *RU-XN* is generally less strict about what is considered relevant. Thus, a highly ranked document that is labeled as "U" instead of "R" will still be considered correct, in turn increasing overall score. Moreover, a real user would be interested in both "R" and "U" labels, which makes this evaluation setup more realistic. For MAP, the situation is reversed. The reason behind this is that MAP, instead of considering only the top of the retrieved list, takes into account the entire list. Consequently, in the *RU-XN* scheme *both* "U" and "R" FAQ-pairs need to be ranked high in order to get a good score. This makes the task more difficult compared to using the *R-UXN* scheme.

6   Conclusion

We described the construction of a domain-specific FAQ retrieval test collection and report a preliminary performance evaluation for several baseline IR models on this collection. We explored how the annotation scheme and individual parts of FAQ-pairs affect the retrieval performance. Results we obtained are as expected and constitute a baseline on which future work using this test collection can build. We make the test collection freely available.[3]

It is our hypothesis that the lessons learned while creating this collection should in general apply to arbitrary domains and languages. Thus, a possible venue of future work is to test this through the creation of a similar collection in a different domain. The main focus of future work, however, will be to use the collection as a development set for more advanced models based on supervised machine learning.

---

[3] http://takelab.fer.hr/data/faqir/.

References

1. Bunescu, R., Huang, Y.: Learning the relative usefulness of questions in community QA. In: Proceedings of the 2010 Conference on Empirical Methods in Natural Language Processing, pp. 97–107. Association for Computational Linguistics (2010)
2. Burke, R.D., Hammond, K.J., Kulyukin, V., Lytinen, S.L., Tomuro, N., Schoenberg, S.: Question answering from frequently asked question files: experiences with the FAQ finder system. AI Mag. **18**(2), 57 (1997)
3. Fleiss, J.L.: Measuring nominal scale agreement among many raters. Psychol. Bull. **76**(5), 378 (1971)
4. Jijkoun, V., de Rijke, M.: Retrieving answers from frequently asked questions pages on the web. In: Proceedings of the 14th ACM International Conference on Information and Knowledge Management, pp. 76–83. ACM (2005)
5. Kim, H., Seo, J.: High-performance FAQ retrieval using an automatic clustering method of query logs. Inf. Process. Manag. **42**(3), 650–661 (2006)
6. Kothari, G., Negi, S., Faruquie, T.A., Chakaravarthy, V.T., Subramaniam, L.V.: SMS based interface for FAQ retrieval. In: Proceedings of the Joint Conference of the 47th Annual Meeting of the ACL and the 4th International Joint Conference on Natural Language Processing of the AFNLP: Volume 2, pp. 852–860. Association for Computational Linguistics (2009)
7. Manning, C.D., Raghavan, P., Schütze, H., et al.: Introduction to Information Retrieval, vol. 1. Cambridge University Press, Cambridge (2008)
8. Mikolov, T., Chen, K., Corrado, G., Dean, J.: Efficient estimation of word representations in vector space. arXiv preprint arXiv:1301.3781 (2013)
9. Robertson, S.E., Walker, S., Jones, S., Hancock-Beaulieu, M.M., Gatford, M., et al.: Okapi at trec-3. NIST SPECIAL PUBLICATION SP, p. 109 (1995)
10. Šarić, F., Glavaš, G., Karan, M., Šnajder, J., Bašić, B.D.: TakeLab: Systems for measuring semantic text similarity. In: Proceedings of the First Joint Conference on Lexical and Computational Semantics-Volume 1: Proceedings of the Main Conference and the Shared Task, and Volume 2: Proceedings of the Sixth International Workshop on Semantic Evaluation, pp. 441–448. Association for Computational Linguistics (2012)
11. Sneiders, E.: Automated FAQ answering with question-specific knowledge representation for web self-service. In: 2009 2nd Conference on Human System Interactions, HSI 2009, pp. 298–305. IEEE (2009)
12. Surdeanu, M., Ciaramita, M., Zaragoza, H.: Learning to rank answers to non-factoid questions from web collections. Comput. Linguist. **37**(2), 351–383 (2011)
13. Voorhees, E.M.: The philosophy of information retrieval evaluation. In: Peters, C., Braschler, M., Gonzalo, J., Kluck, M. (eds.) CLEF 2001. LNCS, vol. 2406, pp. 355–370. Springer, Heidelberg (2002)
14. Wu, C.H., Yeh, J.F., Chen, M.J.: Domain-specific FAQ retrieval using independent aspects. ACM Trans. Asian Lang. Inf. Process. (TALIP) **4**(1), 1–17 (2005)

Combining Dependency Parsers Using Error Rates

Tomáš Jelínek[(⊠)]

Faculty of Arts, Institute of Theoretical and Computational Linguistics,
Charles University, Prague, Czech Republic
tomas.jelinek@ff.cuni.cz

Abstract. In this paper, we present a method of improving dependency
parsing accuracy by combining parsers using error rates. We use four
parsers: MSTParser, MaltParser, TurboParser and MateParser, and the
data of the analytical layer of the Prague Dependency Treebank. We
parse data with each of the parsers and calculate error rates for several parameters such as POS of dependent tokens. These error rates are
then used to determine weights of edges in an oriented graph created by
merging all the parses of a sentence provided by the parsers. We find
the maximum spanning tree in this graph (a dependency tree without
cycles), and achieve a 1.3 % UAS/1.1 % LAS improvement compared to
the best parser in our experiment.

Keywords: Syntax · Czech · Dependency parsing · Ensemble parsing

1   Introduction

In the domain of dependency parsing, a significant improvement has been
achieved in recent years due to the introduction of better parsing algorithms
such as greedy transition-based parsers with dynamic oracle. There is, however,
still room for further improvement. In this paper, we present a method combining
several parsers, which uses pre-computed error rates of the parsers to determine
weights of edges. The approach is based on the observation that each parser
tends to make consistently the same types of errors. With properly adjusted
weights, the parsers' strengths can be boosted which can increase the accuracy
of the parsers in a combination. Based on previously parsed training data, we
calculate error rates of parsers for several coarser or finer parameters and use
them to calculate weights of edges in a graph built by merging all parses of a
sentence. The choice of one dependency parse is then performed by an algorithm
for finding the maximum spanning tree in an oriented graph.

Parser combination is best suited for applications and uses, where parsing
accuracy is of far greater importance than speed.

2   Data and Parsers

In this experiment, we use four dependency parsers: two of them are well-known,
established parsers: MSTParser [6] and MaltParser [7], while the other two

© Springer International Publishing Switzerland 2016
P. Sojka et al. (Eds.): TSD 2016, LNAI 9924, pp. 82–92, 2016.
DOI: 10.1007/978-3-319-45510-5_10

belong to a newer generation of parsers: TurboParser [5] and a dependency parser included in Mate tools [1] (MateParser). The data used in this experiment come from the analytical layer of the Prague Dependency Treebank [4], comprising 1.5 million tokens in 80.000 sentences. The data are morphologically tagged by the Featurama tagger[1] with a precision of 95.3%. Mate tools perform their own tagging, with a slightly lower precision reaching 94.2%. Table 1 shows the accuracy of the four parsers used in a 10-fold cross-validation scheme over the whole PDT a-layer data. Four accuracy measures are shown: UAS and LAS (unlabeled and labeled attachment score for single tokens), SENT_U and SENT_L (unlabeled and labeled attachment score for whole sentences).

Table 1. Accuracy of parsers used in the experiment

	UAS	LAS	SENT_U	SENT_L
MST	86.35	79.25	38.63	24.63
Malt	86.68	81.25	42.19	32.54
Turbo	88.57	82.33	44.63	28.68
Mate	88.50	83.03	45.56	33.59

It is evident from the table that the two older parsers have considerably worse results, both in UAS and in unlabeled attachment score for the whole sentences (SENT_U). Of the two more recent parsers, TurboParser has a lower labeled accuracy (for the whole sentences even lower than MaltParser).

3 Data Partition, Error Rates

The aim of our experiment is to find the best method of combining the parsers using error rates in a 10-fold cross-validation. Error rates are calculated for all values of a range of morphosyntactic parameters (e.g. POS of dependent tokens). Error rates range from 0 to 1, expressing the proportion of correct attachments (dependencies) in occurrences of tokens with a given parameter value. For example, an error rate calculated as 0.79 for the parameter POS:LAS of dependent tokens with the value N (noun) indicates that the parser assigns the correct dependency relation and label to 79% of nouns. In order to calculate error rates for the parsers, we need as much data as possible, excluding test data (10% of data in cross-validation). With enough computing capacity the best method would be to take all the training data in each of the 10 phases of the 10-fold cross-validation, and perform another 10-fold cross-validation (10 times training over 90% and parse 10%) over the training data. In this case, we would obtain parsed data from which we could calculate error rates, then train the parsers on the whole training data, parse the test data and apply the method of parser combination using error rates on the test data, see Fig. 1.

---

[1] See http://sourceforge.net/projects/featurama/.

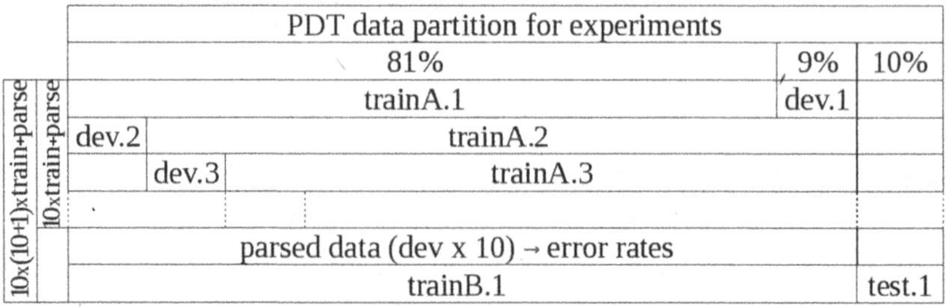

		PDT data partition for experiments			
		81%		9%	10%
		trainA.1		dev.1	
	dev.2	trainA.2			
	dev.3		trainA.3		
	.				
		parsed data (dev x 10) → error rates			
		trainB.1			test.1

Fig. 1. Data distribution (ideal).

However, this approach requires huge computing resources, as it needs 110 cycles of training and parsing for each of the parsers. TurboParser can be trained in a few hours, but MateParser may need up to 15 days on 24 CPU cores. As we have no such resources available, we used an alternative procedure: we split the whole data in 10 splits, parsed the whole data in 10 training and parsing cycles by all parsers and calculated error rates from these parsed data. Then the data were redistributed into another set of 10 splits, and a new 10-fold cross-validation cycle was performed with all parsers, followed by tests of the parser combination, as shown in Fig. 2. This method is less valid, as we use a part of test data to calculate error rates, but it requires much less resources (approx. 5.5 times less).

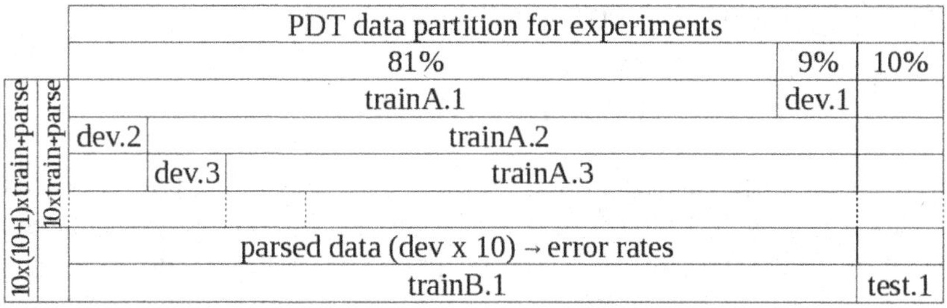

		PDT data partition for experiments			
		81%		9%	10%
		trainA.1		dev.1	
	dev.2	trainA.2			
	dev.3		trainA.3		
		parsed data (dev x 10) → error rates			
		trainB.1			test.1

Fig. 2. Data distribution (alternative)

3.1 Error-Rate Tables

We calculate error-rate tables by comparing parsed data with the gold standard. Error rates are computed for several (linguistic) parameters. For example, the parameter "POS of the dependent token" is calculated for each of the possible values (noun, adjective...) as the proportion of tokens with correct dependencies of a given value (e.g. noun) to the total number of occurrences of such tokens. If the parameters use morphosyntactic values, automatically assigned values are used (rather than the gold standard). The error rates have values between 0 and 1. For each parameter, we calculate error rates for labeled and unlabeled attachment separately. Parameters for which entries in error-rate tables are calculated can be coarse, finer or very detailed. Coarse parameters are overall accuracy of the parsers (LAS/UAS, SENT_L/SENT_U): 1 error rate for each parameter and parser. Medium-fine parameters have approx. 10 error rates for each parameter: POS of the dependent token, POS of the governing token (parameter named 2POS) and the distance between the tokens (parameter DIST, its values are reduced to 5 intervals calculated for dependency relations to the left and to the right of the dependent token, e.g. −3 for tokens dependent on a token situated 4–6 positions to the left). Fine-grained parameters have tens or hundreds of error rates. One of these parameters combines POS of the dependent and governing token (POS2POS); other three parameters use 2 or 3 categories of the morphosyntactic tag of the dependent token: POS; subtype of POS, e.g. personal pronoun, and case (parameters POSSUBPOS, POSCASE, POSSUBPOSCASE).

Table 2 shows a fraction of the error-rate table, calculated for all four parsers used in the experiment. The first column shows the parser, the second indicates the parameter, the third distinguishes between labeled and unlabeled attachments, the fourth gives values of parameters, and the fifth presents the error rates. The parameters and their values are interpreted as follows: SENT indicates the share of whole sentences with correct dependencies; DIST with the "2" value says that the governing token is 2–3 tokens right of the dependent token; POS with the "N" value means that the dependent token is a noun; POSSUB-POSCASE with the "PP4" value implies that the token is a personal pronoun in the accusative case.

Table 2. Example of entries in error table

Parser	Parameter	L/U	Value	E-rate
MALT	SENT	UAS	–	0.4280
MATE	SENT	UAS	–	0.4632
MST	SENT	UAS	–	0.3941
TURBO	SENT	UAS	–	0.4534
MALT	DIST	UAS	2	0.8553
MATE	DIST	UAS	2	0.8688
MST	DIST	UAS	2	0.8460
TURBO	DIST	UAS	2	0.8735
MALT	POS	LAS	N	0.7964
MATE	POS	LAS	N	0.8058
MST	POS	LAS	N	0.7606
TURBO	POS	LAS	N	0.7901
MALT	POSSUBPOSCASE	LAS	PP4	0.9176
MATE	POSSUBPOSCASE	LAS	PP4	0.9192
MST	POSSUBPOSCASE	LAS	PP4	0.9238
TURBO	POSSUBPOSCASE	LAS	PP4	0.9248

The table shows that for all parameters the error rates calculated for various parsers are similar and their order corresponds more or less to the order of the overall accuracy of the parsers. The error-rate table has approx. 5600 entries (1400 entries per parser).

3.2   Error Rate Consistency

The idea of using error rates for parser combination is based on the assumption that parsers using the same parsing algorithm err in a similar way, if training and testing data are similar. Put more simply, their error rates are consistent (if we calculate them on a part of the data, e.g. a training data split, we can use them for the purposes of parser combination on another part of the data, e.g. a test data split). To test this assumption, we used Pearson's correlation of matrices, which we calculated on error-rate tables (matrices) deduced from 10 data splits (dev1 ... dev10). The phenomena with less than 2 occurrences in each split were discarded. Pearson's correlation coefficient has values between -1 and 1, where 1 is a perfect (direct) correlation, 0 denotes independent values, -1 means an inverse correlation. Table 3 presents the correlation matrices calculated for the following three parameters: POS, DIST and POS2POS, for each parser.

Table 3. Error rate consistency

	Mate	Turbo	Malt	MST
POS	0.960	0.994	0.973	0.970
DIST	0.996	0.995	0.999	0.997
POS2POS	0.910	0.766	0.854	0.805

It is clear that error rates calculated on the data splits are really consistent, with only minor differences between the parsers. The most detailed parameter (POS2POS) has a slightly lower consistency.

4   Calculation of Weights

Once we have error-rate tables and the new (test) data parsed by four parsers, we can start experimenting with parser combinations. Parsing results are merged into one file, where sentences can now be represented as oriented graphs including one or several possible trees. First we need to calculate weights of edges (i.e. dependencies), needed for parser combination. For this calculation, it is necessary to choose a parameter (or a combination of parameters) and also an exponent that increases the differences between better and worse rates and that should increase the accuracy of the parser combination, as Green suggests in his thesis [3], when discussing the performance of ensemble parsing (when calculating weights, every numeric value is first raised to the power of the chosen exponent, e.g. 0.46^9). The weight of each edge (e.g. a dependency relation suggested by at least one parser) is calculated as the sum of error rates for every parameter (if more parameters are used). If the parsers agree on a dependency relation, the resulting weight is calculated as the sum of values for all agreeing parsers. We can choose, for example, the parameter POS:LAS with the exponent 2. If a noun having two suggested attachments is encountered, each attached by two parsers (Malt & Mate, MST & Turbo), two weights will be calculated, one for each attachment, using error rates (as shown in Table 2). For the attachment suggested by Malt & Mate parsers, the weight will be calculated as $0.7964^2 + 0.8058^2 = 1.2836$, the attachment suggested by MST & Turbo parsers will be assigned the weight $0.7606^2 + 0.7901^2 = 1.2028$. These weights will then be used for finding the best parse possible, using an algorithm to find the maximum spanning tree.

5   Parser Combination with the MST Algorithm

For the actual combination of parsers we decided to use an approach where syntactic structures proposed by the four parsers are merged into a single oriented graph. All its edges are assigned weights, and we use Chu-Liu-Edmonds' algorithm for finding the maximum spanning tree in a directed graph (see [6], p. 526) to choose the best dependency tree. In simple terms, Chu-Liu-Edmonds' algorithm chooses the incoming edge (dependency) with the highest weight for each node (token). If no cycle is formed, the maximum spanning tree is found. Otherwise, if a local cycle is created, it is treated separately as a subgraph. The incoming edge to the cycle is determined in such a way that the sum of the weight of the incoming edge and the weights of the rest of the (now broken) cycle are the highest possible. Given that the structures generated by the parsers rarely differ to a larger extent or the dependencies are specified in opposing directions, a cycle is rarely formed (less than 1 in 1000 sentences). In most cases, the algorithm just selects edges with the highest weights. When two or three parsers agree on one dependency relation against one parser choosing another one, the edge where more parsers agree is always chosen. Only in case two parsers agree on one edge and the other two parsers on another (or all four parsers disagree), the way the weights are calculated (i.e. which parameters are chosen) starts to be relevant in the parser combination accuracy. Figure 3 shows an example of a successful parser combination in a sentence, where all parsers committed at least one error. The sentence shown is:

Privatizované mlékárny se však zatím mezi sebou nedokázaly domluvit.
Privatized dairies (refl.) however yet among themselves were not able to agree.
'However, the privatized dairies have not been able to come to an agreement yet.'

When parsing this sentence, the parsers did not agree and yielded different results also with respect to the gold standard in three tokens: *zatím* (yet), *mezi* (among) and *se* (refl. particle). Three parsers agreed on the correct attachment

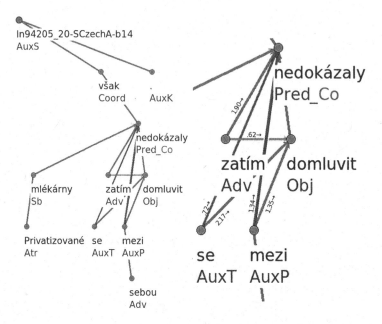

Fig. 3. Parser combination example

of the tokens *se* 'reflexive particle' and *zatím* 'yet', while two parsers (Turbo and MST) agreed on the correct attachment of the preposition *mezi* 'among' and two parsers (Mate and Malt) suggested a different attachment. In the case of *mezi* 'among', the calculated weights were decisive for the outcome but the difference between the weights was very small anyway (1.35 vs. 1.34).

6   Results

With the error-rate tables calculated from the whole data and the data parsed by four parsers (in a different distribution of training and test data splits), we conducted a series of experiments aiming to find the best setting of parameters and the best exponent for all four accuracy measures (UAS, LAS, SENT_U, SENT_L).

6.1   Coarse Parameters

A noticeable improvement against the baseline (MateParser, the best of the four parsers used) was attained by using the coarse parameters. In Table 4 (and in the following two tables), the first column presents the parameter used, the second column shows the distinction between labeled and unlabeled attachments (error rates calculated with or without syntactic labels), while the third column gives the exponent (one exponent with the best results; for one parameter, the results of experiments with two different exponents are shown). The following four columns show the accuracy rates obtained with the parser combination using the parameters indicated. The "ALL" parameter identifies the attachment accuracy for all tokens (i.e. LAS or UAS score).

Table 4. Parser combination accuracy with coarse parameters

	LAS/UAS	Exp.	UAS	LAS	SENT_U	SENT_L
Mate	-	-	88.50	83.03	45.56	33.59
ALL	LAS	1	89.64	84.01	47.11	33.70
ALL	UAS	1	89.60	83.99	47.07	33.66
SENT	LAS	1	89.62	83.99	47.08	33.68
SENT	LAS	9	89.48	83.85	47.28	34.03
SENT	UAS	1	89.58	83.95	46.97	33.62

The best result with the coarse parameters for UAS/LAS was achieved when the "ALL" parameter with labeled score (i.e. LAS of the parser) and a low exponent (1–4, same results) were used, with the improvement reaching 1.1/1.0 UAS/LAS. The best result for the whole sentences was obtained with the "SENT" parameter with labeled score (i.e. SENT_L for the parser), with an improvement achieving 1.7/0.6 SENT_U/SENT_L.

6.2 Fine-Grained Parameters

Table 5 shows the accuracy measures achieved with more detailed parameters. Only the single best combination of LAS/UAS and exponent for each parameter is presented.

Table 5. Parser combination accuracy with fine-grained parameters

	LAS/UAS	Exp.	UAS	LAS	SENT_U	SENT_L
Mate	-	-	88.50	83.03	45.56	33.59
POS	LAS	4	89.75	84.11	47.20	33.69
DIST	UAS	1	89.62	83.99	46.00	33.01
2POS	UAS	1	88.82	83.20	41.64	30.67
POS2POS	LAS	3	89.72	84.09	47.27	33.96
POSCASE	UAS	1	89.75	84.12	47.22	33.68
POSSUBPOS	LAS	1	89.76	84.12	47.19	33.66
POSSUBPOSCASE	LAS	1	89.72	84.09	47.27	33.96

With more fine-grained parameters we achieve a slightly better accuracy than with the coarse-grained ones. The exponent plays rarely any role in the increase of accuracy. The best result (a 1.3/1.1 UAS/LAS increase over the baseline) was achieved with the POSSUBPOS (POS and subtype of POS) parameter, but the differences are mostly marginal (except for the 2POS parameter, i.e. POS of the governing token, which achieved relatively poor results).

6.3   Combination of Parameters

The best result in UAS, LAS and SENT_U can be achieved using a combination of several parameters, as shown in Table 6. The best UAS and LAS scores can be reached with the sum of two fine-grained parameters: POSSUBPOS and POS2POS, both LAS, with exponent 1 (combination 1). To obtain the highest SENT_U score (but not an exceptionally high SENT_L score), several parameters, coarse-grained, finer and very fine-grained are combined: SENT:LAS, POS:LAS, POS:UAS, POS2POS:LAS, POSCASE:LAS i POSSUBPOS:LAS, using the exponent 2.

Table 6. Parser combination accuracy with combined parameters

	LAS/UAS	Exp.	UAS	LAS	SENT_U	SENT_L
Mate	-	-	88.50	83.03	45.56	33.59
Combination 1	LAS	1	89.81	84.17	47.25	33.71
Combination 2	LAS+UAS	2	89.80	84.16	47.37	33.81

6.4   Statistical Significance

We conducted a test of statistical significance for all the results of the experiments using binomial test applied on the number of correctly parsed tokens/sentences. All the differences between results larger than 0.1 % are statistically significant at $p < 0.01$: the differences between the best results among the coarse parameters and the best results among the fine-grained parameters are statistically significant, whereas the differences between the best results among the fine-grained parameters and combined parameters are not statistically significant.

6.5   Conclusion

To conclude this part of the experiment: it is possible to find an optimal set of parameters (and an exponent) to maximize whichever accuracy measure we like (LAS, UAS, SENT_U, SENT_L). We can increase parsing accuracy compared to the baseline (MateParser) by up to 1.31 UAS, 1.14 LAS, 1.81 SENT_U and 0.44 SENT_L. However, the actual choice of parameters alters the results only to some extent, there is only a marginal difference between the coarse parameters (which are much easier to obtain) and the fine-grained parameters (necessitating the use of error-rate tables and the parsing of large data).

7   Future Work

Our next step in the field of combination of parsers will be the introduction of one or two other parsers, e.g. StanfordParser [2] or Parsito [8], and test them in the combination of parsers. Perhaps their use in the combination will prove that using detailed error rates is not necessary. Or, on the contrary, the use of error rates in the combination with 5 or 6 different parsers will result in an even larger accuracy increase. Some of the parsers may be also trained using morphological gold standard, rather than the automatically assigned tags. It will probably decrease their accuracy when using an automatically tagged text, but it may help in a parser combination. As the UAS score approaches 90 %, the precision of morphological tagging at approx. 95 % becomes crucial since a considerable part of syntactic errors is caused by errors in morphological tagging. Therefore we plan to use more sources of morphological annotation (two different taggers, a rule-based disambiguation system etc.) and try to simultaneously choose a correct morphological tag with a correct syntactic structure.

8   Conclusion

In this paper, we have shown that combining parsers can lead to a substantial increase in parsing accuracy. We proposed a system based on the maximum spanning tree algorithm, with weights using detailed information on the errors of parsers. This system has good results (increase in UAS accuracy of 1.3 %), but a simpler system, requiring much less computational resources, which uses only UAS/LAS scores of each parser, has only slightly worse results (1.1 % UAS increase). Our planned addition of other parsers or tests on other languages will show whether detailed error rates are necessary to make the most from the parser combination using weights and the maximum spanning tree algorithm, or a simple overall accuracy of the parsers would do the job as well.

Acknowledgments. This research was supported by Czech Ministry of Education, Youth and Sports through the Czech National Corpus project (LM2015044). A part of the computational resources used in our experiments were provided by the CES-NET project (LM2015042). Both projects are part of the programme *Large Research, Development, and Innovations Infrastructures.*

References

1. Bohnet, B., Nivre, J.: A transition-based system for joint part-of-speech tagging and labeled non-projective dependency parsing. In: Proceedings of EMNLP 2012 (2012)
2. Chen, D., Manning, C.D.: A fast and accurate dependency parser using neural networks. In: Proceedings of EMNLP 2014 (2014)
3. Green, N.D.: Improvements to syntax-based machine translation using ensemble dependency parsers (thesis). Faculty of Mathematics and Physics, Charles University, Prague (2013)

4. Hajič, J.: Complex corpus annotation: the Prague dependency treebank. In: Šimková, M. (ed.) Insight into the Slovak and Czech Corpus Linguistics, pp. 54–73. Veda, Bratislava (2006)

5. Martins, A.F.T., Almeida, M.B., Smith, N.A.: Turning on the turbo: fast third-order non-projective turbo parsers. In: Proceedings of ACL 2013 (2013)

6. McDonald, R., Pereira, F., Ribarov, K., Hajic, J.: Non-projective dependency parsing using spanning tree algorithms. In: Proceedings of EMNLP 2005 (2005)

7. Nivre, J., Hall, J., Nilsson, J.: MaltParser: a data-driven parser-generator for dependency parsing. In: Proceedings of LREC 2006 (2006)

8. Straka, M., Hajič, J., Straková, J., Hajič jr., J.: Parsing universal dependency treebanks using neural networks and search-based Oracle. In: Proceedings of TLT 2015 (2015)

A Modular Chain of NLP Tools for Basque

Arantxa Otegi[✉], Nerea Ezeiza, Iakes Goenaga, and Gorka Labaka

IXA Group, University of the Basque Country, UPV/EHU, San Sebastian, Spain
arantza.otegi@ehu.eus

Abstract. This work describes the initial stage of designing and implementing a modular chain of Natural Language Processing tools for Basque. The main characteristic of this chain is the deep morphosyntactic analysis carried out by the first tool of the chain and the use of these morphologically rich annotations by the following linguistic processing tools of the chain. It is designed following a modular approach, showing high ease of use of its processors. Two tools have been adapted and integrated to the chain so far, and are ready to use and freely available, namely the morphosyntactic analyzer and PoS tagger, and the dependency parser. We have evaluated these tools and obtained competitive results. Furthermore, we have tested the robustness of the tools on an extensive processing of Basque documents in various research projects.

1 Introduction

We live immersed in the Information Society and we have access to a vast amount of information mostly in the form of text. It is increasingly necessary to incorporate the automatic processing of information and consequently, of languages, since the medium in which this information is mostly found is in natural language. Although English is the predominant language in the current globalized environment, information is multilingual, and the presence of minor languages, like Basque, is increasing.

Basque presents a complex intraword morphological structure based on morpheme agglutination. Regarding the agglutinative nature of Basque, it could be said that the affixes are attached to the lemmas, and each attached morpheme carries (ordinarily) only one meaning. For instance, in *mendian* (Basque for 'in the mountain'), *mendi* stands for 'mountain', *-a* for the determiner (translatable as 'the'), and *-n* for the locative case. The determiner, number and declension case morphemes are appended to the last element of the noun phrase and always occur in this order. In addition, often the syntactic function of a word is conveyed by the suffix attached to it as a case marker. In the previous example, the locative case added to the stem *mendi* ('mountain') assigns the verb complement (locative) function to the word. However, not all the lexical words in a phrase are inflected in Basque.

Moreover, word formation is very productive, as it is very usual to create new compounds as well as derivatives. As a result of the wealth of information contained within word forms, complex structures have to be built to represent complete

ⓒ Springer International Publishing Switzerland 2016
P. Sojka et al. (Eds.): TSD 2016, LNAI 9924, pp. 93–100, 2016.
DOI: 10.1007/978-3-319-45510-5_11

morphological information at word level. For example, the compound *oxigeno-hornitzailea* can be segmented into: (a) the sequence *oxigeno* (Basque for 'oxygen'), the hyphen, *hornitzaile* ('provider') and *-a* (meaning 'the person who provides oxygen'); (b) *oxigeno*, the hyphen, *horni* (stem of the verb *hornitu* 'provide'), the lexical suffix *-tzaile* (attached to a verb 'a person or thing that'), and *-a*. This last segmentation is ambiguous in that the derivation affix might be modifying only the verb stem, meaning the same as the previous segmentation, or modifying the compound meaning 'the device that provides oxygen' or 'oxygen bottle'.

The rules to combine the intraword information to represent the final morphosyntactic interpretation are defined via a word grammar [1].

Having all these characteristics in mind, the main characteristic of the chain is the ease to integrate complex processors, such as the deep morphosyntactic analysis carried out by the first tool. Another important feature is the use of morphologically rich annotations to transmit information through the chain. Finally, it is modular to let users decide which processors to include in the chain.

In this paper we introduce the already developed first two processors of the chain: the morphological analyzer and a PoS tagger (ixa-pipe-pos-eu), and the dependency parser (ixa-pipe-dep-eu). In the future, the chain will be extended in order to make it possible to carry out the whole linguistic processing for Basque, as it will provide not only basic processing tools like tokenization, but also more complex tasks as lemmatization, part-of-speech (PoS) tagging, syntactic parsing, already included, as well as Named Entity Recognition and Classification, Named Entity Disambiguation, coreference resolution, semantic role labeling, etc.

The rest of the paper is organized as follows. Next section presents some previous similar work. Section 3 introduces the general description of the modular processing chain for Basque. Section 4 describes the tools so far developed, and we also present their empirical evaluation. Section 5 presents some projects in which these tools have been used. Finally, Sect. 6 discusses some concluding remarks and future work.

2   Related Work

Many NLP toolkits exist providing extensive functionalities, like GATE [7], Stanford CoreNLP [12] and Freeling [13]. They are large and complex system, making difficult the use or the integration of other tools in their chain. Conversely, IXA pipes is intended to be simple, ready to use, modular and portable, as well as efficient, multilingual, accurate and open source [4].

The most similar work to our approach is IXA pipes, as it is a modular set of multilingual NLP tools which is publicly available.[1] It offers robust and efficient linguistic annotation to both researchers and non-NLP experts with the aim of lowering the barriers of using NLP technology either for research purposes or for small industrial developers. It provides several linguistic annotation tools, including, among others, a tokenizer and a module for PoS tagging and lemmatizing, which are the tools we focus on in this work. Additionally, it offers

---

[1] http://ixa2.si.ehu.eus/ixa-pipes/.

third party tools for other linguistic annotations. Currently, it supports a different number of languages for each tool. For instance, it offers tokenization, PoS tagging and lemmatizing for Basque, Dutch, English, French, Galician, German, Italian and Spanish.

The morphosyntactic analysis and PoS tagger tool presented in this work is based on Eustagger, a robust lemmatizer/tagger for Basque [8], which integrates the word grammar described in [1]. This tagger has been extensively used during the last 20 years to process Basque corpora. Among them, we want to highlight EPEC [2], an annotated corpus for Basque, aimed to be a reference corpus for the development and improvement of several NLP tools for Basque. EPEC was initially a 50,000-word sample collection of written standard Basque that has afterwards been extended to 300,000 words. This corpus has been automatically processed and the results have been manually revised. Only half of the complete collection has been annotated with syntactic information yet and part of it (80 % according to [6]) has been automatically converted to Universal Dependencies.[2]

It is worth mentioning that the lemmatizer for Basque in IXA pipes (ixa-pipe-pos) uses a model trained on the mentioned Basque Universal dependency corpus, as opposed to the morphological processor integrated in the chain presented in this work (see Sect. 4.1). Being both approaches different, we intend to compare their performance under the same conditions to see the strengths and weaknesses of each one.

Apart from being modular, our approach has another characteristic in common with IXA pipes: the input/output format of the tools. As all the modules in both set of tools read and write NAF format (see Sect. 3), it is possible the interaction between them. For example, it is possible to extend the modular chain of Basque processing tools with a tool for NERC in IXA pipes (ixa-pipe-nerc), which is ready for Basque (among other several languages) [5].

3   General Description of the Modular Chain

The linguistic processing tools for Basque integrated in the modular chain, and described in Sect. 4, were initially designed for internal use and each of the tools was designed to run independently from the rest. Thus, the tools have been adapted to follow the main characteristics of the modular chain described below.

One of the main features of the chain is its modularity. That is, the tools can be picked and changed, as long as they read and write the required data format via the standard streams. The processors interact like Unix pipes, specifically they all take standard input, do some linguistic processing, and produce standard output which feeds directly the next one.

The data format used to represent and pipe linguistic annotations in such modular chain is NAF [9], a linguistic annotation format designed for complex NLP pipelines.[3] In that way, by default, the input and output of all the tools is formatted in NAF, except for the input of the first one, ixa-pipe-pos-eu, which

---

[2] http://universaldependencies.org/#eu.
[3] http://wordpress.let.vupr.nl/naf/.

```
cat input.txt | sh ixa-pipe-pos-eu/run.sh | sh ixa-pipe-dep-eu/run.sh
```

Fig. 1. Command line invocation for running the two tools of the modular chain.

takes raw text as input. The annotation chain can be applied to any text, such as a single sentence, a paragraph or whole story. All the tools work with UTF-8 character encoding.

Another main feature of the chain is the minimal compilation or installation effort in order to get started using the tools. Besides, running the processing chain is simple and is done by a command-line interface. Once you get the tools, without doing any installation or configuration, doing linguistic processing for a file can be as easy as Fig. 1 shows. Using the command displayed in the example, a raw text file is processed, first, analyzing morphologically and PoS tagging, and next, applying a dependency parser. The linguistic annotations provided by the whole chain will be written through standard output, formatted in NAF. In addition, some tools could have their own properties to allow further customization of their usage.

Additionally, the tools (binary tarballs) are publicly available[4] and are distributed under a free software license, GPL v3.[5]

4   The Tools Integrated in the Modular Chain

4.1   ixa-pipe-pos-eu

ixa-pipe-pos-eu is a robust and wide-coverage morphological analyzer and a PoS tagger, which is an adapted version of Eustagger, a tool lemmatizer/tagger for Basque [8]. It is the first module of the linguistic processing chain. The tool takes a raw text as an input text and outputs the lemma, the PoS tag and the morphological information for each token in NAF format.

It processes a text morphosyntactically following the next steps: tokenization, segmentation, the word grammar, treatment of multiword expressions and morphosyntactic disambiguation.

The morphological segmentation of words is based on a set of two-level rules converted into finite-state transducers. The analysis is performed in two main phases and gives as a result all the possible analyses of each word in the text: on the one hand, the standard analyzer that is able to analyze/generate standard-language words based on a general lexicon and the corresponding rules for morphotactics and morphophonological changes; on the other hand, the guesser or analyzer of words with lemmas not belonging to the previous lexicon. For the guesser the lexicon is simplified by allowing only open categories (nouns, adjectives, verbs, etc.) and any combination of characters as lemmas. This general set of lemmas is combined with affixes related to open categories and general rules in order to capture as many morphologically significant features as possible. In this

---

[4] http://ixa2.si.ehu.es/ixakat/.
[5] http://www.gnu.org/licenses/gpl-3.0.en.html.

two-phase architecture, each word will be processed by the first analyzer that is able to produce at least one valid segmentation following the order defined before. Comparing to Eustagger, this adaptation lacks of the module for linguistic variants, which has been discarded to reduce the complexity of the process and to make it more efficient.

After segmenting the word into its constituent morphemes, the word grammar based processor analyzes and elaborates the sequential intraword information in order to build the information of the word as a whole.

For the detection of multiword expression, we have integrated a reduced version of the processor in Eustagger due to simplicity, which detects only the most common expressions.

The performance of the morphosyntactic analysis in Eustagger is very good, assigning the correct analysis to 99 % of the words despite the high ambiguity (each token has 3.56 reading on average, and 1.56 reading taking only PoS tag into account). Although we still have to confirm the results of this reduced tool in question, we expect to obtain similar results.

Once we have given all possible morphological analysis to each token/multi-token, PoS tagging and lemmatization must be performed in order to assign the correct lemma and grammatical category to each token taking into account the context. The disambiguation is based on linguistic knowledge, as well as statistical information. First, a set of Constraint Grammar rules [11] are used to discard some analysis. After that, a stochastic HMM disambiguation is applied to choose the final analysis. The HMM model has been trained using the training set of the EPEC corpus (100,000 words), and the CG rules have also been defined and tuned based on the same set of the corpus.

Eustagger has been evaluated on the test set of EPEC corpus (50,000 words) obtaining a performance of 95.17 % on PoS tagging accuracy, and 91.89 % when considering all morphological information. As we have noted before, we expect to obtain similar figures for ixa-pipe-pos-eu, because we consider that the differences in the processing (lack of variants and smaller set of multiword expressions) should not distort too much the results of this robust approach.

4.2   ixa-pipe-dep-eu

ixa-pipe-dep-eu is a dependency parser which takes a NAF document containing lemmas, PoS tags and morphological annotations from the output of the previous ixa-pipe-pos-eu tool.

There are two main approaches for dependency parsing: transition-based and graph-based. And these are the state of the art dependency parsers: MaltParser[6], MaltOptimizer[7], MST Parser[8] and Mate.[9] The two former ones follow a transition-based approach, whereas the two latter ones follow a graph-based approach.

---

[6] http://www.maltparser.org/.

[7] http://nil.fdi.ucm.es/maltoptimizer/install.html.

[8] http://www.seas.upenn.edu/~strctlrn/MSTParser/MSTParser.html.

[9] https://code.google.com/archive/p/mate-tools/.

Following a transition-based approach, MaltParser and MaltOptimizer (MaltParser optimization tool) consist of a transition system for deriving dependency trees, coupled with a classifier for deterministically predicting the next transition given a feature representation of the current parser configuration. The main difference between them is that MaltOptimizer first performs an analysis of the training set in order to select a suitable starting point for optimization, and then guides the user through the optimization of parsing algorithm, feature model, and learning algorithm.

In contrast, and following a graph-based approach, MST Parser adopts the second order maximum spanning tree dependency parsing algorithm. A maximum spanning tree dependency based parser decomposes a dependency structure into parts known as factors. The factors of the first order maximum spanning tree parsing algorithm are edges consisting of the head, the dependent (child) and the edge label. The second order parsing algorithm uses the edges to those children which are closest to the dependent, the child of the dependent occurring in the sentence between the head and the dependent, and the edge to a grandchild, in addition to the first order factors.

ixa-pipe-dep-eu tool is based on the graph-based version of Mate parser, which adopts the second order maximum spanning tree dependency parsing algorithm, as does MST Parser. The main two difference between them are the following: (1) Mate parser considers more children and edges in order to create dependency trees and (2) it uses a new parallel parsing and feature extraction algorithm that improves accuracy as well as parsing speed. Our dependency parsing tool has been trained using part of the Basque Dependency Treebank [3] which contains 96,000 tokens and 7,700 sentences.

We have evaluated ixa-pipe-dep-eu using part of the Basque Dependency Treebank (13,851 tokens) and it have obtained 83.00 % LAS (Labeled Attachment Score) [10], which is a good result taking into account that Basque is a morphologically rich language. Having said that, we consider interesting to compare ixa-pipe-dep-eu with the other state of the art statistical parsers mentioned above. The results of all the systems are shown in Table 1. It can be observed that our dependency parsing tool outperforms all the rest of the parsers, yet there is not significant difference between ixa-pipe-dep-eu and MST Parser.

Although the results of ixa-pipe-dep-eu are promising, there is room for improvement. For this reason, on the one hand, we are trying to increase the size of our training set, and on the other hand, we are studying the use of feature engineering in order to select only the most beneficial morphological features for dependency parsing of Basque.

Table 1. Results of different dependency parsers for Basque.

Dependency parser	LAS
MaltParser	77.78 %
MaltParser + MaltOptimizer	80.04 %
MST Parser	82.69 %
ixa-pipe-dep-eu	83.00 %

5   Some Projects Using the Modular Chain

The modular chain of NLP tools for Basque is already being used successfully for the linguistic annotations in several projects.

One of these projects is QTLeap,[10] a FP7 European project on machine translation. QTLeap project explores novel ways for attaining machine translation of higher quality that are opened by a new generation of increasingly sophisticated semantic datasets and by recent advances in deep language processing. Basque is one of the 8 languages (Basque, Bulgarian, Czech, Dutch, English, German, Portuguese and Spanish) involved in the project, and the tools presented in this work are being used for its linguistic processing.

The other project in which these tools have been used is Ber2tek,[11] whose aim is to advance in the research and development of the technologies of analysis of cross-media contents, high quality machine translation and natural spoken multimodal interaction for Basque.

6   Conclusions and Future Work

Many other multilingual NLP toolkits exist, but integrating already developed tools into most of them is not trivial. In this paper, we have presented our approach to adapt and integrate a set of NLP tools for Basque using a modular architecture.

The main characteristics of such tools are the following: to allow the integration of complex tools easily, to transmit information containing morphologically rich annotations among tools, and to be modular.

The robustness of the modular chain of Basque processing tools is already being tested doing extensive processing, mainly in the FP7 European project QTLeap.

Currently we have made publicly available two tools of the chain, namely, the morphosyntactic analyzer and PoS tagger, and the dependency parser. Moreover, tools such as semantic role labeling, named entity disambiguation and coreference resolution are being integrated and will be available soon.

Acknowledgments. This work has received support by the EC's FP7 (FP7/2007-2013) under grant agreement number 610516: "QTLeap: Quality Translation by Deep Language Engineering Approaches".

---

[10] http://qtleap.eu.
[11] http://www.ber2tek.eus/en.

References

1. Aduriz, I., Agirre, E., Aldezabal, I., Alegria, I., Arregi, X., Jose, M.A., Artola, X., Gojenola, K., Sarasola, K., Urkia, M.: A word-grammar based morphological analyzer for agglutinative languages. In: Proceedings of COLING, Saarbrucken, Germany, vol. 1, pp. 1–7 (2000)
2. Aduriz, I., Aranzabe, M., Jose, M.A., Atutxa, A., de Ilarraza, A.D., Ezeiza, N., Gojenola, K., Oronoz, M., Soroa, A., Urizar, R.: Methodology and steps towards the construction of EPEC, a corpus of written Basque tagged at morphological and syntactic levels for the automatic processing. In: Corpus Linguistics Around the World. Language and Computers Series, vol. 56, pp. 1–15. Rodopi, Netherlands (2006)
3. Aduriz, I., et al.: Construction of a basque dependency treebank. In: Proceedings of the Workshop on Treebanks an Linguistic Theories (TLT 2003). Treebanks and Linguistic Theories (2003)
4. Agerri, R., Bermudez, J., Rigau, G.: IXA pipeline: efficient and ready to use multilingual NLP tools. In: Calzolari, N., Choukri, K., Declerck, T., Loftsson, H., Maegaard, B., Mariani, J., Moreno, A., Odijk, J., Piperidis, S. (eds.) Proceedings of LREC 2014, pp. 3823–3828. European Language Resources Association (ELRA) (2014)
5. Agerri, R., Rigau, G.: Robust multilingual named entity recognition with shallow semi-supervised features. Artif. Intell. **238**, 63–82 (2016)
6. Aranzabe, M., Atutxa, A., Bengoetxea, K., de Ilarraza, A.D., Goenaga, I., Gojenola, K., Uria, L.: Automatic conversion of the basque dependency treebank to universal dependencies. In: Proceedings of the Workshop on Treebanks an Linguistic Theories (TLT 2014), pp. 233–241. Institute of Computer Science of the Polish Academy of Sciences, Warszawa (2015)
7. Cunningham, H.: GATE, a general architecture for text engineering. Comput. Humanit. **36**(2), 223–254 (2002)
8. Ezeiza, N., Aduriz, I., Alegria, I., Arriola, M., Urizar, R.: Combining stochastic and rule-based methods for disambiguation in agglutinative languages. In: Proceedings of COLING-ACL 1998, vol. 1, pp. 380–384. Association for Computational Linguistics (1998)
9. Fokkens, A., Soroa, A., Beloki, Z., Ockeloen, N., Rigau, G., van Hage, W.R., Vossen, P.: NAF and GAF: linking linguistic annotations. In: Proceedings 10th Joint ISO-ACL SIGSEM Workshop on Interoperable Semantic Annotation (2014)
10. Goenaga, I., Gojenola, K., Ezeiza, N.: Exploiting the contribution of morphological information to parsing: the BASQUE TEAM system in the SPRML'2013 shared task. In: Proceedings of SPRML-2013 Workshop, ACL, pp. 71–77. Association for Computational Linguistics (2013)
11. Karlsson, F., Voutilainen, A., Heikkila, J., Anttila, A. (eds.): Constraint Grammar: A Language-Independent System for Parsing Unrestricted Text. Mouton de Gruyter, Berlin (1995)
12. Manning, C.D., Surdeanu, M., Bauer, J., Finkel, J., Bethard, S.J., McClosky, D.: The Stanford CoreNLP natural language processing toolkit. In: Proceedings of ACL 2014: System Demonstrations, pp. 55–60. Association for Computational Linguistics (2014)
13. Padro, L., Stanilovsky, E.: FreeLing 3.0: towards wider multilinguality. In: Calzolari, N., Choukri, K., Declerck, T., Doğan, M.U., Maegaard, B., Mariani, J., Moreno, A., Odijk, J., Piperidis, S. (eds.) Proceedings of LREC 2012. European Language Resources Association (ELRA), Istanbul (2012)

Speech-to-Text Summarization Using Automatic Phrase Extraction from Recognized Text

Michal Rott[(✉)] and Petr Červa

Institute of Information Technology and Electronics,
Technical University of Liberec, Studentská 2, 461 17 Liberec, Czech Republic
{michal.rott,petr.cerva}@tul.cz
https://www.ite.tul.cz/speechlab/

Abstract. This paper describes a summarization system that was developed in order to summarize news delivered orally. The system generates text summaries from input audio using three independent components: an automatic speech recognizer, a syntactic analyzer, and a summarizer. The absence of sentence boundaries in the recognized text complicates the summarization process. Therefore, we use a syntactic analyzer to identify continuous segments in the recognized text.

We used 50 reference articles to perform our evaluation. The data are publicly available at http://nlp.ite.tul.cz/sumarizace. The results of the proposed system were compared with the results of sentence summarization in the reference articles. The evaluation was performed using co-occurrence of n-grams in the reference and generated summaries, and by readers mark-ups. The readers marked two aspects of the summaries: readability and information relevance. Experiments confirm that the generated summaries have the same information value as the reference summaries. However, readers state that phrase summaries are hard to read without the whole sentence context.

Keywords: Automatic speech summarization · Phrases extraction · Spoken news · Automatic speech recognition · Syntactic analysis

1 Introduction

Spoken news has gained popularity on Czech news servers over the past few years. These news stories are typically short reports focused on one topic or event. They are usually in the form of a video that contains several video clips with a reporter giving commentary or a studio record of spoken words. Each news page contains an indicative text summary that describes the contents of the news. Nevertheless, the summary is often written in a populist way. It does not contain any real indicative value, but the authors are trying to deceive readers/watchers using buzzwords and attractive words or by using sound bites from the news stories. We propose a system that generates more descriptive summaries. The proposed system performs a Speech-to-Text summarization and generates informative summaries.

© Springer International Publishing Switzerland 2016
P. Sojka et al. (Eds.): TSD 2016, LNAI 9924, pp. 101–108, 2016.
DOI: 10.1007/978-3-319-45510-5_12

We presume that the audio streams of the news contain all necessary information. The proposed system has to transcribe the input audio into text. This step allows for further analysis of the audio contents. Modern Automatic Speech Recognizer (ASR) systems of the Czech language have a greater than 85 % accuracy rate [1, 2]. Out-Of-Vocabulary (OOV) words pose the most serious issue to voice recognition software. ASR replaces OOVs with the most similar word. This replacement creates errors and leads to information being lost.

The inflection of Czech words and the relatively free word order in sentence structure complicate performing an analysis of recognized text. These issues also complicate the detection of sentence boundaries. Experiments in paper [3] show that human annotators agree on only 4.6 % of the sentence boundary placement in a recognized text. How can we expect to automatically detect sentence boundaries when educated specialists are not able to find them?

We propose a way to segment recognized text into something that is similar to sentences. We utilize the syntactic analyzer SET [4] to identify phrases in the text. The phrases are used instead of sentences during summarization. Replacing sentences with phrases allows us to extract even more information from the text because phrases are shorter than sentences. The main disadvantage of this approach is a loss of sentence context. Moreover, phrases can have ambiguous meanings and are often hard to read because they are only fragments of the entire text. For example, the phrase "mělo by mít oficiální informaci první" (word-by-word translation: "should have had official information first"). In this case, the subject is missing and readers cannot know who should have had the official information.

Here are summarizers of English and Chinese broadcast news [5, 6]. Both systems trivialize the problem of sentence boundary detection and detect sentences using probabilistic language models. The detected sentences are summarized and the most informative sentences are extracted and compacted [7]. In contrast, the proposed scheme extracts phrases. Our approach segments and compacts the recognized text at the same time.

The rest of the paper is structured as follows: The next section describes the scheme of the proposed system and each of its subsystems. Section 3 describes the evaluation test set and metrics. An experimental evaluation of the proposed scheme is provided in Sect. 4, where we also compare the recognized text summarization with the original text summarization. The last section concludes this paper.

2   The Proposed Summarization System

The overall scheme of the proposed summarization system is depicted in Fig. 1. It shows that the system is composed of three subsystems: an automatic speech recognizer, a syntactic analyzer and a text summarizer. The output of each subsystem serves as the input for the subsequent one.

The proposed scheme is versatile and allows us to change any subsystem. The change is limited only by the input and output data formats. For example, the ASR system has to read an audio file and produce a one-line text transcription of the input audio.

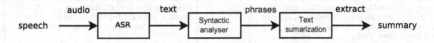

Fig. 1. The scheme of the automatic speech summarization system

An ASR system processes recorded speech and transcribes it into a text. A configuration of the ASR system is described in Sect. 4.1. The Syntactic Engineering Tool (SET) [4] is used as the syntactic analyzer to identify phrases in the recognized text. The SET is able to create phrasal trees of the text. One tree is created when the SET parses the ASR system output, and this tree then must be split into sub-trees for further use. This subsystem is essential for the summarization subsystem because the latter needs the text to be represented by segments[1] in order to work properly. The task of the SET is to segment the ASR system output (represented by a sequence of words) into text segments, which are then used during the summarization.

We utilize the Summarization Engine for Czech (SummEC) as the text summarizer. The SummEC is composed of two modules. The first is a text pre-processing module. This module offers sentence separation, word lemmatization, a stop list, and synonym substitution. Lemmatization is performed by an external tool called Morphodita [8]. The stop list contains over 200 Czech terms, including prepositions, conjunctions, particles and the most frequent Czech lemmas. The synonym dictionary contains 7,443 different groups of synonyms with a total number of 22,856 lemmas.

The second module summarizes the pre-processed text. This module contains five summarization methods that were described and evaluated in papers [9,10]. The TFxIDF method [11] was selected as our preferred method for summarizing automatic speech. This method is not computationally demanding and still produces good extracts [9].

How the proposed system works is shown at Fig. 2. The top left box represents a part of spoken news delivered by a speaker. His speech is recognized by the ASR system and verb phrases are identified in the recognized text by the SET. Finally, the SummEC selects the most informative phrases (the highlighted ones). The text in the Fig. 2 represents only a small part of one reference article.

---

[1] Sentences are used as segments when the text is summarized.

spoken text	recognized text
Americko-australská společnost chce zkoumat podzemí pod dvaadvaceti obcemi na Berounsku. Z velké části pod chráněnou krajinnou oblastí Český kras. Hluboko pod zemí ve vrstvách břidlic může být totiž vázaný hořlavý plyn spolu s naftou a firma Hutton Energy jej umí získat. Česká průzkumná geologická firma, která pro ni pracuje, už obce a Správu CHKO Český kras oslovila. Záměr působí zděšení. "Když si představím, že by kilometr od obce mohl být zkušební vrt, tak mne jímá hrůza," řekl starosta obce Svatý Jan pod Skalou Pavel Vokál.	americko australská společnost chce zkoumat podzemí pod dvaadvaceti obcemi na Berounsku z velké části pod chráněnou krajinnou oblastí český kras hluboko pod zemí ve vrstvách břidlic může být totiž vázaný hořlavý plyn spolu s naftou a firma ho to na energii umí získat česká průzkumná geologická firma, která pro ni pracuje už obce a zprávu jako český kras oslovila záměr působí zděšení, když si představím, že by kilometr od obce mohly být zkušební vrt tak mne jímá hrůza řekl starosta obce Svatý Jan pod Skalou Pavel vokál

ASR

Phrase identification

identified phrases

Americko australská společnost chce zkoumat podzemí pod dvaadvaceti obcemi na Berounsku z velké části pod chráněnou krajinnou oblastí český kras.
Hluboko pod zemí ve vrstvách břidlic může být totiž vázaný hořlavý plyn spolu s naftou.
Firma ho to na energii umí získat česká průzkumná geologická firma.
Která pro ni pracuje už obce a zprávu český kras oslovila záměr.
Zděšeni , když si představím, by kilometr od obce mohly být zkušební vrt.
Tak mne jímá hrůza.
Řekl starosta obce Svatý Jan pod Skalou Pavel vokál.

Fig. 2. An example of spoken news processing.

3 Evaluation Data and Metrics

3.1 Data for Evaluation

The evaluation data set contains informative extracts from 50 newspaper articles. The reference extracts were created by 15 independent annotators. Each annotator created an informative extract for every reference article. The articles contain 7,849 unique words and 92,089 words in total. The articles are selected in columns taken from local and international news, economics and culture. The evaluation test is publicly available at http://nlp.ite.tul.cz/sumarizace.

We expect lower values of recall, precision and f-score in our experiments because the task performed by the annotators was different. The original task was to extract 20 % of the text sentences. This paper's task is to create extracts with a length of up to 25 % of the total number of characters in the original text.

3.2 Tools and Metrics Used

We used the ROUGE toolkit [12] to perform our evaluation. This toolkit supports various summarization evaluation metrics. In this paper, we used the ROUGE-1 metric. The Czech language has a relatively free word order in its sentence structure. We do not use higher n-grams because important text segments do not have to be ordered in the same order in different segments of the text.

ROUGE-1 is defined as:

$$Precision = \frac{TP}{TP + FP} \quad Recall = \frac{TP}{TP + FN} \quad F - score = \frac{2RP}{R + P} \quad (1)$$

where TP, FP and FN are explained in Table 1.

Table 1. The meaning of variables in Eq. (1) for ROUGE-1

# Unigrams	Selected by annotators	Not selected by annotators
Selected by the system	TP	FN
Not selected by the system	FP	TN

Another evaluation metric is based on the subjective opinion of the readers. We provided the sentence extract of the original text and the phrasal extract of the recognized text to volunteer readers. They marked the readability and information relevance of the extracts on a scale from 1 to 5. The original text was provided to the volunteers so that they could mark the relevant information. They had to mark how important the extracted information was.

To make this task easier, we provided volunteers with text representations of the marks meanings. The latter are shown in Table 2. This metric is used in the last experiment that evaluates the final setup of the proposed system, and compares it to the sentence extracts from the original text.

Table 2. Meaning of marks

Mark	Readability	Extract ability
1	Readable and understandable	Fully important
2	Understandable	Mostly important
3	Unclear	Uncertain
4	Not understandable	Mostly unimportant
5	Unreadable	Fully unimportant

4 Experiment

4.1 Experimental Setup

The ASR system uses a hybrid Hidden Markov Model (HMM) and Deep Neural Network (DNN) architecture [13] and a general 550k lexicon. The input audio is split into 25-ms frames with 10-ms overlaps. Each audio frame is described by 39 normalized log filter bank features [14]. The DNN utilizes five hidden layers, with 1,024 neurons per layer, and a learning rate of 0.08. The ReLU function is used as the activation function of neurons. This DNN is trained for 35 epochs using 300 h of speech recordings.

The SET is used to parse the recognized text into a phrasal tree. The verb phrases of the phrasal tree are used in the article summarization. For this purpose, we utilize the newest version of the SET. This version allows us to limit a context within word dependencies that are searched. The context is limited by parameter max-match-len, and the default value of this parameter is 20 words. The average time spent on the parsing using the default value is approx. 2 min.

The length of the generated extract is 25 % of the total characters in the original text. The TFxIDF summarization method requires a stop list and an IDF dictionary. In this work, both of these components are created using 2.2M newspaper articles. The resulting stop list has 283 items and the IDF dictionary has 491k items. The items in the IDF dictionary are inverse document frequencies of lemmas computed as:

$$IDF(t) = log\frac{|D|}{|\{d \in D : t \in d\}|} \tag{2}$$

where $|D|$ is the total number of training documents and $|\{d \in D : t \in d\}|$ is the number of documents that contain the lemma t.

4.2 Comparison of Sentence Extracts and Phrase Extracts of Punctuated Text

This experiment investigates how much information is preserved in phrase extracts. It shows the difference between sentence extracts and phrase extracts of the original text. The ASR system is not used during this experiment. Phrases are detected in the original text without any changes, and the punctuation of the text is preserved. The generated sentence extracts are compared with the phrase extracts of the text in Table 3. The difference in the importance of the extracted information between sentence and phrase extracts was only 0.31 % of the f-score scale. We can presume that phrases are able to contain the same amount of information as sentences.

Table 3. Comparison of sentence and phrase extracts [%]

Unit	Precision	Recall	f-score
Sentence	54.68	47.09	50.29
Phrase	49.53	51.17	49.98

4.3 Comparison of Summaries Between an Ideal and an Actual ASR System

This experiment compares how significant an influence is made by the imperfection of the ASR system. We simulated an ideal ASR system that has no OOV words, and has 100 % accuracy. We joined original article texts into one sentence by removing commas, periods, question marks and exclamation points. The resulting sentence was used as an output of an ideal ASR system. Note that the used ASR system has an accuracy of 89.3 %. We detected 784 Out-Of-Vocabulary words, which represents approx. 10 % of the original text words. The result of this experiment is shown in Table 4. The imperfection of the ASR system has a small effect the summarization results, only 0.67 on the f-score scale.

Table 4. A comparison of summaries between an ideal and an actual ASR system [%]

ASR	Precision	Recall	f-score
Ideal	50.18	45.95	47.64
Real	45.94	48.81	46.97

4.4 Subjective Evaluation – Readability and Relevance

The previous experiments show that phrases have the same informational qualities as sentences. We proposed that the final experiment be based on the subjective evaluation of extracts. Readers marked the readability and extracted information relevance of the sentence extracts and phrase extracts. Both types of extracts have been compared using these criteria. Mark representation is presented in Table 2.

Table 5. Extracts readability and relevance

Extract type	Readability	Relevance
Text sentences	1.88	2.48
Phrases from ASR	2.95	2.44

Following the marks, both sets of extracts have the same information relevance. This is shown in the last column of Table 5. However, extracts composed from the original sentences are easier to read. The phrase extracts contain the same information, but the missing parts of sentences complicate their reading.

5 Conclusions

In this paper, we present an automatic speech summarization system for Czech spoken news. We compare the performance of this system with a text summarization method. All experiments demonstrate that the phrase extracts from the recognized text are equal to the sentence extracts from the original text with regard to the relevance of the information extracted. However, readers consider them harder to read, and some of them are not understandable.

We plan to improve the readability of generated summaries. For this purpose, several techniques can be employed to improve the readability of phrase extracts. For example, reformulating extracted phrases according to the base sentence patterns of Czech that are used in Czech language classes for foreigners.

Resolving anaphora can improve the understanding of phrase extracts. Pronominal anaphora creates greater ambiguity in the text. In the future, we need to resolve pronominal anaphora. Their resolution can boost the score of phrases that contain pronouns and they can more likely be used in extracts.

Acknowledgement. This paper was supported by the Technology Agency of the Czech Republic (Project No. TA04010199) and by the Student Grant Scheme 2016 (SGS) at the Technical University of Liberec.

References

1. Mateju, L., Cerva, P., Zdansky, J.: Investigation into the use of deep neural networks for LVCSR of Czech. In: 2015 IEEE International Workshop of Electronics, Control, Measurement, Signals and their Application to Mechatronics (ECMSM), pp. 1–4. IEEE (2015)
2. Nouza, J., Zdansky, J., Cerva, P., Silovsky, J.: Challenges in speech processing of Slavic languages (case studies in speech recognition of Czech and Slovak). In: Esposito, A., Campbell, N., Vogel, C., Hussain, A., Nijholt, A. (eds.) Second COST 2102. LNCS, vol. 5967, pp. 225–241. Springer, Heidelberg (2010)
3. Bohac, M., Blavka, K., Kucharova, M., Skodova, S.: Post-processing of the recognized speech for web presentation of large audio archive. In: 2012 35th International Conference on Telecommunications and Signal Processing (TSP), pp. 441–445 (2012)
4. Kovář, V., Horák, A., Jakubíček, M.: Syntactic analysis using finite patterns: a new parsing system for Czech. In: Vetulani, Z. (ed.) LTC 2009. LNCS, vol. 6562, pp. 161–171. Springer, Heidelberg (2011)
5. Hori, C., Furui, S., Malkin, R., Yu, H., Waibel, A.: Automatic speech summarization applied to English broadcast news speech. In: 2002 IEEE International Conference on Acoustics, Speech, and Signal Processing (ICASSP), vol. 1, pp. I-9–I-12 (2002)
6. Chen, Y.T., Chen, B., Wang, H.M.: A probabilistic generative framework for extractive broadcast news speech summarization. IEEE Trans. Audio Speech Lang. Process. **17**, 95–106 (2009)
7. Furui, S., Kikuchi, T., Shinnaka, Y., Hori, C.: Speech-to-text and speech-to-speech summarization of spontaneous speech. IEEE Trans. Speech Audio Process. **12**, 401–408 (2004)
8. Straková, J., Straka, M., Hajič, J.: Open-source tools for morphology, lemmatization, POS tagging and named entity recognition. In: Proceedings of 52nd Annual Meeting of the Association for Computational Linguistics: System Demonstrations, pp. 13–18. Association for Computational Linguistics, Baltimore, Maryland (2014)
9. Rott, M., Červa, P.: SummEC: a summarization engine for Czech. In: Habernal, I. (ed.) TSD 2013. LNCS, vol. 8082, pp. 527–535. Springer, Heidelberg (2013)
10. Michal, R.: The initial study of term vector generation methods for news summarization. In: Proceedings of the Ninth Workshop on Recent Advances in Slavonic Natural Languages Processing, pp. 23–30. Tribun EU, Brno (2015)
11. Vanderwende, L., Suzuki, H., Brockett, C., Nenkova, A.: Beyond sumbasic: task-focused summarization with sentence simplification and lexical expansion. Inf. Process. Manag. **43**, 1606–1618 (2007). Text Summarization
12. Lin, C.Y.: Rouge: a package for automatic evaluation of summaries, pp. 25–26 (2004)
13. Dahl, G.E., Yu, D., Deng, L., Acero, A.: Context-dependent pre-trained deep neural networks for large-vocabulary speech recognition. IEEE Trans. Audio Speech Lang. Process. **20**, 30–42 (2012)
14. Svendsen, T., Hamar, J.B.: Combining NDHMM and phonetic feature detection for speech recognition. In: 2015 23rd European Signal Processing Conference (EUSIPCO), pp. 1666–1670 (2015)

Homonymy and Polysemy in the Czech Morphological Dictionary

Jaroslava Hlaváčová[⊠]

Faculty of Mathematics and Physics, Institute of Formal and Applied Linguistics,
Charles University, Prague, Czech Republic
hlavacova@ufal.mff.cuni.cz

Abstract. We focus on a problem of homonymy and polysemy in morphological dictionaries on the example of the Czech morphological dictionary MorfFlex CZ [2]. It is not necessary to distinguish meanings in morphological dictionaries unless the distinction has consequencies in word formation or syntax. The contribution proposes several important rules and principles for achieving consistency.

Keywords: Homonymy · Polysemy · Paradigm · Word formation · Czech

1 Introduction — Morphological Dictionary of Czech

The morphological dictionary of Czech used in Prague was designed by Jan Hajič in 1990s. Despite the fast development in the area of NLP, its main features, including the format, are still in use. It contains almost 450,000 lines of coded information on the basis of which more than 100,000,000 wordforms are generated, belonging to almost 900,000 lemmas. Its format is described in the Chap. 4 of the book [1].

We shall briefly introduce only the main features of the dictionary and rules for its entries.

1.1 Dictionary Format

The morphological dictionary consists of lines, that describe a generation of one or more wordforms, together with their morphological tags, belonging to a single lemma. Each line has the following pattern (taken from [1], simplified):

Technical stem (Root)	Model (Paradigm)	Lemma	Tag	AddInfo

Contrary to the original Hajič's work, we have slightly changed his terminology. In the scheme above, the old terms are in parentheses.

© Springer International Publishing Switzerland 2016
P. Sojka et al. (Eds.): TSD 2016, LNAI 9924, pp. 109–116, 2016.
DOI: 10.1007/978-3-319-45510-5_13

We have replaced the original term Root with the term **Technical stem**, because the term root in its strict linguistic sense is something different. Even the term Stem does not reflect accurately the entity in the dictionary. Moreover, there are many definitions that differ. The technical stem is the beginning part (string) of all wordforms that can be generated from the dictionary line. It may contain all the main types of morphemes: prefix, root or its part as well as suffix or its part. Moreover, it may contain also an ending or its part.

The second replacement concerns the term Paradigm. We replaced it with the term **Model**, as the term paradigm is often used in another sense, namely as a set of all wordforms belonging to a certain lemma. We will use it in this sense in the following text. The Model field contains a name of a derivational model. Each derivational model is connected (in a special table) with at least one inflectional model. For instance, the derivational model mz1 is connected with two inflectional models. The first one, with the same name mz1 generates the wordforms of a single lemma, in this case a soft masculine animate noun. The second inflectional model is named uv and generates all the wordforms of the derived possessive adjective. There is a special model 0, which is used for "exceptions". In that case, the technical stem is always the whole wordform and there must be the unempty field **Tag**, containing all morphological tags belonging to the wordform.

Thus, every dictionary line contains either a non-zero (derivational) model and no tag, or the zero model and a set of morphological tags. Every line may contain also additional information concerning wordforms described by the line. It is often relevant for the whole lemma, even if the line does not describe the whole paradigm.

The **Lemma** field contains a lemma. Especially for human readability, the lemma is just a word in its basic form, there are no precise identifiers. If lemmas were written unambiguously, the word itself could be the unique identifier but it is not so. Different meanings of an ambiguous lemma are distinguished with a number, that becomes part of the lemma itself. Thus, we have the lemma *kohout-1* for an animal (*a cock*), and *kohout-2* for a closure device (*a tap*) for instance of gas or water. If there are more lines for description of a single lemma, the number must be the same for all of them. Moreover, if there is an additional information concerning the lemma (in the dictionary field AddInfo), they all must be the same as well.

The **AddInfo** field contains optional additional information concerning wordforms described by the given line. There might be information concerning derivation, or a semantic explanation (for instance the note "bird" for the lemma *kohout-1*). They are related to the whole paradigm belonging to the given lemma. Another sort of additional information, concerning the style, may be related only to special wordforms. An example is the lemma *téci* (*to flow*) with two forms for the present 3rd person plural: *tekou* and *tečou*. The former form is archaic, which is denoted by a special code on the dictionary line.

1.2   Lemma Numbering

The lines were added to the dictionary mainly manually, by several contributors. They were using data from many various sources, at first from older paper dictionaries of Czech, then from corpora. Despite many different checks, number of errors or inconsistencies entered the dictionary, often due to different opinions of different contributors. One such type of inconsistency is the lemma numbering, described above. There were diligent, punctilious contributors, who tried to put many different senses of individual words to the dictionary, especially from the paper dictionaries [7,8]. However, the morphological dictionary should not be a collection of meanings, especially when some of them are metaphors, or are very close.

A morphological dictionary should contain all the wordforms, not meanings, together with their tags and lemmas, and also information of word formation. If two word paradigms share all the morphological properties, including the word formation consequences, there should be only one lemma, even with different meanings. In this sense, it is not necessary to distinguish between the words *kohoutek* as a name of a flower and *kohoutek* as a tap because both have the same derivational as well as inflectional model, namely for masculine inanimate nouns. On the other hand, there is a different lemma *kohoutek* (*a small cock*), that has different derivational (as well as inflectional) model for masculine animate nouns.

In connection with a project *An Integrated Approach to Derivational and Inflectional Morphology of Czech*, we started a deep inspection of the numbers in lemmas. We found out, that they are used often needlessly. On the other hand, we found cases where they were missing. The inconsistent usage of numbers in lemmas leads to wrong evaluations of relations among some wordforms.

One of the reasons of the inconsistencies is a vague distinction between the two crucial linguistic terms — homonymy and polysemy.

2   Homonymy and Polysemy

The definitions seem to be clear [4]: A word with (at least) two entirely distinct meanings yet sharing a lexical form is said to be **homonymous**, while a word with several related senses is said to be **polysemous**.

It is important to note that in this text, we consider only the homonymy on the level of whole paradigms. We do not deal with homonymous wordforms belonging to the same lemma (e.g. *hrad* as nominative as well as accusative of the lemma *hrad* (*castle*)). Very detailed description of homonymy in Czech is in the recently published work [5].

From the above definitions, it is clear, that both, homonymous as well as polysemous words have different meanings. In this sense, homonymous words[1]

---

[1] The homonymy may be divided into homography for the same spelling and homophony for the same pronouncing. In this work, we use the term homonymy as a synonym for homography, in accordance with the traditional Czech terminology.

are always polysemous. One of the main differences between the two terms is that polysemy always concerns the whole paradigm — a sense of a word is considered to be the same in all its wordforms —, while homonymy might concern only special wordforms. The famous Czech example of homonyms is the word *žeň* as the imperative of the verb *hnát* (*to herd, to chase*), or as the singular nominative of the feminine noun (*a harvest*). The lemmas of those two wordforms are different, there is no homonymy.

However, there are also whole homonymous paradigms. A clear example of such homonymy is the lemma *kolej*, that has two different meanings, namely *a college* and *a track*. These two meanings have different origin and by chance they reached the same spelling after centuries of their development. They have the same inflectional paradigm, but their derivational behaviour differs.

A clear example of polysemy is *průvodce* (*a guide*), that can be a man or a written text. Their origin is the same and it is not surprising that they have the same spelling in their basic form (lemma). Their paradigms differ because of different animateness.

There are many unclear cases that are difficult to distinguish. One of the keys could be a translation to another language. If a word is possible to translate by two different words (like our example of *kolej*), the word is homonymous. If the word is possible to translate with one word, having two meanings even in the other language, like the *guide* from the second example, it is a pure polysemous word. However, this clue is not 100 %.

The distinguishing between homonymy and polysemy is important in the field of NLP, because it has often important consequencies. One of the most important ones concerns derived words. Take for example the homonymous word *kolej*. It is possible to derive an adjective, but each of its meanings will suit another derivation. *Kolej* in the meaning of *college* leads to the adjective *kolejní* (concerning a college, like *kolejní rada* = *college council*). The meaning of *track* has the adjective *kolejový*, like *kolejový jeřáb* = *tracked crane*.

However, the precise distinction is impossible, subjective. Lyons [4] proposes two strategies to avoid the problem. We cite from [6]:[2]

1. "Maximise homonymy — associate every meaning of a word with a distinct lemma."
2. "Maximise polysemy — no two lemmas can be entirely distinct when they are syntactically equivalent and when the set of wordforms they are associated with are identical."

Both approaches have their pros and cons. We decided to adopt the strategy 2 and to postpone the resolution of meanings to upper layers of NLP. It should be stated here, that the principle of maximising polysemy is taken not so strictly as it is stated in the above definition which was aimed probably especially for languages with not very rich inflection. We make exceptions, for instance for animate and inanimate nouns — see later.

---

[2] Available also at ftp://ftp.cogsci.ed.ac.uk/pub/kversp/html/node153.html.

3   Polysemy Within MorfFlex

As stated in the previous section, we decided to maximise polysemy in the morphological dictionary. In other words, if there are two lemmas with the same spelling, they will be considered as one lemma, unless they have different paradigms, or different derived words, or different syntactical behaviour.

For following explanations, we will need a more detailed terminology:
A **lemma** is a triple (L, N, A), where

- L is a pure lemma (word),
- N is a number, or empty string,
- A is additional info concerning the whole paradigm (represented by lemma), or empty string.

There should hold:
For every two lemmas (L, N, A) and (L, M, B) (with the same pure lemma L) the following should be true:
If $M! = N$, then

- $A! = B$.
- both M and N are nonempty.

If $A! = B$ then $M! = N$.

We have extracted all the dictionary lines that contained a numbered lemma. We ignored abbreviations, in other words those lines that contained a model for an abbreviation. These lemmas cannot be used for studying derivatives or other linguistic topics, as they usually are not "normal words", but typically only initial letters. For every numbered lemma (without abbreviations), we checked, if there exists another line in the dictionary, containing the same lemma without any number. We added all such lines to the previous ones. Now, we have a set of dictionary lines where every lemma occurs at least once with a number. We want to repair the set in such a way that they conform to the requirements presented above.

According to the requirements, we checked automatically the lines from our file and found violations of the requirements. The next procedure — error corrections — was manual as it was not possible to make them automatically, their variability was considerable.

In the following paragraphs, we will present several main cases of violation the requirements, together with their solutions.

3.1   Uppercase and Lowercase

In the previous version of the dictionary, the lemmas differing in the case of their initial letter were often labeled with a different number, though it was not necessary. The case of the initial letter is a sufficient distinction to consider the two lemmas to be different.

There are three main groups of such lemmas.

Common and Proper Names. The most frequent are proper names having their counterparts as common words.

Example: The lemma *švanda* (*fun*, feminine) had the lemma *švanda-2*, while the proper family name *Švanda* (masculine animate) had the lemma *Švanda-1*. We have preserved the both lemmas, but removed their numbers, because *švanda* with lowercase *š* is different lemma than *Švanda* with uppercase *Š*, there is no need of an additional number.

Car Brands. The second frequent case are car brands. In normal texts they occur in the both variants — with the uppercase as well as lowercase initial letter. The more common car brand, the more often it appears in lowercase, because it became to be perceived as a common name, opposed to a proper name.

An example is the car brand *Lancia*, where only 17.6 % (out of 2,124) are written with lowercase initial letter in Czech texts, while in case of (in the Czech republic more common) *Trabant* the proportion is opposite — almost 3/4 occurrences (out of 9,567) are written with initial letter lowercase. The figures were counted on the corpus SYN [3].

In the majority of occurences, when speaking about a vehicle, the both variants are interchangable. They have also the same derivational and inflectional models. It is tempting to include them under one lemma. However, when speaking about a factory, or a brand, it is necessary to preserve the lemma with uppercase initial letter as well, we cannot consider the cases variants. The solution is the same as in the previous case — we have two lemmas differing in the initial letters (*trabant* and *Trabant*, *lancia* and *Lancia*).

Latin Names. MorfFlex CZ contains a number of Latin names denoting biological species. The reason of their inclusion into the dictionary is unclear, they probably occured frequently in a text that was to become part of a corpus. In the dictionary, they often occur in two variants, with an upper- and lowercase initial letter, this time with different part of speech. That one with the uppercase is usually a noun, the second one an adjective, both with underspecified other morphological categories (or some of them). The part of speech of such words is disputable — though adjectives in Latin, they do not behave as adjectives in Czech, and as parts of a Latin name they both could be considered (in Czech texts) a noun. One of the solutions could even be their omission from the dictionary, as there are only some of the Latin names included in the dictionary, and the selection is arbitrary. However, we decided not to erase anything that could occur in Czech texts, so we left both lemmas (noun, adjective) in the dictionary, but removed their numbers.

Example: There were two lemmas for the partrige (*Perdix perdix* in Latin): *perdix-1* with the tag `AAXXX----1A----` and *Perdix-2* with `NNMXX-----A----`. After the changes, we have the lemmas *perdix* and *Perdix*. The capitalization is sufficient for the lemma distinction. In the future, we should adopt a better solution, especially concerning their part of speech.

3.2   Fluctuating Declension

There are nouns in the Czech language, the gender of which is not strict. The most of them fluctuate between masculine and feminine, there are 173 animate and 93 inanimate nouns with both masculine and feminine inflecional models. There is a set of expressive words, for instance *šmudla* (*a dirty man/woman*). Another example is *privilej* (*a privilege*) which was interperted as *privilej-1* (fem.), *privilej-2* (masc.). After checking, we have one *privilej*, with feminine as well as masculine declension. Another set fluctuates between feminine and neuter (17 lemmas), for instance the bird *káně* (*a buzzard*).

Until now, we have considered the different genders as a distinguishing mark for different lemmas but this conclusion appears to be wrong. The fluctuating gender has no impact on meaning, there is no homonymy, no polysemy. It is only a matter of inflection. Therefore we removed the numbers from such lemmas. There could arise an objection, that in certain contexts it is not possible to decide the gender. The answer is Yes, that is true. If it is not possible to decide, there are two options — either to decide arbitrarily, or not to try to decide at all. It may depend on an application, what is better, but from the theoretical point of view, the both strategies are good.

3.3   Animate vs Inanimate Nouns

A shift in meaning may happen between an inanimateness and animateness. In our dictionary, there are almost 500 lemmas of this sort. If there is a common meaning for both, we consider them as one lemma. In this case, there could be an objection against merging the two lemmas into one, concerning derivation of possessive adjectives. Naturally, they can be derived only from animate nouns. On the other hand, their common origin and our principle of maximising polysemy speak in the favor of their identity.

Example: *baryton* as a bariton voice or a man singing in bariton voice, *recyklátor* (*a recycler*) as a man or a device, tool. Another example is the lemma *průvodce* (*guide*) from the Introduction section of this paper.

3.4   Predicatives

Predicatives[3] are described as neuter nouns or adverbs. Thus, we have *jasno-1* as noun and *jasno-2* as adverb (both translated into English as *clear*, for instance as in the sentence *Today will be clear.*). It would be natural to have only one lemma for such words, because the meaning is always the same. However, the fluctuation among parts of speech appears to be more severe than fluctuation among genders. Therefore, the situation of predicatives remains unchanged — we have two lemmas differing in number. Again, a better solution should be adopted in future.

---

[3] Czech predicatives form a class of words usually ending in -o. They are problematic in terms of POS classification.

3.5   Figurative Meanings

Virtually all words may be used in their normal or usual context as well as in an unusual one. In the latter case, it may happen, that another — figurative meaning of the original word arises from its usage in an "unusual" context. As there are no strict definitions of "usualness", it is often hard to decide, whether a certain context is "usual", or "newly usual" or completely "unusual". This implies that there is no generally accepted procedure to decide how many meanings a word has. It can be demonstrated by comparisons of some dictionary entries in several explanatory dictionaries. For instance, the dictionary [8] gives an explanation for 8 meanings of the Czech adjective *černý* (*black*), while another dictionary [7] presents only 6 meanings.

If there are no syntactical or derivational differences, it is not reasonable to make any distinction and have a single lemma for all such meanings in the morphological dictionary. The morphological dictionary contained two lemmas for this adjective: *černý* (color) and *černý-2* (*illegal*). We merged them into one lemma *černý*.

4   Conclusion

There are more inconsistencies in MorfFlex CZ, but those concerning homonymy and polysemy seemed to be the most important, as their influence on complex NLP applications may appear crucial. We want to quantify the impact of the actual dictionary cleaning on several results in tasks that use morphological dicionary as one of their bases. And we will continue in fixing other, less visible, errors and inconsistencies that are still present in the dictionary.

We hope that the principles adopted for Czech solution of homonymy and polysemy in morphological dictionaries could be inspirative for other languages.

Acknowledgment. Work on this paper was supported by the grant number 16-18177S of the Grant Agency of the Czech republic (GAČR).

References

1. Hajič, J.: Disambiguation of rich inflection (Computational Morphology of Czech). Nakladatelství Karolinum (2004)
2. Hajič, J., Hlaváčová, J.: MorfFlex CZ, LINDAT/CLARIN digital library at Institute of Formal and Applied Linguistics, Charles University in Prague (2013)
3. Křen, M., Čermák, F., Hlaváčová, J., Hnátková, M., Jelínek, T., Kocek, J., Kopřivová, M., Novotná, R., Petkevič, V., Procházka, P., Schmiedtová, V., Skoumalová, H., Šulc, M.: Corpus SYN, version 3. Institute of the Czech National Corpus FF UK, Prague (2014). http://www.korpus.cz
4. Lyons, J.: Semantics. Cambridge University Press, Cambridge (1977)
5. Petkevič, V.: Morfologická homonymie v současné češtině. Studie z korpusové lingvistiky 22. Nakladatelství Lidové noviny (2016)
6. Verspoor, C.M.: Contextually-dependent lexical semantics. Ph.D. thesis, University of Edinburgh (1997)
7. Slovník spisovného jazyka českého. Nakl. Československé akademie věd (1960)
8. Příruční slovník jazyka českého. Státní nakl (1937)

Cross-Language Dependency Parsing Using Part-of-Speech Patterns

Peter Bednár[(✉)]

Faculty of Electrical Engineering and Informatics,
Department of Cybernetics and Artificial Intelligence,
Technical University of Košice, Letná 9, 04200 Košice, Slovakia
`peter.bednar@tuke.sk`

Abstract. The presented paper describes a simple instance-based learning method for dependency parsing, which is based solely on the part-of-speech n-grams extracted from training data. The presented method is not dependent on any lexical features (i.e. words or lemmas) or other morphological categories so model trained on one language can be directly applied to another similar language with harmonized tagset of coarse-grained part-of-speech categories. Using the instance-based learning allows us to directly evaluate predictive power of part-of-speech patterns on evaluation data from Czech and Slovak treebanks.

Keywords: Dependency parsing · Cross-language parsing · Natural language processing

1 Introduction

Recent state-of-the-art methods for dependency parsing are based on rich set of features extracted from language-dependent properties including word forms and lemmas [1,4,6]. According to empirical experience, extraction of large number of features is one of the main reason why highly accurate parsers have high demands on resources and long parsing time. Training of a parser frequently takes several days and parsing of the sentence can take up to one minute on average. Lexicalized features are also problematic in the cross-domain or cross-language settings where the parser is built on a resource-rich language and then applied to improve accuracy of parsing on resource-poor language.

On the other hand, experiments with delexicalized parsers show that part-of-speech features can be highly predictive for unlabelled attachments [2,3]. In this paper, our main motivation is to experiment with very limited representation of input sentences based solely on the part-of-speech features. Our proposed method is based on the instance-based learning paradigm, where the algorithm directly extracts part-of-speech n-grams and estimates predictive power of the patterns without too much dependency on the parameters of learning algorithm or underlying classification model.

The rest of the paper is organized as follows. Subsequent chapter provides related introduction to dependency parsing problem and graph-based parsers. Section 2 provides details about the proposed method and Sect. 3 describes initial experiments.

© Springer International Publishing Switzerland 2016
P. Sojka et al. (Eds.): TSD 2016, LNAI 9924, pp. 117–124, 2016.
DOI: 10.1007/978-3-319-45510-5_14

1.1 Dependency Parsing

Given an input sentence $x = w_1, \ldots, w_n$, the goal of dependency parsing is to build a dependency tree, which can be represented as a set of directed arcs $\mathbf{d} = \{(h, m, l) : 0 \leq h \leq n; 0 < m \leq n; l \in \mathcal{L}\}$. Directed arc (h, m, l) is labeled with dependency label l from the set of possible labels $\mathcal{L}$ and connects head word w_h to the modifier word w_m. Head with index 0 denotes the root node of the dependency tree.

Current state-of-the-art methods can be divided into two groups which tackle the problem of dependency parsing from different perspectives but achieve comparable accuracy on different languages. Graph-based methods [1,6] view the parsing as the problem of finding maximum spanning tree from the fully-connected weighted directed graph where weights of the graph edges are predicted by statistical model. Transition-based methods transform target tree to the sequence of transitions and statistical model is used directly to predict the sequence of transitions that leads to the legal dependency tree from the sequence of input words. Transition-based parsers typically have a linear or quadratic complexity [4]. In [5] is introduced a transition based non-projective parsing algorithm that has a worst case quadratic complexity and an expected linear parsing time. Depending on factorization, graph-based methods have usually quadratic complexity. When parsing and factorization includes second-order factors and arc labeling, worst case complexity of graph based methods can be $O(n^4)$.

In both cases, statistical model is linear, i.e. parser maximizing score computed as the sum of dot products between the vector of features extracted for the factorized subsequence or subgraph and vector of model parameters – weights optimized on training data. In order to achieve state-of-the-art accuracy, models are build using the reach set of features extracted from the lexical properties of words such as word prefixes, suffixes, forms or lemmas, and general properties such as part-of-speech classes and other morphological categories common for similar languages. Disadvantage of this approach is that the feature space is high dimensional and current methods have high demands of resources and long parsing time. Another disadvantage is that models are dependent on the lexical features of words specific for the given language and they cannot be directly applied to improve parsing accuracy on similar language or domain.

1.2 Graph-Based Dependency Parser

In our approach, we adopted the graph-based method because it allows us to directly derive our weighting scheme for part-of-speech patterns. The graph-based method factors the score of a dependency tree into scores of small subtrees $\mathbf{p}$:

$$score(x, \mathbf{w}, \mathbf{d}) = \mathbf{w} \cdot \mathbf{f}(x, \mathbf{d}) = \sum_{\mathbf{p} \subseteq \mathbf{d}} score(x, \mathbf{w}, \mathbf{p}) \qquad (1)$$

The state-of-the-art models include factors of first and second order [1], i.e. the score of tree is factored as:

$$score(x, \mathbf{w}, \mathbf{d}) = \sum_{\{(h,m)\} \subseteq \mathbf{d}} \mathbf{w}_{dep} \cdot \mathbf{f}_{dep}(x, h, m) +$$
$$\sum_{\{(h,m),(h,s)\} \subseteq \mathbf{d}} \mathbf{w}_{sib} \cdot \mathbf{f}_{sib}(x, h, m, s) \tag{2}$$

where $\mathbf{f}_{dep}(x, h, m)$ and $\mathbf{f}_{sib}(x, h, m, s)$ are feature vectors extracted for single edges between head and modifier nodes or for subtrees of two edges connecting head node to two sibling nodes m and s; and $\mathbf{w}_{dep}$ and $\mathbf{w}_{sib}$ are corresponding vectors of weights inferred from training data using selected learning algorithm, e.g. passive-aggressive perceptron algorithm. For the given input sequence of words, the parsing algorithm finds a dependency tree with the highest score, which corresponds to maximum spanning tree of the weighted fully-connected oriented graph constructed by factorization. Eisner algorithm can be used for parsing of projective trees, meaning that if we put the words in their linear order, preceded by the root, the arcs of the dependency tree can be drawn above the words without crossings, or, equivalently, a word and its descendants form a contiguous substring of the sentence. Complexity of Eisner algorithm is $O(n^3)$. For non-projective trees, the Chu-Liu-Edmonds algorithm can be used to find unconstrained maximum spanning tree. Tarjan proposed efficient implementation of this algorithm for dense graphs with quadratic complexity.

2   Dependency Parsing Using the Part-of-Speech Patterns

Instead of linear statistical model, our approach is based on the instance-based learning referenced also as "lazy" learning since during the training phase, algorithm is not inferring the parameters of the model from the training data. Instead of this, training data are stored as the instance patterns, which are matched during the model application/testing phase on new data. Instances of our model are based directly on the n-gram syntactical patterns. For a given set of part-of-speech labels, length of the pattern n, length of the prefix $0 \leq m$ and postfix $0 \leq k; m + k < n$ parts, the instance is defined as the tuple $(\mathbf{p}, \mathbf{h})$, where $\mathbf{p} = (p_1, \ldots, p_n)$ is the sequence of part-of-speech labels and $\mathbf{h} = (h_{m+1}, \ldots, h_{n-k-1})$ encodes the predicted head nodes of corresponding modifiers at the positions matched by the $\mathbf{p}$ sequence. We will denote $\mathbf{p}$ as the condition part of the instance and $\mathbf{h}$ as the prediction part.

Heads in the prediction part are encoded directly as tuple $h_j = (i_j, y_j)$ where y_j is the head's part-of-speech label and i_j is the index, which counts nodes between the head and the modifier with the head's part-of-speech label. Index i_j also indicates the position of the head relative to the modifier, i.e. if the head node precedes the modifier, value is negative, otherwise value is positive. If the head of the node is root of the sentence, index i_j is equal to 0 and part-of-speech is set to ROOT label.

For training data, we are scanning each sentence in the training set left to right and generating n-gram instances for each scanned position. Sentences are extended with the artificial token ϵ inserted at the beginning and at the end of the sentence. Instances are then stored in the multi map (an associative array with multiple values for one key) where condition part **p** is the key and prediction part **h** is the value. Each instance has associated weight which counts how many times the instance occurred in the training data. The following algorithm summarizes training procedure (Table 1).

For example, for decision tree of sentence "He has good control." on Fig. 1, with pattern length $n = 4$ and prefix/suffix length $= 1$, algorithm generates the following instances:

Algorithm 1. Training phase.

Require: $S = \{(x_j, t_j), j = 1, \ldots, T\}, n, m, k$
Ensure: $patterns[\mathbf{p}][\mathbf{h}], weights[\mathbf{p}][\mathbf{h}]$
 1: **for** $j \leftarrow 1, \ldots, J$ **do**
 2:    $i \leftarrow 1$
 3:    **repeat**
 4:       $\mathbf{p} \leftarrow (p_i, p_{i+1}, \ldots, p_{i+n+1})$
 5:       $\mathbf{h} \leftarrow (h_{i+m}, h_{i+m+1}, \ldots, h_{i+n-k})$
 6:       **if** $patterns[\mathbf{p}][\mathbf{h}]$ exists **then**
 7:          $weights[\mathbf{p}][\mathbf{h}] \leftarrow weights[\mathbf{p}][\mathbf{h}] + 1$
 8:       **else**
 9:          $patterns.\mathrm{add}(\mathbf{p}, \mathbf{h})$
10:          $weights[\mathbf{p}][\mathbf{h}] \leftarrow 1$
11:       **end if**
12:       $i \leftarrow i + 1$
13:    **until** $i \leq |x_j| - n - 1$
14: **end for**

Table 1. Instances generated for English sentence "He has good control."

Index	Instance	
	Condition part	Prediction part
1	(ϵ, PRP, VBZ, JJ)	((1, VBZ), (0, ROOT))
2	(PRP, VBZ, JJ, NN)	((0, ROOT), (1, NN))
3	(VBZ, JJ, NN, PUN)	((1, NN), (−1, VBZ))
4	(JJ, NN, PUN, ϵ)	((−1, VBZ), (−1, VBZ))

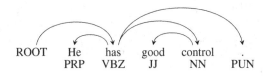

ROOT   He   has   good   control   .
          PRP   VBZ   JJ   NN   PUN

Fig. 1. An example of unlabelled dependency tree for English sentence "He has good control."

During the prediction phase algorithm is constructing fully connected weighted directed graph. Similarly to the training, at first, algorithm scans input sentence from left to right and matches condition part of the instances with the input part-of-speech sequence. For all detected instances, algorithm tries to find heads for each predicted position according to head's part-of-speech and index and increment weight of the edge between the head and the modifier by weight of the instance. Following algorithm summarizes the prediction phase:

Algorithm 2. Prediction phase.

Require: $x, n, m, k, patterns[\mathbf{p}][set(\mathbf{h})], weights[\mathbf{p}][\mathbf{h}]$
Ensure: maximum-spanning-tree($\mathbf{g}$)
  1: $i \leftarrow 1$
  2: $\mathbf{g} \leftarrow 0$
  3: **repeat**
  4:     $\mathbf{p} \leftarrow (p_i, p_{i+1}, \ldots, p_{i+n+1})$
  5:     **if** $patterns[\mathbf{p}][\mathbf{h}]$ exists **then**
  6:        **for all** $\mathbf{h} = (h_m, h_{m+1}, \ldots, h_{n-k}) \leftarrow patterns[\mathbf{p}][set(\mathbf{h})]$ **do**
  7:           **for** $j \leftarrow m, \ldots, n - k$ **do**
  8:             $q \leftarrow \text{head}(h_j, i + j)$
  9:             **if** q exists **then**
10:                $\mathbf{g}[q][i + j] \leftarrow \mathbf{g}[q][i + j] + weights[\mathbf{p}][\mathbf{h}]$
11:             **end if**
12:           **end for**
13:        **end for**
14:     **end if**
15:     $i \leftarrow i + 1$
16: **until** $i \leq |x_j| - n - 1$

We are using Tarjan's algorithm for non-projective trees to find the maximum spanning tree of weighted graph. The predictive part can be applied partially only to all nodes for which it is possible to find the head with specified constrains. Note that the same weight of the instance is added to weight of the edges for all predicted heads.

3   Experiments and Analysis

We tested our proposed algorithm on normalized HamleDT 3.0 dataset [7]. HamleDT provides collection of dependency treebanks with unified set of morphological tags and dependency labels. In order to test cross-language application of part-of-speech patterns, experiments were performed on treebanks for Czech and Slovak languages. Using the part-of-speech patterns we can directly observe to what degree both languages share syntactical structures. This setup also reflects the size of the treebanks in a sence that Czech language is used as the resource-rich language applied to improve accuracy of parsing in Slovak language with smaller resources. In all experiments, data were divided to training, validation and testing set according to a standard division in HamleDT dataset.

In the first experiment, we optimized the length of the patterns (length of the condition part). The patterns were extracted from the training set of Slovak treebank with the length set from 2 to 8-grams and evaluated on the validation set. As the evaluation criteria in all experiments, we used Unlabelled Attachment Score (UAS), i.e. fraction of tokens with correctly assigned head and the number of sentences in which all tokens were parsed correctly.

Similarly, we optimized the length of the prefix and the suffix part of the pattern. In all subsequent experiments, we have fixed the length of the pattern to 5 and the length of prefixes and suffixes to 1, i.e. based on the part-of-speech sequence of 5 tokens, model predicts head assignment of 3 tokens in the middle of the pattern.

In the second experiment, we compared the accuracy in cross-language setting. The results are summarized in Table 2. As the baseline accuracy, we tested patterns extracted from Slovak training set, then we tested patterns extracted from Czech training set and applied to Slovak language and finally, we combined set of patterns extracted from Czech and Slovak language and applied them to Slovak language. For comparison we have also included accuracy on Czech dataset and base-line accuracy of MST parser trained on the unlexicalized Slovak dataset containing part-of-speech features only. For MST parser we have used default settings for non-projective decoding. Altogether, for Slovak dataset, the algorithm generated 283 540 instances with 73 820 unique part-of-speech 5-grams (condition parts). The average ambiguity of part-of-speech patterns, i.e. average number of different prediction parts for the same condition part was 3.84. The longest dependency relation encoded in the prediction part, i.e., highest number of tokens with the head's part-of-speech between head and modifier was 36. For Czech language, algorithm generated 359 768 instances with 89 747 part-of-speech patterns. Average ambiguity increased to 4.01 and longest dependency was 56.

Table 2. Summary of UAS and total correct accuracy for Czech and Slovak language separately and in cross-language setting.

Method/language	sk	cz	cz to sk	(sk + cz) to sk	sk MST
UAS	68.71	70.99	60.27	68.52	74.71
Total correct	19.04	17.46	14.75	18.76	29.61

Since our basic algorithm is using exact matching of patterns and only detected patterns are applied, we have counted number of hits and misses on testing data. For Slovak language, patterns were detected in 721 115 cases and there were 2 979 (about 4 %) undetected training cases, what reflects inferior accuracy of pattern-based algorithm in comparison to MST parser. When the Czech language patterns were applied to Slovak testing set, number of undetected patterns increased to 10 348 (14 %). In conclusion, mixing of patterns from both languages did not increase accuracy of combined parser. This can be explained by coverage of patterns from Czech language in Slovak language, where only 14 373 condition parts (part-of-speech 5-grams) and 174 combinations of condition and prediction parts co-occurred in both datasets.

Regarding the computation time, both training and testing is very fast in comparison to MST parser. Patterns are stored in hashing map so the complexity of pattern lookup is constant. For example, for Slovak language, it took only 5 s to generate patterns from training set and about 1 s for testing including time for loading of data from hard drive. In comparison, on the machine with the same configuration it took about 53 min to build model using MST parser. Each pattern requires approximately n bytes for condition part and $(n-k-m).2$ bytes for prediction part extended with some additional memory for hashing entries.

4 Related Work

Current state-of-the-art parsers (both graph-based or transition-based) are using very rich set of features extracted from language-dependent word properties. Set of features was later extended by latent features (word embeddings) learned on large unannotated corpora. Our research is targeted to build accurate model solely from the part-of-speech sequence and add additional features only when needed to further constrain head/label resolution. Regarding the cross-language setting, our approach is similar to other methods [2,3,8] for learning of delexalized parsers on resource-rich languages for parsing resource-poor languages or domains with harmonized set of morphological features and dependency labels. This approach does not directly require any bitext corpus, but similarly to methods in [9,10], bitext can be incorporated in order to further help transfer of delexicalized parsers.

5 Conclusion

In the presented paper, we have described simple method for dependency parsing using the part-of-speech n-grams. We have presented initial results on Czech and Slovak treebanks and in cross-language setting. The presented method is baseline, which can be extended in many aspects. In future work, we would like to conduct more experiments and analysis of predictive power of part-of-speech n-grams and implement extensions of parsing algorithm, namely partial matching of patterns and boosting optimization of weights for each predicted attachment. Besides these extensions, we would like to investigate methods based on Formal Concept Analysis [11–13] in order to generate frequent set of patterns. We hope that with these extensions we will be able to achieve parsing accuracy comparable with current state-of-the-art methods.

Acknowledgments. The work presented in this paper was supported by the Slovak VEGA grant 1/0493/16 and Slovak KEGA grant 025TUKE-4/2015.

References

1. Bohnet, B.: Top accuracy and fast dependency parsing is not a contradiction. In: Proceedings of the 23rd International Conference on Computational Linguistics, pp. 89–97 (2010)
2. McDonald, R., Petrov, S., Hall, K.: Multi-source transfer of delexicalized dependency parsers. In: Proceedings of the Conference on Empirical Methods in Natural Language Processing, pp. 62–72 (2011)
3. Zeman, D., Resnik, P.: Cross-language parser adaptation between related languages. NLP for Less Privileged Languages, pp. 35–42 (2008)
4. Nivre, J., Hall, J., Nilsson, J.: Memory based dependency parsing. In: Proceedings of the 8th CoNLL, Boston, Massachusetts, pp. 49–56 (2004)
5. Nivre, J.: Non-projective dependency parsing in expected linear time. In: Proceedings of the 47th Annual Meeting of the ACL and the 4th IJCNLP of the AFNLP, Suntec, Singapore, pp. 351–359 (2009)
6. McDonald, R., Crammer, K., Pereira, F.: Online Large-margin Training of Dependency Parsers. In: Proceedings of ACL, 91–98 (2005)
7. Zeman, D., Dušek, O., Mareček, D., Popel, M., Ramasamy, L., Štěpánek, J., Žabokrtský, Z., Hajič, J.: HamleDT: Harmonized Multi-Language Dependency Treebank. In: Language Resources and Evaluation, ISSN 1574–020X, vol. 48, no. 4, Springer, Netherlands, 601–637 (2014)
8. Petrov, S., Das, D., McDonald, R.: A universal part-of-speech tagset. In: arXiv:1104.2086 (2011)
9. Täckström, O., McDonald, R., Nivre, J.: Target language adaptation of discriminative transfer parsers. In: Proceedings of NAACL, pp. 1061–1071 (2013)
10. Täckström, O., McDonald, R., Uszkoreit, J.: Cross-lingual word clusters for direct transfer of linguistic structure. In: Proceedings of NAACL-HLT (2012)
11. Butka, P., Pócs, J., Pócsova, J.: On equivalence of conceptual scaling and generalized one-sided concept lattices. Inf. Sci. **259**, 57–70 (2014)
12. Butka, P., Pócs, J., Pócsová, J., Sarnovský, M.: Multiple data tables processing via one-sided concept lattices. Adv. Intell. Syst. Comput. **183**, 89–98 (2013)
13. Butka, P., Pócs, J., Pócsova, J.: Distributed computation of generalized one-sided concept lattices on sparse data tables. Comput. Inf. **34**(1), 77–98 (2015)

Assessing Context for Extraction of Near Synonyms from Product Reviews in Spanish

Sofía N. Galicia-Haro[1]([⊠]) and Alexander F. Gelbukh[2]

[1] Faculty of Sciences, UNAM, Mexico City, Mexico
sngh@fciencias.unam.mx
[2] Center for Computing Research, National Polytechnic Institute,
Mexico City, Mexico
http://www.Gelbukh.com

Abstract. This paper reports ongoing research on near synonym extraction. The aim of our work is to identify the near synonyms of multiword terms related to an electro domestic product domain. The state of the art approaches for identification of single word synonyms are based on distributional methods. We analyzed for this method different sizes and types of contexts, from a collection of Spanish reviews and from the Web. We present some results and discuss the relations found.

Keywords: Semantic similarity · Product reviews · Vector space model · Context

1 Introduction

The motivation of our work is to analyze if the general methods currently applied on scientific and specialized documents to extract single word synonyms or near-synonyms [1] are the methods to obtain semantically related multiword terms (MWTs) appearing in product reviews. We considered that such MWT synonyms correspond to denominative variants as previously [2] defined: different denominations restricted for example to lexicalized forms, with a minimum of consensus among the users of units in a domain, since such conceptualization reflects more properly the degree of informality expressed in the texts we used and the diversity of terms in daily life products domain.

We found this problem when we were working on a collection of washing machine reviews: There are concepts that have several possible term candidates. For example, the Delayed Start function allows the startup of the washing machine program to be delayed for a number of hours. Some of the variants of the Spanish term related to this concept are: *inicio diferido* 'delayed start', *inicio retardado* 'retarded start', *encendido programable* 'programmable switch on', *preselección de inicio* 'start time-preselection', and others including different word order of the same variants.

We supposed that such terms were generated by the authors of the opinion reviews or maybe by the translators of the washing machine instruction manuals

© Springer International Publishing Switzerland 2016
P. Sojka et al. (Eds.): TSD 2016, LNAI 9924, pp. 125–133, 2016.
DOI: 10.1007/978-3-319-45510-5_15

from English to Spanish, but we found that these variants also have an origin in the instruction manuals of the manufacturers. Automatically grouping such similar MWTs should be useful since there are no glossaries of semantically related terms used by different manufacturers, translated manuals, and users of electro domestic appliances. This type of knowledge should be included in service robots in the future.

For the automatic determination of single word synonyms two main paradigms have been applied: lexicon-based and distributional approach. The former paradigm requires sources of word definitions in general language, i.e. dictionaries or terminology banks are required. The second paradigm relies on the distributional hypothesis of Harris [3]: words with similar meaning tend to occur in similar contexts. The required sources for this approach are corpora. Since sources of word definitions are not available we decided to use the distributional approach. We applied this method first on a collection of washing machine review texts and then on retrieved Web contexts. The work that we describe here is an analysis of the effect of different contexts on MWT synonym extraction.

The paper is organized as follows: in the next section, we present an overview of the method, describing the materials and the details of the method we followed. In Sect. 3 we describe the context sizes and types we defined and some of their statistics. In Sect. 4 we present the results and discuss their interpretation. The final section gives the conclusions.

2   The General Method

We followed the well-known assumption that words are more likely to be semantically related if they share the same contexts. Its common implementation in the Vector Space or Word Space Model [4,5] is based on the computation of a vector for a word w_t. Its dimensions correspond to the close neighbors of w_t, obtained from a corpus.

Seeing that review texts have many linguistic errors we did not consider grammatical relations to select the contexts. We followed other works where close neighbors are computed from one word to the left and one word to the right of w_t to a larger context. Each word neighbor w_i or entry in the vector has a value that measures the tightness between w_i and w_t. In this method, the semantic similarity of two terms is evaluated by applying a similarity measure between their vectors and then making a ranking based on such values. In this work, instead of a word w_t we consider a multiword noun phrase mw_t and its neighbors as single words. We also characterized each MWT mw_t from our collection by a vector computed from its neighbors.

2.1   Multiword Terms

In this work a MWT is a noun phrase of several words including prepositions and articles. They were obtained by the following patterns:

Noun [Noun| Adjective]
Noun Preposition [Article] Noun [Adjective]
Noun Preposition Num Noun

These patterns covered all the names of the functions and the noun phrases in the product characteristics section of the washing machine manuals that we collected. The patterns include prepositions and articles since in colloquial Spanish it is usual to include articles. For example: *bloqueo infantil* 'child lock', *bloqueo para niños* 'child lock', *seguro para niños* 'child safety', *seguridad para niños* 'child safety', *apertura a prueba de niños* 'child-proof opening system', *sistema de bloqueo infantil* 'child lock system', *bloqueo para los niños* 'child lock' are denominative variants for the same concept: the door lock that prevents children putting their hand into the washing machine while it is working.

2.2 Corpora

Corpus of Product Reviews. We used a collection of review texts compiled in a previous study [6], named here Corpus of Product Reviews (CPR), comprising 2,800 reviews extracted from the *ciao.es* website. This site has product reviews in Spanish for diverse electro domestic appliances. The collection was automatically compiled from the washing machine section and it was tagged using Freeling [7].

We wrote a program that executed a sequence of pre-processing steps. From the raw text corpus, the first step split the CPR texts in sentences. Sahlgren [8] stated that the semantic properties of the word space model were determined by the choice of context and that the more linguistically justified definition of context in which to collect syntagmatic information should be a clause or a sentence. Then the program extracted the sentences where the MWT candidates appeared, based on the POS patterns described above. Normalization was not considered so a word with the first letter in upper case was different from that with all letters in lowercase. We did not consider applying any spell-checking correction since Freeling gave the correct tag in many spelling mistakes. In the last step the sentences including MWTs with frequency higher than one were selected since the quantity of examples for each multiword phrase was low. The preprocessing resulted in 34,871 sentences for 5,422 MWTs.

We selected 112 noun phrases that corresponded to 97 MWTs associated with the washing machine functions. Since we did not normalize, *botón de inicio* and *boton de inicio* are different MWTs in our work. We considered that each example was related to a different term. One reason is that we had few examples for most of the MWTs and we could use their results as a base line. Also, we noticed that MWTs may have a distinct behavior since manufacturers, translators and users made them rare cases for the quantity of terms they used for a single concept.

Table 1 shows the distribution of the quantity of examples in the collection for each MWT. For example, in the first row, the first column shows that there were 2 examples for each one of 45 terms and the last column shows that there were 14 examples for only one term.

Table 1. Quantity of examples obtained from the CPR

#Examples	2	3	4	5	6	7	8	9	10	11	13	14
#MUTs	45	20	13	2	6	4	1	2	1	1	2	1
#Examples	15	16	18	19	21	22	24	29	33	34	54	147
#MWTs	1	1	1	1	2	2	1	1	1	1	1	1

Corpus Obtained from the Web. Because of the small number of contexts in the CPR for most of the selected MWT's, we decided to acquire contexts for them from the Web. Two problems inherent in Web searching we tried to avoid: different domain for the MWT's, and incomplete context in the snippets.

We wrote a program to obtain context examples from the Web searching for each one of the selected 97 MWTs, using the Google's asterisk facility [9]. The word *lavadora* 'washing machine' was incorporated in the search to try to limit the context to such domain. For example, for the *inicio diferido* 'delayed start' term the search was launched with the string: "* * *inicio diferido* * *" *lavadora*, where the asterisks substitute for the possible sequences of words around the MWT. The Google search engine tool was limited to the Spanish language. Google returned a quantity of hits where each snippet most probably there would have a string of different words, then the MWT followed by another sequence of words. The program retrieved a maximum of 500 hits for each MWT. The total number of retrieved web contexts were 25,013, from them 7,251 were useful, mainly due to very short contexts and missing words corresponding to the search keywords in context in the snippets.

Figure 1 shows the distribution of the quantity of examples obtained from the Web. The horizontal axis corresponds to the number of multiword terms (from 1 to 3). The vertical axis measures the number of examples. For example, 3 terms have 100 examples, only unique terms have more than 160 examples.

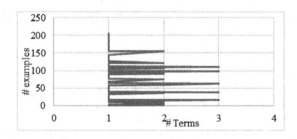

Fig. 1. Distributional quantity of examples obtained from the Web

2.3   Measures for the Distributional Method

In this section we describe the measures that we considered for the tightness between the neighbors and the MWT, and for the semantic similarity of two term vectors.

There are different measures that have been applied in the distributional method, for example Ferret [10] evaluated three measures for neighbor tightness: T-test, Tf-Idf, and Pointwise Mutual Information (PMI), and six measures for term similarity: Cosine, Jaccard, Jaccard†, Dice, Dice†, Lin. He found that Cosine measure with PMI gave the best results. Hazem and Daille [11] also applied diverse measures and they found the results were more contrastive for the Spanish corpus in comparison with a French and an English corpus. The PMI and Cosine measures performed the best for the Spanish corpus. Based on such works we decided to use PMI and Cosine measures.

The general method was applied in three steps. In the first step for each MWT mw_i its vector was built gathering all the words that co-occurred with mw_i within each specific window that we detail in the next section. In the second step the program computed the value based on PMI [12] that measures the tightness of the mw_i with each word in its specific window. This value was computed by the following equation:

$$\text{PMI}(mw_i, w_j) = \log \frac{P(mw_i, w_j)}{P(mw_i), P(w_j)}$$

where

$$P(mw_i) = P(w_{pmi_1})P(w_{pmi_2})\ldots P(w_{pmi_n})$$

$P(mw_i, w_j)$ is the coocurrence of mw_i with the word w_j appearing in the window of mw_i and $P(w_i)$ is the ocurrence of w_i in the collection. The formula for $P(mw_i)$ is the model of independence of McInnes [13] for n-gram of any size n. We considered the lemma of each word to group those that differ in gender and number, for example: *lavadora* and *lavadoras* were gathered and represented by the *lavadora* lemma. For the corpus obtained from the Web, $P(mw_i, w_j)$ and $P(w_i)$ corresponded to the number of hits retrieved by the Google search engine in the same sense described above.

In the third step the cosine similarity [14] was computed for each pair of vectors v_k and v_l by the following equation:

$$\text{cosine}_{v_l}^{v_k} = \frac{\sum_i \text{PMI}(mw_i, l)\text{PMI}(mw_i, k)}{\sqrt{\sum_i \text{PMI}(mw_i, l)^2}\sqrt{\sum_i \text{PMI}(mw_i, k)^2}}$$

The candidate denominative variants of the MWT mw_i are the MWTs best ranked following their cosine value.

3   Contexts

We experimented on the selection of the quantity of word neighbors of the MWT, i.e. on the context size. But we also experimented with the restrictions imposed by the review authors according to the punctuation marks they included. Finally, we experimented delimiting the context by means of eliminating specific kind of words.

3.1 Context Sizes

Different window sizes have been defined in the distributional method. For example, Ferret [10] analyzed a measure that performed well on an extended TOEFL test, it was applied for synonym extraction. The measures were tested with window sizes between 1 and 5 words. He found the best results for the window size of 1 word on a corpus made of around 380 million words from news articles.

Rosner and Sultana [15] investigated methods for extending dictionaries using non-aligned corpora by finding translations through context similarity. The contexts were converted into vectors and then compared using the cosine measure. They used news text as the main source of comparable text for English and Maltese. The authors tested different window sizes from 1 to 5 words, and the window size of 3 was found to be the optimal.

Hazem and Daille [11] applied a 7-window size. Their experiments were carried out on a French/English/Spanish specialized corpus from the domain of wind energy of 400,000 words. Their work was devoted to extracting synonyms of MWTs by means of a semi-compositional method.

Seeing that the best size for the window differ from one work to another we decided to use two window sizes: 12 words around the MWT, named CT12, and 6 words around the MWT, named CT6, for the contexts obtained from the CPR. Regarding the contexts extracted from the Web, since we used the snippets retrieved by the Google search engine we did not consider experimenting on sizes for no-clear-cut contexts.

3.2 Context Types

Sahlgren [8] considered that clauses and sentences or at least the functional equivalent to such entities seem to be linguistic universals, i.e. some sequence delimited by some kind of delimiter. We followed this idea considering that delimiters in the CPR texts should be taken into account to restrict the window size since users used punctuation marks with a specific purpose. We proposed to reduce the contexts according to the following punctuation marks: points, quotes, parenthesis, exclamation mark, slash, semicolon, and hyphen as delimiters of the left and right contexts.

We wrote a program to obtain two reduced contexts from the previous CT12 and CT6, delimited by the indicated punctuation marks and named CR12 and CR6 respectively. Table 2 shows the cosine values obtained for some MWTs from the CPR contexts for the various sizes and types described above. We could observe that the highest values corresponded to the taxonomic relation between *sistema de bloqueo* 'lock system' and *sistema de bloqueo infantil* 'child lock system' while the lowest values corresponded to the first row for another semantic relation across taxonomic relation links between *seguro para niños* 'child safety locks' and *sistemas de seguridad* 'security systems'. The rest of the MWTs corresponded to near synonyms or denominative variants.

For the Web contexts we defined 3 types of context delimitation. We wrote a program to obtain the context delimited by the indicated punctuation marks

Table 2. Cosine values for some MWTs obtained from CPR contexts

Multiword terms	CT12	CT6	CR12	CR6
Seguro para niños VS *sistemas de seguridad*	0.5000	0.4376	0.3574	0.2989
Boton de on VS *botón de inicio*	0.5645	0.8629	0.4823	0.5215
Tiempo restante VS *tiempo remanente*	0.85438	0.79748	0.89059	0.5139
Programación de fin VS *fin diferido*	0.9473	**0.8929**	0.6349	0.6349
Bloqueo de seguridad VS *sistema de bloqueo*	0.9485	0.8825	0.8752	0.7812
Sistema de bloqueo VS *sistema de bloqueo infantil*	**0.9762**	0.8428	**0.9781**	**0.7888**

and by deleting the following function words: determinants, pronouns, numbers, conjunctions, prepositions and auxiliary verbs, named MW1. We also obtained another context delimited as MW1 and reduced additionally by deleting adverbs from them, named MW2. We supposed that attributes could not be useful for context similarity. The third context type named MW3 was delimited by deleting in addition to the previous ones the short words (1–2 letters) with unknown POS.

Table 3. Cosine values for some MWTs obtained from Web contexts

Multiword terms	MW1	MW2	MW3
Boton de on VS *botón de inicio*	0.1462	0.1477	0.1409
Seguro para niños VS *sistemas de seguridad*	0.1553	0.1511	0.1779
Programación de fin VS *fin diferido*	0.2357	0.2418	0.2569
Bloqueo de seguridad VS *sistema de bloqueo*	0.2715	0.2639	0.2560
Tiempo restante VS *tiempo remanente*	0.3146	0.3079	0.2940
Sistema de bloqueo VS *sistema de bloqueo infantil*	**0.3369**	**0.3384**	**0.3285**

Table 3 shows the cosine values obtained for some MWTs according to their Web contexts. We could observe that the cosine values are lower than those obtained from the CPR collection since the number of occurrences and co-occurrences were taken from the total hits reported by the Google search engine.

4   Results and Discussion

One method for evaluating the performance of an extraction system is to compare the similarity scores assigned by the system to the results given by human judges. Since we do not have such a golden standard we manually analyzed the first 100 top results for each one of the several context types we defined. Two students that manually analyzed the first 100 top results for each kind of context were required to search in the Web to clarify the specific meaning of many MWTs

Table 4. Precision for top 100 results

Context	CT12	CR6	CR12	CT6	MW1	MW2	MW3
Precision	0.33	0.43	0.45	0.48	0.55	0.60	0.60

since initially their agreement rate (kappa statistic [16]) was 69. The precision for the 100 top values of similarity is shown in Table 4.

We observe that we obtained some differences in results among the various contexts applied to the collection of product reviews. The delimitation by punctuation marks was more useful on the 12-word window increasing their results by 12 %. This delimitation has an adverse effect on the 6-word context where the complete context scored 5 % percent higher than its reduced counterpart.

Regarding the results obtained for the 3 context types defined for Web contexts, we observe that the precision did not change for the MW2 and MW3 context types and that short words elimination had no effect on results. The MW2 type obtained 5 % better results than the MW1 type where elimination of adverbs was the only difference between them. The attributes elimination has more sense if applied to product reviews since the texts include personal experiences, personal thoughts, opinions about anything, etc. but we wanted to analyze their effect on the Web contexts.

Despite the 60 % precision obtained for the better results, we obtained several MWT groups related to a concept, we show two of such groups:

delayed start: *comienzo retrasado, marcha diferida, programación diferida, preselección de fin, función de inicio, inicio diferido, retardo horario, tiempo diferido*

on/off button: *botón de arranque, botón inicio, botón de encendido, tecla de encendido, botón de inicio*

5   Conclusions

As Sahlgren [8] stated, the distributional models are not only grounded in empirical observation, but they also rest on a solid theoretical foundation. Despite the lower quantity of examples used in this work we concluded that the results are useful according to the task complexity. We present in this work experiments to analyze the adequacy of several kind of contexts to extract denominative variants of MWTs applying the distributional method first to a collection of Spanish reviews for washing machines and then to contexts retrieved from the Web for the MWTs obtained from such product reviews.

We manually tested the results for multiword terms associated to different concepts and the best results were obtained for the Web contexts delimited by punctuation marks, function words and attributes.

Acknowledgments. The second author recognizes the support of the Instituto Politécnico Nacional, grants SIP-20161958 and SIP-20162064.

References

1. Edmonds, P., Hirst, G.: Near-synonymy and lexical choice. Comput. Linguist. **28**(2), 105–144 (2002)
2. Freixa, J.: Causes of denominative variation in terminology: a typology proposal. Terminology **12**(1), 51–77 (2006)
3. Harris, Z.: Distributional structure. Word **10**(23), 146–162 (1954)
4. Schütze, H.: Dimensions of meaning. In: Proceedings of the ACM/IEEE Conference on Supercomputing, pp. 787–796. IEEE Computer Society Press (1992)
5. Schütze, H.: Automatic word sense discrimination. Comput. Linguist. **24**(1), 97–123 (1998)
6. Galicia-Haro, S.N., Gelbukh, A.: Extraction of semantic relations from opinion reviews in Spanish. In: Gelbukh, A., Espinoza, F.C., Galicia-Haro, S.N. (eds.) MICAI 2014, Part I. LNCS, vol. 8856, pp. 175–190. Springer, Heidelberg (2014)
7. Padró, L., Stanilovsky, E.: Freeling 3.0: towards wider multilinguality. In: Proceedings of the Language Resources and Evaluation Conference, LREC 2012. ELRA (2012)
8. Sahlgren, M.: The word-space model: using distributional analysis to represent syntagmatic and paradigmatic relations between words in high-dimensional vector spaces. Ph.D. thesis, Department of Linguistics, Stockholm University (2006)
9. Gelbukh, A.F., Bolshakov, I.A.: Internet, a true friend of translator: the Google wildcard operator. Int. J. Trans. **18**(1–2), 41–48 (2006)
10. Ferret, O.: Testing semantic similarity measures for extracting synonyms from a corpus. In: Proceedings of the 7th International Conference on Language Resources and Evaluation, LREC 2010, pp. 3338–3343 (2010)
11. Hazem, A., Daille, B.: Semi-compositional method for synonym extraction of multi word terms. In: Proceedings of the 9th International Conference on Language Resources and Evaluation, LREC 2014, pp. 1202–1207 (2014)
12. Manning, C., Schütze, H.: Foundations of Statistical Natural Language Processing. MIT Press, Cambridge (1999)
13. McInnes, B.T.: Extending the log-likelihood measure to improve collocation identification. Master thesis, University of Minnesota (2004)
14. Salton, G., Lesk, M.E.: Computer evaluation of indexing and text processing. J. Assoc. Comput. Mach. **15**(1), 8–36 (1968)
15. Rosner, M., Sultana, K.: Automatic methods for the extension of a bilingual dictionary using comparable corpora. In: Proceedings of the 9th International Conference on Language Resources and Evaluation, LREC 2014, pp. 3790–3797 (2014)
16. Manning, C.D., Raghavan, P., Schütze, H.: Introduction to Information Retrieval. Cambridge University Press, Cambridge (2008)

Gathering Information About Word Similarity from Neighbor Sentences

Natalia Loukachevitch[(✉)] and Aleksei Alekseev

Research Computing Center of Lomonosov Moscow State University, Moscow, Russia
louk_nat@mail.ru, a.a.alekseevv@gmail.com

Abstract. In this paper we present the first results of detecting word semantic similarity on the Russian translations of Miller-Charles and Rubenstein-Goodenough sets prepared for the first Russian word semantic evaluation Russe-2015. The experiments were carried out on three text collections: Russian Wikipedia, a news collection, and their united collection. We found that the best results in detection of lexical paradigmatic relations are achieved using the combination of word2vec with the new type of features based on word co-occurrences in neighbor sentences.

Keywords: Russian word semantic similarity · Evaluation · Neighbor sentences · Word2vec · Spearman's correlation

1 Introduction

The knowledge about semantic word similarity can be used in various tasks of natural language processing and information retrieval including word sense disambiguation, text clustering, textual entailment, search query expansion, etc.

To check the possibility of current methods to automatically determine word semantic similarity, various datasets were created. Most known gold standards for this task include the Rubenstein-Goodenough (RG) dataset [21], the Miller-Charles (MC) dataset [18] and WordSim-353 [7]. These datasets contain word pairs annotated by human subjects according to lexical semantic similarity. Results of automatic approaches are compared with the datasets using Pearson or Spearman's correlations [1].

Some researchers distinguish proper similarity between word senses that is mainly based on paradigmatic relations (synonyms, hyponyms, or hyperonyms), and relatedness that is established on other types of lexical relations. Agirre et al. [1] subdivided the WordSim-353 dataset into two subsets: the WordSim-353 similarity set and the WordSim-353 relatedness set. The former set consists of word pairs classified as synonyms, antonyms, identical, or hyponym-hyperonym and unrelated pairs. The latter set contains word-pairs connected with other relations and unrelated pairs.

To provide word similarity research in other languages, the English similarity tests have been translated. Gurevych translated the RG and MC datasets into

© Springer International Publishing Switzerland 2016
P. Sojka et al. (Eds.): TSD 2016, LNAI 9924, pp. 134–141, 2016.
DOI: 10.1007/978-3-319-45510-5_16

German [9]; Hassan and Mihalcea translated them into Spanish, Arabic and Romanian [11]; Postma and Vossen translated the datasets into Dutch [20].

In 2015 the first Russian evaluation of word semantic similarity was organized [19][1]. One of the datasets created for the evaluation was the word pair sets with the scores of their similarity obtained by human judgments (hj-dataset). The Russe hj-dataset was prepared by translation of the existing English datasets MC, RG, and WordSim-353.

One of traditional approaches to detect semantic word similarity are distributional approaches that are based on similar word neighbors in sentences. In this paper, we present a study of a possible contribution of features based on location of words in neighbor sentences. We suppose that co-occurrence of words in the same sentences can indicate that they are components of a collocation or are in the relatedness relation [22], but frequent co-occurrence in neighbor sentences indicates namely paradigmatic similarity between words. We conduct our experiments on the Russe hj test set. Then we work with the Russian translations of RG and MC datasets prepared during the Russe evaluation. We show that newly introduced features allow us to achieve the best results on the Russian translations of RG and MC datasets.

2   Related Work

The existing approaches to calculate semantic similarity between words are based on various resources and their features. One of well-known approaches utilizes the lexical relations described in manually created thesauri such as WordNet or Roget's thesaurus [1,5] or such online resource as Wikipedia [8].

Another type of approaches to detect semantic word similarity are distributional approaches that account for shared neighbors of words [14]. In these approaches a matrix, where each row represents a word w and each column corresponds to contexts of w, is constructed. The value of each matrix cell represents the association between the word w_i and the context c_j. A popular measure of this association is pointwise mutual information (PMI) and especially positive PMI (PPMI), in which all negative values of PMI are replaced by 0 [2,15]. Similarity between two words can be calculated, for example, as the scalar product between their context vectors.

Recently, neural-network based approaches, in which words are "embedded" into a low-dimensional space, appeared and became to be used in lexical semantic tasks. Especially, word2vec tool[2], a program for creating word embedding, became popular. In [2,15] traditional distributional methods are compared with new neural-network approaches in several lexical tasks. Baroni et al. [2] found that new approaches consistently outperform traditional approaches in most lexical tasks. For example, on the RG test traditional approaches achieve 0.74 Spearman's correlation rank, the neural approach obtains 0.84.

---

[1] http://russe.nlpub.ru/.
[2] https://code.google.com/archive/p/word2vec/.

Levy et al. [15] tried to find the best parameters for each approach and showed that on the WordSim-353 subsets, the word2vec similarity is better on the WordSim-353 similarity subset (0.773 vs. 0.755), while traditional distributional models based on PPMI weighting achieve better results on the WordSim-353 relatedness subset (0.688 vs. 0.623). They conclude that word2vec provides a robust baseline: "while it might not be the best method for every task, it does not significantly underperform in any scenario."

Some authors suppose to combine various features to improve detection of word semantic similarity. Agirre et al. [1] use a supervised combination of WordNet-based similarity, distributional similarity, syntactic similarity, and context window similarity, and achieve 0.96 Spearman's correlation on the RG set and 0.83 on the WordSim-353 similarity subset.

The best result achieved on the Russe-2016 hj test set [19], 0.7625 Spearman's correlation, was obtained with a supervised approach combining word2vec similarity scores calculated on a united text collection (consisting of the ruwac web corpus, the lib.ru fiction collection, and Russian Wikipedia), synonym database, prefix dictionary, and orthographic similarity [16]. The second result (0.7187) was obtained with the application of word2vec to two text corpora: Russian National Corpus and a news corpus. The word2vec scores from the news corpus were used if target words were absent in RNC [13]. The best result achieved with a traditional distributional approach was 0.7029 [19].

3 Datasets and Text Collections

In this study we use similarity datasets created for the first Russian evaluation on word semantic similarity Russe [19]. The Russe hj data set was constructed by translation of three English datasets: Miller-Charles set, Rubenstein-Goodenough set, and WordSim-353.

The RG set consists of 65 word pairs ranging from synonymy pairs (e.g., car – automobile) to completely unrelated terms (e.g., noon – string). The MC set is a subset of the RG dataset, consisting of 30 word pairs. The WordSim-353 set consists of 353 word pairs. The Miller-Charles set is also a subset in the WordSim-353 data set. The noun pairs in each set were annotated by human subjects, using scales of semantic similarity.

These datasets were joined together and translated into Russian in a consistent way. The word similarity scores for this joined dataset were obtained by crowdsourcing from Russian native speakers. Each annotator has been asked to assess the similarity of each pair using four possible values of similarity.

All estimated word pairs were subdivided to the training set (66 word pairs) for adapting the participants' methods and the test set, which was used for evaluation (335 word pairs) (Russe hj test set). Correlations with human judgments were calculated in terms of Spearman's rank correlation.

In this study we present our results on the Russe hj test set (335 word pairs). Besides, we singled out the Russian translations of the MC and RG sets from the whole Russe hj similarity dataset and show our results on each dataset.

We use two texts collections: Russian Wikipedia (0.24 B tokens) and Russian news collection (0.45 B tokens). Besides, we experimented on the united collection of the above-mentioned collections.

4   Using Neighbor Sentences for Word Similarity Task

Analysing the variety of features used for word semantic similarity detection, we could not find any work utilizing co-occurrence of words in neigbor sentences.

However, repetitions of semantically related words in natural language texts bear a very important role providing the cohesion of text fragments [10], that is device for "sticking together" different parts of the text. Cohesion can be based on the use of semantically related words, reference, ellipsis and conjunctions. Among these types the most frequent type is lexical cohesion, which is created with sense related words such as repetitions, synonyms, hyponyms, etc.

When analyzing a specific text, the location of words in sentences can be employed for constructing lexical chains. In classical works [3,12], the location of words in neighbor sentences is used for adding new elements in the current lexical chain. The distance in sentences between referential candidates is an important feature in tasks such as anaphora resolution and coreference chains construction [6].

Thus, in the text next sentences are often lexically attached to previous sentences. It allows us to suppose that the analysis of word distribution in neighbor sentences can give additional information about their semantic similarity. If two words frequently co-occur near each other it often indicates that they are components of the same collocation. Frequent word co-occurrence in the same sentences often mean that they are participants of the same situations [22]. But if words are frequently met in neighbor sentences it could show that are semantically or thematically similar.

In this study, we calculate co-occurrence of words in two sentences that are direct neighbors to each other and analyze its importance for lexical similarity detection in form of two basic features:

- frequency of co-occurrence of words in neighbor sentences (NS) where words should be located in different sentences: one word should be in the first sentence and the second word in the second sentences,
- frequency of co-occurrence of words in a large window equal to two sentences (TS).

We consider these features in two variant weightings: pointwise mutual information (pmiNS, pmiTS) and normalized pointwise mutual information (npmiNS, npmiTS). It should be also mentioned that both types of features do not require any tuning because they are based on sentence boundaries, that is they are self-adaptive to a text collection.

To compare these features with features extracted from a window within the same sentences, we calculated the inside-sentence window feature (IS), with its variants pmiIS, npmiIS. These features show co-occurrence of a word pair in a specific word window within a sentence. We experimented with various sizes of word windows and presents only the best results usually based on a large window of 20 words (10 words to the left and 10 words to the right within a sentence). In fact, in most cases IS-window is equal to the whole sentence. Besides, we calculated npmi for word pair co-occurrences within documents (npmiDoc).

To compare with state-of-the art approaches, we utilized the word2vec tool for processing our collections (w2v). For preprocessing, the text collections were lemmatized, all function words were removed and, thus, the word2vec similarity between words was calculated only on the contexts containing nouns, verbs, adjectives, and adverbs. We used CBOW model with 500 dimensions, window sizes were varied to produce the best results.

We also considered simple combinations between pairs of features calculated as summation of the ranks of word pairs in the ordering determined with those features. Table 1 presents the results on three collections: the news collection, Ru-wiki, and the united collection for the Russe hj test set.

One can see that the best results obtained from a single feature for all three corpora are based on the inside sentence co-occurrence feature npmiIS. The best results for two collections also correlate with the inside-sentence features.

Table 1. The results obtained on the Russe hj test set. The best single features and pairs of features are highlighted. Best w2v window sizes for all collections - 5

Features	News collection	Ru-Wiki	United collection
NS	0.519	0.494	0.543
pmiNS	0.632	0.598	0.660
npmiNS	0.638	0.612	0.677
TS	0.604	0.580	0.607
pmiTS	0.645	0.627	0.657
npmiTS	0.648	0.636	0.670
IS	0.577	0.601	0.594
pmiIS	0.645	0.633	0.652
npmiIS	**0.663**	**0.658**	**0.681**
npmiDOC	0.611	0.567	0.641
best of w2v	0.651	0.618	0.680
NS+w2v	0.659	0.633	0.674
pmiNS+w2v	0.687	0.647	**0.712**
npmiNS+w2v	0.661	0.652	0.704
TS+w2v	0.679	0.662	0.689
pmiTS+w2v	**0.694**	0.663	0.710
npmiTS+w2v	0.679	0.662	0.697
pmiIS+w2v	0.693	0.670	0.710
npmiIS+w2v	0.689	**0.674**	0.705
npmiDoc+w2v	0.659	0.619	0.692

In our opinion, it is because the Russe hj test set contains a lot of related (non-paradigmatic) word pairs. However, the best result (0.712) for this set is achieved on summation of the word rankings obtained with two features word2vec and pmiNS that is pointwise mutual information calculated on co-occurrence of words in neighbor sentences. This result is very close to the second result achieved during the Russe evaluation (0.717) but in that approach a three times larger news collection and balanced Russian National Corpus (RNC) were used [13].

Thus, on the Russe test set, the best result was obtained with additional accounting for co-occurrence of words in neighbor sentences. Though it should be noted that combination of word2vec with an inside-sentence feature generated a very similar result (0.710). The specificity of this set is that it contains paradigmatic relations (proper semantic similarity) together with relatedness relations from the translated the WordSim-353 set.

Therefore we decided to study the proposed features on the RG and MC sets, which are constructed for measuring paradigmatic relations between words. Table 2 presents the results obtained for the Russian translations of the RG and MC sets. Only best features and its pairs are shown. The presented results are the first results obtained for these Russian datasets.

Table 2. The results obtained on the Russian RG and MC sets. The best single features and pairs of features are highlighted. Best windows of word2vec are indicated

Features	Dataset	News collection	Ru-Wiki	United collection
npmiNS	RG	0.759	0.751	0.813
	MC	0.803	**0.833**	0.827
npmiTS	RG	0.779	**0.808**	0.820
	MC	**0.834**	0.832	0.823
npmiIS	RG	0.744	0.803	0.811
	MC	0.805	0.824	0.817
(Best of w2v)	RG (wind=2–5)	**0.791**	0.805	**0.858**
	MC (wind=1–3)	0.812	**0.833**	**0.848**
NS+w2v	RG	0.833	0.798	0.860
	MC	0.841	0.785	0.848
npmiNS+w2v	RG	0.835	**0.850**	0.868
	MC	0.837	**0.861**	0.868
TS+w2v	RG	**0.841**	0.842	**0.870**
	MC	**0.872**	0.816	**0.881**
npmiTS+w2v	RG	0.831	0.807	0.863
	MC	0.858	0.850	0.863
npmiIS+w2v	RG	0.826	**0.851**	0.857
	MC	0.852	0.843	0.867

We can see a quite different picture on these sets than on the Russe hj test set. In both cases the maximal values of Spearman's correlation rank for all three collections are obtained with the combination of word2vec and features based on co-occurrence of words in neigbor sentences. The best single features are mainly associated with neighbor-sentence features.

Our best result for the Russian MC set (0.881) is achieved with the pair of features (word2vec, TS – the co-occurrence of words in the window of two sentences) on the united text collection of Ru-Wiki and the news collection. The best result for Russian RG set (0.870) is obtained with the same combination of features on the same collection. The achieved results are much better than the results for each single feature. In our opinion, it means that the frequent co-occurrence of words in neighbor sentences bears additional useful information for detecting paradigmatic lexical similarity.

5   Conclusion

In this paper we presented the first results of detecting word semantic similarity on the Russian translations of MC and RG sets prepared for the first Russian word semantic evaluation Russe-2015. The experiments were performed on three text collections: Russian Wikipedia, a news collection, and the united collection of Ru-Wiki and the news collection.

We found that the best results in detection of lexical paradigmatic relations are achieved using the combination of word2vec with the new type of features proposed in this paper – the co-occurrence of words in neighbor sentences. We suppose that the found feature combination is especially useful for improvement of sense-related word extraction on medium-sized collections.

We plan to continue the study of the proposed features on other datasets including the Russian translation of the WordSim-353 dataset and original English datasets. In addition, we suppose to test possible contributions of these neighbor sentence features in probabilistic topic modeling [4].

Acknowledgments. This work was partially supported by Russian Foundation for Basic Research, grant N14-07-00383.

References

1. Agirre, E., Alfonseca, E., Hall, K., Kravalova, J., Paşca, M., Soroa, A.: A study on similarity and relatedness using distributional and wordnet-based approaches. In: The 2009 Annual Conference of the North American Chapter of the Association for Computational Linguistics, pp. 19–27, May 2009
2. Baroni, M., Dinu, G., Kruszewski, G.: Don't count, predict! A systematic comparison of context-counting vs. context-predicting semantic vectors. In: Proceedings of ACL-2014, pp. 238–247 (2014)
3. Barzilay, R., Elhadad, M.: Using lexical chains for text summarization. In: Advances in Automatic Text Summarization, pp. 111–121 (1999)

4. Blei, D.M., Ng, A.Y., Jordan, M.I.: Latent Dirichlet allocation. J. Mach. Learn. Res. **3**, 993–1022 (2003)
5. Budanitsky, A., Hirst, G.: Evaluating wordnet-based measures of lexical semantic relatedness. Comput. Linguist. **32**(1), 13–47 (2006)
6. Fernandes, E.R., dos Santos, C.N., Milidiú, R.L.: Latent trees for coreference resolution. Comput. Linguist. **40**, 801–835 (2014)
7. Finkelstein, L., Gabrilovich, E., Matias, Y., Rivlin, E., Solan, Z., Wolfman, G., Ruppin, E.: Placing search in context: the concept revisited. In: Proceedings of the 10th International Conference on World Wide Web, pp. 406–414 (2001)
8. Gabrilovich, E., Markovitch, S.: Computing semantic relatedness using Wikipedia-based explicit semantic analysis. In: Proceedings of IJCAI, pp. 6–12 (2007)
9. Gurevych, I.: Using the structure of a conceptual network in computing semantic relatedness. In: Proceedings of the 2nd International Joint Conference on Natural Language Processing, Jeju Island, South Korea, pp. 767–778 (2005)
10. Halliday, M., Hasan, R.: Cohesion in English. Routledge, London (2014)
11. Hassan, S., Mihalcea, R.: Cross-lingual semantic relatedness using encyclopedic knowledge. In: Proceedings of the 2009 Conference on Empirical Methods in Natural Language Processing, Singapore, vol. 3, pp. 1192–1201 (2009)
12. Hirst, G., St-Onge, D.: Lexical chains as representations of context for the detection and correction of malapropisms. In: WordNet: An Electronic Lexical Database, pp. 305–332 (1998)
13. Kutuzov, A., Kuzmenko, E.: Comparing neural lexical models of a classic national corpus and a web corpus: the case for Russian. In: Gelbukh, A. (ed.) CICLing 2015. LNCS, vol. 9041, pp. 47–58. Springer, Heidelberg (2015)
14. Lapesa, G., Evert, S.: A large scale evaluation of distributional semantic models: parameters, interactions and model selection. Trans. Assoc. Comput. Linguist. **2**, 531–545 (2014)
15. Levy, O., Goldberg, Y., Dagan, I.: Improving distributional similarity with lessons learned from word embeddings. Trans. Assoc. Comput. Linguist. **3**, 211–225 (2015)
16. Lopukhin K.A., Lopukhina A.A., Nosyrev G.V.: The impact of different vector space models and supplementary techniques on Russian semantic similarity task. In: Computational Linguistics and Intellectual Technologies: Papers from the Annual Conference, Dialogue, vol. 2, pp. 145–153 (2015)
17. Mikolov, T., Sutskever, I., Chen, K., Corrado, G., Dean, J.: Distributed representations of words and phrases and their compositionality. In: Advances in Neural Information Processing Systems, pp. 3111–3119 (2013)
18. Miller, G.A., Charles, W.G.: Contextual correlates of semantic similarity. Lang. Cogn. Process. **6**(1), 1–28 (1991)
19. Panchenko, A., Loukachevitch, N., Ustalov, D., Paperno, D., Meyer, C., Konstantinova, N.: RUSSE: the first workshop on Russian semantic similarity. In: Computational Linguistics and Intellectual Technologies: Papers from the Annual Conference, Dialogue, vol. 2, pp. 89–105 (2015)
20. Postma, M., Vossen, P.: What implementation and translation teach us: the case of semantic similarity measures in wordnets. In: Proceedings of Global Word-Net Conference GWC-2014, Tartu, Estonia, pp. 133–141 (2014)
21. Rubenstein, H., Goodenough, J.B.: Contextual correlates of synonymy. Commun. ACM **8**(10), 627–633 (1965)
22. Sahlgren, M.: The word-space model: using distributional analysis to represent syntagmatic and paradigmatic relations between words in highdimensional vector spaces. Ph.D. thesis, University of Stockolm (2006)

Short Messages Spam Filtering Using Sentiment Analysis

Enaitz Ezpeleta[1]([✉]), Urko Zurutuza[1], and José María Gómez Hidalgo[2]

[1] Electronics and Computing Department, Mondragon University,
Goiru Kalea, 2, 20500 Arrasate-mondragón, Spain
{eezpeleta,uzurutuza}@mondragon.edu
[2] Pragsis Technologies, Manuel Tovar, 43-53, Fuencarral, 28034 Madrid, Spain
jmgomez@pragsis.com

Abstract. In the same way that short instant messages are more and more used, spam and non-legitimate campaigns through this type of communication systems are growing up. Those campaigns, besides being an illegal online activity, are a direct threat to the privacy of the users. Previous short messages spam filtering techniques focus on automatic text classification and do not take message polarity into account. Focusing on phone SMS messages, this work demonstrates that it is possible to improve spam filtering in short message services using sentiment analysis techniques. Using a publicly available labelled (spam/legitimate) SMS dataset, we calculate the polarity of each message and aggregate the polarity score to the original dataset, creating new datasets. We compare the results of the best classifiers and filters over the different datasets (with and without polarity) in order to demonstrate the influence of the polarity. Experiments show that polarity score improves the SMS spam classification, on the one hand, reaching to a 98.91 % of accuracy. And on the other hand, obtaining a result of 0 false positives with 98.67 % of accuracy.

Keywords: SMS · Spam · Polarity · Sentiment analysis · Security

1 Introduction

In the era of smartphones and online social networks, instant short message communication tools are growing up very fast. For instance, one of the famous instant messaging applications, WhatsApp, reached 1 billion users on February 2016[1]. Another example is that 6.1 billion people all over the world use an SMS-capable mobile phone on June 2015, so SMS messages can reach more than 6 billion consumers[2]. Such a growth turned those systems in a very attractive objective to malicious companies and groups. Because of this, more and more illegal activities are being carried out through these communication methods.

---

[1] https://blog.whatsapp.com/616/One-billion/.
[2] http://goo.gl/yqzMDz.

© Springer International Publishing Switzerland 2016
P. Sojka et al. (Eds.): TSD 2016, LNAI 9924, pp. 142–153, 2016.
DOI: 10.1007/978-3-319-45510-5_17

For example, a gang made at least 5 million Euros over the last decade from a premium-rate SMS messaging scam[3]. Attackers know that there is a huge number of users whose privacy can be threatened in an easy way, sending a direct instant message, such as SMS or WhatsApp messages. Additionally, what makes SMS appropriate for illegitimate activities is the open rate (how many SMSs are opened (or viewed) in a particular SMS campaign) of 98 % (for instance, email marketing reports a 22 % open rate)[4]. Currently, with a spam proportion of 20–20 % of all SMS traffic in China and India, SMS spam is an emerging problem specially in the Middle East and Asia[5].

To deal with this type of problems several tools are designed and developed by researchers all over the word. Those systems are mostly focussed on automatic text classification, but do not take message sentiment into account. It could be considered that a spam message aims at selling products, thus the message should tend to be positive.

The main objective of this paper is to analyze the influence of the polarity in short instant messages spam filtering, testing if sentiment analysis can help on this task. It also aims to provides means to validate the hypothesis that sentiment feature of the short messages can improve the results obtained using common short messages filtering classifiers. Taking into account the publicly available datasets, we focus on SMS messages, which are structurally similar to other instant short messages.

Using a publicly available labelled (spam/legitimate) SMS dataset, we calculate the polarity of each message and aggregate the polarity score to the original dataset, creating new datasets. We compare the results of the best classifiers and filters over the different datasets (with and without polarity).

The remainder of this paper is organized as follows. Section 2 describes the previous work conducted in the area of SMS spam, natural language processing and sentiment analysis. Section 3 describes the process of the aforementioned experiments, regarding sentiment analysis of SMS in terms of polarity and SMS spam filtering. In Sect. 4, the obtained results are described, presenting the results of the descriptive experiment, and making a comparison between the actual filtering results and the filtering results using the polarity of the messages. Finally, we summarize our findings and give conclusions in Sect. 5.

2   Related Work

2.1   SMS Spam

During the last years attacker have detected that instant message systems are good place to perform malicious activities, specially attracted by the huge amount of users. In this study we are going to focus specially on SMS messages because structurally are similar to other currently more used short messages

---

[3] http://elpais.com/elpais/2015/04/20/inenglish/1429529298_001329.html.

[4] http://goo.gl/CaxweY.

[5] https://goo.gl/g6R7uW.

applications in our area such as Whatsapp or Twitter. Our decision to focus on certain messages is principally based on the public access to a tagged datasets needed to perform our experiments. This give us the possibility to compare our results with previous works. But also we take into account SMS is an emerging problem in the Middle East and Asia, with SMS spam contributing to 20–30 % of all SMS traffic in China and India[6].

In [3] authors presented a survey of work on filtering SMS spam and reviewed recent developments in SMS spam filtering. Also a brief discussion about publicly available corpora and availability for future research in the area is shown.

The authors in [1] compare different machine learning methods and indicated that Support Vector Machine was the best. They obtained an accuracy of 97.64 % using this method. Furthermore, they offer a public and non-encoded SMS spam collection that can be used by the research community. This study gives us the possibility to test with the same dataset and to compare results.

In other recent studies such as [13,15] two-level classifiers are used to obtain better results filtering the spam. But in this study we are going to focus on improving one-level learning-based classifiers.

2.2 Sentiment Analysis

Natural Language Processing (NLP) techniques are becoming more and more useful for spam filtering, as it is demonstrated in [7] using sender information and text content based NLP techniques.

Researchers in [4,10] confirmed that it is possible to create an application or a system to detect spam in different formats using text mining techniques and semantic language models respectively.

Among all NLP techniques we focus on the use of Sentiment Analysis (SA) to improve the detection of illegitimate short instant messages. This is a different strategy if we compare with the traditional short spam detection techniques which focus on automatic text classification, but do not take SA into account.

During the last years SA has been used in several research areas, although there has been a continued interest for a while. In [11] the most important research opportunities related to SA are described. Based on that, we select document sentiment classification topic as a possible option to short messages filtering.

As it is presented in [16] this area aims at defining if a document is positive or negative based on the its content. In order to improve the classification into positive, negative or neutral, other studies propose supervised learning techniques [17] or unsupervised learning techniques based on opinion words or phrases [18].

Different tools with the objective of helping during the sentiment classification have been proposed in the last years. Lexicon-based methods are interesting tools for our work. Those methods are used to extract the polarity of a certain

---

[6] https://goo.gl/g6R7uW.

word or phrase. In [8] a comparison between 8 popular sentiment analysis methods is presented and the author develops a combined method to improve the results. Centered on short messages, another comparison between lexicon-based approaches is described in [12].

Taking into account those comparisons, we decided to use the publicly available dictionary called *SentiWordNet*. The last version of this tool was presented in [2], which is an improved version of the first dictionary introduced by [5].

3   SMS Spam Filtering Using Sentiment Analysis

This study has been carried out following the procedure showed in Fig. 1, where two main experiments are implemented during the work.

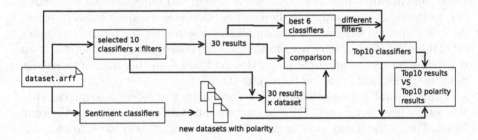

Fig. 1. SMS spam filtering using sentiment analysis

The process is divided in two phases. On the one hand, a sentiment analysis of the dataset is done to create new datasets including the measure of polarity.

On the other hand, two experiments are carried out applying different classifiers to the datasets in order to demonstrate the influence of the polarity over the SMS spam filtering.

Those experiments are carried out using the 10-fold cross-validation technique and the results are analyzed in terms of false positive rate and accuracy, being the percentage of testing set examples correctly classified by the classifier.

3.1   Datasets

During this study two different publicly available dataset are used:

– *SemEval-2013*[7]: Introduced in [14]. This dataset contains labelled (positive, negative and neutral) mobile phone messages, and we use it to evaluate the effectiveness of each sentiment classifier during the first phase. Specifically we use positive (492 SMS) and negative (394 SMS) messages.

---

[7] https://www.cs.york.ac.uk/semeval-2013/task2/.

– *SMS Spam Collection v.1*[8]: Published in [1], it is composed by 5,574 English, real and non-enconded messages, tagged according being legitimate (ham) or spam. Specifically, it contains 747 spam messages and 4,827 ham messages. This dataset is used to carry out the two spam filtering experiments.

3.2 Sentiment Analysis

The main objective of this part is to add the polarity of each message to the original dataset *SMS Spam Collection v.1* in order to carry out the experiments. To do that, we analyze different possibilities for sentiment classification of text messages, we evaluate those classifiers using the *SemEval-2013* dataset, we choose the best three classifiers and we apply those to the dataset.

Own Sentiment Classifier. In order to design and implement an own classifier, we decided to use sentiment dictionaries to develop a lexicon based classifier. As it is explained in Sect. 2.2, among all the possibilities the publicly available *SentiWordNet* is used. It returns to the user the polarity score of a certain word depending on its grammatical properties, and based on this score, the average polarity of the SMS messages is calculated.

Using several settings offered by the dictionary, five different sentiment classifiers have been developed. The first four are called Adjective, Adverb, Verb and Noun. The methodology of each classifier is to consider every word to be a certain part of speech (represented by the name), so we have obtained the polarity of those words that have that grammatical property. For example, in the Noun classifier every word is considered to be a noun, so the polarity of the words is extracted from the dictionary. And the last classifier is called AllPosition where every part of speech per each word is considered to obtain the polarity scores.

TextBlob Classifier. Another way to do sentiment analysis is to use publicly available resources. In this case, we focus on a simple API for diving into common NLP tasks called TextBlob[9]. More specifically, this tool returns a float value within the range $[-1.0, 1.0]$ for the polarity giving a certain string. The system calculates the average score of the sentence using a lexicon based score of each word.

To improve the effectiveness of those classifiers we change settings and select different thresholds (−0.05, −0.1, 0, 0.1, 0.05). The threshold means the point were we consider the polarity score positive or negative, and we use it in the name of the classifier to differentiate from each other.

Comparison Between Classifiers. Once the classifiers have been defined, a tagged (in terms of polarity) publicly available dataset is required to evaluate the effectiveness of the classifiers. Taking into account that SMS messages are

---

[8] http://www.dt.fee.unicamp.br/~tiago/smsspamcollection/.
[9] http://textblob.readthedocs.org/.

the final objective, we decided to choose a dataset composed by SMS messages in order to obtain more reliable results: *SemEval-2013* dataset.

We apply the different sentiment classifiers to the dataset, and we analyze the accuracy of correctly classified messages. In Table 1 a comparison between different classifiers and thresholds is shown.

Table 1. Comparison in terms of accuracy between classifiers

Classifier	Accuracy	Classifier	Accuracy
TextBlob 0.05	0.78	Adjectives	0.66
TextBlob 0.1	0.76	Nouns	0.58
TextBlob -0.05	0.73	TextBlob 0	0.56
AllPositions	0.72	Adverb	0.53
TextBlob -0.1	0.71	Verb	0.52

Based on the accuracies offered by the table, the best three classifiers are selected (*TextBlob 0.05*, *TextBlob 0.1* and *TextBlob −0.05*) in order to use those ones to annotate the messages included in *SMS Spam Collection v.1* which has not been annotated for sentiment. As a result, we obtain three new datasets (one per each classifier). The original one and the new three ones are used in the next experiments.

3.3 Spam Filtering

To analyze the influence of the polarity over the filtering of SMS messages, the first step is to select 10 representative classifiers and some of the best filter settings for natural language processing. To do that the results presented in [6] are taken into account. Consequently the best five classifiers from the mentioned paper are used in this study. Also the best three settings of the filters are chosen. Moreover, we added more classifiers to the list based on other research studies such as [9]. Final list: Large-scale Bayesian logistic regression for text categorization, discriminative parameter learning for Bayesian networks, complement class Naive Bayes classifier, multi-nominal Naive Bayes classifier, updateable multi-nominal Naive Bayes classifier, decision tree (C4.5), random tree, forest of random trees, Support Vector Machine (SMO) and adaptive boosting meta-algorithm with Naive Bayes.

The next step is to apply those classifiers, combined with the best three filters and settings, to the datasets (with and without polarity) and compare the results.

This first step provides the best classifier for text messages, so in the following phase, the best six classifiers are picked. And the second experiment is carried out applying those classifiers with different combination of filters and settings (56 combinations per classifier) to the datasets. The objective of the combination of this filters and settings is to follow a text mining process in order to compare results and identified the best ones, and those are some of its main details:

- A filter to convert a string to feature vector containing the words. We combine different options: words are converted to lower case, special characters ($.,;: \$\& \times \% = \_@()?! + -\#[]$) are removed using tokenizers; n-gram with min and max grams are created; roots of the words are obtained using a stemmer...
- Attribute Selection: a ranker to evaluate the worth of an attribute by measuring the information gain with respect to the class (spam/ham) is used.

Using those combinations we identify the best ten settings and classifiers for SMS spam classification, and those are applied to the dataset with polarity to compare the results.

4   Experimental Results

In this section the results obtained during the previously explained experiments are shown. To carry out those experiment the dataset called *SMS Spam Collection v.1* is used.

4.1   Descriptive Experiment

Once the dataset is selected, we perform a descriptive experiment of the dataset. The objective of this step is to analyze the polarity of the messages applying our previously selected (Sect. 3.2) sentiment classifiers. This is the point where the polarity extracted during the analysis is inserted in the dataset, creating three new datasets (one per each classifier) and where statistics about the polarity are calculated.

In Table 2 the results of the experiment are presented (*Tb 005* means *TextBlob 0.05*, *Tb 01* means *TextBlob 0.1* and *Tb-005* means *TextBlob −0.05*).

Table 2. Sentiment analysis of SMS messages

	Number of messages						Percentage (%)					
	Tb 005		Tb 01		Tb-005		Tb 005		Tb 01		Tb-005	
	P	N	P	N	P	N	P	N	P	N	P	N
Spam	430	317	408	339	688	53	58	42	55	45	92	7
Ham	1,960	2,867	1,859	2,968	4,121	687	41	59	39	61	85	14

According to the classifiers it is possible to see, specially in the first two sentiment classifiers, that spam messages are mostly positive while ham messages are more negative. This means that there is a difference between spam and ham messages in terms of polarity, so it can be helpful for improving SMS spam filtering.

Two experiments to see the real influence are carried out.

4.2 First Experiment: Finding the Best SMS Spam Filtering Classifiers

This experiments aims to identify the best SMS spam filtering classifiers in order to use them in the next experiment with more filters and settings. As it is mentioned in Sect. 4 we choose 10 classifiers and the following filter combinations per each one. Those filters are used to obtain the results presented in the Table 4. The explanation is based on the results obtained in [6].

1. *stwv.go.wtok*: the best result.
2. *i.t.c.stwv.go.ngtok.stemmer.igain*: into the best two algorithms this settings obtained the best result in one, and the second result in the other.
3. *i.t.c.stwv.go.wtok*: appeared in the top ten results and is was the first with NGrams and information gain filter.

The nomenclature used in this list and in the following tables is explained in Table 3, where idft means Inverse Document Frequency (IDF) Transformation, tft means Term Frequency score (TF) Transformation and outwc counts the words occurrences.

Table 3. Nomenclatures

	Meaning		Meaning
NBMU	Naive Bayes Multinominal Updatable	.stwv	String to Word Vector
NBM	Naive Bayes Multinominal	.go	general options
CNB	Complement Naive	.wtok	Word Tokenizer
BLR	Bayesian Logistic Regression	.ngtok	NGram Tokenizer 1-3
.c	idft False, tft False, outwc True	.stemmer	Stemmer
.i.c	idft True, tft False, outwc True	.igain	Attribute selection using
.i.t.c	idft True, tft True, outwc True		InfoGainAttributeEval

In Table 4 the results of the ten classifiers with the three listed filters are presented. The number in the name represents the type of the filter used.

Analyzing the table we can see that applying classifiers to the original dataset the best one in terms of accuracy is SMO with the third settings. In this case the polarity does not help to improve the result. But applying the Discriminative Parameter Learning for Bayesian Network (DMNBtext) classifier and the first filter to the dataset created using the sentiment analyzer *TextBlob01* the top result is improved. Specifically, an accuracy of 98.76 % is obtained. In other two cases the top result is also improved.

Another important information shown in the table is that although it is not the best result in terms of accuracy there is case that must be highlighted. This case is the Bayesian Logistic Regression with the second filter applied to the dataset *TextBlob01*, which obtained an accuracy of 98.67 % and 0 false positives.

Table 4. Comparison between results

Spam classifier	Normal		Tb 005		Tb 01		Tb -005	
	FP	Acc	FP	Acc	FP	Acc	FP	Acc
SMO.3	3	98.73	5	98.71	4	98.67	3	98.73
NaiveBayesMultinomial.3	12	98.69	12	98.62	12	98.69	8	98.69
NaiveBayesMultinomialUpdateable.3	12	98.69	12	98.62	12	98.69	8	98.69
BayesianLogisticRegression.3	5	98.64	6	98.60	5	98.67	4	98.64
DMNBtext.1	10	98.62	7	**98.74**	7	**98.76**	6	98.69
BayesianLogisticRegression.2	2	98.60	3	98.49	0	**98.67**	0	**98.58**
NaiveBayesMultinomial.1	23	98.53	23	98.51	22	98.53	16	98.64
SMO.2	4	98.53	4	98.58	5	98.58	4	98.53
NaiveBayesMultinomialUpdateable.2	36	98.51	36	98.49	35	98.51	26	98.67
NaiveBayesMultinomialUpdateable.1	29	98.49	25	98.56	26	98.55	16	98.73
ComplementNaiveBayes.1	31	98.44	32	98.40	32	98.39	18	98.64
NaiveBayesMultinomial.2	52	98.37	52	98.33	52	98.31	46	98.48
DMNBtext.2	4	98.28	3	98.31	3	98.37	2	98.31
DMNBtext.3	4	98.28	3	98.31	3	98.37	2	98.31
ComplementNaiveBayes.2	64	98.19	64	98.15	62	98.19	58	98.30
ComplementNaiveBayes.3	56	98.17	48	98.30	48	98.26	19	**98.74**
BayesianLogisticRegression.1	1	97.45	0	96.41	0	96.59	0	96.18
SMO.1	0	97.45	0	97.54	0	97.56	0	97.45
J48.3	54	97.02	58	96.68	56	96.72	54	96.97
J48.2	58	96.90	62	96.56	62	96.54	58	96.86
J48.1	42	96.86	43	96.90	43	96.90	42	96.86
RandomForest.2	0	96.38	0	95.82	2	96.05	0	96.39
RandomForest.3	0	96.21	0	96.27	1	95.91	0	96.29
RandomTree.1	25	95.60	18	95.59	24	95.71	17	95.95
RandomForest.1	0	95.19	0	94.76	0	94.94	0	95.03
RandomTree.3	84	95.16	79	95.43	92	94.90	95	95.25
RandomTree.2	88	95.07	73	95.35	79	95.28	93	95.32
AdaBoostM1.2	167	91.44	166	91.46	166	91.46	167	91.44
AdaBoostM1.3	167	91.44	166	91.46	166	91.46	167	91.44
AdaBoostM1.1	188	91.32	139	91.59	139	91.59	188	91.32

4.3   Second Experiment: SMS Spam Filtering with Polarity Score

The second experiment is based on the results obtained in the first one. While the previous aims to search the best algorithms, this one aims to explore most of the possible filter combinations with the best classifiers.

On this way, we identify the best 6 classifiers in Table 4 and combined each one with 56 different filter settings. We analyze the achieved results and we get the classifiers that obtained the best ten results in terms of accuracy. The next step is to apply those classifiers to the new datasets that we created using the sentiment classifiers. Those results are shown in Table 5.

Table 5. Comparison between Top10 results

Spam classifier	Sentiment analyzer							
	None		Tb 005		Tb 01		Tb -005	
	FP	Acc	FP	Acc	FP	Acc	FP	Acc
NBMU.i.c.stwv.go.ngtok	28	98.85	36	98.73	36	98.73	35	98.74
NBMU.i.t.c.stwv.go.ngtok	27	98.82	17	98.60	16	98.71	8	98.76
NBM.i.t.c.stwv.go.ngtok	32	98.78	37	98.74	37	98.74	33	98.78
NBMU.i.t.c.stwv.go.ngtok.stemmer	23	98.78	36	98.71	36	98.71	34	98.74
NBM.c.stwv.go.wtok	13	98.76	33	98.78	32	98.80	28	98.85
NBM.i.t.c.stwv.go.ngtok.stemmer	34	98.76	34	98.74	33	98.74	32	98.76
NBMU.c.stwv.go.wtok	13	98.76	17	98.60	16	98.71	8	98.76
CNB.i.t.c.stwv.go.ngtok.stemmer	37	98.73	28	98.85	28	98.85	27	98.82
NBM.i.c.stwv.go.ngtok	37	98.73	26	98.85	25	98.87	22	**98.91**
NBM.i.c.stwv.go.ngtok.stemmer	36	98.73	23	98.80	22	98.82	19	98.82

The table shows that a higher accuracy than in the previous experiment is obtained applying new settings of the filters to the original SMS dataset.

Analyzing the data we realize that in half of cases polarity helps to improve the accuracy, and also that by applying the Bayesian Logistic Regression classifier to the dataset created by *TextBlob-005* classifier we improve the best result. While without polarity the best result is 98.85 %, using the polarity a 98.91 % of accuracy is obtained.

Furthermore, in same cases where better accuracy is not obtained, polarity helps to reduce the number of false positives. Obtaining a percentage of 98.76 % and 8 false positives in two cases, reducing from 27 false positives in one case and from 13 in the other.

5   Conclusions

This work shows that sentiment analysis can help improving short messages spam filtering. We have demonstrated that, adding the polarity obtained during a sentiment analysis of the short text messages, in most of the cases the result is improved in terms of accuracy. Moreover, we have proved our hypothesis obtaining better results with the polarity score than the top result without polarity. (98.91 % versus 98.58 %). Despite the difference in the percentage does not seem to be relevant, if we take into account the amount of real SMS traffic the improvement is significant.

In addition, a substantial improvement in terms of the number of false positive messages have been achieved in this work. For instance, during the first experiment the best accuracy with 0 false positives is obtained using polarity: 98.67 %.

Acknowledgments. This work has been partially funded by the Basque Department of Education, Language policy and Culture under the project SocialSPAM (PI_2014_1_102).

References

1. Almeida, T.A., Gómez Hidalgo, J.M., Yamakami, A.: Contributions to the study of SMS spam filtering: new collection and results. In: Proceedings of the 11th ACM Symposium on Document Engineering, pp. 259–262. ACM (2011)
2. Baccianella, S., Esuli, A., Sebastiani, F.: Sentiwordnet 3.0: an enhanced lexical resource for sentiment analysis and opinion mining. In: LREC, vol. 10, pp. 2200–2204 (2010)
3. Delany, S.J., Buckley, M., Greene, D.: SMS spam filtering: methods and data. Expert Syst. Appl. **39**(10), 9899–9908 (2012)
4. Echeverria Briones, P.F., Altamirano Valarezo, Z.V., Pinto Astudillo, A.B., Sanchez Guerrero, J.D.C.: Text mining aplicado a la clasificación y distribución automática de correo electrónico y detección de correo spam (2009)
5. Esuli, A., Sebastiani, F.: Sentiwordnet: a publicly available lexical resource for opinion mining. In: Proceedings of LREC, vol. 6, pp. 417–422. Citeseer (2006)
6. Ezpeleta, E., Zurutuza, U., Gómez Hidalgo, J.M.: Does sentiment analysis help in Bayesian spam filtering? In: Martínez-Álvarez, F., Troncoso, A., Quintián, H., Corchado, E. (eds.) HAIS 2016. LNCS, vol. 9648, pp. 79–90. Springer, Heidelberg (2016). doi:10.1007/978-3-319-32034-2_7
7. Giyanani, R., Desai, M.: Spam detection using natural language processing. Int. J. Comput. Sci. Res. Technol. **1**, 55–58 (2013)
8. Gonçalves, P., Araújo, M., Benevenuto, F., Cha, M.: Comparing and combining sentiment analysis methods. In: Proceedings of the First ACM Conference on Online Social Networks, pp. 27–38. ACM (2013)
9. Kumar, R.K., Poonkuzhali, G., Sudhakar, P.: Comparative study on email spam classifier using data mining techniques. In: Proceedings of the International MultiConference of Engineers and Computer Scientists, vol. 1, pp. 14–16 (2012)
10. Lau, R.Y.K., Liao, S.Y., Kwok, R.C.W., Xu, K., Xia, Y., Li, Y.: Text mining and probabilistic language modeling for online review spam detection. ACM Trans. Manag. Inf. Syst. **2**(4), 25:1–25:30 (2012). http://doi.acm.org/10.1145/2070710.2070716
11. Liu, B., Zhang, L.: A survey of opinion mining and sentiment analysis. In: Aggarwal, C.C., Zhai, C. (eds.) Mining Text Data, pp. 415–463. Springer, Berlin (2012). http://scholar.google.de/scholar.bib?q=info:CEE7xsbkW6cJ:scholar.google.com/&output=citation&hl=de&as_sdt=0&as_ylo=2012&ct=citation&cd=1

12. Musto, C., Semeraro, G., Polignano, M.: A comparison of lexicon-based approaches for sentiment analysis of microblog posts. In: Information Filtering and Retrieval, p. 59 (2014)
13. Nagwani, N.K., Sharaff, A.: SMS spam filtering and thread identification using bi-level text classification and clustering techniques. J. Inf. Sci. 1–13, 3 December 2015. doi:10.1177/0165551515616310
14. Nakov, P., Kozareva, Z., Ritter, A., Rosenthal, S., Stoyanov, V., Wilson, T.: Semeval-2013 task 2: Sentiment analysis in Twitter (2013)
15. Narayan, A., Saxena, P.: The curse of 140 characters: evaluating the efficacy of SMS spam detection on android. In: Proceedings of the Third ACM Workshop on Security and Privacy in Smartphones and Mobile Devices, pp. 33–42. ACM (2013)
16. Pang, B., Lee, L.: Opinion mining and sentiment analysis. Found. Trends Inf. Retr. **2**(1–2), 1–135 (2008)
17. Pang, B., Lee, L., Vaithyanathan, S.: Thumbs up? Sentiment classification using machine learning techniques. In: Proceedings of the ACL-02 Conference on Empirical Methods in Natural Language Processing, vol. 10, pp. 79–86, EMNLP 2002, Association for Computational Linguistics, Stroudsburg, PA, USA (2002). http://dx.doi.org/10.3115/1118693.1118704
18. Turney, P.D.: Thumbs up or thumbs down? Semantic orientation applied to unsupervised classification of reviews. In: Proceedings of the 40th Annual Meeting on Association for Computational Linguistics, pp. 417–424, ACL 2002, Association for Computational Linguistics, Stroudsburg, PA, USA (2002). http://dx.doi.org/10.3115/1073083.1073153

Preliminary Study on Automatic Recognition of Spatial Expressions in Polish Texts

Michał Marcińczuk[✉], Marcin Oleksy, and Jan Wieczorek

G4.19 Research Group, Department of Computational Intelligence,
Faculty of Computer Science and Management,
Wrocław University of Technology, Wrocław, Poland
{michal.marcinczuk,marcin.oleksy,jan.wieczorek}@pwr.edu.pl

Abstract. In the paper we cover the problem of spatial expression recognition in text for Polish language. A spatial expression is a text fragment which describes a relative location of two or more physical objects to each other. The first part of the paper treats about a Polish corpus annotated with spatial expressions and annotators agreement. In the second part we analyse the feasibility of spatial expression recognition by overviewing relevant tools and resources for text processing for Polish. Then we present a knowledge-based approach which utilizes the existing tools and resources for Polish, including: a morpho-syntactic tagger, shallow parsers, a dependency parser, a named entity recognizer, a general ontology, a wordnet and a wordnet to ontology mapping. We also present a dedicated set of manually created syntactic and semantic patterns for generating and filtering candidates of spatial expressions. In the last part we discuss the results obtained on the reference corpus with the proposed method and present detailed error analysis.

Keywords: Information extraction · Spatial expressions · Spatial relations

1 Introduction

Spatial information describes a physical location of an object in a space. The location of the object can be encoded using some absolute values in a coordinate system or by relative references to other entities in the space. The latter are called spatial relations. The spatial relations can be expressed directly by spatial expressions [1] or indirectly by a chain of semantic relations [2]. A comprehensive recognition of spatial relations between objects described in a text requires a complex chain of processing and reasoning, including a morphological analysis of text, a recognition of object mentions, a text parsing, a named entity recognition and classification, a coreference resolution, a semantic relation recognition and interpretation. Thus, the feasibility and the quality relies on the availability of certain tools and their performance.

In the article we focus on spatial expressions in which the spatial information is encoded using a spatial preposition. According to our preliminary research this type of spatial relation is predominant in Polish texts (more than 53 % of all expressions), while the second dominant group for which the spatial information is contained by a verb covers only 24 % of all instances.

© Springer International Publishing Switzerland 2016
P. Sojka et al. (Eds.): TSD 2016, LNAI 9924, pp. 154–162, 2016.
DOI: 10.1007/978-3-319-45510-5_18

2   Reference Corpora

To annotate the corpora we followed the guideline for *Task 3: Spatial Role Labeling* [1]. In the current research we annotated only four types of elements: trajector (localized object, henceforth TR), landmark (object of reference, henceforth LM), spatial indicator (henceforth SI) and region (henceforth RE). The remaining elements (i.e. path, direction, motion) will be included in the future. Below is a brief description of the two sets of documents and Table 1 contains their statistics.

WGT — a set of 50 geographical texts from Polish Wikipedia (Wikipedia Geographical Texts). This type of articles contains many spatial relations between objects. The set contains 17,407 tokens and 484 spatial expressions (one expression for every 36 tokens). This set was annotated by two linguists independently and the inter-annotator agreement was measured by means of the Dice coefficient. The agreement was 82%. This set of documents was used to define syntactic patterns (see Sect. 3.3) and semantic constraints (see Sect. 3.4).

KPWr — 1,526 document from the KPWr corpus [3]. The set contains 419,769 tokens and 2,581 spatial expressions (one relation for every 162 tokens). The documents were annotated only by one linguist. The set was used to evaluate our method and to perform an error analysis.

Table 1. Statistic of corpora annotated with spatial expressions

	WGT	KPWr
Documents	50	1,526
Tokens	17,407	419,769
Spatial expressions	484	2,581
Annotations		
Spatial object	1212	5050
Spatial indicator	743	2615
Region	114	149

3   Recognition of Spatial Expressions

3.1   Procedure

We assume that the text will be preprocessed with a morphological tagger, a shallow parser, a dependency parser, a named entity recognizer and a word sense disambiguation (see Sect. 3.2). We will also use a wordnet for Polish, an ontology and a mapping between the wordnet and the ontology. We will use a set of syntactic patterns to identify spatial expression candidates, i.e., tuples containing a trajector (TR), a spatial preposition (SI), a landmark (LM) and optionally a region (RE) (see Sect. 3.3). In the last step, the set of generated candidates will be tested against a set of semantic constrains (see Sect. 3.4).

3.2   Preprocessing

Different ways of expressing spatial relations require specialized tools and resources to make the task feasible. The basic text processing, which includes text segmentation, morphological analysis and disambiguation, can be easily performed with any of the existing taggers for Polish, i.e., WCRFT [4], Concraft [5] or Pantera [6]. The accuracy of the taggers is satisfactory and varies between 89–91 %.

First we need to identify relevant entity mentions. The mentions can be: named entities, nominal phrases, pronouns and null verbs (verbs which do not have an explicit subject cf. [7]). The spans of entity mentions can be recognized using a shallow parser for Polish, i.e., Spejd [8] with a NKJP grammar [9] or IOBBER [10]. Spejd recognizes a flat structure of nominal groups (NG) with their semantic and syntactic heads. A noun group preceded by a preposition is marked with the preposition as a prepositional nominal group (PrepNG). Every noun and pronoun creates a separate nominal group. The only exception is a sequence of nouns that is annotated as a single nominal group. IOBBER also recognizes a flat structure of nominal phrases (NP). A nominal phrase is defined as a phrase which is a subject or an object of a predicative-argument structure. This means, that some NP can contain several NGs. For example "m czyzna siedz cy w piwnicy" (*a man sitting in the basement*) is a single NP that contains two NGs: "m czyzna" (*a man*) and "piwnicy" (*the basement*) as a part of the PrepNG "w piwnicy". Spejd combined with IOBBER can be used to identify expressions with a spatial preposition within a single NP. According to [11] the NKJP grammars evaluated on the NKJP corpus obtained 78 % of precision and 81 % of recall in recognition of NGs, PrepNGs, NumNGs and PrepNumNGs. IOBBER evaluated on the KPWr corpus obtained 74 % of precision and 74 % of recall in recognition of NPs [10].

Second we need to categorize the entities into physical and non-physical. For nominal phrases this can done using a mapping between plWordNet [12] and the SUMO ontology [13]. The mapping contains more than 175000 links between synsets from plWordNet and SUMO concepts. Other types of mentions (i.e., named entities, pronouns and null verbs) require additional processing. Most named entities are not present in the plWordNet so they cannot be mapped onto SUMO through the mapping. However, they can be mapped by their categories which can be recognized using one of the named entity recognition tools for Polish, i.e., Liner2 [14] or Nerf [5]. Liner2 for a coarse-grained model recognizing top 9 categories obtained 73 % of precision and 69 % of recall, and for a fine-grained model with 82 categories 67 % and 59 %, respectively. Nevertheless, a mapping of named entity categories onto SUMO is required.

3.3   Spatial Expression Patterns

Using the WGT dataset we generated a list of most frequent syntactis patterns for spatial expressions. We used phrases recognized by Spejd and IOBBER, i.e. noun groups and noun phrases. We have identified the following types of patterns:

1. TR and LM appear in the same noun phrase:
   - TR is followed directly by SI and LM, i.e. "a cat on the roof",
   - TR and LM are arguments of a participle, i.e. "a cat sitting on the roof (...)"
2. TR and LM appear in different noun phrases — they are arguments of the same verb:
   - TR and LM are single objects, i.e. "a cat is sitting on the roof",
   - TR and/or LM are lists of objects.

For the first group of patters we consider every noun phrases (NP) containing all the elements in a certain order, i.e. a noun group (NG) as a TR, an optional participle, a preposition (SI), a noun group as a LM with potential RE. For the second group of patterns we used a dependency parser for Polish [15] and select verbs with attached a noun group as a subject (TR) and a preposition with a noun group (LM).

3.4   Semantic Constraints

The last step is categorization of the expressions into spatial and non-spatial. For English, a common approach is to categorize the expressions on the basis of preposition categorization [16]. Since there are no resources nor tools to recognize spatial prepositions for Polish, we decided to apply a knowledge-based approach, i.e., the candidate expressions are tested against a set of semantic constraints.

Information about the type of a spatial relation comes not only from the meaning of a preposition (spatial indicator). Lexemes referring to a TR and to a LM also influence the identification of the relation denoted in a text. We can use the same preposition (in a formal sense, i.e., in a combination with the same grammatical case of a noun) to introduce information about spatial or non-spatial relations (e.g. time)). For example:

1. Piotr siedział przed domem. *(Piotr was sitting in front of the house.)*
2. Piotr siedział przed godziną w biurze. *(Piotr was sitting in the office an hour ago.)*

Preposition: on (Pol. *na*)
Interpretation: Object TR is outside the LM, typically in contact with external limit of LM by applying pressure with its weight.
Example of usage: "ksi ka le y na stole" (*a book is on the table*)
TR's semantic restrictions (SUMO classes): artifact, contentbearingobject, device, animal, plant, pottery, meat, preparedfood, chain,
LM's semantic restrictions (SUMO classes): artifact, LandTransitway, boardorblock, boatdeck, shipdeck, stationaryartifact

Fig. 1. Schema #1 for preposition ON (Pol. "NA")

The semantic restrictions of TR and LM can be used to distinguish a specific meaning of the preposition due to a specific spatial cognitive pattern [17]. We described them using classes from the SUMO ontology trying to capture the prototypical conceptualization of the patterns. The set of consists contains over 170 cognitive schemes for spatial relations (including the specificity of the objects in the relation). For example there are 18 schemes for preposition "NA" (*on*). A sample schema for preposition "NA" is presented on Fig. 1.

4   Evaluation

The evaluation was performed on the KPWr corpus presented in Sect. 2.

4.1   Generation of Spatial Expression Candidates

The candidates of spatial expressions are recognized using syntactic patterns presented in Sect. 3.3. We were able to recognize 44.58 % of all expressions with the precision of 11.12 %. At this stage the precision is not an issue (the candidates will be filtered in the second step). The problem was a low recall, however, we do not cover this problem in this article and we have left it for the future research.

4.2   Semantic Filtering of Candidates

Table 2 shows the impact of semantic filtering of spatial expressions. The number of false positives was dramatically reduced and the precision increased from 11.12 % to 66.67 %. At the same time the number of true positives was lowered and the recall dropped from 44.58 % to 29.81 %. In the next two sections we discuss the reasons of false positives and false negatives.

Table 2. Complete evaluation

Semantic filtering	Precision	Recall	F-measure
No	11.12 %	44.58 %	17.80 %
Yes	66.67 %	29.81 %	41.20 %

4.3   Analysis of False Positives

We have carefully analysed more than 200 false positives in order to identify the source of errors. We have identified the following types of errors grouped into two categories (external and procedure):

- external errors — errors committed by the external tools:
  - WSD error (ca. 19 %) — the tool for word sense disambiguation assigned an incorrect meaning of the trajector or landmark what caused an error of mapping the noun to a SUMO concept. For example "szczyt długiej przerwy" (Eng. *peak of the playtime*) was interpreted as a mountain peak and mapped onto the LandArea concept.

- Spejd error (ca. 13 %) — two adjacent noun groups were incorrectly joined into a single noun group. For example *"przejmowanie przez [banki przestrzeni publicznej]"* (Eng. acquiring by [banks the public space]) should be recognized as two separate noun groups, i.e. *"przejmowanie przez [banki] [przestrzeni publicznej]"* (Eng. *acquiring by [banks] [the public space]*). This error leads to incorrect assignment of phrase head.
- Liner2 error (ca. 10 %) — the named entity recognition tool assigned an incorrect category of named entity, for example a person was marked as a city.
- dependency parser error (ca. 3 %) — trajector and/or landmark were assigned to an incorrect verb,
- WCRFT error (ca. 1 %) — incorrect morphological disambiguation.
- procedure errors — errors committed by our procedure:
  - the meaning of the verb is required to interpret the expression (ca. 17 %) — this problem affects candidates generated with dependency patterns (a verb with arguments). The interpretation of the expression depends on the meaning of the verb, for example "Adam stoi przed Martą" (Eng. *Adam is standing in front of Marta*) reflects a spatial relation but *"Adam schował zeszyt przed Martą" (Eng. Adam hid a notebook from Marta)* does not.
  - semantic filtering error (ca. 9 %) — this type of error is caused by too general cognitive schemes.
  - motion expressions (ca. 10 %) — semantic constrains do not distinguish between static and motion expressions. In our experiment we focused only on static expressions and motion ones were not annotated in the corpus.
  - TR is the phrase head's modifier, not the head itself (ca. 10 %) — for example in the phrase "[szef rady osiedla] w [Sokolnikach]" (Eng. *[the head of council estate] in [Sokolniki]*) the *council estate* should be the TR, not the *head*,
  - non-physical objects (ca. 5 %) — a non-physical object is recognized as a TR, for example "napis na ścianie" (Eng. *writing on the wall*).
  - inverse order of the TR and LM (ca. 3 %) — in most cases the TR is followed by a preposition and a LM. In some cases the order is shifted, for example "[ławki] w [galeriach handlowych] pod [schodami ruchomymi]" (Eng. *[benches] in [shopping malls] under [the escalator]*).

Near 46 % of false positives were caused by errors committed by the external tools used in the text preprocessing (tagging, chunking, dependency parsing, named entity recognition, word sense disambiguation). It might be a laborious task to improve them as it requires an improvement of the tools which are already state-of-the-art.

The remaining 54 % of false positives are caused by our procedure of spatial expression recognition and there is still a room for improvement. The largest group of errors are caused by the fact that we did not consider verbs in the semantic filtering. The preliminary experiment proofed that the verbs should be included. The second largest groups of errors are committed by the current set of semantic schemes which are in same cases to general. The set of schemes need to be revised.

4.4 Analysis of False Negatives

We also carefully analysed about 130 false negatives (candidates incorrectly discarded by the semantic filtering) to identify the main sources of errors. We identified the following groups of errors:

- WSD error (ca. 50 %) — a candidate was discarded because the WSD tool assigned incorrect sense which was latter mapped on a SUMO concept not present in the schemas,
- missing mapping to SUMO (ca. 15 %) — a candidate was discarded because trajector and/or landmark were not mapped onto SUMO and the semantic filtering could not be applied,
- missing schema for semantic filtering (ca. 14 %) — a candidate was discarded due to missing semantic schema.
- Liner2 error (ca. 11 %) — a candidate was discarded due to incorrect proper name category assignment.

In the case of false negatives (candidates incorrectly discarded by semantic filtering) the majority (76 %) were caused by errors committed by external tools. Only 15 % of candidates were discarded due to missing semantic schemas.

5 Conclusions and Future Work

In the paper we discussed the problem of spatial expression recognition for Polish language. We presented and evaluated a proof of concept for recognition of spatial expressions for Polish using a knowledge-based two-stage approach. We focused on expressions containing a spatial preposition. The preliminary results are promising in terms for precision — 66.67 %. There is still a room for improvement by revising the set of semantic schemes and including semantic of the verbs. The main problem which still needs to be addressed is the recall of spatial expression candidates. The current set of patterns without semantic filtering was able to discover only 44.58 % of all expressions. The next 15 % of expressions are lost due to missing schemes for semantic filtering. The other way to improve the performance of semantic expression recognition is to improve the tools used in the preprocessing. However, this is a laborious task as the tools already have state-of-the-art performance and the small errors committed by every single tools cumulate to a large number.

Acknowledgements. Work financed as part of the investment in the CLARIN-PL research infrastructure funded by the Polish Ministry of Science and Higher Education.

References

1. Kolomiyets, O., Kordjamshidi, P., Bethard, S., Moens, M.: SemEval-2013 task 3: spatial role labeling. In: Second Joint Conference on Lexical and Computational Semantics (SEM). Proceedings of the Seventh International Workshop on Semantic Evaluation (SemEval 2013), Atlanta, USA. ACL, East Stroudsburg (2013)
2. LDC: ACE (Automatic Content Extraction) English Annotation Guidelines for Relations. Argument (2008)
3. Broda, B., Marcińczuk, M., Maziarz, M., Radziszewski, A., Wardyński, A.: KPWr: towards a free corpus of Polish. In: Calzolari, N., Choukri, K., Declerck, T., Doğan, M.U., Maegaard, B., Mariani, J., Odijk, J., Piperidis, S. (eds.) Proceedings of the Eight International Conference on Language Resources and Evaluation (LREC 2012), Istanbul, Turkey. European Language Resources Association (ELRA), May 2012
4. Radziszewski, A.: A tiered CRF tagger for Polish. In: Bembenik, R., Skonieczny, Ł., Rybiński, H., Kryszkiewicz, M., Niezgódka, M. (eds.) Intelligent Tools for Building a Scientific Information. SCI, vol. 467, pp. 215–230. Springer, Heidelberg (2013)
5. Waszczuk, J.: Harnessing the CRF complexity with domain-specific constraints. The case of morphosyntactic tagging of a highly inflected language. In: Proceedings of COLING 2012, no. December 2012, pp. 2789–2804 (2012)
6. Acedański, S.: A morphosyntactic Brill tagger for inflectional languages. In: Loftsson, H., Rögnvaldsson, E., Helgadóttir, S. (eds.) IceTAL 2010. LNCS, vol. 6233, pp. 3–14. Springer, Heidelberg (2010)
7. Kaczmarek, A., Marcińczuk, M.: Heuristic algorithm for zero subject detection in Polish. In: Král, P., Matoušek, V. (eds.) TSD 2015. LNCS, vol. 9302, pp. 378–386. Springer, Heidelberg (2015). doi:10.1007/978-3-319-24033-6_43
8. Przepiórkowski, A.: Powierzchniowe przetwarzanie języka polskiego. Problemy współczesnej nauki, teoria i zastosowania: Inżynieria lingwistyczna. Akademicka Oficyna Wydawnicza "Exit" (2008)
9. Głowińska, K.: Anotacja składniowa NKJP. In: Przepiórkowski, A., Bańko, M., Górski, R.L., Lewandowska-Tomaszczyk, B. (eds.) Narodowy Korpus Języka Polskiego, pp. 107–127. Wydawnictwo Naukowe PWN, Warsaw (2012)
10. Radziszewski, A., Pawlaczek, A.: Large-scale experiments with NP chunking of Polish. In: Sojka, P., Horák, A., Kopeček, I., Pala, K. (eds.) TSD 2012. LNCS, vol. 7499, pp. 143–149. Springer, Heidelberg (2012)
11. Radziszewski, A.: Metody znakowania morfosyntaktycznego i automatycznej płytkiej analizy składniowej języka polski. Ph.D. thesis, Politechnika Wrocławska, Wrocław (2012)
12. Maziarz, M., Piasecki, M., Szpakowicz, S.: Approaching plWordNet 2.0. In: Proceedings of the 6th Global Wordnet Conference, Matsue, Japan, January 2012
13. Pease, A., Niles, I., Li, J.: The suggested upper merged ontology: a large ontology for the semantic web and its applications. In: Working Notes of the AAAI-2002 Workshop on Ontologies and the Semantic Web (2002)
14. Marcińczuk, M., Kocoń, J., Janicki, M.: Liner2 — a customizable framework for proper names recognition for Polish. In: Bembenik, R., Skonieczny, Ł., Rybiński, H., Kryszkiewicz, M., Niezgódka, M. (eds.) Intelligent Tools for Building a Scientific Information. SCI, vol. 467, pp. 231–254. Springer, Heidelberg (2013)
15. Wróblewska, A., Woliński, M.: Preliminary experiments in Polish dependency parsing. In: Bouvry, P., Kłopotek, M.A., Leprévost, F., Marciniak, M., Mykowiecka, A., Rybiński, H. (eds.) SIIS 2011. LNCS, vol. 7053, pp. 279–292. Springer, Heidelberg (2012)

16. Kordjamshidi, P., Van Otterlo, M., Moens, M.F.: Spatial role labeling: towards extraction of spatial relations from natural language. ACM Trans. Speech Lang. Process. **8**(3), 1–36 (2011)
17. Przybylska, R.: Polisemia przyimków polskich w świetle semantyki kognitywnej. Universitas, Kraków (2002)

Topic Modeling over Text Streams
from Social Media

Miroslav Smatana[✉], Ján Paralič, and Peter Butka

Department of Cybernetics and Artificial Intelligence,
Faculty of Electrical Engineering and Informatics,
Technical University of Košice, Letná 9, 04200 Košice, Slovakia
{miroslav.smatana,jan.paralic,peter.butka}@tuke.sk

Abstract. Topic modeling becomes a popular research area which shows us new way to search, browse and summarize large amount of texts. Methods of topic modeling try to uncover the hidden thematic structure in document collections. Topic modeling in connection with social networks, which are one of the strongest communication tool and produces large amount of opinions and attitudes on world events, can be useful for analysis in case of crisis situations, elections, launching a new product on the market etc. For that reason we pro-pose a tool for topic modeling over text streams from social networks in this paper. Description of proposed tool is extended with practical experiments. Realized experiments shown promising results when using our tool on real data in comparison to state-of-the-art methods.

Keywords: Topic modeling · Social media · Natural language processing · Modularity clustering

1 Introduction

In recent years social networks became one of the strongest communication tool. Users publish hundreds of millions of contributions every day. These contributions reflect the opinions and attitudes of users on world events. It follows that social networks are an interesting source of large amounts of information. They can be used in case of crisis situations (wars, earthquakes, tornados, rebellions), elections, but also when e.g. launching a new product on the market etc. The importance of these contributions also highlights the fact that search engines such as Google and Bing include them into their search results.

The manual analysis of contributions from social networks is very difficult and time consuming. From that reason there is a strong need to automate this process. It can be achieved by using the methods of natural language processing.

In this paper we propose a system for automatic analysis of short messages from Twitter social network (https://twitter.com/). Proposed system is looking for the most interesting topics over streams of data and monitors their evolution over time, finds keywords, key tweets and key hashtags.

© Springer International Publishing Switzerland 2016
P. Sojka et al. (Eds.): TSD 2016, LNAI 9924, pp. 163–172, 2016.
DOI: 10.1007/978-3-319-45510-5_19

When analyzing Twitter's short messages one might encounter the following three main problems:

1. Short messages have text with maximal length about 140 characters, which contains small amount of information for processing by computer.
2. There is often used Internet slang.
3. A lot of these messages contain only link to other website.

The proposed system described in this paper deals with the first two problems. The rest of the paper is organized as follows. In next section some details regarding the basic idea of topic modeling and related work are described. In Sect. 3 we provide the details on our proposed solution, including the applied methods, architecture and processing steps. In the following section we provide setup and results of experiments based on the selected data from Twitter.

2   Topic Modeling and Related Work

Topic modeling tries to uncover the hidden semantic structures in document collections. It shows us new way to search, browse and summarize large amount of texts, so it became increasingly important step in text processing area. Simple example of topic modeling application can be demonstrated on traditional search engines. Most of these engines use match of their query with words or keywords in documents and return those with best match. This approach does not provide enough information so many times it returns irrelevant documents. With topic modeling we can obtain enough information (which topic(s) is (are) most important in given document, relations between topics, topic evolution in time) to achieve better results of search [1]. For example if a user is browsing documents for the query "best football players in 90s", (s)he will focus on documents from topic "football" and years from 1990 to 1999.

Topic modeling is highly correlated and often confused with document clustering, which aims to organize documents with similar content to same groups. Most of the classic clustering methods for documents are usually based on bag-of-words model which uses only raw data and does not have additional information about semantic structure. Topic models are able to put words with similar semantic into the same group. Although document clustering does not fulfill the same role as topic modeling in some naive way it can be transformed to this task. To make clustering promote topic modeling, we first use classic clustering method to obtain groups of similar documents (clusters) and after that we use the method described in [2] to obtain clusters' labels and importance of words in each cluster.

As we already mentioned in the previous section, topic model tries to uncover hidden semantic structure from given corpus of documents. Latent Semantic Analysis (LSA) [3] can be seen as a type of effort to uncover semantic structure from texts. There were created several extensions and improvements of LSA, e.g., probabilistic latent semantic analysis described in [4], which was a basis for Latent Dirichlet Allocation (LDA) [5]. LDA became dominantly used method for topic modeling.

Area of topic modeling has been studied by many researchers in recent years. There have been proposed many extensions of LDA, e.g. Petterson et al. [6] extend LDA to explicitly allow the encoding of side information in the distribution of words which lead to the improvement of topic alignment quality and also allows synchronize topics over different languages. Zhai and Boyd-Graber [7] proposed online version of LDA which is able to incorporate new words to model. Also other methods as hierarchical Dirichlet processes (HDP) [8], moving average stochastic variational inference (MASVI) [9], or stochastic variational inference [10] have been proposed. All described models only deal with longer texts; they do not consider problems with topic modeling of short texts.

However recent studies in topic modeling show increasing interest in short texts, especially Twitter. For that reason there have been also proposed a few models focused on that part of topic modeling area. For example in [11] authors proposed a method using conventional topic model based on large-scale external data. The disadvantage of this method is that external data must be closely related to modeled data. Sridhar [12] use Gaussian mixture models in his work, which does not need aggregation of short texts as in the conventional topics models. Cheng et al. [13] proposed Biterm Topic Model (BTM) in their work. BTM explicitly models word co-occurrence patterns in the whole corpus, which makes inference more effective. Quan et al. [14] applied heuristic aggregation for aggregating short text into long pseudo-documents before using standard conventional topic model.

3   Proposed Solution

3.1   Modularity-Based Clustering

Modularity-based clustering [15] belongs to the graph methods, where objects from which we want to make clusters are represented as nodes and edges represent similarity between objects. Main goal of this method is to identify sub-networks of objects which are highly correlated. This sub-networks expose hidden dependencies between objects and usually represent objects from the same community or about the same topic. Quality of created communities is calculated by modularity. Modularity reflects intensity of connections of objects in individual communities in comparison to connections between other communities, and can be calculated by the formula:

$$Q = \frac{1}{2m} \sum_{i,j} \left[A_{ij} - \frac{k_i k_j}{2m} \right] \delta(c_i, c_j) \tag{1}$$

where A_{ij} represents weight between nodes i and j, k_i is sum of weights of edges connected with node i, c_i is community of node i, $\delta(u, v)$ is equal to 1 if $u = v$ (otherwise is 0), and $m = \frac{1}{2} \sum_{i,j} A_{ij}$.

Algorithm of modularity-based clustering method consists of two main phases, which are repeated again until modularity rises. In the first phase every node is assigned to its own community. The algorithm is trying to merge every pair of communities to new one and if the modularity rises, the change will be kept. Rise of modularity can be calculated as:

$$\Delta Q = \left[\frac{N_{in} + k_{i,in}}{2M} - \left(\frac{N_{tot} + k_i}{2M} \right)^2 \right] - \left[\frac{N_{in}}{2M} - \left(\frac{N_{tot}}{2M} \right)^2 - \left(\frac{k_i}{2M} \right)^2 \right] \quad (2)$$

where N_{in} is the sum of connections inside the community, N_{tot} is the sum of all connections, which occurred within the community nodes, k_i is the sum of connections with node i, $k_{i,in}$ is the sum of connections with node i inside the community, M is sum of all connections in the graph.

In the second phase a new graph is created from communities, which were found in the first phase. In this new graph every community is represented as one node and weight of the edge between two nodes is calculated as the sum of values of edges' weights between all nodes in these communities. Internal connections inside the community lead to self-connection.

3.2 Modified PageRank

The PageRank algorithm was originally invented to rank bibliographic references and (later) web pages. PageRank of a web page A is calculated by the formula [16]:

$$PR(A) = (1 - d) + d \sum_{i=1}^{n} \frac{PR(T_i)}{C(T_i)} \quad (3)$$

where $PR(A)$ is PageRank of page A, d is damping factor, T_i is one of n pages (with their PageRank values $PR(T_i)$) that link to the page A, $C(T_i)$ is the number of outbound links from T_i. In our tool we use original PageRank with modified weighting of connections between nodes. Weights are calculated by the following formula:

$$w_{ij} = \frac{n_{ij}}{n_t} \quad (4)$$

where w_{ij} is weight of connection from node i to node j, n_t is total number of words in the topic, n_{ij} is number of co-occurrences of words i and j. If some word occurs in message more than once it leads to a self-connection of the respective node.

3.3 Proposed Tool

The main purpose of our work was to develop the tool which will be able to analyze stream data consisting of short texts without user intervention and will be language independent. The proposed tool is created as a library in Java, so it is free to use in any other application. It is based on two frameworks, Gate (https://gate.ac.uk/) and Gephi (http://gephi.github.io/). Gate is open-source tool for language processing and annotation problems. Gephi is support framework for analysis, searching and visualization of networks and graphs.

The architecture of proposed tool is shown in Fig. 1. The process performed by the proposed tool consists of six main steps: preprocessing, clustering, data extraction, cluster labeling, comparison with topics in memory and feature selection. All steps are described in the following subsections.

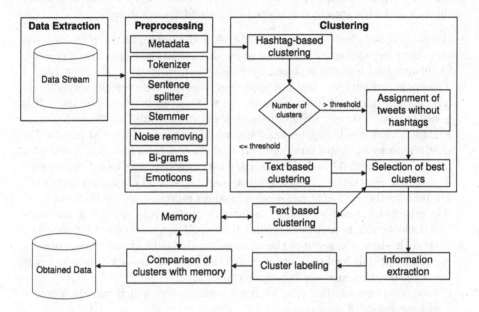

Fig. 1. Architecture of the proposed tool for Twitter stream topic modeling.

Preprocessing. In this step unstructured messages from the input are processed to understandable form for our tool. Because many tasks of preprocessing step depend on language of messages we decided to use methods from Gate which supports multilingual texts. This step consists of seven tasks:

- Metadata - we crawl basic data such as time, date, hashtags from input message metadata.
- Tokenizer - is used to split input text to tokens (in our case: words, emoticons).
- Sentence splitter - splits text messages into sentences.
- Stemmer - removes the common morphological and inflexional endings and beginnings from words.
- Bi-grams - candidates for phrases are extracted here in form of bi-grams (for example the following sentence "A great hotel in good location" contains bi-grams "A great", "great hotel", "hotel in", "in good", "good location").
- Emoticons - we extract graphic symbols for emoticons which often represent sentiment, attitudes of author.
- Noise removal - we remove words which length is less than 3 characters, punctuation, numbers, and webpage links.

Clustering. This step represents first part of topic modeling, where clusters are created from input messages based on their similarity using modularity-based clustering presented in Sect. 3.1. This method offers several advantages over other methods, e.g., no initial adjustment of parameters is needed, it has better computation complexity, and also better noise identification than other methods. Clustering step in our system consists of these subtasks:

- Hashtags-based clustering - in this subtask nodes are represented by hashtags and edges by number of hashtags co-occurrence in messages. After creation of clusters hashtags are replaced by messages in which they are occurring.
- Threshold comparison - this task represents a switch between clustering methods, because usage of hashtags is not so widespread in some countries. It is possible that if we get million messages input and only 10,000 will contain hashtags, then clustering by hashtags would give very distorted results. For that reason we compare number of created clusters from hashtags-based clustering with threshold and if it is satisfactory, we skip text based clustering.
- Text-based clustering - it is similar as hashtags-based clustering but nodes are represented by whole messages and edges represent message similarity.
- Assignment of message without hashtags - this step is applied if hashtags-based clustering is satisfactory and it is used for assignment of messages without hashtags to some of the clusters. In this subtask we iterate over all messages without hashtags and if more than 50 % of message tokens occur in some cluster, this message is assigned to it.
- Cluster selection - in this step we remove all clusters which contain a below-average number of messages.

Data Extraction. In data extraction step we aim to extract keywords, key tweets and key hashtags from clusters. As first, the informativeness of every word and hashtag in every cluster is computed by modified PageRank approach presented in Sect. 3.2. Then, informativeness of words is computed using formula:

$$S_i = \sum_{j=1}^{n} \frac{I_j}{n} \tag{5}$$

where S_i is informativeness of message i; I_j is informativeness of word j, n is the number of words in message. Last step is selection of words, hashtags and messages with the largest value of their informativeness are selected.

Cluster Labeling. Cluster labeling is performed using information about key hashtags extracted in previous step, where label of cluster is represented as hashtag with the highest value of informativeness.

Comparison with Topics in Memory. In this step the actual topic is compared with topics saved in memory. Comparison is based on the similarity between topics. If similarity is greater than defined threshold then topics are declared as the same, in other case actual topic is added to memory. We use simple similarity between topic i and topic j given by equation:

$$sim_{ij} = \frac{t_j}{V} \tag{6}$$

where t_{ij} is number of identical words in topic i and topic j, V is size of vocabulary of these topics.

Feature Selection. The disadvantage of working with the large amount of data is growing computational complexity and size of feature space. For that reason we used feature selection methods, which remove words with low informativeness from feature space. In work [17] five basic methods for feature selection are presented, we decided to use method based on information gain selection.

Word informativeness calculated by information gain is based on presence or absence of term in documents (messages). Information gain is computed by the following formula:

$$IG(t) = -\sum_{i=1}^{N_c} P_r(c_i)logP_r(c_i) + P_r(t)\sum_{i=1}^{N_c} P_r(c_i|t)logP_r(c_i|t)$$
$$+ P_r(\bar{t})\sum_{i=1}^{N_c} P_r(c_i|\bar{t})logP_r(c_i|\bar{t}) \tag{7}$$

where $IG(t)$ is informativeness of term t, N_c is the number of categories, c_i is i-th category, t means presence and $\bar{t}$ absence of particular term and $P_r(c_i)$ is probability of category c_i.

4 Experiments

In the presented series of experiments we compared the proposed clustering and labeling solution to the baseline solutions which were implemented within the same architecture as the proposed solution.

4.1 Sample Data

The dataset described in this section is available as free from page of the Sanders Analytics (http://www.sananalytics.com/lab/twitter-sentiment/). It consists of manually annotated short messages from Twitter. The original dataset has 5513 samples, where every sample has assigned topic (Twitter, Google, Microsoft, Apple) and sentiment (positive, negative, neutral, irrelevant). Due to fact that Twitter license terms dataset contains only identification key of each message and we need to download them from Twitter, some of the messages were private or no longer available, therefore we used 2945 samples from original dataset. The distribution of messages by assigned topic was 23 % Twitter, 32 % Apple, 24 % Microsoft and 23 % Google. The distribution of messages by assigned sentiment was 63 % of neutral messages, 15 % positive, 17 % negative and 5 % irrelevant.

4.2   Experimental Tasks and Evaluation Metrics

In our experiments we followed two main tasks, which we used to evaluate quality of our proposed solution:

1. Clustering similar samples to same cluster
2. Information extraction for individual topics

In clustering task we evaluated quality of clusters by three evaluation metrics:

- **Coverage** – it is represented by the ratio of samples which were assigned to the same topic to all samples in dataset.
- **Purity** – for calculation of purity most frequent label is used for cluster and then average purity of objects within clusters is computed [18].
- **Normalized mutual information (NMI)** – information-theoretic metric similar to purity, where value 1.0 means that clustering results match exactly with the category labels of clusters [18].

For evaluation of information extraction task we used questionnaire method, because of missing annotations for keywords, key hashtags and key tweets. The questionnaire was based on the list of keywords (hashtags, tweets) for every evaluated method and each of 20 respondents independently rated every method with value from 1 to 5, where 1 means very bad and 5 means very good.

4.3   Results

The results achieved in the experiments for clustering evaluation are presented in the Table 1. From this table one can see that methods which we used in our system achieved better quality than other evaluated methods and generate acceptable number of clusters. K-means and LDA achieved better quality only in coverage metric, because our proposed methods (clustering using hashtags and clustering using the whole text of messages) can annotate some samples as trash (not assigned to any cluster).

Table 1. The results of experiments with clustering methods - comparison of our proposed clustering approaches (using hashtags and using the whole text of messages) with K-means and LDA methods according to three evaluation metrics. Critical N values.

Method	Number of clusters	Coverage	Purity	NMI
K-means	4	1.0	0.766	0.728
LDA	4	1.0	0.699	0.438
Clustering (hashtags)	5	0.925	0.892	0.768
Clustering (text of messages)	4	0.991	0.956	0.848

Table 2. The results of experiments with information extraction methods - comparison of methods used for extraction of keywords, key hashtags and key tweets based on ratings from users. PageRank is our modified version of extraction and is compared to standard approaches.

Rating for ...	TF-IDF + Chi-square	TF-IDF	Chi-square	PageRank
Keywords	1.13	3.73	1.13	1.4
Key hashtags	1.06	3.6	1.13	3.2
Key tweets	2.6	2.46	3.3	4

Table 2 presents the results for information extraction experiments for particular evaluated methods (standard methods – TF-IDF, Chi-square, their combination, our method – modified PageRank) with their ratings from respondents, all according to the particular goal (type) of extraction – keywords, key hashtags and key tweets. The best result for keywords extraction was achieved by standard TF-IDF method. Our implemented modification of PageRank did not offer very good keywords but it is still better than other evaluated methods. In case of key hashtags extraction our modified PageRank approach achieved results of comparable quality with TF-IDF method. In last row of table we can see that for key tweets modified PageRank achieved the best quality from all evaluated methods. Moreover, another good feature of proposed method is the ability to pick out very varied messages as key tweets. In the future work we would like to extend this approach using methods from the area of Formal Concept Analysis, especially heterogeneous fuzzy approaches [19,20]. Also, our vision is also to extend our approach for distributed clustering of larger datasets [21].

5   Conclusion

In this paper we describe a system for topic modeling over short text streams from social media, which can be also easily included as supportive library to already existing systems. The achieved results showed that used methods in proposed system have comparable or better quality than other evaluated methods. However more detailed evaluation on more and bigger datasets are needed.

Acknowledgments. The work presented in this paper was supported by the Slovak VEGA grant 1/0493/16 and Slovak KEGA grant 025TUKE-4/2015.

References

1. Blei, D.: Probabilistic topic models. Commun. ACM **55**(4), 77–84 (2012)
2. Xie, P., Xing, E.: Integrating document clustering and topic modeling. In: Proceedings of 29th Conference Uncertainty in Artificial Intelligence, Bellevue, US, pp. 694–703 (2013)

3. Landauer, T.K., Foltz, P.W., Laham, D.: An introduction to latent semantic analysis. Discourse Process **25**(2–3), 259–284 (1998)
4. Hofmann, T.: Probabilistic latent semantic analysis. In: Proceedings of 15th Conference Uncertainty in Artificial Intelligence, Stockholm, Sweden, pp. 289–296 (1999)
5. Blei, D., Ng, A., Jordan, M.: Latent Dirichlet allocation. J. Mach. Learn. Res. **3**, 694–703 (2003)
6. Petterson, J., Buntine, W., Narayanamurthy, S., Caetano, T., Smola, A.: Word features for latent Dirichlet allocation. Adv. Neural Inf. Process. Syst. **23**, 1921–1929 (2010)
7. Zhai, K., Boyd-Graber, J.: Online latent Dirichlet allocation with infine vocabulary. In: Proceedings of 30th International Conference on Machine Learning, Atlanta, US, pp. 561–569 (2013)
8. Teh, Y., Jordan, M., Beal, M., Blei, D.: Hierarchical Dirichlet processes. J. Am. Stat. Assoc. **101**(476), 1566–1581 (2006)
9. Li, X., Ouyang, J., Lu, Y.: Topic modeling for large-scale text data. Front. Electr. Electron. Eng. **16**(6), 457–465 (2015)
10. Hoffman, M., Blei, D., Wang, C., Paisley, D.: Stochastic variational inference. J. Mach. Learn. Res. **14**, 1303–1347 (2013)
11. Phan, X., Nguyen, L., Horiguchi, S.: Learning to classify short and sparse text & web with hidden topics from large-scale data collections. In: Proceedings of 17th International Conference on World Wide Web, Beijing, China, pp. 91–99 (2008)
12. Sridhar, V.: Unsupervised topic modeling for short texts using distributed representations of words. In: Proceedings of NAACL-HLT 2015, Denver, US, pp. 192–200 (2015)
13. Cheng, S., Yan, X., Lan, Y., Guo, J.: BTM - topic modeling over short texts. IEEE Trans. Knowl. Data Eng. **26**(12), 2928–2941 (2014)
14. Quan, X., Kit, C., Ge, Y., Pan, S.: Short and sparse text topic modeling via self-aggregation. In: Proceedings of 24th International Conference on Artificial Intelligence, Buenos Aires, Argentina, pp. 2270–2276 (2015)
15. Blondel, V., Guillaume, J., Lambiotte, R., Lefebvre, E.: Fast unfolding of communities in large networks. J. Stat. Mech. Theory Exp. **2008**(10), P10008 (2008). (pp. 1–12)
16. Page, L., Brin, S., Motwani, R., Winograd, T.: The PageRank citation ranking: bringing order to the Web. Technical report, Stanford Digital Libraries (1998)
17. Yang, Y., Pedersen, J.: A comparative study of feature selection in text categorizations. In: Proceedings of 14th International Conference on Machine Learning, San Francisco, US, pp. 412–420 (1997)
18. Manning, C., Raghavan, P., Schütze, H.: Introduction to Information Retrieval. Cambridge University Press, New York (2008)
19. Pocs, J., Pocsova, J.: Basic theorem as representation of heterogeneous concept lattices. Front. Comput. Sci. **9**(4), 636–642 (2015)
20. Pocs, J., Pocsova, J.: Bipolarized extension of heterogeneous concept lattices. Appl. Math. Sci. **8**(125–128), 6359–6365 (2014)
21. Sarnovsky, M., Carnoka, N.: Distributed algorithm for text documents clustering based on k-means approach. Adv. Intell. Syst. Comput. **430**, 165–174 (2016)

Neural Networks for Featureless Named Entity Recognition in Czech

Jana Straková[✉], Milan Straka, and Jan Hajič

Faculty of Mathematics and Physics,
Institute of Formal and Applied Linguistics, Charles University in Prague,
Malostranské náměstí 25, 118 00 Prague, Czech Republic
{strakova,straka,hajic}@ufal.mff.cuni.cz
http://ufal.mff.cuni.cz

Abstract. We present a completely featureless, language agnostic named entity recognition system. Following recent advances in artificial neural network research, the recognizer employs parametric rectified linear units (PReLU), word embeddings and character-level embeddings based on gated linear units (GRU). Without any feature engineering, only with surface forms, lemmas and tags as input, the network achieves excellent results in Czech NER and surpasses the current state of the art of previously published Czech NER systems, which use manually designed rule-based orthographic classification features. Furthermore, the neural network achieves robust results even when only surface forms are available as input. In addition, the proposed neural network can use the manually designed rule-based orthographic classification features and in such combination, it exceeds the current state of the art by a wide margin.

Keywords: Neural networks · Named entity recognition · Czech · Word embeddings · Character-level embeddings · Parametric rectified linear unit (PReLU) · Gated linear unit (GRU)

1 Introduction

Recent years have seen a dramatic progress in the field of artificial neural networks. The publication of word embeddings [19] opened reliable and computationally affordable ways of using tokens as classification features in artificial neural networks. For morphologically rich languages, word embeddings appear rather too coarse for many tasks, especially those where the inner structure of the word such as prefixes and suffixes, is crucial. Therefore, the ideas go further in publication of character-level embeddings [21], which recently improved the state of the art in POS-tagging [17]. One of the advantages of word- and character-level embeddings is that they are learned automatically from large raw corpora.

Another paradigm-changing publication introduces the long short-term memory units (LSTMs, [10]). In simple words, LSTMs are specially shaped units

© Springer International Publishing Switzerland 2016
P. Sojka et al. (Eds.): TSD 2016, LNAI 9924, pp. 173–181, 2016.
DOI: 10.1007/978-3-319-45510-5_20

of artificial neural networks designed to process whole sequences. LSTMs have been shown to capture non-linear and non-local dynamics in sequences [10] and have been used to obtain many state-of-the-art results in sequence classification [15, 17].

Recently, a gated linear unit (GRU) was proposed by [3] as an alternative to LSTM, and was shown to have similar performance, while being less computationally demanding.

In this work, we use artificial neural networks employing parametric rectified linear units (PReLU), word embeddings and character-level embeddings based on gated linear units (GRU). We describe our methodology in Sect. 3. We report our results and discussion in Sect. 4 and we conclude in Sect. 5.

2   Related Work

Czech named entity recognition (NER) has become a well-established field. Following the publication of the Czech Named Entity Corpus [14, 26], a selection of named entity recognizers for Czech has been published: [11–14, 23] and even a publicly available Czech named entity recognition exists (NameTag,[1] [9]).

All these works use manually selected rule-based orthographic classification features, such as first character capitalization, existence of special characters in the word, regular expressions designed to reveal particular named entity types. Also gazetteers are extensively utilized. The authors employed a wide selection of machine learning techniques (decision trees [26], SVMs [14], maximum entropy classifier [23], CRFs [12]), clustering techniques [11] and stemming approaches [13].

The contribution of our work is that we use artificial neural networks with parametric rectified linear units, word embeddings and character-level embeddings, which do not need manually designed classification features or gazetteers, and still surpass the current state of the art.

In [5], the authors present a semi-supervised learning approach based on neural networks for Czech and Turkish NER utilizing word embeddings [19], but there are some differences in the neural network design and in classification features used. Instead regularized averaged perceptron, we use parametric rectified linear units, character-level embeddings and dropout. The NER system in [5] does not use morphological analysis, it is therefore similar to our experiments with only surface forms as input. However, the system does use "type information of the window c_i, i.e. is-capitalized, all-capitalized, all-digits, ..." etc. Our system surpasses these results even without using such features.

English named entity recognition has a successful tradition in computational linguistics and the state of the art [20] has recently been pushed forward by [2, 15, 16, 18]. We present a comparison with these works in Sect. 4. The most similar to our proposed design is [15], which was accepted to NAACL 2016 the exact month of this paper submission. The authors propose a very similar network with LSTMs, word embeddings and character-level embeddings. However,

---

[1] http://ufal.mff.cuni.cz/nametag.

while we classify each word separately and use Viterbi to perform the final decoding, [15] employs LSTMs combined with CRF layer to decode whole sentences, which brings a determining advantage over our framework as we show in Sect. 4.

3   Methodology

We conduct our experiments on all available Czech NER corpora, so that we are able to compare with all available related work in Czech NER: Czech Named Entity Corpus (CNEC) 1.0 [14,26], CNEC 2.0,[2] CoNLL-based Extended CNEC 1.1 [12], CoNLL-based Extended CNEC 2.0.[3]

The named entities in CNEC 1.0 and 2.0 are hierarchically organized in two hierarchies (fine-grained "types" and coarse "supertypes", [14]), may be nested and labeled with more than one label.[4]

CoNLL-based Extended CNEC 1.1 and 2.0 are based on the respective original CNEC corpora, but they use only the coarser 7 classes, are flattened and assume that entities are non-nested and labeled with one label.

For comparison with the English state of the art, we evaluated our NER system on CoNLL-2003 shared task dataset [25]. In this task, four classes are predicted: PER (person), LOC (location), ORG (organization) and MISC (miscellaneous). The named entities are non-nested, non-overlapping and annotated with exactly one label.

In case of the original CNEC 1.0 and CNEC 2.0, we present results for both fine-grained and coarse-grained classes hierarchy ("types" and "supertypes", [14]) and we evaluate our results with the script provided with the corpora, which computes F-measure of selected types [14].

In case of the CoNLL-based Extended CNEC 1.1 and 2.0, we present results for the 7 classes present in these corpora and evaluate our results with the standard CoNLL evaluation script `conlleval`.

Similarly, the English CoNLL-2003 dataset is evaluated with CoNLL evaluation script `conlleval`.

3.1   The Network Classifier

For each word (and its context), we compute the probability distribution of labeling this word with BILOU-encoded [20] named entities. We then determine the best consistent assignment of BILOU-encoded entities to the words in the sentence using the Viterbi algorithm.

We compute the probability distribution for each word using an artificial neural network. The input layer consists of representations of surface forms (and

---

[2] http://ufal.mff.cuni.cz/cnec/cnec2.0.

[3] http://home.zcu.cz/~konkol/cnec2.0.php.

[4] Our system learns and predicts only outermost entities and is thus penalized for every misted nested named entity during evaluation.

optionally lemmas, tags, characters, character-level embeddings and classification features) of the word and W previous and W following words. The input layer is connected to a hidden layer of parametric rectified linear units [8] and the hidden layer is connected to the output layer which is a softmax layer producing probability distribution for all possible named entity classes in BILOU encoding.

We represent each word using a combination of the following:

- *word embedding:* Word embeddings are vector representations of low dimension [19]. We generated the word embeddings using word2vec [19] and we chose the Skip-gram model with negative sampling.[5]
- *character-level embedding:* To overcome drawbacks of word embeddings (embeddings for different words are independent; unknown words cannot be handled), several orthography aware models have been proposed [17,21], which compute word representation from the characters of the word.
  We hypothesized that character-level embeddings such as published in [17] have the potential to increase the performance of Czech NER system. Our assumption was that Czech as a morphologically rich language would benefit from character-level embeddings rather than word embeddings especially in cases where no morphological analysis is available.
  We use bidirectional GRUs [3,7] in line with [17]: we represent every Unicode character with a vector of C real numbers, and we use GRUs to compute two outputs, a sequence of word characters and a sequence of reversed word characters, and we then concatenate the two outputs, as shown in Fig. 1.
- *prefix and suffix:* For comparison with character-level embeddings, we also include "poor man's" character-level embeddings – we encode first two and last two characters encoded as one-hot vectors. We hypothesize that character-level embeddings as a more sophisticated means should perform better.
- *tag:* We encode part-of-speech tags as one-hot vectors.
- *manually designed classification features:* We also publish a combination of our neural network framework with traditional manually designed rule-based orthographic classification features. We use quite a limited set of classification features inspired by [23]: capitalization information, punctuation information, number information and Brown clusters [1]. We do not use gazetteers, context aggregation, prediction history nor two-stage decoding.

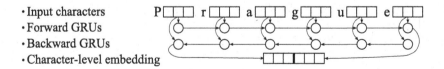

- Input characters
- Forward GRUs
- Backward GRUs
- Character-level embedding

Fig. 1. Neural network for character-level embedding computation.

---

[5] We used the following options: `-cbow 0 -window 5 -negative 5 -iter 1`.

The network is trained with AdaGrad [6] and we use dropout [22] on the hidden layer. We implemented our neural network in Torch7 [4], a scientific computing framework with wide support for machine learning algorithms.

We tuned most of the hyperparameters on development portion of CNEC 1.0 and used them for all other corpora. Notably, we utilize window size $W = 2$, hidden layer of 200 nodes, dropout 0.5, minibatches of size 100 and learning rate 0.02 with decay. We tune the dimension C of the character-level embeddings for every corpus separately, choosing either 32 or 64. All reported experiments use an ensemble of 5 networks, each using different random seed, with the resulting distributions being an average of individual networks distributions. The training of a single network took half a day on a single CPU to stabilize performance on development data. During evaluation of testing data, we add the development data to the training data, a technique proposed in context of NER by [20].

We trained the word embeddings of dimension 200 on English Gigaword Fifth Edition corpus and on Czech SYN [9]. We also lemmatized the corpora with MorphoDiTa [24] in order to train the lemma embeddings.

4   Results and Discussion

We present two groups of experiments with low and high complexity depending on the available network input: experiments where only surface forms were used, a putatively more difficult task as no linguistic knowledge is available to the NER system; and experiments with morphologically analyzed and POS-tagged text. We automatically generate lemmas and POS-tags from surface forms with MorphoDiTa [24], an open source tagger and lemmatizer.

Table 1 presents all results of this work. Our baseline is an artificial neural network with only surface forms encoded as word embeddings. We then add more computational complexity to the network: WE stands for word embeddings of forms and lemmas, CLE stands for character-level embeddings of forms and lemmas, 2CH stands for first two and last two characters of forms, lemmas and POS tags, and CF stands for experiments with traditional classification features.

4.1   Experiments with Surface Forms in Czech

This group of experiments dealt with situations when only surface forms are available as input. Since most of the previous literature heavily depends on manually selected language-dependent features, as well as gazetteers and more or less linguistically motivated variants of lemmatization of stemming, the only work to be directly compared with is [5]. The authors of [5] use a similar, semi-supervised neural network based approach. Their final system, which uses word embeddings, capitalization and punctuation information, prefixes, suffixes, context aggregation and prediction history, achieves CoNLL F-measure 75.61 for CoNLL-based Extended CNEC 1.1. We surpass these results with CoNLL F-measure 76.72, using only word embeddings, character-level embeddings and first two and last two characters. If the traditional features are added, we even achieve CoNLL F-measure 78.21.

Table 1. Experiment results and comparison with related work. Columns denote corpora, rows our experiments or related work. First group of rows describes our experiments with surface forms only (*f*), second group our experiments with forms, lemmas and POS-tags (*f,l,t*). *WE* stands for word embeddings, *CLE* for character-level embeddings, *2CH* for first two and last two characters, *CF* for traditional classification features. Third group of rows describes related work in Czech NER, and fourth group related work in English NER.

Experiment/related work	Corpus						
	Original CNEC 1.0		Original CNEC 2.0		Extended CNEC 1.1	Extended CNEC 2.0	English CoNLL-2003
	Types	Supt.	Types	Supt.	Classes	Classes	Classes
f+WE (baseline)	63.24	69.61	63.33	68.87	63.48	63.91	67.99
f+CLE	71.43	76.13	70.50	75.80	69.59	70.06	82.65
f+WE+2CH	69.73	74.49	69.44	74.31	75.15	74.36	79.40
f+WE+CLE	73.30	78.11	73.10	77.89	73.33	73.80	84.08
f+WE+CLE+2CH	73.71	78.32	72.81	77.87	76.72	77.18	84.29
f+WE+CLE+2CH+CF	73.73	78.50	72.91	77.65	78.21	78.20	86.06
f,l,t+WE	80.07	83.21	77.45	80.92	78.42	78.18	87.92
f,l,t+CLE	75.63	80.88	74.38	79.85	75.32	76.02	83.70
f,l,t+WE+2CH	80.46	83.85	78.32	82.09	79.68	79.48	89.37
f,l,t+WE+CLE	80.64	84.06	78.62	82.48	80.11	80.41	89.74
f,l,t+WE+CLE+2CH	80.92	84.18	78.63	82.41	**80.88**	**80.79**	89.71
f,l,t+WE+CLE+2CH+CF	**81.20**	**84.68**	**79.23**	**82.78**	80.73	80.73	**89.92**
Kravalová and Žabokrtský [14]	*68.00*	*71.00*	–	–	–	–	–
Konkol and Konopík [12]	–	*79.00*	–	–	*74.08*	–	*83.24*
Straková et al. [23]	*79.23*	*82.82*	–	–	–	–	–
Konkol and Konopík [13]	–	–	–	–	*74.23*	*74.37*	–
Demir and Özgür [5]	–	–	–	–	*75.61*	–	–
Konkol et al. [11]	–	–	–	–	*74.08*	–	*89.44*
Ratinov and Roth [20]	–	–	–	–	–	–	*90.80*
Lin and Wu [16]	–	–	–	–	–	–	*90.90*
Chiu and Nichols [2]	–	–	–	–	–	–	*90.77*
Luo et al. [18]	–	–	–	–	–	–	*91.20*
Lample et al. [15]	–	–	–	–	–	–	*90.94*

4.2 Experiments with Lemmas and POS Tags in Czech

Table 1 presents a comparison with related work on all available Czech NER corpora. The row denoted *f,l,t+WE+CLE+2CH+CF* is our best setting, including manually selected classification features. Our proposed network clearly exceeds the current state of the art on all Czech corpora in measures selected by the authors of the respective literature.

We shall now focus our discussion on featureless neural networks. Our system exceeds the current Czech state of the art solely with automatically obtained word embeddings (see row $f,l,t+WE$ in Table 1), without requiring manually designed rule-based orthographic features, gazetteers, context aggregation, prediction history or two-stage decoding. The effect is even stronger with character-level embeddings and optionally first two and last two characters.

4.3 English Experiments

Our best result (row $f,l,t+WE+CLE+2CH+CF$) is F-measure 89.92, which is near the English state of the art. A work most similar to ours [15], also proposed neural network architecture with word embeddings and character-level embeddings. Nevertheless, in [15] sentence-level decoding using bidirectional LSTMs with additional CRF layer is used, while our framework decodes the entities using Viterbi algorithm on probability distributions of named entity classes.

5 Conclusions

We presented an artificial neural network based NER system which achieves excellent results in Czech NER and near state-of-the-art results in English NER without manually designed rule-based orthographic classification features, gazetteers, context aggregation, prediction history or two-stage decoding. Our proposed architecture exceeds all known Czech published results only with forms, lemmas and POS tags encoded as word embeddings and achieves even better results in combination with character-level embeddings, prefixes and suffixes. Finally, it surpasses the current state of the art of Czech NER in combination with traditional classification features by a wide margin. The proposed neural network also yields very robust results without morphologic analysis or POS-tagging, when only surface forms are available. As our future work, we plan to improve our decoding in line with [15].

Acknowledgments. This work has been partially supported and has been using language resources and tools developed, stored and distributed by the LINDAT/CLARIN project of the Ministry of Education, Youth and Sports of the Czech Republic (project LM2015071). This research was also partially supported by SVV project number 260 333.

References

1. Brown, P.F., deSouza, P.V., Mercer, R.L., Pietra, V.J.D., Lai, J.C.: Class-based n-gram models of natural language. Comput. Linguist. **18**(4), 467–479 (1992)
2. Chiu, J.P.C., Nichols, E.: Named Entity Recognition with Bidirectional LSTM-CNNs. CoRR abs/1511.08308 (2015). http://arXiv.org/abs/1511.08308
3. Cho, K., van Merrienboer, B., Bahdanau, D., Bengio, Y.: On the Properties of Neural Machine Translation: Encoder-Decoder Approaches. CoRR abs/1409.1259 (2014). http://arXiv.org/abs/1409.1259

4. Collobert, R., Kavukcuoglu, K., Farabet, C.: Torch7: a matlab-like environment for machine learning. In: BigLearn, NIPS Workshop (2011)
5. Demir, H., Özgür, A.: Improving named entity recognition for morphologically rich languages using word embeddings. In: 2014 13th International Conference on Machine Learning and Applications (ICMLA), pp. 117–122, December 2014
6. Duchi, J., Hazan, E., Singer, Y.: Adaptive subgradient methods for online learning and stochastic optimization. J. Mach. Learn. Res. **12**, 2121–2159 (2011)
7. Graves, A., Schmidhuber, J.: Framewise phoneme classification with bidirectional LSTM and other neural network architectures. Neural Netw. **18**, 5–6 (2005)
8. He, K., Zhang, X., Ren, S., Sun, J.: Delving deep into rectifiers: Surpassing human-level performance on imagenet classification. CoRR abs/1502.01852 (2015). http://arXiv.org/abs/1502.01852
9. Hnátková, M., Křen, M., Procházka, P., Skoumalová, H.: The SYN-series corpora of written Czech. In: Proceedings of the Ninth International Conference on Language Resources and Evaluation (LREC 2014). ELRA, Reykjavik, May 2014
10. Hochreiter, S., Schmidhuber, J.: Long short-term memory. Neural Comput. **9**(8), 1735–1780 (1997). http://dx.doi.org/10.1162/neco.1997.9.8.1735
11. Konkol, M., Brychcín, T., Konopík, M.: Latent semantics in named entity recognition. Expert Syst. Appl. **42**(7), 3470–3479 (2015)
12. Konkol, M., Konopík, M.: CRF-based Czech named entity recognizer and consolidation of Czech NER research. In: Habernal, I. (ed.) TSD 2013. LNCS, vol. 8082, pp. 153–160. Springer, Heidelberg (2013)
13. Konkol, M., Konopík, M.: Named entity recognition for highly inflectional languages: effects of various lemmatization and stemming approaches. In: Sojka, P., Horák, A., Kopeček, I., Pala, K. (eds.) TSD 2014. LNCS, vol. 8655, pp. 267–274. Springer, Heidelberg (2014)
14. Kravalová, J., Žabokrtský, Z.: Czech named entity corpus and SVM-based recognizer. In: Proceedings of the 2009 Named Entities Workshop: Shared Task on Transliteration, NEWS 2009, ACL, pp. 194–201 (2009)
15. Lample, G., Ballesteros, M., Kawakami, K., Subramanian, S., Dyer, C.: Neural architectures for named entity recognition. CoRR abs/1603.01360v1 (2016). To appear at NAACL 2016
16. Lin, D., Wu, X.: Phrase clustering for discriminative learning. In: Proceedings of the Joint Conference of the 47th Annual Meeting of the ACL and the 4th International Joint Conference on Natural Language Processing of the AFNLP: Volume 2, pp. 1030–1038. Association for Computational Linguistics (2009)
17. Ling, W., Luís, T., Marujo, L., Astudillo, R.F., Amir, S., Dyer, C., Black, A.W., Trancoso, I.: Finding Function in Form: Compositional Character Models for Open Vocabulary Word Representation. CoRR abs/1508.02096 (2015). http://arXiv.org/abs/1508.02096
18. Luo, G., Huang, X., Lin, C.Y., Nie, Z.: Joint named entity recognition and disambiguation. In: Proceedings of the 2015 Conference on Empirical Methods in Natural Language Processing, pp. 879–888. ACL (2015)
19. Mikolov, T., Sutskever, I., Chen, K., Corrado, G.S., Dean, J.: Distributed representations of words and phrases and their compositionality. In: Advances in Neural Information Processing Systems, vol. 26, pp. 3111–3119. Curran Associates, Inc. (2013)
20. Ratinov, L., Roth, D.: Design challenges and misconceptions in named entity recognition. In: CoNLL 2009: Proceedings of the Thirteenth Conference on Computational Natural Language Learning, pp. 147–155. ACL (2009)

21. Santos, C.D., Zadrozny, B.: Learning character-level representations for part-of-speech tagging. In: Proceedings of the 31st International Conference on Machine Learning, pp. 1818–1826. JMLR Workshop and Conference Proceedings (2014)
22. Srivastava, N., Hinton, G., Krizhevsky, A., Sutskever, I., Salakhutdinov, R.: Dropout: a simple way to prevent neural networks from overfitting. J. Mach. Learn. Res. **15**, 1929–1958 (2014)
23. Straková, J., Straka, M., Hajič, J.: A new state-of-the-art Czech named entity recognizer. In: Habernal, I. (ed.) TSD 2013. LNCS, vol. 8082, pp. 68–75. Springer, Heidelberg (2013). http://dx.doi.org/10.1007/978-3-642-40585-3_10
24. Straková, J., Straka, M., Hajič, J.: Open-source tools for morphology, lemmatization, POS tagging and named entity recognition. In: Proceedings of 52nd Annual Meeting of the Association for Computational Linguistics: System Demonstrations, pp. 13–18. ACL, Baltimore. http://www.aclweb.org/anthology/P/P14/P14-5003.pdf
25. Tjong Kim Sang, E.F., De Meulder, F.: Introduction to the CoNLL-2003 shared task: language-independent named entity recognition. In: Proceedings of CoNLL-2003, pp. 142–147, Edmonton, Canada (2003)
26. Ševčíková, M., Žabokrtský, Z., Krůza, O.: Named entities in Czech: annotating data and developing NE tagger. In: Matoušek, V., Mautner, P. (eds.) TSD 2007. LNCS (LNAI), vol. 4629, pp. 188–195. Springer, Heidelberg (2007)

SubGram: Extending Skip-Gram Word Representation with Substrings

Tom Kocmi[(⊠)] and Ondřej Bojar

Faculty of Mathematics and Physics, Institute of Formal and Applied Linguistics,
Charles University in Prague, Prague, Czech Republic
{kocmi,bojar}@ufal.mff.cuni.cz

Abstract. Skip-gram (word2vec) is a recent method for creating vector representations of words ("distributed word representations") using a neural network. The representation gained popularity in various areas of natural language processing, because it seems to capture syntactic and semantic information about words without any explicit supervision in this respect.

We propose SubGram, a refinement of the Skip-gram model to consider also the word structure during the training process, achieving large gains on the Skip-gram original test set.

Keywords: Distributed word representations · Unsupervised learning of morphological relations

1 Introduction

Vector representations of words learned using neural networks (NN) have proven helpful in many algorithms for image annotation [1] or [2], language modeling [3–5] or other natural language processing (NLP) tasks [6] or [7].

Traditionally, every input word of an NN is stored in the "one-hot" representation, where the vector has only one element set to one and the rest of the vector are zeros. The size of the vector equals to the size of the vocabulary. The NN is trained to perform some prediction, e.g. to predict surrounding words given a word of interest. Instead of using this prediction capacity in some task, the practice is to extract the output of NN's hidden layer of each word (called *distributed representation*) and directly use this deterministic mapping $vec(\cdot)$ of word forms to the vectors of real numbers as the word representation.

The input one-hot representation of words has two weaknesses: the bloat of the size of the vector with more words in vocabulary and the inability to provide any explicit semantic or syntactic information to the NN.

The learned distributed representation of words relies on much shorter vectors (e.g. vocabularies containing millions words are represented in vectors of a few hundred elements) and semantic or syntactic information is often found to be implicitly present ("embedded") in the vector space. For example, the Euclidean distance between two words in the vector space may be related to semantic or syntactic similarity between them.

© Springer International Publishing Switzerland 2016
P. Sojka et al. (Eds.): TSD 2016, LNAI 9924, pp. 182–189, 2016.
DOI: 10.1007/978-3-319-45510-5_21

1.1 Skip-Gram Model

The authors of [8] created a model called Skip-gram, in which linear vector operations allow to find related words with surprisingly good results. For instance $vec(king) - vec(man) + vec(woman)$ gives a value close to $vec(queen)$.

In this paper, we extend Skip-gram model with the internal word structure and show how it improves the performance on embedding morpho-syntactic information.

The Skip-gram model defined in [8] is trained to predict context words of the input word. Given a corpus T of words w and their context words $c \in C(w)$ (i.e. individual words c appearing close the original word w), it considers the conditional probabilities $p(c|w)$. The training finds the parameters θ of $p(c|w;\theta)$ to maximize the corpus probability:

$$\arg\max_{\theta} \prod_{w \in T} \prod_{c \in C(w)} p(c|w;\theta) \tag{1}$$

The Skip-gram model is a classic NN, where activation functions are removed and hierarchical soft-max [9] is used instead of soft-max normalization. The input representation is one-hot so the activation function is not needed on hidden layer, there is nothing to be summed up. This way, the model is learned much faster than comparable non-linear NNs and lends itself to linear vector operations possibly useful for finding semantically or syntactically related words.

2 Related Work

In [10] was proposed to append part-of-speech (POS) tags to each word and train Skip-gram model on the new vocabulary. This avoided conflating, e.g. nouns and verbs, leading to a better performance, at the cost of (1) the reliance on POS tags and their accurate estimation and (2) the increased sparsity of the data due to the larger vocabulary.

The authors in [11] used character-level input to train language models using a complex setup of NNs of several types. Their model was able to assign meaning to out-of-vocabulary words based on the closest neighbor. One disadvantage of the model is its need to run the computation on a GPU for a long time.

The authors of [12] proposed an extension of Skip-gram model which uses character similarity of words to improve performance on syntactic and semantic tasks. They are using a set of similar words as additional features for the NN. Various similarity measures are tested: Levenshtein, longest common substring, morpheme and syllable similarity.

The authors of [13] added the information about word's root, affixes, syllables, synonyms, antonyms and POS tags to continuous bag-of-words model (CBOW) proposed by [8] and showed how these types of knowledge lead to better word embeddings. The CBOW model is a simpler model with usually worse performance than Skip-gram.

3 SubGram

We propose a substring-oriented extension of Skip-gram model which induces vector embeddings from character-level structure of individual words. This approach gives the NN more information about the examined word with no drawbacks in data sparsity or reliance on explicit linguistic annotation.

We append the characters ^ and $ to the word to indicate its beginning and end. In order to generate the vector of substrings, we take all character bigrams, trigrams etc. up to the length of the word. This way, even the word itself is represented as one of the substrings. For the NN, each input word is then represented as a binary vector indicating which substrings appear in the word.

The original Skip-gram model [8] uses one-hot representation of a word in vocabulary as the input vector. This representation makes training fast because no summation or normalization is needed. The weights w_i of the input word i can be directly used as the output of hidden layer h (and as the distributed word representation): $h_j = w_i j$.

In our approach, we provide the network with a binary vector representing all substrings of the word. To compute the input of hidden layer we decided to use mean value as it is computationally simpler than sigmoid:

$$h_j = \frac{\sum_{i=1}^{|X|} x_i * w_{ij}}{|S|} \tag{2}$$

where $|S|$ is the number of substrings of the word x.

4 Evaluation and Data Sets

We train our NN on words and their contexts extracted from the English wikipedia dump from May 2015. We have cleaned the data by replacing all numbers with 0 and removing special characters except those usually present in the English text like dots, brackets, apostrophes etc. For the final training data we have randomly selected only 2.5M segments (mostly sentences). It consist of 96M running words with the vocabulary size of 1.09M distinct word forms.

We consider only the 141K most frequent word forms to simplify the training. The remaining word forms fall out of vocabulary (OOV), so the original Skip-gram cannot provide them with any vector representation. Our SubGram relies on known substrings and always provides at least some approximation.

We test our model on the original test set [8]. The test set consists of 19544 "questions", of which 8869 are called "semantic" and 10675 are called "syntactic" and further divided into 14 types, see Table 1. Each question contains two pairs of words (x_1, x_2, y_1, y_2) and captures relations like "What is to 'woman' (y_1) as 'king' (x_2) is to 'man' (x_1)?", together with the expected answer 'queen' (y_2). The model is evaluated by finding the word whose representation is the nearest (cosine similarity) to the vector $vec(king) - vec(man) + vec(woman)$. If the nearest neighbor is $vec(queen)$, we consider the question answered correctly.

Table 1. Mikolov's test set question types, the upper part are "semantic" questions, the lower part are "syntactic".

Question type	Sample pair
Capital-countries	Athens – Greece
Capital-world	Abuja – Nigeria
Currency	Algeria – dinar
City-in-state	Houston – Texas
Family	boy – girl
Adjective-to-adverb	calm – calmly
Opposite	aware – unaware
Comparative	bad – worse
Superlative	bad – worst
Present-participle	code – coding
Nationality-adjective	Albania – Albanian
Past-tense	dancing – danced
Plural	banana – bananas
Plural-verbs	decrease – decreases

In this work, we use Mikolov's test set which is used in many papers. After a closer examination we came to the conclusion, that it does not test what the broad terms "syntactic" and "semantic relations" suggest. "Semantics" is covered by questions of only 3 types: predict a city based on a country or state, currency name from the country and the feminine variant of nouns denoting family relations. The authors of [14] showed, that many other semantic relationships could be tested, e.g. walk-run, dog-puppy, bark-dog, cook-eat and others.

"Syntactic" questions cover a wider range of relations at the boundary of morphology and syntax. The problem is that all questions of a given type are constructed from just a few dozens of word pairs, comparing pairs with each other. Overall, there are 313 distinct pairs throughout the whole syntactic test set of 10675 questions, which means only around 35 different pairs per question set. Moreover, of the 313 pairs, 286 pairs are regularly formed (e.g. by adding the suffix 'ly' to change an adjective into the corresponding adverb). Though it has to be mentioned that original model could not use this kind of information.

We find such a small test set unreliable to answer the question whether the embedding captures semantic and syntactic properties of words.

4.1 Rule-Based Baseline Approach

Although the original test set has been used to compare results in several papers, no-one tried to process it with some baseline approach. Therefore, we created a very simple set of rules for comparison on the syntactic part of the test set. The rules cover only the most frequent grammatical phenomenona.

- adjective-to-adverb: Add *ly* at the end of the adjective.
- opposite: Add *un* at the beginning of positive form.
- comparative: If the adjective ends with *y*, replace it with *ier*. If it ends with *e*, add *r*. Otherwise add *er* at the end.
- superlative: If the adjective ends with *y*, replace it with *iest*. If it ends with *e*, add *st*. Otherwise add *est* at the end.
- present-participle: If the verb ends with *e*, replace it with *ing*, otherwise add *ing* at the end.
- nationality-adjective: Add *n* at the end, e.g. *Russia → Russian*.
- past-tense: Remove *ing* and add *ed* at the end of the verb.
- plural: Add *s* at the end of the word.
- plural-verbs: If the word ends with a vowel, add *es* at the end, else add *s*.

4.2 Our Test Set

We have decided to create more general test set which would consider more than 35 pairs per question set. Since we are interested in morphosyntactic relations, we extended only the questions of the "syntactic" type with exception of nationality adjectives which is already covered completely in original test set.

We constructed the pairs more or less manually, taking inspiration in the Czech side of the CzEng corpus [15], where explicit morphological annotation allows to identify various pairs of Czech words (different grades of adjectives, words and their negations, etc.). The word-aligned English words often shared the same properties. Another sources of pairs were acquired from various webpages usually written for learners of English. For example for verb tense, we relied on a freely available list of English verbs and their morphological variations. We have included 100–1000 different pairs for each question set. The questions were constructed from the pairs similarly as by Mikolov: generating all possible pairs of pairs. This leads to millions of questions, so we randomly selected 1000 instances per question set, to keep the test set in the same order of magnitude. Additionally, we decided to extend set of questions on opposites to cover not only opposites of adjectives but also of nouns and verbs.

In order to test our extension of Skip-gram on out-of-vocabulary words, we created an additional subset of our test set with questions where at least one of x_1, x_2 and y_1 is not among the known word forms. Note that the last word y_2 must be in vocabulary in order to check if the output vector is correct.

5   Experiments and Results

We used a Python implementation of word2vec[1] as the basis for our SubGram, which we have made freely available[2].

---

[1] http://radimrehurek.com/gensim

Gensim implements the model twice, in Python and an optimized version in C. For our prototype, we opted to modify the Python version, which unfortunately resulted in a code about 100 times slower and forced us to train the model only on the 96M word corpus as opposed to Mikolov's 100,000M word2vec training data used in training of the released model.

[2] https://github.com/tomkocmi/SubGram.

We limit the vocabulary, requiring each word form to appear at least 10 times in the corpus and each substring to appear at least 500 times in the corpus. This way, we get the 141K unique words mentioned above and 170K unique substrings (+141K words, as we downsample words separately).

Our word vectors have the size of 100. The size of the context window is 5.

The accuracy is computed as the number of correctly answered questions divided by the total number of questions in the set. Because the Skip-gram cannot answer questions containing OOV words, we also provide results with such questions excluded from the test set (scores in brackets).

Tables 2 and 3 report the results. The first column shows the rule-based approach. The column "Released Skip-gram" shows results of the model released by Mikolov[3] and was trained on a 100 billion word corpus from Google News and generates 300 dimensional vector representation. The third column shows Skip-gram model trained on our training data, the same data as used for the training of the SubGram. Last column shows the results obtained from our SubGram model.

Comparing Skip-gram and SubGram on the original test set (Table 2), we see that our SubGram outperforms Skip-gram in several morpho-syntactic question sets but over all performs similarly (42.5 % vs. 42.3 %). On the other hand, it does not capture the tested semantic relations at all, getting a zero score on average.

Table 2. Results on original test set questions. The values in brackets are based on questions without any OOVs.

	Rule based	Released skip-gram	Our skip-gram	SubGram
Capital-countries	0 %	18.6 % (24.7 %)	71.9 % (71.9 %)	0 % (0 %)
Capital-world	0 %	2.2 % (15.0 %)	53.6 % (54.6 %)	0 % (0 %)
Currency	0 %	7 % (12.2 %)	3 % (4.7 %)	0.1 % (0.2 %)
City-in-state	0 %	9.2 % (14 %)	40.5 % (40.5 %)	0.1 % (0.1 %)
Family	0 %	84.6 % (84.6 %)	82.6 % (82.6 %)	5.9 % (5.9 %)
Overall semantic	0 %	10.2 % (24.8 %)	47.7 % (50 %)	0 % (0 %)
Adjective-to-adverb	90.6 %	28.5 % (28.5 %)	16.3 % (16.3 %)	73.7 % (73.7 %)
Opposite	65.5 %	42.7 % (42.7 %)	9.4 % (10.1 %)	43.1 % (46.3 %)
Comparative	89.2 %	90.8 % (90.8 %)	72.1 % (72.1 %)	46.5 % (46.5 %)
Superlative	88.2 %	87.3 % (87.3 %)	24.4 % (25.9 %)	45.9 % (48.8 %)
Present-participle	87.9 %	78.1 % (78.1 %)	44.2 % (44.2 %)	43.5 % (43.5 %)
Nationality-adjective	31.7 %	13.3 % (21.9 %)	60.4 % (60.4 %)	21.8 % (21.8 %)
Past-tense	42.5 %	66 % (66 %)	35.6 % (35.6 %)	15.8 % (15.8 %)
Plural	86.5 %	89.9 % (89.9 %)	46.8 % (46.8 %)	44.7 % (44.7 %)
Plural-verbs	93.3 %	67.9 % (67.9 %)	51.5 % (51.5 %)	74.3 % (74.3 %)
Overall syntactic	71.9 %	62.5 % (66.5 %)	42.5 % (43 %)	42.3 % (42.7 %)

---

[3] https://code.google.com/archive/p/word2vec/.

Table 3. Results on our test set questions.

Type	Rule based	Released skip-gram	Our skip-gram	SubGram	OOV
	Our test set				
Adjective-to-adverb	68.4 %	18.8 % (20.9 %)	1.9 % (3.7 %)	32.3 % (62.7 %)	2.3 %
Opposite	3.7 %	6.3 % (6.4 %)	5.3 % (5.6 %)	0.6 % (0.6 %)	0.7 %
Comparative	90.2 %	67.1 % (68.9 %)	12 % (31.5 %)	14.4 % (37.8 %)	0 %
Superlative	92.5 %	57 % (59.9 %)	4.4 % (16.1 %)	12.2 % (44.7 %)	0.5 %
Present-participle	88.7 %	50.2 % (53 %)	12.8 % (16.2 %)	37.3 % (47.3 %)	4.8 %
Past-tense	75 %	53.5 % (56.5 %)	17.1 % (22.8 %)	24.2 % (32.3 %)	0.5 %
Plural	26.8 %	39.1 % (42.1 %)	8.9 % (13.8 %)	13.6 % (21.1 %)	1.7 %
Plural-verbs	85.8 %	56.1 % (59 %)	15.4 % (20.3 %)	44.3 % (58.5 %)	2 %
Overall syntactic	66.4 %	43.5 % (45.9 %)	9.7 % (15.4 %)	22.4 % (35.4 %)	1.6 %

When comparing models on our test set (Table 3), we see that given the same training set, SubGram significantly outperforms Skip-gram model (22.4 % vs. 9.7 %). The performance of Skip-gram trained on the much larger dataset is higher (43.5 %) and it would be interesting to see the SubGram model, if we could get access to such training data. Note however, that the Rule-based baseline is significantly better on both test sets.

The last column suggests that the performance of our model on OOV words is not very high, but it is still an improvement over flat zero of the Skip-gram model. The performance on OOVs is expected to be lower, since the model has no knowledge of exceptions and can only benefit from regularities in substrings.

6   Future Work

We are working on a better test set for word embeddings which would include many more relations over a larger vocabulary especially semantics relations. We want to extend the test set with Czech and perhaps other languages, to see what word embeddings can bring to languages morphologically richer than English.

As shown in the results, the rule based approach outperform NN approach on this type of task, therefore we would like to create a hybrid system which could use rules and part-of-speech tags. We will also include morphological tags in the model as proposed in [10] but without making the data sparse.

Finally, we plan to reimplement SubGram to scale up to larger training data.

7   Conclusion

We described SubGram, an extension of the Skip-gram model that considers also substrings of input words. The learned embeddings then better capture almost all morpho-syntactic relations tested on test set which we extended from original described in [8]. This test set is released for the public use[4].

---

[4] https://ufal.mff.cuni.cz/tom-kocmi/syntactic-questions.

An useful feature of our model is the ability to generate vector embeddings even for unseen words. This could be exploited by NNs also in different tasks.

Acknowledgment. This work has received funding from the European Union's Horizon 2020 research and innovation programme under grant agreement no. 645452 (QT21), the grant GAUK 8502/2016, and SVV project number 260 333.

This work has been using language resources developed, stored and distributed by the LINDAT/CLARIN project of the Ministry of Education, Youth and Sports of the Czech Republic (project LM2015071).

References

1. Lazaridou, A., Pham, N.T., Baroni, M.: Combining language and vision with a multimodal skip-gram model (2015). arXiv preprint arXiv:1501.02598
2. Weston, J., Bengio, S., Usunier, N.: Wsabie: scaling up to large vocabulary image annotation. In: IJCAI, vol. 11 (2011)
3. Schwenk, H., Gauvain, J.L.: Neural network language models for conversational speech recognition. In: INTERSPEECH (2004)
4. Schwenk, H., Dchelotte, D., Gauvain, J.L.: Continuous space language models for statistical machine translation. In: Proceedings of the COLING/ACL on Main Conference Poster Sessions (2006)
5. Mnih, A., Hinton, G.: Three new graphical models for statistical language modelling. In: Proceedings of the 24th International Conference on Machine Learning (2007)
6. Soricut, R., Och, F.: Unsupervised morphology induction using word embeddings. In: Proceedings of NAACL (2015)
7. Wang, Z., Zhang, J., Feng, J., Chen, Z.: Knowledge graph and text jointly embedding. In: Proceedings of the 2014 Conference on Empirical Methods in Natural Language Processing (EMNLP). Association for Computational Linguistics (2014)
8. Mikolov, T., Chen, K., Corrado, G., Dean., J.: Efficient estimation of word representations in vector space (2013). arXiv preprint arXiv:1301.3781
9. Morin, F., Bengio, Y.: Hierarchical probabilistic neural network language model. In: Proceedings of the International Workshop on AI and Statistics (2005)
10. Lin, Q., Cao, Y., Nie, Z., Rui, Y.: Learning word representation considering proximity and ambiguity. In: Twenty-Eighth AAAI Conference on Artificial Intelligence (2014)
11. Yoon, K., Jernite, Y., Sontag, D., Rush, A.M.: Character-aware neural language models (2015). arXiv preprint arXiv:1508.06615
12. Cui, Q., Gao, B., Bian, J., Qiu, S., Liu, T.Y.: A framework for learning knowledge-powered word embedding (2014)
13. Bian, J., Gao, B., Liu, T.Y.: Knowledge-powered deep learning for word embedding. In: Machine Learning and Knowledge Discovery in Databases (2014)
14. Vylomova, E., Rimmel, L., Cohn, T., Baldwin, T.: Take and took, gaggle and goose, book and read: evaluating the utility of vector differences for lexical relation learning (2015). arXiv preprint arXiv:1509.01692
15. Bojar, O., Dušek, O., Kocmi, T., Libovický, J., Novák, M., Popel, M., Sudarikov, R., Variš, D.: Czeng 1.6: enlarged Czech-English parallel corpus with processing tools dockered. In: Sojka, P., et al. (eds.) TSD 2016. LNAI, vol. 9924, pp. 231–238. Springer International Publishing, Heidelberg (2016)

WordSim353 for Czech

Silvie Cinková[(✉)]

Faculty of Mathematics and Physics, Institute of Formal and Applied Linguistics,
Charles University, Malostranské náměstí 25, Praha 1, Prague, Czech Republic
cinkova@ufal.mff.cuni.cz
http://ufal.mff.cuni.cz

Abstract. Human judgments of lexical similarity/relatedness are used
as evaluation data for Vector Space Models, helping to judge how the distributional similarity captured by a given Vector Space Model correlates
with human intuitions. A well established data set for the evaluation of
lexical similarity/relatedness is WordSim353, along with its translations
into several other languages. This paper presents its Czech translation
and annotation, which is publicly available via the LINDAT-CLARIN
repository at hdl.handle.net/11234/1-1713.

Keywords: Word similarity · Word relatedness · Czech · WordSim353 ·
Language resource

1 Introduction

1.1 Human Judgments of Lexical Similarity/Relatedness in NLP

Recent years have seen a substantial interest in vector space modeling applied to
lexical similarity or lexical relatedness[1], also in multilingual terms. A number of
human judgment datasets have been created to evaluate the available semantic
metrics to make sure that the metrics would simulate the human lexical simi-
arity/relatedness reasoning. One of the first and best-known datasets of this
kind is WordSim353 [3]. It has also been translated into several other languages:
Arabic, Spanish, Romanian [2], and most recently also to Russian, Italian, and
German [1].

To the best of our knowledge, there has not been any WordSim353 translation
to Czech so far – therefore we have created one. In the following sections, we
will describe the resource as well as the methodology we used, along with the
results.

I thank Jan Hajič and Jana Straková for allowing me to use the Czech translations
of WordSim353 they had gathered earlier.

[1] See Sect. 1.3 for a terminological clarification.

© Springer International Publishing Switzerland 2016
P. Sojka et al. (Eds.): TSD 2016, LNAI 9924, pp. 190–197, 2016.
DOI: 10.1007/978-3-319-45510-5_22

1.2   The Original WordSim353

The original WordSim533 was created as an evaluation dataset for context-based keyword search [3]. It consists of 353 noun pairs, some of which are multiword expressions. It includes an older lexical resource – Miller and Charles' word list [4] of 30 noun pairs. The noun pairs represent diverse degrees of lexical similarity - from totally unrelated concepts to frequently co-occurring words and synonyms or antonyms in both a narrow and a broad sense. The dataset is partially related to the English WordNet. (The authors note that 82 word pairs contain at least one word not captured by WordNet.)

WordSim353 contains human similarity judgments. Human annotators were to rank the "similarity"[2] between the words in a pair, using a continuous scale from 0 ("completely unrelated") to 10 ("identical or extremely similar"). 153 word pairs were annotated by 13 annotators, the rest by 16 annotators[3].

1.3   Similarity vs. Relatedness

The semantic relations between the paired words in the WordSim353 data set are of different kinds:

1. synonyms and broader synonyms *(midday-noon, asylum-madhouse, money-cash, football-soccer)*;
2. antonyms *(smart-stupid)*;
3. hyponym + hyperonym *(bird-cock)*;
4. words frequently co-occurring in the same domain *(doctor-nurse, arrival-hotel)*;
5. parts of a multiword expression *(soap-opera)*;
6. unrelated words *(professor-cucumber)*.

Agirre et al. [6] appointed human annotators to classify the word pairs according to the semantic relation between their members, achieving a high interannotator agreement. As a result of this manual classification, WordSim353 was divided into two subsets with pairs containing mutually **similar** words and pairs containing mutually **related** words, respectively. The similarity group contained words paired by synonymy, antonymy, and hypo/hyperonymy, whereas the relatedness group contained words with relations numbered 4–6 in our list above; i.e. including words with low scores given by the original WordSim353 annotators. Agirre et al. demonstrated on this data that the similarity-relatedness distinction plays an important role in distributional semantic modeling, as the modeling of each requires different methods.

---

[2] Although the relation between the pair members is often referred to as "relatedness", which we also find more appropriate, the annotator instructions consistently use the word "similarity". See also Sect. 1.3.

[3] See http://www.cs.technion.ac.il/~gabr/resources/data/wordsim353/.

1.4 Multilingual WordSim353

A multilingual release of WordSim353 [2] comprises Arabic, Spanish, and Romanian. This dataset was created to evaluate vector-space representations of these languages obtained by extracting information from the interlanguage links between concept definitions in Wikipedia by means of Explicit Semantic Analysis.

The translations were obtained from native speakers of the respective languages with high proficiency in English. The translators could see the English similarity scores for each word pair and were familiar with the similarity rating task. They were instructed to provide equivalent pairs that would possibly obtain the same similarity scores as their English originals in the original task.

For Spanish, the pilot language, five independent translations were gathered, which were merged into a single selection by an adjudicator. This Spanish list of word pairs was then rescored by five independent human annotators using the same scale as the English WordSim353. The average scores for English and Spanish reached a high degree of correlation (0.86), which indicates that the translations have closely followed the original relatedness degree. Also the agreement among the translators was high (at least three annotators agreed on the same translation for a word pair). Since the pilot English-to-Spanish translation was successful, the Arabic and Romanian translations were only provided by a single translator.

1.5 Judgment Language Effect

Most recently, Leviant and Reichart [1] published a study on the effect of the judgment language on the human scoring decisions as well as on the performance of Vector Space Models, for which they built a new lexical resource – a multiligual translation of WordSim353 and SimLex999 [5]. To ensure that all their translations as well as scoring decisions followed the same policy, they rescored the English data and retranslated language versions that had already been translated for earlier lexical resources of this type. To investigate the effect of the judgment language on human decisions on lexical relatedness/similarity, Leviant and Reichart compared the interannotator agreement within each judgment language and across the judgment languages. The observed interannotator agreement was higher between than across the judgment languages, which suggests an effect of the judgment language on the human relatedness/similarity scoring. To assess the judgment language effect on the Vector Space Models, Leviant and Reichart trained two models on identical parts of the multilingual data for each language separately. Also the test data comprised identical terms across the language versions. The performance of the Vector Space Models also turned out to be language-specific. Besides, the models were also trained on the entire multilingual data. The performance of the multilingual Vector Space Models was higher than that of the monolingual ones. Leviant and Reichart draw the conclusion that both humans and Vector Space Models are affected by the judgment language, although each in a different way. The judgment-language effect appears to decrease the performance of the Vector Space Models and can be partially alleviated by multilingual setup.

Their multilingual data set comprises Italian, German and Russian as representatives of three different Indo-European language families. Two native speakers were in charge of the translation into each language. The translators were given several additional instructions beforehand to keep the translation strategy as consistent as possible. The inter-translator agreement for WordSim was 87.9 % for Russian, 91.9 % in Italian and 83.9 % in German (calculated on single words, not the entire pairs).

The translation instructions concerned the following aspects:

1. gender disambiguation (e.g. all the languages have gender-specific expressions for *cat*),
2. sense proximity (in homonymous and polysemous words, use the other word pair member to select the sense that is semantically closest to it. When the other pair member provided no hint, the translation manager selected one meaning randomly, and this interpretation was used across all three languages.)
3. POS proximity (when the part of speech of the English original was ambiguous, the translators were supposed to use an equivalent whose part of speech conforms to the other pair member).

2 The Czech WordSim353

2.1 Translation Instructions

The Czech translations were collected from four independent native speakers with high proficiency in English, largely following the translation instructions in [2]. This mainly means that i.e. multi-word expressions were not allowed and ambiguity was to be preserved whenever possible. The translation was performed long before [1] was published, hence any similarities in their translation instructions are coincidental.

The Czech translators were shown the English mean scores of each original word pair and were asked to provide a translation as close as possible in terms of the lexical similarity score: for scores between 5 and 10 the preferred equivalents should be synonyms[4] or be as closely semantically associated as possible, whereas the equivalents with low scores should make the semantic distance of the pair members as obvious as possible. On the other hand, disambiguation was not required. On the contrary, the ideal equivalent was supposed to preserve the ambiguity (knowing this was rarely possible). As Czech is rich in (almost synonymous) derivatives, the translators were also asked to use the same translation equivalent of a given word throughout the dataset. In some cases where only one of the derivatives was possible (preserving the semantic distance/proximity) in

---

[4] Cf. [6] and Sect. 1.3.

one pair, while making no difference in another pair, the equivalent from the more restricted pair was preferred.

An external adjudicator preliminarily merged the translations into one dataset. However, we decided to let the annotators score all sensible translation equivalents, eliminating only typos and clear mistranslations. We indexed each Czech translation equivalent with the corresponding English original.

3   Scoring Instructions

The Czech scoring instructions were obtained by translation from the English original WSim353 instructions. The document differed only by offering the respondents three different file formats for their convenience (.xls, .xlsx, and .csv). All respondents were individually contacted and their responses collected by e-mail. They were randomly chosen with respect to gender and age. The group is likely to be biased towards people with higher education and interest in language(s), but we prohibited participation to linguists (both scholars and computer scientists). The scorers did not know that the word-pair collection had an English counterpart, and naturally they were neither shown the original English word pairs nor the original English scores.

The online description of WordSim353[5] uses the word "relatedness" to describe what the annotators were rating. Nevertheless, the `instruction.txt` file containing the original annotator instructions consistently refers to "similarity", and therefore the Czech instructions also used "similarity", although "relatedness" seemed more appropriate with respect to the fact that "similarity" evokes "synonymy".

Five of the 25 Czech annotators even reacted to the word "similarity" in the instructions as soon as they had started their annotation. They suggested "relatedness" instead and asked for confirmation that "related" words should also get high ratings. When directly asked, we told them they were right but had to decide the scores according to their own intuition. We were always giving only individual answers; that is, we never intervened in the rest of the group. A few annotators complained after they had submitted their work that "there were many more similar words that were not synonyms" and hence the instructions were misleading. However, all annotators were eventually rating what they perceived as "relatedness" beyond "similarity" in the strict sense of synonymy and achieved a high pairwise correlation.

As we wanted to prevent the annotators from creating anchors based on too many close pairs that have arisen due to the translation variants, we displayed the word pairs in random order. According to the feedback delivered after submitting their responses, the annotators had not recognized that some selected words were different translations of the same English original. Pairs identified as similar to some already seen were believed to be "bogus items" to detect careless responding.

---

[5] http://www.cs.technion.ac.il/~gabr/resources/data/wordsim353/.

4   Format

Unlike [2], we had our annotators score all translation variants of each original English word pair. This complete annotation comes as WordSim-cs-Multi, corresponding to a file named `WordSim-cs-Multi.csv`. It comprises 633 Czech word pairs (mapped on the original 353 English word pairs). From this dataset we have made a selection of Czech equivalents (WordSim353-cs, file `WordSim353-cs.csv`), whose judgments were most similar to the judgments of the English originals, compared by the absolute value of the difference between the means over all annotators in each language counterpart. In one case (*psychology-cognition*), two Czech equivalent pairs had identical means as well as confidence intervals, so we randomly selected one. In both data sets, we preserved all 25 individual scores. In the WordSim353-cs data set, we added a column with their Czech means as well as a column containing the original English means and 95 % confidence for each mean (computed by the CI function in the Rmisc R package)[6]. The WordSim-cs-Multi data set contains only the Czech means and confidence intervals. For the most convenient lexical search, we provided separate columns with the respective Czech and English single words, entire word pairs, and eventually an English-Czech quadruple in both data sets. The lexical resource also contains an .xls file with the four translation variants and adjudication before the annotation. The entire resource including the annotation and translation instructions (in Czech) is publicly available via the LINDAT-CLARIN repository at hdl.handle.net/11234/1-1713.

5   Evaluation

5.1   Correlation Between WordSim353 and WordSim353-Cs

As we were able to optimize the selection of the Czech equivalent pairs by minimizing the difference between the Czech and English mean values for each pair, the correlation between these samples does not say much about actual differences between judgment languages with respect to lexical similarity/relatedness. Nevertheless – since correlation was reported for the earlier translated WordSim-based sets, we will use the same evaluation measure: the correlation between WordSim353 and WordSim353-cs lies at 0.9 (almost identical for both Spearman's ρ and Pearson's product-moment correlation, with the confidence interval of Pearson's product-moment correlation at 0.895–0.930 and with p-value $< 2.2e - 16$).

We have observed a rather symmetric distribution of the differences between the corresponding means for each language with the mean at approx. -0.1 (Czech minus English) and the standard deviation at approx. 0.9.

---

[6] [7] and [8].

5.2 Intertranslator and Adjudicator-Translator Agreement

We have analyzed the translations as well as the decisions of the adjudicator with Fleiss kappa. The intertranslator agreement on single words was 0.785. In 37.8 % of the words (267 of 706), at least 3 annotators agreed, of which 244 reached full agreement. Pairwise, the annotators had similar agreement (0.622–0.676); there was no outlier. According to the pairwise agreement observations of the adjudicator with each translator, the adjudicator did neither seem to strongly prefer nor disprefer any translator (0.719–0.787). A manual analysis of intertranslator disagreements revealed several distinct types:

- orthographical variants, e.g., *chléb-chleba (bread)*, *maraton-maratón (marathon)*, *brambora-brambor (potato)*
- typing errors or omited translations (rare), e.g. *arichtektura ("arichtecture")*, *projekt-project, hudba-music*
- synonyms in the narrow sense (rare), e.g. *šukat-šoustat (fuck)*, *šálek-hrnek (cup-mug)*
- synonyms - stylistic variation (formal/less formal), e.g. *chlapec-hoch (boy)*, *auto-automobil (car)*, *doktor-lékař (physician)*
- original Czech words vs. loan words not necessarily belonging to different registers, e.g. *teritorium-území(territory)*, *menšina-minorita (minority)*, *třída-avenue (avenue)*, *katastrofa-pohroma (disaster)*
- POS - disagreement (rare), e.g. *mýdlo-mýdlová* (*soap* noun vs. derived adjective), *pít-pití* (*drink* verb vs. noun, meaning either the event of drinking or a drink), *akcie-akciový* (*share/stock* noun vs. derived adjective)
- synonyms in a broader sense, derivation difference, e.g. *vývoj-rozvoj (evolution-development)*, *vejce-vajíčko*(*egg* vs. the diminutive form of *egg*, with partially different connotations)
- disambiguation guesses (*closet* in the pair *closet-clothes*: *wardrobe* vs. *lavatory*).

Broader synonyms sometimes differ in concept granularity and in register at the same time (e.g. *dítě-batole-miminko* for *baby*. More precisely: *child-toddler-baby*). The most interesting part of the disagreements are certainly the disambiguation guesses as well as failed or successful attempts to preserve the original ambiguity. The instructions regarding ambiguity preservation, potential similarity score matching, and consistency across the entire data set were sometimes conflicting, e.g. the first two in *alkohol-líh (alcohol-spirit)*, which occurred both in the chemical domain as well as together with alcoholic beverages, and, presumably, the association of *alkohol (alcohol)* with different types of alcoholic beverages is stronger than that of *líh* (*spirit* but not *spirits*)[7], although, technically speaking, it is the same substance, since *alkohol*, unlike *líh*, is also used as a shortcut for *alcoholic beverages*.

In a few cases, some translators did not notice (or at least not process) the fact that the pair members were parts of a multiword expression, such as

---

[7] A margin note: there is, though, a Czech derivative of *líh* equivalent to *spirits*, also typically used in plural: *lihoviny*, which the translators did not consider.

cell in *cell phone*. The context-independent translation *buňka (cell)* evokes no association with *telefon (phone)*, while the American English association between *cell* and *phone* is naturally tight.

In some cases, however, the English concept was so blurred, that the translation decision was necessarily random and in 8 cases led to total disagreement; e.g. *pojištění (insurance) - riziko (risk) - náchylnost (predisposition, proneness) - ručení (guarantee)* (as translations of *liability*), *zaměření (focus, specialization) - pozornost (attention) - zaostření (focus, visual accommodation, zoom) - smysl (sense, meaning)* (as translations of *focus*), *sklad (storage) - zásobit (provide, supply) - zásoba (supplies) - dobytek (livestock, cattle)* as translations of *stock*).

6   Conclusion

We have created a multiple-equivalent Czech translation and annotation of WordSim353, one of the standard datasets for evauations of semantic vector space representations. The resource is publicly available at hdl.handle.net/11 234/1-1713 via the LINDAT-CLARIN repository.

Acknowledgments. This project was supported by the Czech Science Foundation grant GA-15-20031S and has been using language resources developed and/or stored and/or distributed by the LINDAT/CLARIN project of the Ministry of Education, Youth, and Sports of the Czech Republic (project LM2015071).

References

1. Leviant, I., Reichart, R.: Separated by an Un-common Language: Towards Judgment Language Informed Vector Space Modeling. arXiv:508.00106v5 [cs.CL], 6 December 2015
2. Hassan, S., Mihalcea, R.: Cross-lingual semantic relatedness using encyclopedic knowledge. In: Proceedings of the Conference on Empirical Methods in Natural Language Processing, Singapore (2009)
3. Finkelstein, L., Gabrilovich, E., Matias, Y., Rivlin, E., Solan, Z., Wolfman, G., Ruppin, E.: Placing search in context: the concept revisited. ACM Trans. Inf. Syst. **20**(1), 116–131 (2002)
4. Miller, G.A., Charles, W.G.: Contextual correlates of semantic similarity. Lang. Cogn. Process. **6**(1), 1–28 (1991)
5. Hill, F., Reichart, R., Korhonen, A.: Simlex-999: evaluating semantic models with (genuine) similarity estimation. [cs.CL]. arXiv:1408.3456
6. Agirre, E., Alfonseca, E., Hall, K., Kravalova, J., Pasca, M., Soroa, A.: A study on similarity and relatedness using distributional and wordnet-based approaches. In: Proceedings of NAACL-HLT 2009 (2011)
7. Hope, R.M.: Rmisc: Ryan Miscellaneous. R package version 1.5. https://CRAN.R-project.org/package=Rmisc
8. R Core Team: A language and environment for statistical computing. R Foundation for Statistical Computing, Vienna, Austria. https://www.R-project.org/

Automatic Restoration of Diacritics for Igbo Language

Ignatius Ezeani[✉], Mark Hepple, and Ikechukwu Onyenwe

NLP Group, Department of Computer Science,
The University of Sheffield, Sheffield, UK
ignatius.ezeani@sheffield.ac.uk

Abstract. Igbo is a low-resource African language with orthographic and tonal diacritics, which capture distinctions between words that are important for both meaning and pronunciation, and hence of potential value for a range of language processing tasks. Such diacritics, however, are often largely absent from the electronic texts we might want to process, or assemble into corpora, and so the need arises for effective methods for automatic diacritic restoration for Igbo. In this paper, we experiment using an Igbo bible corpus, which is extensively marked for vowel distinctions, and partially for tonal distinctions, and attempt the task of reinstating these diacritics when they have been deleted. We investigate a number of word-level diacritic restoration methods, based on n-grams, under a closed-world assumption, achieving an accuracy of 98.83 % with our most effective method.

Keywords: Diacritic restoration · Sense disambiguation · Low resourced languages · Igbo language

1 Introduction

Diacritics are simply defined as marks placed over, under, or through a letter in some languages to indicate a different sound value from the same letter without the diacritics[1]. The word "diacritics" was derived from the Greek word *diakritikós*, meaning "distinguishing".

Although English does not have diacritics (apart from some few borrowed words), many of the worlds language groups (Germanic, Celtic, Romance, Slavic, Baltic, Finno-Ugric, Turkic etc.), as well as many African languages, use a wide range of diacritized letters in their orthography.

Automatic Diacritic Restoration Systems (ADRS) are tools that enable the restoration of missing diacritics in texts. Many forms of such tools have been proposed, designed and developed. Some ADRS restore diacritics on existing texts while others insert appropriate diacritics "on-the-fly" during text creation [9] but not much has been done for Igbo language.

---

[1] http://www.merriam-webster.com/dictionary/diacritic.

© Springer International Publishing Switzerland 2016
P. Sojka et al. (Eds.): TSD 2016, LNAI 9924, pp. 198–205, 2016.
DOI: 10.1007/978-3-319-45510-5_23

1.1 Igbo Writing System and Diacritics

Igbo, one of the three major Nigerian languages and the primary native language of the Igbo people of southeastern Nigeria, is spoken by over 30 million people mostly resident in Nigeria and are of predominantly Igbo descent. It is written with the Latin scripts and has many dialects. Most written works, however, use the official orthography produced by the Ọnwụ Committee[2].

The orthography has 8 vowels (*a, i, o, u, ị, ọ, ụ*) and 28 consonants (*b, gb, ch, d, f, g, gw, gh, h, j, k, kw, kp, l, m, n, nw, ny, ṅ, p, r, s, sh, t, v, w, y, z*). Some researchers, however, consider the Ọnwụ orthography inadequate because of the inability to represent many dialectal sounds with it [1].

In Table 1, the Igbo letters with orthographic or tonal (or both) diacritics are presented with their diacritic forms and some examples of how they can change the meanings of the words they appear in[3].

Table 1. Diacritics in Igbo language

Char	Ortho	Tonal	Examples
a	–	à,á, ā	*ákwà*(cloth), *àkwà*(bed/bridge), *ákwá*(cry), *àkwá*(egg)
e	–	è,é, ē	*égbè*(gun), *égbé*(kite),
i	ị	ì, í, ī, ì, í, ī	*ísí*(head), *ísì*(smell), *ísī*(to cook), *ísī*(to say)
o	ọ	ò, ó, ō, ọ̀, ọ́, ọ̄	*ólù*(neck), *ọ́lù*(work); *ódō*(pestle), *ọ̀dọ́*(pool)
u	ụ	ù, ú, ū, ụ̀, ụ́, ụ̄	*égwú*(dance/song), *égwù*(fear)
m	–	m̀,ḿ, m̄,	*ḿmádù*(a person), *m̀bèrèdé* (accident)
n	ṅ	ǹ,ń, n̄,	*ńdụ̀*(life), *ǹdò* (shelter)

2 Problem Definition

Lack of diacritics can often lead to some semantic ambiguity in written Igbo sentences. Although a human reader can, in most cases, infer the intended meaning from context, the machine may not. Consider the following statements and their literal translations:

Missing Orthographic Diacritics

1. *Nwanyi ahu banyere n'**ugbo** ya.* (The woman entered her [**farm**|**boat/craft**])
2. *O kwuru banyere **olu** ya.* (He/she talked about his/her [**neck/voice**|**work/job**])

---

[2] http://www.columbia.edu/itc/mealac/pritchett/00fwp/igbo/txt_onwu_1961.pdf.

[3] Observe that *m* and *n*, nasal consonants, are sometimes treated as tone marked vowels.

Missing Tonal Diacritics

1. *Nwoke ahu nwere* ***egbe*** *n'ulo ya.* (That man has a [**gun|kite**] in his house)
2. *O dina n'elu* ***akwa.*** (He/she is lying on the [**cloth|bed,bridge|egg|cry**])
3. ***Egwu*** *ji ya aka.* (He/she is held/gripped by [**fear|song/dance/music**])

As seen above, ambiguities may arise when diacritics – orthographic or tonal – are omitted in Igbo texts. In the first examples, **ugbo**(*farm*) and **ụgbọ**(*boat/craft*) as well as **olu**(*neck/voice*)and **ọlụ**(*work/job*) were candidates in their sentences.

Also, in the second examples, **égbé**(*kite*) and **égbè**(gun); **ákwà**(cloth), **àkwà** (bed or bridge), **àkwá**(egg) (or even **ákwá**(cry) in a philosophical or artistic sense); as well as **égwù**(fear) and **égwú**(music) are all qualified to replace the ambiguous word in their respective sentences.

The examples above incidentally showed words that belong to the same class i.e. nouns. However, instances abound of keywords (i.e. non-diacritic variant of a word) that represent actual forms that span different classes. For example, in the first two sentences, **banyere** could mean *bànyèrè* (*entered*, a verb) or *bànyéré* (*about*, a preposition). A high-level description of the proposed system is presented in Fig. 1 below.

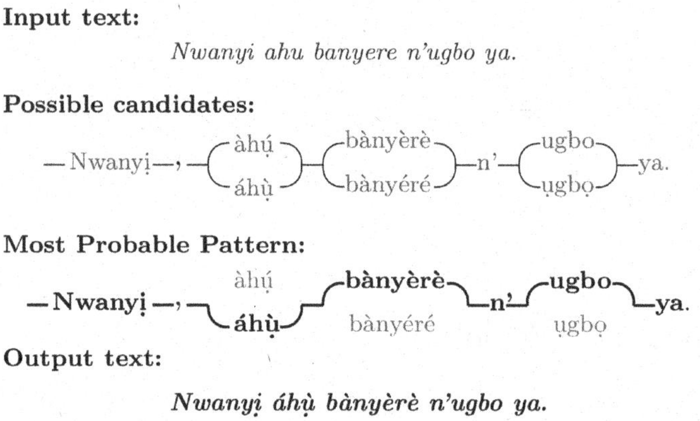
Input text:
Nwanyi ahu banyere n'ugbo ya.
Possible candidates:
Nwanyị
àhụ́
áhụ̀
bànyèrè
bànyéré
n'
ugbo
ụgbọ
ya.
Most Probable Pattern:
Nwanyị
àhụ́
áhụ̀
bànyèrè
bànyéré
n'
ugbo
ụgbọ
ya.
Output text:
Nwanyị áhụ̀ bànyèrè n'ugbo ya.

Fig. 1. Illustrative view of the diacritic restoration process

3 Related Literature

Two common approaches to diacritic restoration are highlighted in this paper: *word based* and *character based.*

3.1 Word Level Diacritic Restoration

Different implementation schemes of the word-based approach have been described e.g. Simard [9] adopted successive operations of *segmentation*, *hypothesis generation* and *disambiguation* using POS-tags and HMM language models for French texts. The work was later extended to an accent insertion tool capable of "on-the-fly-accentuation" (OTFA), on text editors. On the Croatian language, Šantić *et al.* [7] applied the use of substitution schemes, a dictionary and language models in implementing a similar architecture. It involves *tokenisation*, *candidate generation*, and *correct form selection*.

Yarowsky [12] also used dictionaries with decision lists, Bayesian classification and Viterbi decoding based on the surrounding context. Crandall [3], using Bayesian approach, HMM and a hybrid of both, as well as different evaluation method, attempted to improve on Yarowsky's work on Spanish text and reported a restoration accuracy of 7.6 errors per 1000 words. Cocks and Keegan [2], with a multi-genre corpus of about 4.2 million Maori words used naive Bayes classifiers and extracted word-based n-grams relating to the target word as instance features. Tufiş and Chiţu [10] also applied POS tagging to restore diacritics in Romanian texts and subsequently dealt with "unknown words" using character based back-off.

3.2 Grapheme or Letter Level Diacritic Restoration

Mihalcea [5], with their work on Romanian, presented an argument for letter based diacritic restoration for low resource languages. With a 3 million word corpus from the online Romanian newspaper, they implemented an instance based and a decision tree classifiers which gave a letter-level accuracy of 99 % implying a much lower accuracy at the word-level.

This approach became popular among developers of language tools for low resourced languages. However, Wagacha *et al.* [11], replaced the evaluation method in Mihalcea's work, with a more appropriate word-based approach while De Pauw et al. [4] used, the "lexical diffusion" (*LexDif*)[4] metric to quantify the disambiguation challenges on language bases. Each of these works recorded an accuracy level over 98 %.

3.3 Igbo Diacritic Restoration

The only attempt we know of at restoring Igbo diacritics is reported by Scannell [8] in which a combination of word- and character-level models were applied. For the word-level, they used two lexicon lookup methods, *LL* which replaces ambiguous words with the most frequent word and *LL2* that uses a bigram model to determine the output. They reported accuracies of 88.6 % and 89.5 % for Igbo language on the *LL* and *LL2* models respectively.

4   Experimental Setup

We present the preliminary report on the implementation of some word-level Igbo ADRS models based on ideas from the work of Scannell [8]. We developed two broad categories of n-gram - bigrams and trigrams - models with their variants.

---

[4] *LexDif* is the average number of candidates per wordkey, calculated by dividing the total word types with the unique wordkeys. A wordkey is gotten by stripping the diacritics off a word.

4.1 Experimental Methods

In this work, we aimed at extending the work of Scannell [8] on the word-level diacritic restoration of Igbo language. Some key distinctions of the approach we used from their works are highlighted below:

Data Size: Their experiment used a data size of 31k tokens with 4.3k word types while ours used 1.07m with 14.4k unique tokens. Table 2 shows percentage distributions of some ambiguous words in the text.

Preprocessing: Our tokenizer considered certain language information[5] that their method may not have considered.

Baseline Model: Their work reported a baseline measure on the stripped version which is our *bottom line* measure. Our baseline model is similar to their first lexicon lookup *LL* model.

Bigram Models: Our work extended their best model (bigram) with different smoothing techniques, *words vs keys* approach, and *backward replacement* features.

Trigram Models: Trigram models were not included in their work but we implemented different trigram models with similar structures as the bigram models.

Table 2. Sample percentage distribution of some of the ambiguous words

Key	Words	# of occurrences	%age
na	**na**	28041	95.41 %
	ná	1349	4.59 %
o	**o**	7477	24.75 %
	ó	8	0.03 %
	ọ́	252	0.83 %
	ò	83	0.27 %
	ọ̀	1053	3.49 %
	ọ	21339	70.63 %
ruru	**ruru**	225	49.34 %
	rụrụ	231	50.66 %
agbago	**agbago**	49	51.58 %
	agbagọ	46	48.42 %
bu	**bu:**	241	1.49 %
	bú:	2	0.01 %
	bụ́ :	6050	37.39 %
	bụ:	9887	61.11 %

---

[5] For example: strings like *"na"*, (mostly conjunction), *"na-"* (auxiliary) or *"n' "* (preposition) are treated as valid tokens due to the special roles the symbols play in distinguishing the word classes.

4.2 Experimental Data

The Igbo Bible corpus[6] used for this work contains 1,070,429 tokens (including punctuations and numbers) with 14,422 unique word types out of which 5,580 are unambiguous (i.e. appeared only in one diacritic form) in the text. The lexical diffusion on the text is 1.0398. A bible verse (not a sentence) was used as the unit of data entry.

Bigram and trigram counts were generated while a non-diacritic version of the corpus was created by stripping off all diacritics from the main bible text. An outright evaluation of the stripped text yields a bottom line accuracy of 71.30 %.

Our task involves creating the non-diacritic version, generating a look-up dictionary of unique key entries and their diacritic variants based on a closed-world[7] assumption, applying a restoration model to the stripped version one line at a time and keeping a count of the correctly restored tokens. The performance evaluation is measured by computing the percentage of the entire tokens that are correctly restored.

4.3 Model Descriptions

Baseline Models: *a01* and *a02*. As stated earlier, the bottom line model (a01) compared every diacritic token with its corresponding non-diacritic key from the stripped text. The baseline (a02) applied a simple unigram model that picks the most occurring candidate from the data.

Bigram Models: *b01* ... *b06*. Given a stripped word (or wordkey), these models use maximum likelihood estimation (MLE) to select the most probable word given the previous word. *b01* and *b02* are the "one key" variants. They use only the wordkey of the current word to generate candidates while retaining the previously restored preceding word. They differ in the smoothing techniques (Add 1 and Backoff). *b03* to *b06*" are the "two key" variants i.e. for every bigram, the two wordkeys will be used to find all possible candidates of both words and then select the most probable combination. This is motivated by the "assumption" that an error may have occurred in the previous step and should not be carried along. Also a technique called *backward replacement* was introduced to provide an opportunity to "step back and correct"[8] any assumed error as we go along i.e. if the most probable bigram suggest a different diacritic form for the preceding word, then it will be replaced with the current word.

Trigram Models: *t07* ... *t12*. These are the trigram versions of the models described above. The "one key" variants, *t07* and *t08* use the last two restored

---

[6] This corpus was originally processed by Onyenwe *et al.* [6].

[7] Since we did not deal with unknown words, we simplified our models by assuming that words not found in our dictionary do not exist.

[8] We recognise that this might be counter productive as correctly restored words in the previous step may be wrongly replaced again in the next.

words and the candidates of the given key to get the most probable trigram from the data while the rest generate fresh candidates with the current and the two preceding keys. The smoothing techniques as well as the backward replacement methods were also tested on these models.

4.4   Results, Conclusion and Future Work

This paper describes a knowledge-light language independent method for diacritic restoration for Igbo texts using n-gram models with a comparison of smoothing and replacement techniques under a closed-world assumption. The baseline method used a unigram model with an accuracy of 96.23 %. The results show that the trigram models generally outperform the bigram models (Table 3).

Table 3. Results of experiments using the different models

Models	Accuracy	Amb Acc	%Impr
a01:Bottomline: Non-diac Text	71.30 %	–	–
a02:Baseline: Most frequent	96.23 %	91.52 %	86.86 %
b01:Bigram-1Key+Add1:	97.21 %	94.45 %	90.28 %
b02:Bigram-1Key+Backoff:	97.75 %	94.18 %	92.16 %
b03:Bigram-2Key+Add 1:	97.36 %	94.60 %	90.80 %
b04:Bigram-2Key+Add 1-BR:	97.48 %	94.85 %	91.22 %
b05:Bigram-2Key+Backoff:	97.77 %	94.36 %	92.23 %
b06:Bigram-2Key+Backoff-BR:	96.17 %	90.14 %	86.66 %
t07:Trigram-1Key+Add 1:	97.66 %	92.94 %	91.85 %
t08:Trigram-1Key+Backoff:	92.01 %	77.96 %	72.16 %
t09:Trigram-3Key+Add 1:	98.46 %	95.91 %	94.63 %
t10:Trigram-3Key+Backoff:	92.86 %	81.37 %	75.12 %
t11:Trigram-3Key+Add 1-BR:	**98.83 %**	**97.57 %**	**95.92 %**
t12:Trigram-3Key+Backoff-BR:	92.29 %	77.99 %	73.14 %

While the Add 1 smoothing technique improves as the experiment progressed, the Backoff seems inconsistent, beating the Add 1 with the bigrams and dropping in performance with the trigrams. Backward replacement is introduced and it seems to work though it is not yet clear why it does. However, while it has boosted the performance of the Add 1 at each stage, it has clearly deteriorated that of the Backoff model.

The "three key" trigram model with the *Add 1* and backward replacement is the most effective method with a performance accuracy of 98.83 %. They outperformed the best models in literature but future works will attempt to improve on the robustness with an open-world assumption while exploring the backward replacement and other smoothing techniques. Expanding the data size across multiple genre and handling "unknown words" will form the main focus of the next experiments. We also intend to investigate the effects of POS-tagging and a morphological analysis on the performance of the models and explore the connections between this work and the broader field of word sense disambiguation.

Acknowledgments. Many thanks to Nnamdi Azikiwe University & TETFund Nigeria for the funding, my colleagues at the IgboNLP Project, University of Sheffield, UK and Prof. Kelvin P. Scannell, St Louis University, USA.

References

1. Achebe, I., Ikekeonwu, C., Eme, C., Emenanjo, N., Wanjiku, N.: A Composite Synchronic Alphabet of Igbo Dialects (CSAID). IADP, New York (2011)
2. Cocks, J., Keegan, T.: A word-based approach for diacritic restoration in Māori. In: 2011 Proceedings of the Australasian Language Technology Association Workshop, Canberra, Australia, pp. 126–130, December 2011
3. Crandall, D.: Automatic Accent Restoration in Spanish Text (2016). http://www.cs.indiana.edu/~djcran/projects/674_final.pdf. Accessed 7 Jan 2016
4. De Pauw, G., De Schryver, G.M., Pretorius, L., Levin, L.: Introduction to the special issue on African language technology. Lang. Resour. Eval. **45**, 263–269 (2011). Springer Online
5. Mihalcea, R.F.: Diacritics restoration: learning from letters versus learning from words. In: Gelbukh, A. (ed.) CICLing 2002. LNCS, vol. 2276, pp. 339–348. Springer, Heidelberg (2002)
6. Onyenwe, I.E., Uchechukwu, C., Hepple, M.R.: Part-of-speech tagset and corpus development for Igbo, an African language. In: LAW VIII - The 8th Linguistic Annotation Workshop, pp. 93–98. ACL, Dublin (2014)
7. Šantić, N., Šnajder, J., Dalbelo Bašić, B.: Automatic diacritics restoration in Croatian texts. In: Stančić, H., Seljan, S., Bawden, D., Lasić-Lazić, J., Slavić, A. (eds.) The Future of Information Sciences, Digital Resources and Knowledge Sharing, pp. 126–130 (2009). ISBN 978-953-175-355-5
8. Scannell, K.P.: Statistical unicodification of African languages. Lang. Resour. Eval. **45**(3), 375–386 (2011). Springer New York Inc., Secaucus, NJ, USA
9. Simard, M.: Automatic insertion of accents in French texts. In: Proceedings of the Third Conference on Empirical Methods in Natural, Language Processing, pp. 27–35 (1998)
10. Tufiş, D., Chiţu, A.: Automatic diacritics insertion in Romanian texts. In: Proceedings of International Conference on Computational Lexicography, Pecs, Hungary, pp. 185–194 (1999)
11. Wagacha, P.W., De Pauw, G., Githinji, P.W.: A grapheme-based approach to accent restoration in Gĩkũyũ. In: Proceedings of 5th International Conference on Language Resources and Evaluation (2006)
12. Yarowsky, D.: Corpus-based techniques for restoring accents in Spanish and French text. In: Armstrong, S., Church, K., Isabelle, P., Manzi, S., Tzoukermann, E., Yarowsky, D. (eds.) National Language Processing Using Very Large Corpora. Text, Speech and Language Technology, vol. 11, pp. 99–120. Springer, Netherlands (1999). Kluwer Academic Publishers

Predicting Morphologically-Complex Unknown Words in Igbo

Ikechukwu E. Onyenwe[(⊠)] and Mark Hepple

NLP Group, Computer Science Department, University of Sheffield, Sheffield, UK
{i.onyenwe,m.hepple}@sheffield.ac.uk

Abstract. The effective handling of previously unseen words is an important factor in the performance of part-of-speech taggers. Some trainable POS taggers use suffix (sometimes prefix) strings as cues in handling unknown words (in effect serving as a proxy for actual linguistic affixes). In the context of creating a tagger for the African language Igbo, we compare the performance of some existing taggers, implementing such an approach, to a novel method for handling morphologically complex unknown words, based on morphological reconstruction (i.e. a linguistically-informed segmentation into root and affixes). The novel method outperforms these other systems by several percentage points, achieving accuracies of around 92 % on morphologically-complex unknown words.

Keywords: Morphology · Morphological reconstruction · Igbo · Unknown words prediction · Part-of-speech tagging

1 Introduction

The handling of unknown words is an important task in NLP, which can be assisted by morphological analysis, i.e. decomposing inflected words into their stem and associated affixes. In this paper, we address the handling of unknown words in POS tagging for Igbo, an *agglutinative* African language. We present a novel method for handling morphologically-complex unknown words of Igbo, based on morphological reconstruction (i.e. a linguistically-informed segmentation into root and affixes), and show that it outperforms standard methods using arbitrary suffix strings as cues.

In the rest of the paper, we first note prior work on unknown word handling in POS tagging, and consider the suitability of these methods to Igbo, as an agglutinative language. We then present some experiments using morphological reconstruction in unknown word handling for Igbo, and discuss our results.

2 Related Literature

Previous work on POS tagging unknown words has used features such as prefix and suffix strings of the word, spelling cues like capitalization, and the word/tag

© Springer International Publishing Switzerland 2016
P. Sojka et al. (Eds.): TSD 2016, LNAI 9924, pp. 206–214, 2016.
DOI: 10.1007/978-3-319-45510-5_24

values of neighbouring words [1,4,9,11]. The HMM method of Kupiec [5] assigns probabilities and state transformations to a set of suffixes of unknown words. Samuelsson [10] used starting/ending n-length letter sequences as predictive features of unknown words. Brants [1] showed that word endings can predict POS, e.g. −*able* is likely to be adjective in English. Toutanova *et al.* [11] uses variables of length up to n for extracting word *endings*, such that $n = 4$ for *negotiable* generates feature list [e,le,ble,able]. These methods have worked well in languages like English and German whose derivational and inflectional affixes reveal much about the grammatical classes of words in question.

3   Problem Description

Igbo is an agglutinative language, with many frequent suffixes and prefixes [3]. A single stem can yield many word-forms by addition of affixes, that extend its original meaning. Suffixes have different grammatical classes, and may concatenate with a stem in variable order, as in e.g.: *abịakwa* "*a-bịa-kwa*", *bịakwaghị* "*bịa-kwa-ghị*", *bịaghịkwa* "*bịa-ghị-kwa*", *bịaghachiri* "*bịa-gha-chi-ri*", *bịachighara* "*bịa-chi-gha-ra*", *bịaghachirịrị* "*bịa-gha-chi-rịrị*", etc. Methods to automatically identify suffix-string cues (e.g. for use in POS tagging), based on extracting the last n letters of words, seem likely to be challenged by such complexity, e.g. that *bịaghachirịrị* "must come back" has 3 suffixes of length 3 or 4, to a total length of 10, which may elsewhere appear in a different order.

The Igbo POS tagset of Onyenwe *et al.* [7] uses "_XS" to indicate *extensional suffixes*, e.g. tag "VSI_XS" applies to a word that is VSI (verb simple) and includes ≥1 extensional suffix. In our experiments, using 10-fold cross validation and deriving the lexicon from the training data, we find that the majority of unknown words encountered arise due to agglutination (see Table 1).

4   Experiments

Our experiments compare methods using automatically-identified n-character suffix (and prefix) strings to methods based on (linguistically-informed) morphological reconstruction, in regard to their performance for handling morphological-complex Igbo words previously unseen in the training data during POS tagging.

4.1   Experimental Data

There are two sets of corpus data used in this research, the selected books from new testament Bible[1] that represents Igbo tagged religious texts (IgbTNT) and novel[2] for modern Igbo tagged texts genre (IgbTMT). The corpus data and the tagset used were developed in [7,8].

---

[1] Obtained from jw.org.
[2] "Mmadụ Ka A Na-Arịa" written in 2013.

4.2   Experimental Tools

POS Taggers. We chose POS tagging tools that generally perform well, and have parameters to control word feature extraction for unknown word handling: Stanford Log-linear POS Tagger (SLLT) [11], Trigrams'n'Tagger (TnT) [1], HunPOS [4] (a reimplementation of TnT), and FnTBL [6] which uses the *transformation-based learning* (TBL) method of [2], adapted for speed. TBL starts with an initial state (where known words are assigned their most common tag, or a default) and applies *tranformation rules*, to correct errors in the initial state based on context. The training process compares the initial state tags to the true tags of the training data, and iteratively acquires an list of rules correcting errors in the initial state, until it sufficiently resembles the truth.

Morphological Reconstruction. We used morphological reconstruction to segment morphologically-complex words into stems and affixes, so that patterns can be learnt over these sequences, which are used to predict the tags of unknown words. Items in these sequences classified as stem (ROOT), prefix (PRE) and suffix (SUF), i.e. ignoring finer distinctions of their grammatical function. For example, the word *enwechaghị* tagged "VPP_XS" in the IgbTC will have the form "e/PRE nwe/ROOT cha/SUF ghị/SUF" after morphological reconstruction. The idea is to use these morphological clues to predict the tag "VPP_XS", should the word appear as an unknown word.

For an inflected word w, morphological reconstruction involves extracting the stem cv and all n possible affixes attached to it. An Igbo stem is a formation of cv, starting with a consonant c and ending with a vowel v [3], where c could be a single letter or a digraph. Digraphs are two character strings pronounced as one sound and are non split (e.g. "gh", "ch"). We used a list of suffixes from [3] as a dictionary to search for valid morphological forms.

4.3   Experimental Setup

In our experiments, unknown words arise due to our use of 10-fold cross validation, i.e. the unknown words of any test fold are the words that were not present anywhere in the corresponding training set (i.e. the other 9 folds). Table 1 shows the unknown word ratios for our different data sets (listed under experiment1).

Table 1. Average sizes of train, test, and unknown words ratio of the experimental corpus data. Train2 and test2 are morphologically inflected data.

Corpus	Data for experiment1			Data for experiment2	
	Train	Test	Unknown ratio	Train2	Test2
IgbTNT	35938	3993	3.18 %	4120	088
IgbTMT	35965	3996	4.90 %	4855	134
IgbTC	71902	7989	3.39 %	8975	222

Our experiments compare the effectiveness of different methods for tagging such unknown words, and specifically the inflected ones. In our first experiment (experiment1), we apply standard taggers to the data, and score their performance on the inflected unknown words. Our second experiment (experiment2) handles these same unknown words via morphological reconstruction. For this, we extract only the inflected unknown words from the data of experiment1, giving rise (under 10-fold cross validation) to the data set sizes listed under experiment2 of Table 1. (Note that these numbers might seem to be less than as implied by the "Unknown Ratio" column, as only the *inflected* unknown words are extracted, which correspond to around 70 % of all unknown words.)

4.4   Experiment 1: Using Original Word-Forms

HunPOS, TnT and SLLT taggers were used because they have robust methods for extracting last/first letters of words for use as cues in handling unknown words. We chose $n = 5$ and $n = 1$ for extracting last and first letters of a word because the longest suffixes and prefixes in Igbo so far are of these lengths, and the taggers performed well at these settings. These systems also use the context of neighbouring words/tags to help in handling the unknown words. Table 2 shows the performance of these systems for the correct tagging of *only* the inflected unknown words (listed under experiment1).

Table 2. Average statistics and accuracy scores on the inflected tokens based on different approaches.

Corpus	Size	1st experiment			2nd experiment				
		HunPOS	TnT	SLLT	Taggers	PRE+SUF	PRE+SUF +rV	All	All(-PRE)
IgbTNT	88	70.73 %	73.94 %	83.77 %	FnTBL	78.03 %	82.81 %	90.44 %	82.78 %
					SLLT2	66.31 %	67.11 %	66.53 %	70.87 %
IgbTMT	134	67.17 %	70.37 %	86.48 %	FnTBL	78.96 %	86.03 %	91.99 %	85.95 %
					SLLT2	74.45 %	75.27 %	76.01 %	77.15 %
IgbTC	193	70.28 %	73.16 %	84.67 %	FnTBL	83.75 %	86.23 %	88.46 %	83.27 %
					SLLT2	76.41 %	77.62 %	76.09 %	76.54 %

4.5   Experiment 2: Using Morphologically Reconstructed Forms

Our morphology segmentation module was used to perform morphological reconstruction of the data listed under experiment2 of Table 1. In the representation produced, the *correct* tag of an unknown word is marked on its *stem* within the stem/affix sequence. For example, abịakwara has tag VPP_XS, and so, after reconstruction, would be represented as "a/PRE bịa/VPP_XS kwa/SUF ra/SUF".

Four variants of the method were used, differing mostly in the extent to which the grammatical function of affixes were distinguished. In *Pattern1*, all affixes were classed as only either SUF (suffix) or PRE (prefix). In *Pattern2*, an "rV" tag was used for past tense suffixes.[3] In *Pattern3*, more morph-tags for suffixes were added to indicate grammatical functions (see Table 4 for a list of the morph-tags). In *Pattern4*, prefix and stem were collapsed to form one part (e.g. changing "a/PRE bịa/VSI_XS kwa/LSUF" to "abịa/VSI_XS kwa/LSUF"), eliminating the "PRE" tag. Morph-tags serve as important clues for disambiguation (Tables 3 and 4).

Table 3. Some samples of morphological reconstructed words into stems and affixes.

Word form	FnTBL initial state	FnTBL truth state
	Pattern1 PRE+SUF	
nwukwasị	nwu/ROOT kwasị/SUF	nwu/VSI_XS kwasị/SUF
nwukwara	nwu/ROOT kwa/SUF ra/SUF	nwu/VrV_XS kwa/SUF ra/SUF
nwukwasịrị	nwu/ROOT kwasị/SUF rị/SUF	nwu/VrV_XS kwasị/SUF rị/SUF
ịnọdonwu	ị/PRE nọ/ROOT do/SUF nwu/SUF	ị/PRE nọ/VIF_XS do/SUF nwu/SUF
abịakwara	a/PRE bịa/ROOT kwa/SUF ra/SUF	a/PRE bịa/VPP_XS kwa/SUF ra/SUF
nụrụkwanụ	nụ/ROOT rụ/SUF kwa/SUF nụ/SUF	nụ/VSI_XS rụ/SUF kwa/SUF nụ/SUF
enwechaghị	e/PRE nwe/ROOT cha/SUF ghị/SUF	e/PRE nwe/VSI_XS cha/SUF ghị/SUF
	Pattern2 added "rV" to pattern1 and pattern3 added all Morpho-tags	
nwukwasị	nwu/ROOT kwasị/LSUF	nwu/VSI_XS kwasị/LSUF
nwukwara	nwu/ROOT kwa/rSUF ra/rV	nwu/VrV_XS kwa/rSUF ra/rV
nwukwasịrị	nwu/ROOT kwasị/rSUF rị/rV	nwu/VrV_XS kwasị/rSUF rị/rV
ịnọdonwu	ị/PRE nọ/ROOT do/iSUF nwu/iSUF	ị/PRE nọ/VIF_XS do/iSUF nwu/iSUF
abịakwara	a/PRE bịa/ROOT kwa/eSUF ra/APP	a/PRE bịa/VPP_XS kwa/eSUF ra/APP
nụrụkwanụ	nụ/ROOT rụ/xSUF kwa/xSUF nụ/LSUF	nụ/VSI_XS rụ/xSUF kwa/xSUF nụ/LSUF
enwechaghị	e/PRE nwe/ROOT cha/xSUF ghị/NEG	e/PRE nwe/VSI_XS cha/xSUF ghị/NEG

We applied FnTBL and SLLT to the morphologically reconstructed data (here referring to the latter as SLLT2, to differentiate from its earlier use in experiment 1). Note that the reconstructed representations for individual words are presented in isolation, i.e. so the systems cannot exploit contextual information of neighbouring words/tags (in contrast to experiment 1). FnTBL was chosen due to its effective pattern induction method, and SLLT because it outperformed the other systems in experiment 1. SLLT2 was simply trained directly over the reconstructed data. For FnTBL, we intervene to specify a particular initial state for TBL, in which the stem is given the initial tag "ROOT". Hence, TBL should generate only rules that, based on the morphological context, replace a ROOT tag with a final tag, the latter being a POS tag for a complete inflected unknown word. Results are shown in Table 2 under experiment2.

---

[3] Here, "rV" means letter *r* and any vowel $(a,e,u,o,i,ị,ọ,ụ)$ attached to a word in Igbo like "bịara" *came*, "kọrọ" *told*, "riri" *ate*, "nwuru" *shone*, etc. It is a past tense marker if attached to active verb or indicate stative/passive meaning if attached to a stative verb [3]. Therefore, it is an important cue in predicting past tense verbs or verbs having applicative meaning "APP".

Table 4. Morph-tags and meanings

Tag/marker	Meaning
APP	Applicative
NEG	Negative
INFL	Inflection for perfect tense
rV	Inflection for past tense
LSUF	Last suffix marker for morphologically-inflected simple verb
xSUF	Suffix within morphologically-inflected simple verb
eSUF	Suffixes within morphologically-inflected participle
iSUF	Suffixes within morphologically-inflected infinitive
rSUF	Suffixes within morphologically-inflected past tense verb

5 Discussion

Table 5 illustrates how the root+affixes have served as important cues for predicting the tags of morphological-complex unknown words. "Initial Tag" column is the FnTBL initial state, "Transformation Process" is predicted tag after applying transformational rules (adjacent to the stems are rules indexes that fired) and "Final Tag" is the FnTBL predicted tags returned as the tags for morphologically-complex unknown words. In Example 1 of Table 5, the word "begorochaa" refers to "perching activity of a group of birds", and is an inflected simple verb (VSI_XS) with "be" as the stem. Two transformational rules fired to transform its initial tag "ROOT" to the final tag "VSI_XS". The first change is made by Rule 0, which is a generic rule that changes ROOT tag to VrV (past tense verb) tag, provided there is a suffix within the [+1,+2] window. This rule is ordered first in the rule list, as it has the highest correction score over the training data. Rule 2 applies next, changing VrV to "VSI_XS" because xSUF and LSUF occur with inflected simple verbs. In other examples, Rule 2 changes VrV to VrV_XS (inflected past tense verbs) because of rSUF that occur in past tense verbs, Rules 3 and 4 change VrV and VSI_XS to VPP_XS (inflected participle) tag whenever the previous tag after stem is PRE, Rule 5 changes VPP_XS to VPERF (perfect tense verbs) due to presence of INFL, Rule 6 changes VPP_XS to VIF_XS (inflected infinitive verb) due to i prefix, and Rule 36 changes VPERF to VPERF_XS (inflected perfect tense) due to xSUF.

The accuracy scores of both experiments are shown in 1st and 2nd experiment columns of Table 2. "PRE+SUF" column is for *Pattern1* variation, the accuracy scores are substantive, FnTBL did better than SLLT2 in all cases and performed better than other taggers in experiment 1 except SLLT. Column "PRE+SUF+rV" shows *Pattern2* variation, SLLT2 and FnTBL performances generally improve and FnTBL scored better than majority in 1st experiment. "All" column is for *Pattern3*, here is to test the prospect of paradigmatic tagging where meaningful tags for affixes are added to indicate their grammatical

Table 5. Examples of transformational rules generated by FnTBL. The numbers are the rule identity numbers that fired.

Initial tag	Transformation process	Final tag
Example 1		
be ROOT VSI_XS	be VSI_XS VSI_XS | 0 1	begorochaa/VSI_XS
go xSUF xSUF	go xSUF xSUF	
ro APP APP	ro APP APP	
chaa LSUF LSUF	chaa LSUF LSUF	
Example 2		
kpọ ROOT VrV_XS	kpọ VrV_XS VrV_XS | 0 2	kpọchibidoro/VrV_XS
chi rSUF rSUF	chi rSUF rSUF	
bi rSUF rSUF	bi rSUF rSUF	
do rSUF rSUF	do rSUF rSUF	
ro rV rV	ro rV rV	
Example 3		
e PRE PRE	e PRE PRE	ekpochapụ/VPP_XS
kpo ROOT VPP_XS	kpo VPP_XS VPP_XS | 0 3	
cha eSUF eSUF	cha eSUF eSUF	
pụ eSUF eSUF	eSUF eSUF	
Example 4		
ị PRE PRE	ị PRE PRE	ịkpọcha/VIF_XS
kpọ ROOT VIF_XS	kp VIF_XS VIF_XS | 0 1 4 6	
cha LSUF LSUF	cha LSUF LSUF	
Example 5		
e PRE PRE	e PRE PRE	echekwala/VPERF_XS
che ROOT VPERF_XS	che VPERF_XS VPERF_XS | 0 3 5 36	
kwa xSUF xSUF	kwa xSUF xSUF	
la INFL INFL	la INFL INFL	

functions. This gave best scores of 90.44 %, 91.99 % and 88.46 % for FnTBL and these scores are several points better than scores achieved by the taggers used in the 1st experiment (see Table 2).

Finally, column "All(-PRE)" for *Pattern4* is to verify the strength of prefix as unknown word predictive feature considering it is only one character length. Comparing columns "All(-PRE)" and "All", shows that there are lost in accuracies of column "All" for FnTBL (e.g. about 9.0 in IgbTNT). This is contrary to English where addition of prefix as feature caused negative effect on the accuracy of unknown words [11]. Surprisingly, SLLT2 increased in its accuracy against decrease in FnTBL scores. But an experiment on IgbTMT using SLLT tagger's technique for handling unknown words shows that using only suffix features gave accuracy of 77.26 % and addition of prefix features improved the accuracy on the morphologically-complex words by 9.22 %. The reason for SLLT2's accuracy increment can be explained in regard with "PRE" ambiguity. "PRE" tag is used to indicate prefix whether it is "i̩/i" for infinitive or "a/e" for participle and simple verbs, therefore, collapsing it with the stem removes this ambiguity. Statistical taggers will require large data size to properly disambiguate this case.

6   Conclusion

We have shown that use of actual linguistically-informed segmentation into stems and associated affixes are good for predicting unknown inflected words in Igbo. Through morphological reconstruction, inflected words are represented in machine learnable pattern that exploits morphological characteristics during tagging process for handling unknown words. The performance of FnTBL that inductively learns linguistic patterns reveals that our method is better than methods that automatically identify suffix-string cues (e.g. for use in POS tagging), based on extracting the last n letters of words to serve as proxy for actual linguistic affixes. The standard method using arbitrary suffix strings as cues is challenged by complexity associated with morphologically-complex unknown words of the language. In Igbo language, a single root can produce as many possible word-forms as possible through the use of affixes of varying lengths ranging from 1 to 5, which may concatenate with a stem in variable orders.

In the future work, it is important to perform full morphological analysis on Igbo. This experiment excludes some inflected classes (like nouns) as it will lead to full morphological analysis which is beyond the research scope. Also, morphological analysis on the compound verbs and exploiting n neighbouring words information are ignored. These lapses will hide some important information required for NLP task. Of course, this is pointing towards building a large-scale computational morphologies for Igbo.

Acknowledgments. We acknowledge the financial support of Tertiary Education Trust Fund Nigeria and Nnamdi Azikiwe University (NAU) Nigeria. Many thanks to Dr. Uchechukwu Chinedu of linguistic department, NAU for his very helpful discussion.

References

1. Brants, T.: TnT: a statistical part-of-speech tagger. In: Proceedings of the Sixth Conference on Applied Natural Language Processing, pp. 224–231 (2000)
2. Brill, E.: Transformation-based error-driven learning and natural language processing: a case study in part-of-speech tagging. Comput. Linguist. **21**, 543–565 (1995). MIT Press, Cambridge
3. Emenanjo, N.E.: Elements of Modern Igbo Grammar: A Descriptive Approach. Oxford University Press, Ibadan (1978)

4. Halácsy, P., Kornai, A., Oravecz, C.: HunPos: an open source trigram tagger. In: Proceedings of the 45th Annual Meeting of the ACL on Interactive Poster and Demonstration Sessions, pp. 209–212 (2007)
5. Kupiec, J.: Robust part-of-speech tagging using a hidden Markov model. J. Comput. Speech Lang. **6**(3), 225–242 (1992)
6. Ngai, G., Florian, R.: Transformation-based learning in the fast lane. In: Proceedings of the Second Meeting of the North American Chapter of the Association for Computational Linguistics on Language Technologies, pp. 1–8 (2001)
7. Onyenwe, I.E., Uchechukwu, C., Hepple, M.: Part-of-speech tagset and corpus development for Igbo, an African. In: Proceedings of LAW VIII-8th Linguistic Annotation, Workshop 2014 in conjuction with COLING 2014, Dublin, Ireland 23–24 August 2014, pp. 93–98. Association for Computational Linguistics (2014)
8. Onyenwe, I.E., Hepple, M., Uchechukwu, C., Ezeani, I.: Use of transformation-based learning in annotation pipeline of Igbo, an African language. In: Joint Workshop on Language Technology for Closely Related Languages, Varieties and Dialects, p. 24 (2015)
9. Ratnaparkhi, A., et al.: A maximum entropy model for part-of-speech tagging. In: Proceedings of the Conference on Empirical Methods in Natural Language Processing, vol. 1, pp. 133–142 (1996)
10. Samuelsson, C.: Morphological tagging based entirely on Bayesian inference. In: 9th Nordic Conference on Computational Linguistics (2013)
11. Toutanova, K., Klein, D., Manning, C.D., Singer, Y.: Feature-rich part-of-speech tagging with a cyclic dependency network. In: Proceedings of the 2003 Conference of the North American Chapter of the Association for Computational Linguistics on Human Language Technology, vol. 1, pp. 173–180 (2003)

Morphosyntactic Analyzer for the Tibetan Language: Aspects of Structural Ambiguity

Alexei Dobrov[1], Anastasia Dobrova[2], Pavel Grokhovskiy[1(✉)],
Nikolay Soms[2], and Victor Zakharov[1]

[1] Saint-Petersburg State University, Saint-Petersburg, Russia
{a.dobrov,p.grokhovskiy,v.zakharov}@spbu.ru
[2] LLC "AIIRE", Saint-Petersburg, Russia
{adobrova,nsoms}@aiire.org

Abstract. The paper deals with the development of a morphosyntactic analyzer for the Tibetan language. It aims to create a consistent formal grammatical description (formal grammar) of the Tibetan language, including all grammar levels of the language system from morphosyntax (syntactics of morphemes) to the syntax of composite sentences and supra-phrasal entities. Syntactic annotation was created on the basis of morphologically tagged corpora of Tibetan texts. The peculiarity of the annotation consists in combining both the immediate constituents structure and the dependency one. An individual (basic) grammar module of Tibetan grammatical categories, its possible values, and restrictions on their combination are created. Types of tokens and their grammatical features form the basis of the formal grammar being produced, allowing linguistic processor to build syntactic trees of various kinds. Methods of avoiding redundant structural ambiguity are proposed.

Keywords: Corpus linguistics · Tibetan language · Morphosyntactic analyzer · Tokenization · Immediate constituents · Dependency grammar · Natural language processing

1 Introduction

In order to build a morphosyntactic analyzer of Tibetan texts it is necessary to create a formal grammar, which includes all levels of the grammatical system of the Tibetan language from morphosyntax (syntactics of morphemes) to the syntax of sentences and supra-phrasal units.

A few currently available studies of the Tibetan language analyze mainly Tibetan morphology, the only notable exception being "The Classical Tibetan language" by Stephen Beyer [1], which also includes an extensive presentation of Tibetan syntax. Still, this work does not fully describe the Tibetan system of syntactic units and often has a speculative character, since the conclusions are not supported by textual corpora.

The current project has the following objectives:

© Springer International Publishing Switzerland 2016
P. Sojka et al. (Eds.): TSD 2016, LNAI 9924, pp. 215–222, 2016.
DOI: 10.1007/978-3-319-45510-5_25

1. To create a system of syntactical annotation of Tibetan texts, including the information about Tibetan grammatical categories, their possible values, and restrictions on their combinations;
2. To develop a formal grammatical module of the open natural language processing system, which is able to perform a complete morphological and syntactic analysis of Tibetan texts;
3. To annotate a corpus of Tibetan texts syntactically.

The developed tools of language processing allow automatic markup procedures for further extension of the corpus.

The project uses an innovative approach to syntactic analysis, combining the immediate constituents structure (CS) and the dependency structure (DS). Such combination was proposed in [2] for the first time, but the available mathematical model did not allow to implement it in an algorithm. This study takes advantage of the AIIRE linguistic processor (Artificial Intelligence-based Information Retrieval Engine), which is one of the most successful computer realizations of combined CS and DS analysis [3]. Still, in order to be apllied to the Tibetan language, it requires a new research on Tibetan syntax.

2   The Project's Corpora Resources and Software

The project's database comprises two corpora of the Tibetan language developed at the Saint-Petersburg University. The Basic Corpus of the Tibetan Classical Language includes texts in a variety of classical Tibetan literary genres. The Corpus of Indigenous Tibetan Grammar Treatises consists of the most influential grammar works, the earliest of them proposedly dating back to 7th-8th centuries. Both corpora are provided with metadata and morphological annotation.

The corpora comprise 34,000 and 48,000 tokens, respectively. Tibetan texts are represented both in a Tibetan Unicode script and in a standard Latin transliteration [4].

The AIIRE linguistic processor with an open code is used for the project. AIIRE implements the method of inter-level interaction proposed by Tseitin in 1985 [5], which ensures the effective ambiguity resolution, based on the rules.

The principle of inter-level interaction helps to minimize the combinatorial explosion, which is very important for NLP software. The formal grammar analysis produces a considerable rate of ambiguity, especially when ellipsis is possible. The principle of inter-level interaction, implemented in the AIIRE linguistic processor, allows to apply upper-level constraints to lower-level ambiguity, and thus reduces the number of produced combinations.

The architecture of AIIRE and the developed algorithms of text analysis allow to apply this technology to languages of different types in the form of independent language modules, while the analysis algorithms are independent of the language. Besides the modules for the Russian language, modules for Arabic and Abkhaz languages were previously created, and the present project aims at developing a module for the Tibetan language, which is well known for the absence of formally marked word boundaries and ambiguity of word segmentation as such.

3 Representation of Tibetan Morphological Structures in AIIRE

The linguistic processor needs to recognize all the relevant linguistic units in the input text. For inflectional languages the input units are easy to identify as word forms, separated by space, punctuation marks etc. It is not the case for the Tibetan language, as there are no universal symbols to separate the input string into words or morphemes.

The developed module for the Tibetan language performs the segmentation of the input string into morphemes by using the Aho-Corasick algorithm (by Aho and Corasick), that allows to find all possible substrings of the input string according to a given dictionary. The algorithm builds a tree, describing a finite state machine with terminal nodes corresponding to completed character strings of elements (in this case, morphemes) from the input dictionary.

Language module contains a dictionary of morphemes, which allows the machine to create the tree in advance at the build stage of the language module, while in the runtime of the linguistic processor the tree is being loaded as a component of an executable module which brings its initialization time to minimum.

Two special files were created in order to analyze Tibetan morphology and morphonemics: the grammarDefines.py file determines types of tokens, their properties and restrictions, while the atoms.txt file (the allomorphs dictionary) specifies the morpheme, the token type and properties for each allomorph, also in accordance with grammarDefines.py file. For example, the following entry in the allomorphs dictionary དགའ་|morpheme=དགའ་|type=v_root|mood=ind|has_tense=False|fin_phoneme=vowel indicates that the དགའ་ (dga') allomorph is the basic allomorph of the དགའ་ (dga') morpheme, that is the verb root in the indicative mood, having no tense property and ending in a vowel.

The materials processed on the pilot stage allow to identify the following token types: v_suff (verbal suffix), punct (punctuation mark), p_dem_root (the root of the demonstrative pronoun), n_root (noun root), p_pers_root (the root of the personal pronoun), case_marker, v_root (verbal root), num_root (numeral root), p_def_root (attributive pronoun), fin (statement end marker). All these types of tokens have their possible morphological and morphonemic features indicated in the grammarDefines.py file. For example, the verbal root has such potential properties indicated as the mood (indicative, imperative), the tense (present, past, future), the availability of tense category (true/false) and the type of final phonemes defining the compatibility of the verbal root with suffix allomorphs. The restrictions for the verbal root require that the category of tense is available only if respective parameter "has tense" is set to "true", and the parameter of "mood" is set to "indicative".

These types of tokens and their grammatical features form the basis of the formal grammar being developed, allowing the linguistic processor to build syntactic treebanks of various structure.

Case markers of the Tibetan language, unlike inflected languages, function as postpositions rather than as suffix morphemes; and the most appropriate model to correspond to Tibetan morpheme order is seen as representing the nominal phrases followed by a case marker in postposition. Case marker takes the final position after all the modifiers of the nominal phrases, including numerals and pronouns. In this case, the order of English morphemes is opposite: the phrase *dus gcig na* is translated as at (*na*, locative) one (*gcig*) time (*dus*). Both the numeral and the nominal phrase modified by it may be further modified: the numeral may be complex, and the nominal phrase may be modified by an adjective or a participle etc. Thus, it seems to interpret the case marker of the Tibetan morphosyntax as a major constituent, and not as a dependent one, that corresponds, for example, to prepositions in prepositional groups in English and Russian languages.

Nominal phrases followed by a postpositive case marker may have structures of any complexity, including those modified by complex participle clauses, sometimes without a head, as shown in Fig. 1. Such nominal phrases are often proper names or epithets (in this case it is the Tibetan morpheme-for-morpheme rendering of the Sanskrit name of the Indian city of Śravasti); the head constituent

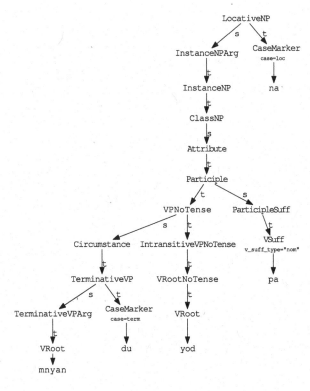

Fig. 1. Locative NP exemplified by a participle clause with a terminative adverbial modifier

"city" being omitted due to its semantic redundancy. In this example, there is a participle verbal phrase, modified by a circumstance, which is expressed by a terminative nominal phrase (TerminativeNP), where the termenative case marker follows the masdar nominal phrase (MasdarNP), that is expressed by a verbal root (V_Root). The masdar nominal phrase omits the nominalizer, that is typical for Tibetan complex verbal nouns, including proper names: the nominalizer may be omitted both by participles and masdars, and for the time being current authors have not identified the precise rules of such omission. Literally translated, the given example reads as follows:

> *to hear* + nominalizer omitted (missing in the tree) +
> *for* (terminative) +
> *to exist* + nominalizer +
> *in* (locative)

That is, *in Existing-To Be-Heard* (where *Existing-To Be-Heard* is a name of the city).

The above mentioned features of the Tibetan morphosyntax cause a considerable rate of ambiguity in Tibetan text while being processed by a computer: due to ellipsis every verbal root can be treated as a participle or a masdar within a personal name, and each modifer can be treated as a separate proper name. As in other languages, circumstances and complements can get ambiguous interpretations if there are several recursive verbal phrases.

4 Avoiding Redundant Structural Ambiguity: Undocumented Restrictions on Tibetan Syntax

Ambiguity of formal syntactic structures is often produced not merely by intrinsic linguistic units' polysemy, but rather by combinatorial redundancy of the formal grammar itself. Nevertheless, exactly in these fairly frequent cases, ambiguity of formal structures shows lack of accuracy in conventional informal descriptions of language, and works as a clue to choose one of several possible ways to specify these descriptions.

As for Tibetan grammar, this on-going study has already shown some formal ambiguity cases of this kind. The examples below show how only three cases of description opacity can produce a combinatorial explosion in quite a short sentence (འདི་སྐད་བདག་གིས་ཐོས་པ, The story about this is heard by me/The one who told this is heard by me/Those who told this are heard by me).

First off, it is not strictly specified in any of existing Tibetan grammar descriptions, including the most detailed one [1], if Tibetan predicates can be omitted. It is known that link-verbs are omitted in composite nominal predicates, but it is not clear, whether the whole predicative VP can be omitted like in Russian, or it is obligatory in any sentence like in English.

As Fig. 2 shows, allowing predicate ellipsis makes the analyzed sentence about 5.3 times more ambiguous (333 vs. 63 versions of parsing). Supposing predicate ellipsis produces not only obvious versions like 'The story about this that I heard (zero predicate)', but also quite weird hypotheses like 'This (zero predicate), the

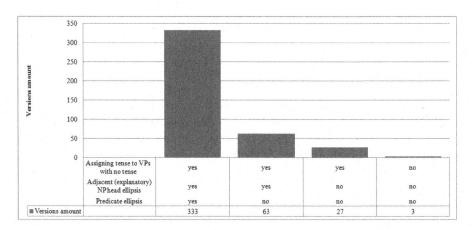

Fig. 2. Amount of versions depending on ways to formalize Tibetan syntax

story (zero predicate), (something) heard by me (zero predicate)', which seem to be ungrammatical, and a lot of their possible combinations.

Another problem of Tibetan syntax formalization that has proven to be quite important is the question about NP head ellipsis limitations. Nouns are very often omitted in Tibetan, especially in proper names, but it is absolutely unclear from existing linguistic descriptions, if such ellipsis is possible in adjacent NPs (e.g., རྒྱལ་ བུ་ རྒྱལ་ བྱེད་ཀྱི་ ཚལ་ མགོན་ མེད་ ཟས་ རྗིན་ གྱི་ ཀུན་ དགའ་ ར་ བ,, king Making-Victory's grove - the (merchant) Giving-Food-To-the-Unprotected-Ones's amusement park). Figure 2 shows that allowing ellipsis in adjacent NPs makes ambiguity level more than 2.3 times higher (63 vs. 27 options). This ellipsis, when allowed, produces ambiguity for each attribute of an NP, as attributes are postpositional in Tibetan and never have any markers to distinguish them from adjacent NPs. Prohibiting ellipsis in such NPs can be achieved by creating a separate constituent class for them. Generally, we can say that ellipsis and related issues in Tibetan require more theoretical research than is currently available in [1,6,7], to give a few examples.

One of the most important difficulties in Tibetan grammar formalization, however, is the problem of verbal tense. Tense is not expressed by any separate marker, but is denoted by verbal root allomorph itself. The problem is that not nearly all Tibetan verbal roots have different allomorphs for different tenses; it is the case for many verbs that tense remains unexpressed in the sentence at all. There are two options to deal with this phenomenon in formal grammar: (1) to build hypotheses for all three possible tenses for each verb root (2) to create different constituent classes (with tense feature and without tense feature) for sentences, predicates, VPs, participle and masdar phrases, etc. First option may seem attractive, as it allows to make grammar shorter, but, as Fig. 2 shows, it makes the above-mentioned sentence 9 times more ambiguous (3 versions for 2 verbs make 3*3 = 9 combinations). The conclusion is therefore obvious, that the second option, which means that such verbs and their phrases are not ambiguous in terms of tense, but rather have no tense at all, is far more plausible for Tibetan.

5   Evaluation

Analysis of 94,814 versions of parsing a sample of 374 different morpheme combinations (from 2-morphemic, and up to 36-morphemic) from the developed part of corpus has shown that all versions of parsing are correct from the formal point of view, except for 18361 versions that violate the requirements of verb-meaning dative object government model (which are pretty much semantic, and not syntactic). This result corresponds to 100 % recall, and, at least, 80 % precision of analysis. However, the average number of formally correct parsing versions for each combination was 974, with an exponential dependence on the number of morphemes, approximated by (1).

$$2,0403\,e^{0,2336\,x} \tag{1}$$

E.g.,   there   are   3,020   formally   correct   versions   of   parsing, for   example,   for   a   compound   sentence
འདི་སྐད་བདག་གིས་ཐོས་པ་དུས་གཅིག་ན་བཅོམ་ལྡན་འདས་མཉན་དུ་ཡོད་པ་ན་རྒྱལ་བུ་རྒྱལ་བྱེད་ཀྱི་ཚལ་མགོན་མེད་ཟས་སྦྱིན་གྱི་ཀུན་དགའ་ར་བ་ན་བཞུགས་སོ།། .
Analysis of these 3020 versions has shown the urgent need for different semantic restrictions, and the impossibility to resolve this ambiguity at the level of syntax.

6   Conclusion and Further Work

Computational linguists dealing with Tibetan language data face new kinds of challenges which are characteristic of this language combining isolation with agglutination. It turns out that many traditional techniques and concepts are not directly applicable for this data, and new ways of text processing should be developed.

We have identified some problems which arise during development of a morphosyntactic analyzer for Tibetan texts, and offered some solutions, such as adjacent NP head and predicate ellipsis prohibition, and discarding artificial assignment of tense to VPs with no tense.

Further prospects imply that (1) more theoretical research should be done on ellipsis in Tibetan, (2) more available corpus material should be syntactically annotated to reveal the maximum scope of syntactic constructions of Tibetan, (3) semantic dimension is to be added to the current version of Tibetan language module of the morphosyntactic analyzer in order to reduce ambiguity of Tibetan syntax further at later stages of this research.

Acknowledgment. Development of the morphosyntactic analyzer for the Tibetan language is supported by the grant No. 16-06-00578 "Morphosyntactic Analyser of Texts in the Tibetan language" by Russian Foundation for Basic Research. Authors also acknowledge Saint-Petersburg State University for a research grant 2.38.293.2014 "Modernizing the Tibetan Literary Tradition" which enabled the conceptual study of the original Tibetan texts.

References

1. Beyer, S.: The Classical Tibetan Language. State University of New York, New York (1992)
2. Gladkii, A.V.: Syntactic Structures of Natural Language in Automated Communication Systems [Sintaksicheskie struktury estestvennogo jazyka v avtomatizirovannyh sistemah obshhenija]. Nauka, Moscow (1985)
3. Dobrov, A.V.: Automatic Classification of News by Means of Syntactic Semantics [Avtomaticheskaja rubrikacija novostnyh soobshhenij sredstvami sintaksicheskoj semantiki]. Doctoral thesis, Saint-Petersburg State University (2014)
4. Grokhovskiy, P., Khokhlova, M., Smirnova, M., Zakharov, V.: Tibetan linguistic terminology on the base of the Tibetan traditional grammar treatises corpus. In: Král, P., et al. (eds.) TSD 2015. LNCS, vol. 9302, pp. 299–306. Springer, Heidelberg (2015). doi:10.1007/978-3-319-24033-6_34
5. Tseitin, G.S.: Programming in Associative Networks [Programmirovanie na associativnyh setjah], Computers in Designing and Manufacturing [EVM v proektirovanii i proizvodstve] (2). Mashinostroenie, Leningrad (1985)
6. Andersen, P.K.: Zero-anaphora and related phenomena in classical Tibetan. Stud. Lang. **11**, 279–312 (1987)
7. Denwood, P.: Tibetan. John Benjamins Publishing, Amsterdam (1999)

Automatic Question Generation Based on Analysis of Sentence Structure

Miroslav Blšták[✉] and Viera Rozinajová

Institute of Informatics, Information Systems and Software Engineering,
Faculty of Informatics and Information Technologies,
Slovak University of Technology in Bratislava, Ilkovičova 2, Bratislava, Slovakia
{miroslav.blstak,viera.rozinajova}@stuba.sk

Abstract. This paper presents a novel approach to the area of automated factual question generation. We propose a template-based method which uses the structure of sentences to create multiple sentence patterns on various levels of abstraction. The pattern is used to classify the sentences and to generate questions. Our approach allows to create questions on different levels of difficulty and generality e.g. from general questions to specific ones. Other advantages lie in simple expansion of patterns and in increasing the text coverage. We also suggest a new way of storing patterns which significantly improves pattern matching process. Our first results indicate that the proposed method can be an interesting direction in the research of automated question generation.

Keywords: Automatic question generation · Template-based methods

1 Introduction

Nowadays, there are many opportunities to find information about various topics of interest. This fact contributes also to the area of education. Verification of knowledge is also an important part of educational process. Here we need a human expert (teacher), who is able to examine the students. It would be useful to support or even substitute people by machines in this activity. In this paper we focus on automatic question generation (AQG) task. The input is educational text and the output is a set of factual questions from it. In comparison to question answering task, this is the complementary problem. The most common usage of AQG is in the field of education [1,2,4,5,7]. It is useful in many other domains, for example in question answering systems [13], in creating database of frequently asked questions [10] or in applying personalization in Online Courses. The significance and importance of this topic was stressed by workshop Question Generation Shared Task and Evaluation Challenge [8,9].

Research in AQG is an actual topic. Various approaches are used to solve this task. We focus on template-based approaches and approaches based on sentence structure analysis. Our contribution is to use multiple sentence patterns in order to improve this process by generating larger number of questions of different types. The remaining part of this document is organized as follows: in the second section we mention related works, the third section contains summarization of

© Springer International Publishing Switzerland 2016
P. Sojka et al. (Eds.): TSD 2016, LNAI 9924, pp. 223–230, 2016.
DOI: 10.1007/978-3-319-45510-5_26

template-based methods and in the fourth part we introduce our approach of combining patterns. The fifth section is devoted to evaluation of our approach and in the last section we make conclusions and suggest future work.

2   Related Work

When we look back, one of the first system specialized to question generation task was AUTOQUEST [11]. It was a console application which generated questions from sentences utilizing sentence structure. The quality of questions was low, but many works adopted ideas from this work. Actual works oriented on factual questions generation [2–4] used methods based on templates and rules. In [4] authors presents overgenerate-and-rank algorithm which consists of three steps: the input sentences are simplified to reduce number of words, then the declarative sentences are turned into a set of questions by syntactic transformation rules (e.g. inserting wh-word[1], subject-auxiliary inversion etc.) and generated questions are ranked by score representing expected quality of generated question. In [2], the sentence patterns consist of three parts: subject, object and preposition. Each is classified into some coarse class (e.g. human, entity, time). They created 90 predefined rules based on classes and each class contains set of possible questions. In [3], they used lexico-syntactic patterns consisting from part-of-speech (POS) tags sequence and learned from existing set of question/answer pairs. Software system based on template driven scheme has been proposed by [6]. User imports the text and system identifies the sequence of words and POS tags and create question based on this sentence template (if exists in database) or asks the user to create question manually. We mentioned approaches based on lexical and syntactic sentence patterns. Semantic information about words can be also used [2,4,12]. Typically, the semantic information about nouns are used to recognize, if the noun is object or human. This information also helps in coreference resolution or word sense disambiguation.

3   Template-Based Methods

Generally, template-based methods work with a set of sentence patterns into which are assigned the sentences from input text. Questions are generated by pattern each pattern handles predefined set of possible questions. Differences in various approaches lie in types of pattern (e.g. pattern consists of triplet subject-verb-object or sequence of POS classes or something else). When we analyzed various AQG methods which are usable to create factual questions from educational text, we found out that methods based on templates look advisable, but have some limitations. These limitations were for us the challenges to solve. First, there is a task to find out compromise, how to choose suitable level of sentence pattern specificity. When the patterns are too general (e.g. consist of triplets like subject-object-preposition [2]), they cover different sentences, but the

---

[1] An interrogative or relative pronoun (what, why, where, which, who, or how).

generated questions are also too general (e.g. questions like: *What is X, When occurs X*). On the other hand, when the patterns are more specific (e.g. pattern represented by sequence of POS tags), one pattern covers less sentences and the amount of patterns necessary for covering comparable set of sentences as before, is significantly larger. In area of factual questions, we often need additional information about words (e.g. if the word is entity or location). Patterns enriched by these information are more accurate on one hand, but also more specific, so they cover less types of sentences. Let us go over the mentioned problems. When more specific patterns are used, the total number of them increases rapidly. It created some problems: which pattern to choose in case that there are found more usable patterns or what to do, if no pattern is found. It is also difficult to create large set of patterns manually. In related work, these problems are not solved. Authors mention that the quality of system will increase with creation of new patterns. When we look at simple patterns used in [2] consisting of three parts (subject-object-preposition) and five possible classes for each part, the maximum number of patterns is 150. In works [3,4,6] where the pattern consists of POS sequence, each word can be classified into one of about 38 word classes[2]. So when we want to cover sentences up to 10 words, we need millions of patterns. And the variety of patterns will be even greater, if we want to take into account semantic information about nouns (e.g. locations, humans etc.). We consider template-based methods as perspective, but it is necessary to solve limitations, when we want to use them. It has some limitations and our approach looks advisable to solve them.

4 Our Approach

Our primary goal is to design template-based AQG process with intent to generate factual questions with some traits: to generate general and also specific questions, to generate questions about various topics of interests and to cover various types of sentences.

4.1 Text Preprocessing

Firstly, the text is split into sentences and tokens are annotated by tools able to recognize POS tags and semantic meaning (e.g. named entities). Sequences of annotated tokens are used as patterns. In comparison to other works mentioned in the previous section, sentence could have more patterns at various levels of details (e.g. sequence of POS tags or sequence of named entity tokens). Output of the preprocessing step is a set of annotated sentences. As the last step of preprocessing, we try to simplify the sentences with intent to reduce number of tokens, but with constraint not to remove words. We merge the related tokens terms consisting of more tokens are merged into one (e.g. term *United States of America* has four tokens of different tags which we merge to one atomic token).

---

[2] Classifications of tokens by Stanford POS Tagger.

4.2 Structures of Sentence as Patterns

We use combination of patterns to classify the sentence and to generate questions based on this classification. One sentence could have more patterns according to various levels of details of its tokens. Examples of these levels are: POS tags, named entity tags (NER), super sense tags (SST) obtained from WordNet dictionary or linked data categories. In Table 1 there are examples of sentences and types of patterns.

Table 1. Structures of sentence on various levels of abstraction

Sentence	The	first	president	of	Slovakia	was	Michal Kovac
POS	DT	JJ	NN	IN	NNP	VBZ	NNP
NER					Location		Person
SST			noun.role		noun.location		
Linked Data					country		Person

As a base pattern we use POS sequence. Let's consider these sentences:

- Michael is a nice boy.
- Slovakia is a small country.

At POS level, the sentences are same. In the first sentence, the noun represents person and in the second sentence it refers to location. Using named entity tags, we can distinguish what type of interrogative pronoun it is better to use (who or what). This type of differentiation by named entity recognition was already used in some related work, but only as conditional rule. We combine NER, POS and other sentence patterns together. This allows us to generate different questions for sentences with the same POS pattern, so they could be more specific. Pattern acquired from WordNet (marked as SST) reveals that token *president* has semantic meaning *role*. So we can create more specified question (e.g. *What is role of person X?*). Also, by help of linked data concept, we get the information that Slovakia is a country. Default question (*Where is Bratislava?*) should be then replaced by more specific question (*In which country is Bratislava?*). This approach also resolves the problem of extracting information about token, if a particular tool does not recognize it. Named entity recognizers usually identify only a few word categories (e.g. person or location). Combination of patterns helps us to overcome this lack of information, because information from one pattern can replace missing information in the other.

4.3 Working with Multiple Patterns

It is obvious that larger amount of patterns allows to generate more questions. Adding new patterns also causes other problems like how to set up priority or how to search for pattern effectively. Let's consider some sentences (Table 2).

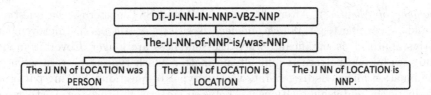

Fig. 1. Hierarchy of sentence patterns based on level of abstraction.

Sentences have the same POS pattern. In sentence number 2, the last two tokens were merged into one in preprocessing step. When we analyze named entities, the sentence patterns are different in some words of type NNP. There are several types of entities into which are words classified: location (Slovakia, Bratislava, Albania, Slovak Republic) or person (Michal Kovac). Some entities (Albanian in this case) are unknown for entity recognizer tools. In these cases, the tokens inherit tag from upper pattern. In Fig. 1, there are patterns of sentences and its classification in hierarchy.

The combination of multiple patterns causes that the process will be computationally more expensive, but the pattern matching should use hierarchy class is searched by traversing through hierarchy. We start at the top of hierarchy and after some patterns are matched, we go deeper to search for more detailed patterns (child nodes). This allows us to effectively find general and specific patterns (Table 3).

We have demonstrated our approach on simple sentences. The combination of multiple patterns allows us to create more specific questions. If we had only POS pattern, we would be limited to ask simple questions (e.g.: *What is the official*

Table 2. Example of sentences and its syntactic pattern

POS	DT	JJ	NN	IN	NNP	VBZ	NNP
1	The	capital	city	of	Slovakia	is	Bratislava
2	The	first	president	of	Slovakia	was	Michal Kovac
3	The	official	language	of	Albania	is	Albanian
4	The	alternative	name	of	Slovakia	is	Slovak Republic

Table 3. Question generation by POS sentence pattern

POS pattern	DT	JJ	NN	IN	NNP1	VBZ	NNP2
Sentence 1	The	capital	city	of	Slovakia	is	Bratislava
Sentence 2	The	official	language	of	Albania	is	Albanian
Question pattern	What	VBZ	DT	JJ	NN	IN	NNP1
Question 1	What	is	the	capital	city	of	Slovakia?
Question 2	What	is	the	oficial	language	of	Albania?

language of Albania?). After obtaining NER and SST patterns, the question should be specific (e.g.: *Which language is spoken in Albania?*). Improvement of this approach is significant, when the sentences are longer. Covering many sentences by one pattern is difficult, no matter which type of pattern we use. Combination allows us to generate various types of questions. Storing patterns into hierarchy and using relationships between patterns improve the effectiveness of generating questions.

5   Evaluation

Typically, the AQG system evaluation is focused on scoring the number of generated questions and its quality [2,4,8,9]: number of generated and acceptable questions from the given text. Our objective is to improve the coverage of text and to generate questions of various level of abstraction. In order to compare the methods according to these criteria, we proposed an experiment: we took a piece of text and we investigated the number of patterns and the extent of necessary browsing over them. Input sentences contain four tokens and each can be classified into subset of classes (we use subset due to the endeavour to make explanation and calculation more understandable). Tokens in experiment can be classified only into these categories:

- POS: proper noun (NNP), noun (NN), verb (VBZ) or adjective (JJ).
- NER (for proper noun tokens): location, person or organization.
- SST (for adjective tokens): color or property.

In Table 4, there are sample sentences. In Table 5, there are questions generated from these sentences. In case of questions focused on adjectives (last two rows marked as JJ), there are two possibilities of specificity.

We need 24 patterns to cover all example sentences (combination without repetition). If NER patterns are used, there are 3 possible categories for NNP, so there are 72 patterns totally. In addition, if SST pattern is used (two categories for JJ tag) there are 144 patterns in total. Adding SST pattern allows us to generate specified question for adjective (so the number of possible questions is 5). In our approach, we search through POS patterns in first cycle and after the POS pattern is matched, we searched in NER patterns assigned to this POS pattern (3 possible patterns). We need only 27 patterns to search (instead of 72) (Table 6).

Table 4. Example sentences and their patterns.

	John ate green apple	Rome won famous medal
POS	NNP VBZ JJ NN	NNP VBZ JJ NN
NER	Person VBZ JJ NN	Location VBZ JJ NN
SST	NNP VBZ color NN	NNP VBZ property NN

Table 5. Example sentences and their patterns.

Question type	John ate green apple	Rome won famous medal
NNP	Who ate green Apple?	Which location won famous medal?
NN	What did John eat?	What did Rome win?
JJ (general)	How was apple?	How was medal?
JJ (specific)	What color was apple?	What property has the medal?

Table 6. Number of patterns used for question generation process

	Basic approaches		Our approach	
	POS + NER	POS + NER + SST	POS + NER	POS + NER + SST
# Questions	4	5	4	5
Without token repetition				
# Patterns	72	144	72	144
# Searched p.	72	144	27	29
With token repetition				
# Patterns	768	1536	768	1536
# Searched p.	768	1536	337	353

If we take into consideration that the number of categories is significantly larger (e.g. there are 38 POS categories instead of 4 and 5–7 NER categories), we can state that state-of-the-art approaches are not well applicable. The pattern hierarchy allows to manage greater number of them and use them in combination.

6   Conclusion and Future Work

We presented a novel approach to using template-based methods in AQG task. We use combination of multiple patterns. Combination causes that the process will be more complex, but it allows to generate various types of questions and covers various types of sentences. Whereas there are relations between patterns, it is possible to classify them into hierarchy. This concept allows to search for patterns more effectively.

We compared our approach to approaches used in related works and found out that we match the patterns more efficiently. Without storing patterns into hierarchy, it is hardly possible to use combination of multiple patterns, because the total number of searched patterns grows rapidly.

There are some directions for further improvement. Quantity and quality of questions depends on number of transformation patterns. We suggest leveraging existing pairs of declarative sentence related to generated question. We can obtain these pairs from existing datasets (e.g. question answering task on Text REtrieval Conferences) or create computer interface similarly as it is used in parallel text editors. We also consider to use another natural language tools for obtaining more detailed patterns, e.g. hypernym words or abstracted terms.

Acknowledgments. The work reported here was supported by the Scientific Grant Agency of Slovak Republic (VEGA) under the grants No. VG 1/0752/14, VG 1/0646/15 and ITMS 26240120039.

References

1. Ali, H., Chali, Y., Hasan, S.A.: Automation of question generation from sentences. In: Proceedings of QG2010: The Third Workshop on Question Generation, Pittsburgh, USA, pp. 58–67 (2010)
2. Ali, H.D.A.D.: Automatic question generation: a syntactical approach to the sentence-to-question generation case. Ph.D. thesis, Lethbridge, Alta.: University of Lethbridge, Deparmant of Mathematics and Computer Science (2012)
3. Curto, S., Mendes, A.C., Coheur, L.: Question generation based on lexico-syntactic patterns learned from the web. Dialogue Discourse **3**(2), 147–175 (2012)
4. Heilman, M.: Automatic factual question generation from text. Ph.D. thesis, Carnegie Mellon University (2011)
5. Huang, Y.T., Tseng, Y.M., Sun, Y.S., Chen, M.C.: Tedquiz: automatic quiz generation for ted talks video clips to assess listening comprehension. In: 2014 IEEE 14th International Conference on Advanced Learning Technologies, pp. 350–354, July 2014
6. Hussein, H., Elmogy, M., Guirguis, S.: Automatic English question generation system based on template driven scheme. Int. J. Comput. Sci. Issues (IJCSI) **11**(6), 45–53 (2014)
7. Mostow, J., Jang, H.: Generating diagnostic multiple choice comprehension cloze questions. In: Proceedings of the 7th Workshop on Building Educational Applications Using NLP, pp. 136–146. Association for Computational Linguistics (2012)
8. Rus, V., Graessar, A.: The question generation shared task and evaluation challenge (workshop report) (2009)
9. Rus, V., Wyse, B., Piwek, P., Lintean, M., Stoyanchev, S., Moldovan, C.: A detailed account of the first question generation shared task evaluation challenge. Dialogue Discourse **3**(2), 177–204 (2012)
10. Sneiders, E.: Automated FAQ answering with question-specific knowledge representation for web self-service. In: 2nd Conference on Human System Interactions, 2009, pp. 298–305. IEEE (2009)
11. Wolfe, J.H.: Automatic question generation from text-an aid to independent study. ACM SIGCSE Bull. **8**(1), 104–112 (1976)
12. Yi-Ting, L., Chen, M.C., Yeali, S.: Automatic text-coherence question generation based on coreference resolution. In: 2009 ICCE 17th International Conference on Computers in Education, pp. 5–9 (2009)
13. Zhao, S., Wang, H., Li, C., Liu, T., Guan, Y.: Automatically generating questions from queries for community-based question answering. In: IJCNLP, pp. 929–937 (2011)

CzEng 1.6: Enlarged Czech-English Parallel Corpus with Processing Tools Dockered

Ondřej Bojar[⊠], Ondřej Dušek, Tom Kocmi, Jindřich Libovický,
Michal Novák, Martin Popel, Roman Sudarikov, and Dušan Variš

Faculty of Mathematics and Physics, Institute of Formal and Applied Linguistics,
Charles University in Prague, Prague, Czech Republic
{bojar,odusek,kocmi,libovicky,mnovak,popel,
sudarikov,varis}@ufal.mff.cuni.cz

Abstract. We present a new release of the Czech-English parallel corpus CzEng. CzEng 1.6 consists of about 0.5 billion words ("gigaword") in each language. The corpus is equipped with automatic annotation at a deep syntactic level of representation and alternatively in Universal Dependencies. Additionally, we release the complete annotation pipeline as a virtual machine in the Docker virtualization toolkit.

Keywords: Parallel corpus · Automatic annotation · Machine translation

1 Introduction

We present the new release of our Czech-English parallel corpus with rich automatic annotation, CzEng 1.6. The version number is aligned with the year of the release, 2016.

CzEng 1.6 is the fifth release of the corpus and serves as a replacement for the previous version, CzEng 1.0 [1]. The parallel corpus CzEng was successfully used in multiple NLP experiments, most notably in the WMT shared translation tasks since 2010, see [2] through [3].[1] A pre-release of CzEng 1.6 has been already released and this year's WMT shared task is based on it.

CzEng releases are freely available for research and educational purposes and restricted versions of CzEng have been separately licensed for commercial use.

The main aim of the current release is to update and enlarge the collection of sources and to provide CzEng users with all tools needed to replicate the automatic annotation on other data.

2 CzEng 1.6 Data

CzEng 1.6 primarily uses the same data sources as the previous versions. Most of the sources grow in time and some can be better exploited. Table 1 summarizes the number of parallel sentences and surface (a-layer) and deep-syntactic (t-layer) nodes from each source for both languages. The a-layer nodes correspond to words

---

[1] http://www.statmt.org/wmt10 through http://www.statmt.org/wmt15.

© Springer International Publishing Switzerland 2016
P. Sojka et al. (Eds.): TSD 2016, LNAI 9924, pp. 231–238, 2016.
DOI: 10.1007/978-3-319-45510-5_27

Table 1. CzEng 1.6 data size.

Section	Sentence Pairs	Czech		English	
		a-layer	t-layer	a-layer	t-layer
Subtitles	39.44 M	286.70 M	211.49 M	325.20 M	208.78 M
EU legislation	10.18 M	296.19 M	219.79 M	324.11 M	200.47 M
Fiction	6.06 M	80.65 M	57.68 M	89.37 M	54.20 M
Parallel web pages	2.35 M	37.08 M	28.07 M	41.26 M	26.45 M
Technical documentation	2.00 M	12.92 M	10.10 M	13.82 M	9.65 M
Medical	1.53 M	22.30 M	16.67 M	23.08 M	15.29 M
PDFs from web	0.64 M	9.64 M	7.51 M	10.32 M	6.61 M
News	0.26 M	5.65 M	4.20 M	6.22 M	3.93 M
Navajo	35.29 k	501.01 k	371.70 k	566.33 k	352.23 k
Tweets	0.52 k	9.55 k	6.97 k	10.19 k	6.78 k
Total	62.49 M	751.65 M	555.88 M	833.96 M	525.73 M

and punctuation symbols, with English sentences being by about 10 % longer due to articles and other auxiliaries.

CzEng 1.6 is shuffled at the level of sequences of not more than 15 consecutive sentences. The original texts cannot be reconstructed but some inter-sentential relations are retained for automatic processing of coreference (Sect. 3.3). Only sentences aligned 1-1 and passing the threshold of 0.3 of our filtering pipeline were included, leading to (indicated) gaps in the sequences. The filtering pipeline reduces the overall corpus size from 75 M sentence pairs to the 62 M reported in Table 1.

We prefer to de-duplicate each source at the level of documents, where available, which necessarily leads to duplicated sentences. Comparing the overall size with the previous release (de-duplicated in the same manner), we see a substantial growth in size: 62 M vs. 15 M sentence pairs.

The largest portions of CzEng 1.6 come from movie subtitles, European legislation and fiction, as it was the case in the past. In this release, we also attempt to improve the coverage of the medical domain. In the following, we list changes since the last release for specific data sources:

European legislation was previously based on DGT-Acquis[2] in one of its preliminary versions, spanning the years 2004–2010 of the Official Journal of the European Union. Since there is no recent update of the corpus, we now use the search facility of EUR-Lex[3] to get access to more recent documents. The added benefit is that we can also obtain other documents in the collection, not only issues of the Official Journal. Particularly interesting are the Summaries of EU legislation,[4] which are written in less formal style and intended for general audience.

---

[2] https://ec.europa.eu/jrc/en/language-technologies/dgt-acquis.

[3] http://eur-lex.europa.eu/.

[4] http://eur-lex.europa.eu/browse/summaries.html.

Movie subtitles are available in the OPUS corpus [4] and since the previous CzEng release, the collection has been significantly extended with Open-Subtitles.[5] Very recently, OPUS released yet another update[6] but it did not make it in time to be included in CzEng 1.6.

Subtitles of educational videos can be obtained from other sources and represent a rather different genre than movie subtitles. Not only are the topics slightly different (and mostly, there is one clear topic for each video), but the register is different: the sentences are longer and there are nearly no dialogues. CzEng 1.6 includes translated subtitles coming from Khan Academy[7] and TED talks.[8]

Medical domain is of special interest of several European research and innovation projects. We try to extend CzEng in this direction by specifically crawling some parallel health-related web sites using Bitextor [5] and also by re-crawling EMEA (European Medicines Agency)[9] corpus because its OPUS version[10] suffers from tokenization issues (e.g., decimal numbers split) and it is probably smaller than what can be currently obtained from the database.

3   Rich Annotation in CzEng 1.6

As in previous versions, CzEng is automatically annotated in the multi-purpose NLP processing framework Treex [6][11] based on the theory of Functional Generative Description [7]. The core of the platform is available on CPAN[12] and the various NLP models get downloaded automatically.

Treex integrates many processing tools including morphological taggers, lemmatizers, named entity recognizers, dependency and constituency parsers, coreference resolvers, and dictionaries of various kinds. Many of these tools are well-known third-party solutions, such as McDonald's MST parser [8]; Treex wraps them into a unified shape. The heart of the integration is the Treex data format, where each of the processing modules (called "blocks" in Treex terminology) modifies the common data. Complete NLP applications such as a dialogue system or a transfer-based MT system are implemented using sequences of processing blocks, called "scenarios".

Figure 1 illustrates the core annotation of CzEng. The left-hand trees represent the Czech (top) and English (bottom) sentences at the surface-syntactic layer of representation (analytical, a-tree), and include morphological analysis (shown in teal). The dashed links between the trees show one of the automatic

---

[5] http://www.opensubtitles.org/.
[6] http://opus.lingfil.uu.se/OpenSubtitles2016.php.
[7] http://www.khanacademy.org/ and http://www.khanovaskola.cz/.
[8] http://www.ted.com/.
[9] http://www.ema.europa.eu/.
[10] http://opus.lingfil.uu.se/EMEA.php.
[11] http://ufal.mff.cuni.cz/treex, a web demo is available at http://lindat.mff.cuni.cz/services/treex-web/run.
[12] https://metacpan.org/release/Treex-Core.

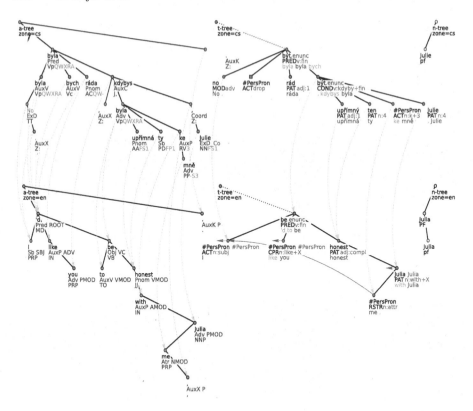

Fig. 1. One sentence pair from CzEng with the core parts of the automatic annotation.

word alignments provided in the data. The right-hand pair of trees are the deep-syntactic (tectogrammatical, t-tree) analyses and include again cross-language alignment. The blue arrows are co-reference links. The rightmost tree (n-tree) captures named entities in the sentence, as annotated by NameTag [9].

For the purposes of CzEng 1.6, we introduced several improvements into the pipeline, mostly on the deep-syntactic layer. They concern semantic role labeling (Sect. 3.1), word sense disambiguation (Sect. 3.2), and co-reference (Sect. 3.3).

3.1 Semantic Role Labels – Functors

The t-tree annotation involves assigning to each node its semantic role, *functor* [10], similar to PropBank [11] labels (shown in capitals in t-trees in Fig. 1). Functor assignment in CzEng 1.0 was handled by a logistic regression classifier [12] based on the LibLINEAR package [13]. Using Prague Dependency Treebank 2.5 [10] for Czech and Prague Czech-English Dependency Treebank (PCEDT) 2.0 [14] for English, we trained a new linear classifier using the VowpalWabbit library [15] and features based on syntactic and topological context. Automatic analysis with projected gold-standard labels is used for training to make the method more robust. We achieve about 2 % accuracy gain in comparison to the previous method; classification accuracy on the evaluation sections of the treebanks used is 80.16 % and 78.12 % for Czech and English, respectively.

3.2 Verbal Word Sense Disambiguation

CzEng 1.6 includes word sense disambiguation in verbs [16], providing links to senses in valency lexicons for Czech and English. The setup used in parallel analysis exploits information from both languages (lemmas of aligned nodes) and the CzEngVallex bilingual valency lexicon [17] to gain better annotation accuracy.

3.3 Coreference Resolution

All the documents also contain annotation of coreference. In Czech, this is performed by the same Treex internal resolvers that were used in annotating CzEng 1.0. It covers coreference of pronouns and zeros. For English, coreference annotation has been provided mostly by BART 2.0 [18,19]. BART is a modular end-to-end toolkit for coreference resolution. It covers almost all kinds of anaphoric expressions including nominal phrases. Only relative pronouns must be processed by the Treex internal resolver. To smooth down the processing pipeline, we set limits on BART's time and memory usage, which may cause that some documents are excluded from coreference annotation. However, it happens only in around 1 % of CzEng documents. In addition, anaphoricity detection by the NADA tool [20] is applied to instances of the pronoun *it*. For an instance declared as non-anaphoric, a possible coreferential link assigned by BART is deleted.

Furthermore, coreferential expressions are exploited to improve word alignment. We use the approach presented in [21]. It is based on a supervised model trained on 1,000 sentences from PCEDT 2.0 [14] with manual annotation of word alignment for coreferential expressions. The only difference is that we do the analysis of PCEDT completely automatically in order to obtain the features distributed similarly to CzEng. Using this approach the alignment quality rises from 78 % to 85 % and from 71 % to 85 % for English and Czech coreferential expressions, respectively.

4 Analysis Dockered

While the whole Treex platform is in principle freely available and a lot of effort is invested in lowering the entry barrier, it is still not very easy to get the pipeline running, especially with all the processing blocks utilized in CzEng.

If a part of CzEng rich annotation is used as training data for an NLP task, CzEng users naturally want to apply such models to new sentences. Indeed, we have received several requests to analyze some data with the same pipeline as CzEng was annotated.

With the current release, we want to remove this unfortunate obstacle, and we release CzEng 1.6 with the complete monolingual analysis pipeline (for both Czech and English) wrapped as a Docker[13] container. Docker is a software bundle designed to provide a standardized environment for software development and execution through container virtualization. Docker's standardized containers allow us to make Treex installation automatic, even without the knowledge of the user's physical machine environment. This should make it very easy to replicate the analysis on most operating systems.

An important added benefit is that the whole processing pipeline will be frozen in the version used for CzEng. This strong replicability is very hard to achieve even with current solid versioning, because some of the processing blocks depend on models that cannot be included in the repository for space or licensing reasons. Dockering CzEng analysis will thus help also ourselves.

We release two Treex-related Docker images: *ufal/treex*,[14] which creates a container with the latest release of Treex, and *ufal/czeng16*,[15] which contains Treex frozen on the revision that was used to process CzEng 1.6. Both images are aimed at simplifying the process of Treex installation; the latter providing the means to easily reproduce the CzEng 1.6 monolingual analysis scenario.

The Dockerfile that is used to build the *ufal/czeng16* image simply specifies all the dependencies that have to be installed to run the Treex modules correctly, then it clones the Treex repository from GitHub[16] and configures the necessary environment variables. It also downloads and installs dependencies that are not available in the repository (mainly Morče tagger and NADA tool).

To run the analysis pipeline, you just need to pull the CzEng 1.6 Docker image from the Docker Hub repository and follow the instructions in the Readme file. The pipeline is able to process data as a filter (read standard input and write into standard output) as well as process multiple input files and save the results into all CzEng 1.6 export formats.

5 Availability

CzEng 1.6 is freely available for educational and research non-commercial uses and can be downloaded from the following website:

$$\text{http://ufal.mff.cuni.cz/czeng}$$

It is available in the following file formats; the first three are identical with the previous release, the last one is new.

Plain text format is very simple and has only four tab-separated columns on each line: sentence pair ID, filter score, Czech and English sentence.

---

[13] https://www.docker.com/.
[14] https://hub.docker.com/r/ufal/treex/.
[15] https://hub.docker.com/r/ufal/czeng16/.
[16] https://github.com/ufal/treex.

Treex XML format is the format used internally in the Treex NLP toolkit. It can be viewed and searched with the TrEd tool.[17]

Export format is a line-oriented format that contains most of the annotation. It uses one sentence per line, with tab-separated data columns (see the website and [1] for details).

CoNLL-U is a new text-based format with one token per line, introduced within the Universal Dependencies project[18] and further enriched with word alignment and multilingual support within the Udapi project.[19] The a-trees were automatically converted to the Universal Dependencies style, the t-trees are missing in CoNLL-U.

6   Conclusion

We introduced a new release of the Czech-English parallel corpus CzEng, version 1.6. We hope that the new release will follow the success and popularity of the previous version, CzEng 1.0.

CzEng 1.6 is enlarged, contains a slightly improved and extended linguistic annotation, and the whole annotation pipeline is now available for simple installation using the Docker tool. This makes it much easier to annotate other data than what is already provided in the corpus, which has been one of the major drawbacks of the previous CzEng release. We hope that by wrapping the analysis pipeline as a Docker container, we make an important step to making the annotation widely usable.

Acknowledgement. We would like to thank to Christian Buck for providing us with raw crawled PDFs. This project received funding from the European Union's Horizon 2020 research and innovation programme under grant agreement 645452 (QT21) and 644402 (HimL), and also from FP7-ICT-2013-10-610516 (QTLeap), GA15-10472S (Manyla), GA16-05394S, GAUK 338915, and SVV 260 333. This work has been using language resources and tools developed and/or stored and/or distributed by the LINDAT/CLARIN project of the Ministry of Education, Youth and Sports of the Czech Republic (project LM2015071).

References

1. Bojar, O., Žabokrtský, Z., Dušek, O., Galuščáková, P., Majliš, M., Mareček, D., Maršík, J., Novák, M., Popel, M., Tamchyna, A.: The joy of parallelism with CzEng 1.0. In: LREC, Istanbul, Turkey, pp. 3921–3928 (2012)
2. Callison-Burch, C., Koehn, P., Monz, C., Peterson, K., Przybocki, M., Zaidan, O.: Findings of the 2010 joint workshop on statistical machine translation and metrics for machine translation. In: Joint WMT and MetricsMATR, pp. 17–53 (2010)

---

[17] http://ufal.mff.cuni.cz/tred.

[18] http://universaldependencies.org/.

[19] http://udapi.github.io/.

3. Bojar, O., Chatterjee, R., Federmann, C., Haddow, B., Huck, M., Hokamp, C., Koehn, P., Logacheva, V., Monz, C., Negri, M., Post, M., Scarton, C., Specia, L., Turchi, M.: Findings of the 2015 workshop on statistical machine translation. In: WMT, Lisboa, Portugal, pp. 1–46 (2015)
4. Tiedemann, J.: News from OPUS - a collection of multilingual parallel corpora with tools and interfaces. In: RANLP, pp. 237–248 (2009)
5. Esplà-Gomis, M., Forcada, M.L.: Combining Content-based and URL-Based Heuristics to Harvest Aligned Bitexts from Multilingual Sites With Bitextor. Prague Bulletin of Mathematical Linguistics, vol. 93. Charles University (2010)
6. Popel, M., Žabokrtský, Z.: TectoMT: modular NLP framework. In: Loftsson, H., Rögnvaldsson, E., Helgadóttir, S. (eds.) IceTAL 2010. LNCS, vol. 6233, pp. 293–304. Springer, Heidelberg (2010)
7. Sgall, P., Hajičová, E., Panevová, J.: The Meaning of the Sentence and Its Semantic and Pragmatic Aspects. Academia/Reidel Publishing Company, Prague (1986)
8. McDonald, R., Pereira, F., Ribarov, K., Hajič, J.: Non-projective dependency parsing using spanning tree algorithms. In: HLT/EMNLP, pp. 523–530 (2005)
9. Straková, J., Straka, M., Hajič, J.: Open-source tools for morphology, lemmatization, POS tagging and named entity recognition. In: Proceedings of ACL: System Demonstrations, Baltimore, Maryland, pp. 13–18. ACL (2014)
10. Bejček, E., Panevová, J., Popelka, J., Straňák, P., Ševčíková, M., Štěpánek, J., Žabokrtský, Z.: Prague dependency treebank 2.5 a revisited version of PDT 2.0. In: Coling, pp. 231–246 (2012)
11. Palmer, M., Gildea, D., Kingsbury, P.: The proposition bank: an annotated corpus of semantic roles. Comput. Linguist. 31, 71–106 (2005)
12. Mareek, D., Duek, O., Rosa, R.: Progress report on translation with deep generation. Project FAUST deliverable D5.5 (2012)
13. Fan, R.E., Chang, K.W., Hsieh, C.J., Wang, X.R., Lin, C.J.: LIBLINEAR: a library for large linear classification. JMLR 9, 1871–1874 (2008)
14. Hajič, J., Hajičová, E., Panevová, J., Sgall, P., Bojar, O., Cinková, S., Fučíková, E., Mikulová, M., Pajas, P., Popelka, J., Semecký, J., Šindlerová, J., Štěpánek, J., Toman, J., Urešová, Z., Žabokrtský, Z.: Announcing Prague Czech-english dependency treebank 2.0. In: LREC, Istanbul, Turkey (2012)
15. Langford, J., Li, L., Strehl, A.: Vowpal Wabbit online learning project. Technical report (2007)
16. Dušek, O., Fučíková, E., Hajič, J., Popel, M., Šindlerová, J., Urešová, Z.: Using parallel texts and lexicons for verbal word sense disambiguation. In: Depling, Uppsala, Sweden, pp. 82–90 (2015)
17. Urešová, Z., Dušek, O., Fučíková, E., Hajič, J., Šindlerová, J.: Bilingual English-Czech valency lexicon linked to a parallel corpus. In: LAW IX, Denver, Colorado, pp. 124–128 (2015)
18. Versley, Y., Ponzetto, S.P., Poesio, M., Eidelman, V., Jern, A., Smith, J., Yang, X., Moschitti, A.: BART: a modular toolkit for coreference resolution. In: ACL-HLT, pp. 9–12 (2008)
19. Uryupina, O., Moschitti, A., Poesio, M.: BART goes multilingual: the UniTN/Essex submission to the CoNLL-2012 shared task. In: EMNLP-CoNLL (2012)
20. Bergsma, S., Yarowsky, D.: NADA: a robust system for non-referential pronoun detection. In: Hendrickx, I., Lalitha Devi, S., Branco, A., Mitkov, R. (eds.) DAARC 2011. LNCS, vol. 7099, pp. 12–23. Springer, Heidelberg (2011)
21. Nedoluzhko, A., Novák, M., Cinková, S., Mikulová, M., Mírovský, J.: Coreference in Prague Czech-English dependency treebank. In: LREC (2016)

Using Alliteration in Authorship Attribution of Historical Texts

Lubomir Ivanov[✉]

Computer Science Department, Iona College,
715 North Avenue, New Rochelle, NY 10801, USA
livanov@iona.edu

Abstract. The paper describes the use of alliteration, by itself or in combination with other features, in training machine learning algorithms to perform attribution of texts of unknown/disputed authorship. The methodology is applied to a corpus of 18th century political writings, and used to improve the attribution accuracy.

Keywords: Authorship attribution · Alliteration · Lexical stress · Machine learning

1 Introduction

Authorship attribution is an interdisciplinary field, aimed at developing methodologies for identifying the writer(s) of texts of unknown or disputed authorship. Historically, authorship attribution has been carried out by human experts. Based on their knowledge, experience, and, sometimes, intuition, the experts put forth hypotheses, and attempt to prove them through an exhaustive analysis of the literary style and techniques used by the author, as well as by analyzing the content of the work in the context of the historical and political realities of the specific time period and the philosophical, political, and socio-economic views of the author. Human expert attribution, however, is tedious and difficult to perform on large texts or corpora. It often misses subtle nuances of authors' styles, and may be tainted by the personal beliefs of the attribution expert. With the rapid advance of data mining, machine learning, and natural language processing, novel authorship attribution methods have been developed, which can perform accurate computer analyses of digitized texts. There are many advantages to automated attribution: Analysis of the large texts/corpora can be carried out significantly faster and in greater depth, uncovering inconspicuous stylistic features used by authors consistently and without a conscious thought. Automated attribution is not influenced by the subjective beliefs of the attribution expert, and the results can be verified independently.

Automated authorship attribution has found a wide-spread application in areas as diverse as literature, digital rights, plagiarism detection, forensic linguistics, and anti-terrorism investigation [1–5]. A variety of techniques have emerged to handle specific types of attribution – poetry, long and short prose, historical

© Springer International Publishing Switzerland 2016
P. Sojka et al. (Eds.): TSD 2016, LNAI 9924, pp. 239–248, 2016.
DOI: 10.1007/978-3-319-45510-5_28

documents, electronic communication (e.g. email, Twitter messages), etc. The most well-known authorship attribution success is the work of Mosteller and Wallace on the Federalist papers [6]. Other notable attributions include works by William Shakespeare [7,8], Jane Austen [9], and Greek prose [10]. A number of important attribution studies of American and British writings of the late 18th century include works of Thomas Paine, Anthony Benezet, and other political writers from the time of the American and French Revolutions [11–13].

A new direction in authorship attribution is the use of prosodic features. This paper focuses on the use of alliteration. We present an algorithm for extracting alliteration patterns from written text and using them for training machine learning algorithms to perform author attribution. We also present results from combining alliteration-based attribution with traditional methods. The results are analyzed, pointing out the strengths and weaknesses of alliteration-based attribution in contrast to lexical-stress attribution. Finally, the paper presents future directions based on the use of other prosodic features.

2 Alliteration

2.1 Motivation and Definition

When analyzing the writings of Thomas Paine, a colleague commented that "there is a distinct rhythm – almost a melody – to his writing that nobody else has". This casual observation led to our effort to quantify the notion of "melody in text" and to use it for authorship attribution. In [12] we demonstrated how lexical stress can be used for authorship attribution of 18th century historical texts. Lexical stress, however, is confined to individual words, and, thus, has a relatively small influence on the flow of melody in text. Lexical stress also requires a significant literary sophistication on behalf of the author for expressing emotion in text. Other prosodic features – intonation, alliteration – are more commonly employed for providing emotional emphasis.

Alliteration is a prosodic literary technique used to emphasize and strengthen the emotive effect of a group of words in a phrase, a sentence, or a paragraph. Among the different definitions of alliteration [14–16] the most linguistically-appropriate one is that alliteration is the repetition of the same consonant sound in the primary-stressed syllables of nearby words. Some resources [17] indicate that it is also appropriate to consider the repetition of initial vowel sounds in nearby words as alliteration.

Simple examples of alliteration are common tongue-twisters such as "Peter Piper Picked a Peck of Pickled Peppers" and "But a better butter makes a batter better" as well as "catchy" company names like "Dunkin' Donuts" and "Bed, Bath, and Beyond". Aside from children's rhymes and popular advertising, alliteration has traditionally been used both in poetry and prose, dating as far back as 8th century and possibly much earlier.

For a technique so widely used in literature, alliteration has been analyzed very little with modern computer-based methodologies. One study of alliteration in "Beowulf' is presented in [18]. More recently, alliteration was mentioned in the context of forensic linguistics [19]. An analysis of alliteration's usefulness as a stand-alone measure or in combination with other features was not presented. The goal of this research is to investigate the appropriateness of using alliteration for attribution. We consider alliteration with different factors of interleaving and analyze the effectiveness of combining alliteration with other features for training classifiers to perform author attribution.

2.2 Extracting Alliteration from Text

For alliteration to be usable as an attribution feature, we must show that authors tend to use alliteration consistently and uniquely relative to other authors. Our first task was to extract all alliteration patterns from our corpus. At present, our text collection consists 224 attributed (Table 1) and 50 unattributed documents. The documents' attribution is fairly certain, although the age and nature of the material must allow for a percentage of misattributions. Other issues include the unequal number of documents per author, the different document lengths (950 to 20,000 words), and problems with the digitization of the original manuscripts.

To extract the alliteration patterns we use the Carnegie Mellon University (CMU) pronunciation dictionary [20]. Each word in the dictionary is transcribed for pronunciation into syllables with lexical stress indicated by three numeric values: 0 – no stress, 1 – primary stress, 2 – secondary stress. We use the CMU dictionary to extract the main stress syllable, and, from it, its leading consonant, for every word in each text in our corpus. Connective function words are

Table 1. Authors of attributed documents

Author	Num. of docs	Author	Num. of docs
John Adams	10	James Mackintosh	7
Joel Barlow	4	William Moore	5
Anthony Benezet	5	William Ogilvie	4
James Boswell	5	Thomas Paine	11
James Burgh	7	Richard Price	4
Edmund Burke	6	Joseph Priestley	5
Charles Carroll	3	Benjamin Rush	6
John Cartwright	13	George Sackville	2
Cassandra (pseud. of J. Cannon)	4	Granville Sharp	8
Earl of Chatham (W. Pitt Sr.)	3	Earl of Shelburne (William Petty)	3
John Dickinson	4	Thomas Spence	6
Philip Francis	4	Charles Stanhope	2
Benjamin Franklin	9	Sir Richard Temple	2
George Grenville	3	John Horne Tooke	4
Samuel Hopkins	5	John Wesley	4
Francis Hopkinson	21	John Wilkes	1
Thomas Jefferson	7	John Witherspoon	8
Marquis de Lafayette	5	Mary Wollstonecraft	7
Thomas Macaulay	7	John Woolman	6

not processed. For example, the following excerpt from [21]: "Up the aisle, the moans and screams merged with the sickening smell of woolen black clothes worn in summer weather and green leaves wilting over yellow flowers." (1) yields the string "U A M S M S S W B K W S W G L W O Y F". Each input file produces a similar string of main-stress leading consonants. These strings are then processed for alliteration using the algorithm described below.

2.3 Alliteration Algorithm

Notice that the excerpt (1) above contains alliteration, which is far more complex than the simple tongue-twister alliterations presented earlier. The "moans...merged" alliteration is interleaved with the "screams...sickening smell ...summer" alliteration, which, in turn, is interleaved with the "worn...weather ...wilting" alliteration. It is possible for one or more alliteration patterns to begin in the middle of another alliteration pattern. The new pattern may be entirely contained within the original pattern (a special case is referred to as symmetric alliteration, e.g. "...better find friends brave..."), or may continue past the end of the original pattern. To handle complex alliteration interleaving, we developed the alliteration extraction algorithm below:

```
Create an empty list of <letter, alliterationCount> pairs
Create a queue, Q, for search start positions and enqueue position 0
while (Q is not empty) {
    Dequeue Q: get next search position, i, and set it to"processed"
    Set count = 1
    for(all position j = i+1 to end of the ''Text'' string ) {
        if (Text[j] equals Text[i]) {
            count ++
            Set position j as ''processed''
        }
        else {
            Enqueue j as a new search position in Q
            for(all positions p = j+1 to j+skipRange) {
                if (Text[p] equals Text[i])  Set i=p-1
            if (no letter[i] found in the skip range) {
                Add <Text[i],count> to the alliteration list
                Continue with the next iteration of the main while loop
                }
        }
    }
    if (not already added) Add <Text[i], count> to the alliter. list
    }
}
```

Starting with the character in position 0, the algorithm searches for alliterated characters in the input string "Text". One by one, the subsequent characters of the string are examined. If the next character matches the current search character, the alliteration count is incremented. Otherwise either the alliteration sequence is complete or the next alliterated word is not immediately adjacent. The skip range value indicates the maximum distance, which words with the same main stressed syllable leading consonant can be apart before they are no longer considered alliterated. Thus, when a different character is encountered, its position is enqueued as a new starting search position for a different alliteration sequence. Then the characters up to the length of the skip range are examined. If a character equal to the original search character is encountered, the original search resumes from that position. If no matching character is found up to the skip range, the current alliteration search is marked complete and the <character, alliter.count> is recorded. The next search position is dequeued, and the processing continues with the next alliteration sequence.

For example, when sentence (1) is processed with a skip range set to 3, it yields the alliteration patterns: <M, 2>, <S, 3>, and <W, 4>, i.e. there is a sequence of two alliterated words with the leading stressed consonant 'M', a sequence of three alliterated words with a leading stressed consonant 'S', and a sequence of four alliterated words with a leading stressed consonant 'W'. On the other hand, setting the skip range to 4 yields <M, 2>, <S, 4>, and <W, 4>. Once all alliteration patterns are recorded, the number of occurrences of each pattern is counted, and the counts are normalized, yielding the alliteration patterns' frequencies. The vector of all alliteration patterns and their frequencies is written to a file, and can be used for attribution using different classifiers.

3 Machine Learning Algorithms

3.1 Learning Methods

The success of the attribution task depends on the selection of appropriate machine learning models. In this study we used support vector machines with sequential minimal optimization (SMO) and multilayer perceptrons (MLP) implemented in WEKA [23]. The inputs are the normalized frequencies of all alliteration patterns across all files in the corpus. Our output nodes represent classification categories – one for each author.

3.2 Features

Initially, we used the frequencies of alliteration patterns. In further experiments, we combined alliteration with several traditional features: function words (fw), parts-of-speech (PoS) [24], and most frequent words (mfw). We also used feature N-grams.

3.3 Evaluating Performance

To evaluate the accuracy of a learning method, the documents are divided into training and testing sets. We adopted a "leave-one-out" validation: $n - 1$ documents are used for training, and validation is done using the remaining document. This is repeated n times, so each document is used at some point for validation. The percentage of correctly classified documents constitutes the "leave-one-out" accuracy of the method. Two other measures of performance are precision and recall. The precision of a method is the fraction of documents attributed to an author, which are indeed his/her work. The recall of a method is the fraction of an author's documents attributed to him/her.

3.4 Weighted Voting

When an expert attempts to attribute a text of unknown or disputed authorship, he/she usually examines a broad set of features. We use same approach –
a methodology, which combines different learning methods and features through
weighted voting [13]. Each method independently supports one author propor-
tionally to the method's leave-one-out accuracy. The overall support for an
author is the sum of the supports the author received from methods that choose
him/her. The accuracy-weighted method selects the author with the highest
overall support. Only methods that have an accuracy above a preset threshold
are considered.

4   Using Alliteration for Authorship Attribution

4.1 Experiments: Stand-Alone Alliteration

In the first set of experiments we wanted to test the usefulness of alliteration
as a stand alone feature for authorship attribution. We began with the full set
of 38 authors and 224 attributed documents, extracting all alliteration patterns
and their associated frequencies from the texts. There were 135 patterns, which
determined the size of our input vectors. The size of the output vectors was 38
(the number of authors). The learning methods used were SMO and MLP. The
MLP was set up with a single hidden layer using either the average (MLP-A) or
the sum (MLP-S) of the number of inputs and outputs. We varied the alliteration
skip range between 0 and 4 words.

The results of our initial experiments indicated that alliteration, as an attribution method, works well only for some authors. They scored high for both
recall and precision, while other authors performed poorly (Table 2). Across all
experiments with the full set of authors/documents, between 42 % and 50 %
of the authors scored very low in both precision and recall. This reduced the
accuracy in all full-corpus experiments.

Next, we selected a sample of 22 authors corresponding to the experiments
presented in [12] and repeated all experiments. The accuracy improved (33.58 %),
but not significantly. It is worth noting that all authors who performed well in
the full-corpus experiments continued to do so in the new set of experiments.
Some of the poorly performing authors from the first experiments (e.g. Benezet,
Franklin) did significantly better, while others (Boswell, Priestley) continued to
perform poorly.

In our third set of experiments we wanted to find out if alliteration can
distinguish among those authors who performed well in the original experiments – Burgh, Burke, Hopkins, Hopkinson, Paine, Wollstonecraft, and Woolman. Depending on the learning method and the skip range, the accuracy varied
between 60 % and 83.33 % (Table 3).

In all of the above experiments, the maximum length of the alliteration
sequences was not limited. There were documents, which contained alliteration
sequences as long as 11 alliterated words. We surmised that we can use such long

Table 2. Precision and recall results for a sample of authors (experiment with 38 authors, 224 docs, skip-Range 3, and SMO)

Author	Precision	Recall
Mary Wollstonecraft	71.4%	83.3%
John Woolman	75.0%	60.0%
Edmund Burke	57.1%	66.7%
Samuel Hopkins	44.4%	80.0%
Edmund Burke	57.1%	66.7%
Thomas Paine	41.2%	63.6%
...	...	...
John Witherspoon	25.0%	12.5%
Philip Francis	12.5%	25.0%
George Grenville	0%	0%
Richard Price	0%	0%

Table 3. Accuracy for seven authors using SMO, MLP-A, MLP-S, and skip range 1–4

Learning method	Skip range	Avg. accuracy
SMO	1	70.00%
SMO	2	83.33%
SMO	3	76.79%
SMO	4	64.82%
MLP-A	1	60.00%
MLP-A	2	80.93%
MLP-A	3	80.36%
MLP-A	4	72.22%
MLP-S	1	65.00%
MLP-S	2	83.33%
MLP-S	3	80.36%
MLP-S	4	72.22%

alliteration sequences as a style discriminator for authorship. However, experiments with alliteration sequences of length 5 or more produced poor results. Much more significant results were obtained by limiting alliteration sequence length to alliterations sequences of length between 2 and 5 words. Thus, it appears that the more frequent, shorter alliteration patterns play a more significant role for authorship attribution than the outlier long sequences.

4.2 Comparison with Lexical Stress

It is interesting to compare the performance of alliteration and lexical stress as stand-alone features for authorship attribution. Table 4 lists the maximum recorded accuracies from all experiments using lexical stress and alliteration:

Table 4. Traditional features and classifiers

Num. authors (allit/L.S.)	Num. docs (allit/L.S.)	Max accuracy (allit.)	Max accuracy (L.S.)
38/38	224/224	27.93%	30.80%
22/20	134/140	33.58%	33.57%
7/7	64/65	83.33%	73.85%

Notice that, despite the different nature of the two literary techniques, the results correlate well: The average accuracy is low when a large number of authors with diverse writing styles is considered, but improves dramatically if limited to authors with a more "melodic" writing style. It is important to note, however, that the set of top-performing authors in each case is different: For alliteration it includes Burgh, Burke, Hopkins, Hopkinson, Paine, Wollstonecraft, and Woolman. For lexical stress, the set consists of Adams, Benezet, Hopkinson,

Jefferson, Moore, Paine, Wollstonecraft. The intersection of the two sets includes Hopkinson, Paine, and Wollstonecraft, who apparently skillfully employ both lexical stress and alliteration in their writings for maximum emotive effect.

The experiments appear to support our original conjecture that some authors have a more unique, "melodic" writing style. Examining the works of Paine, Wollstonecraft, and Hopkins clearly reveals their extensive use of alliteration. Whether subconsciously or in full awareness, these authors use alliteration skillfully and consistently in order to strengthen the impact of their arguments. Moreover, the alliteration patterns they choose to use are unique, and, therefore, relatively easy to classify using machine learning. On the other hand, the writing styles of authors who either do not use alliteration or use non-distinct alliteration patterns are more difficult to capture using alliteration as the sole stylistic feature. Thus, it is clear that, just like lexical stress, alliteration, by itself, is insufficient to differentiate between arbitrary author styles.

4.3   Combining Alliteration with Other Features

In [12], it was shown that combining lexical stress with other features improves the accuracy of attribution. We wanted to find out if the same holds true for alliteration. We carried out experiments pairing the traditional features and classifiers listed in Table 5.

Table 5. Traditional features and classifiers

Traditional features	Classifiers
Character N-grams	Linear SVM
Course part-of-speech tagger	WEKA SMO
First word in sentence	WEKA MLP-A
MW function words	WEKA MLP-S
PoS	Nearest neighbor (cosine distance)
PoS N-grams	Nearest neighbor (histogram distance)
Prepositions	Centroid (cosine distance)
Suffices	Centroid (histogram distance)
Vowel-initiated words	
Word stems	
Rare words	

The average author precision and recall recorded were 86.7% and 84.32% respectively. We then added alliteration and lexical stress to the set of traditional features. The average author precision increased to 90.07% (4.65% increase), while the average recall remained unchanged. While not dramatic, these improvements indicate that alliteration and lexical stress, in combination with other traditional stylistic features, do positively affect the outcome of the authorship attribution task.

5   Conclusion and Future Work

This paper presented some results from research on the use of alliteration for authorship attribution. The work is a part of a larger project, which investigates the usefulness of prosodic features for attribution. We have demonstrated that the writing style of some authors can be readily recognized by attribution techniques based on the writers' use of alliteration. However, other authors either make no use alliteration or do not use distinctive alliteration patterns. Thus, alliteration, by itself, is not sufficient to carry out authorship attribution. However, combined with lexical stress and with traditional stylistic features, alliteration does improve the accuracy of authorship attribution and the recall and precision of the individual authors. Our work on the use of prosodic features for authorship attribution faces a number of challenges and limitations: The most significant of those is the fact that there exists no accurate time-period pronunciation dictionary. Since the first voice recordings of spoken English did not appear until the mid-19th century, there is really no way of knowing the historically accurate pronunciation and intonation patterns used during the 18th century in England and the United States. Thus, our work is based on present-day pronunciations of written words from a much earlier historical period. Moreover, many words encountered in the 18th century texts are not in the CMU pronunciation dictionary. We, therefore, rely on the expertise of humanities colleagues to augment our pronunciation dictionary with the proper pronunciation patterns of missing 18th century words. Another issue stems from the fact that some words have multiple pronunciations. For some words, the stress placement may vary between dialects of the same language, and/or affect the semantics of the word. One direction of our research is finding a robust approach for dealing with multiple pronunciations. Other future research directions include investigating the usefulness of other prosodic features such as assonance, consonance, and intonation for authorship attribution.

Acknowledgements. We would like to acknowledge the help of Dr. S. Petrovic and G. Berton, and the assistance of S. Campbell, who carried out some of the weighted voting experiments.

References

1. Stamatatos, E.: A survey of modern authorship attribution methods. J. Am. Soc. Inform. Sci. Technol. **60**(3), 538–556 (2009)
2. Abbasi, A., Chen, H.: Applying authorship analysis to extremist-group web forum messages. IEEE Intell. Syst. **20**(5), 67–75 (2005)
3. Argamon, S., Saric, M., Stein, S.: Style mining of electronic messages for multiple authorship discrimination. In: Proceedings of the 9th ACM SIGKDD, pp. 475–480 (2003)
4. de Vel, O., Anderson, A., Corney, M., Mohay, G.M.: Mining e-mail content for author identification forensics. SIGMOD Rec. **30**(4), 55–64 (2001)

5. Zheng, R., Li, J., Chen, H., Huang, Z.: A framework for authorship identification of online messages: writing style features and classification techniques. J. Am. Soc. Inf. Sci. Technol. **57**(3), 378–393 (2006)

6. Mosteller, F., Wallace, D.: Inference and Disputed Authorship: The Federalist. Addison-Wesley, Reading (1964)

7. Lowe, D., Matthews, R.: Shakespeare vs. Fletcher: a stylometric analysis by radial basis functions. Comput. Humanit. **29**, 449–461 (1995)

8. Matthews, R., Merriam, T.: Neural computation in stylometry: an application to the works of Shakespeare and Fletcher. Literary Linguist. Comput. **8**(4), 203–209 (1993)

9. Burrows, J.: Computation into Criticism: A Study of Jane Austen's Novels and an Experiment in Method. Clarendon Press, Oxford (1987)

10. Morton, A.Q.: The Authorship of Greek Prose. J. Roy. Stat. Soc. (A) **128**, 169–233 (1965)

11. Petrovic, S., Berton, G., Campbell, S., Ivanov, L.: Attribution of 18th century political writings using machine learning. J. Technol. Soc. **11**(3), 1–13 (2015)

12. Ivanov, L., Petrovic, S.: Using lexical stress in authorship attribution of historical texts. In: Král, P., Matoušek, V. (eds.) TSD 2015. LNCS, vol. 9302, pp. 105–113. Springer, Heidelberg (2015). doi:10.1007/978-3-319-24033-6_12

13. Petrovic, S., Berton, G., Schiaffino, R., Ivanov, L.: Authorship attribution of Thomas Paine works. In: Proceedings of the International Conference on Data Mining, DMIN 2014, 12–24 July 2014, pp. 182–188. CSREA Press, ISBN: 1- 60132-267-4

14. Kricka, L.: Alliteration, Again and Again. Xlibris Publication (2013). ISBN: 978-1479776467. https://www.amazon.com/Alliteration-Again-Larry-J-Kricka/dp/1479776467/ref=tmm_pap_swatch_0?_encoding=UTF8&qid=&sr=

15. Internet resource. http://literarydevices.net/alliteration/

16. Internet resource. http://www.dailywritingtips.com/alliteration/

17. Internet resource. http://www.britannica.com/art/alliteration/

18. Barquist, C., Shie, D.: Computer analysis of alliteration in beowulf using distinctive feature theory. Lit Linguist Comput. **6**(4), 274–280 (1991). doi:10.1093/llc/6.4.274

19. Kotzé, E.: Author identification from opposing perspectives in forensic linguistics. South. Afr. Linguist. Appl. Lang. Stud. **28**(2), 185–197 (2010). http://dx.doi.org/10.2989/16073614.2010.519111

20. Internet resource. http://www.speech.cs.cmu.edu/cgi-bin/cmudict

21. Angelou, M.: The Collected Autobiographies of Maya. Modern Library, New York (2004). ISBN: 978-0679643258

22. Hall, M., Frank, E., Holmes, G., Pfahringer, B., Reutemann, P., Witten, L.: The WEKA data mining software: an update. In: ACM SIGKDD Explorations Newsletter, vol. 11(1), pp. 10–18. ACM, New York (2009). doi:10.1145/1656274.1656278

23. Juola, P.: Authorship attribution. Found. Trends Inf. Retrieval **3**, 233–334 (2006)

24. Toutanova, K., Klein, D., Manning, C.: Feature-rich part-of-speech tagging with a cyclic dependency network. In: HLT-NAACL, pp. 252–259 (2003)

Collecting Facebook Posts and WhatsApp Chats
Corpus Compilation of Private Social Media Messages

Lieke Verheijen[✉] and Wessel Stoop

Radboud University, Nijmegen, The Netherlands
{lieke.verheijen,w.stoop}@let.ru.nl

Abstract. This paper describes the compilation of a social media corpus with Facebook posts and WhatsApp chats. Authentic messages were voluntarily donated by Dutch youths between 12 and 23 years old. Social media nowadays constitute a fundamental part of youths' private lives, constantly connecting them to friends and family via computer-mediated communication (CMC). The social networking site Facebook and mobile phone chat application WhatsApp are currently quite popular in the Netherlands. Several relevant issues concerning corpus compilation are discussed, including website creation, promotion, metadata collection, and intellectual property rights/ethical approval. The application that was created for scraping Facebook posts from users' timelines, of course with their consent, can serve as an example for future data collection. The Facebook and WhatsApp messages are collected for a sociolinguistic study into Dutch youths' written CMC, of which a preliminary analysis is presented, but also present a valuable data source for further research.

Keywords: Computer-mediated communication · Social media · New media · Facebook · WhatsApp · Corpus compilation · Data collection

1 Introduction

Increasingly more youths around the world, including the Netherlands, are in the habit of using social media such as SMS text messaging, chat, instant messaging, microblogging, and networking sites in their private lives on a regular and frequent basis. This has raised worries among parents and teachers alike that the informal, non-standard lingo used by youngsters while communicating via social media may have a (negative) impact upon their traditional literacy skills, i.e. writing and reading [1,2]. Before studying the possible effect of unconventional language use in social media on literacy, it is paramount to know what that language actually looks like. Yet little is known so far about the exact linguistic manifestation of Dutch social media texts, in terms of key features of writing such as orthography (spelling), syntax (grammar and sentence structure), and lexis (vocabulary). As such, a linguistic analysis into Dutch youths' written computer-mediated communication is an urgent matter for research. To conduct such a study, an up-to-date corpus of social media texts is of the utmost

© Springer International Publishing Switzerland 2016
P. Sojka et al. (Eds.): TSD 2016, LNAI 9924, pp. 249–258, 2016.
DOI: 10.1007/978-3-319-45510-5_29

importance. This paper describes the compilation of such a social media corpus, specifically of WhatsApp chats and Facebook posts. Ultimately, the corpus can help to answer the following questions: how does Dutch youths' language use on WhatsApp and Facebook differ from Standard Dutch? And how do WhatsApp and Facebook messages differ, linguistically speaking, from other new media genres, such as SMS text messages and tweets?

First, we collected WhatsApp chats. These are private online chats, which involve typed spontaneous communication in real time between two or more users of the mobile phone application WhatsApp Messenger. This instant messaging client, whose name is a contraction of 'what's up' and 'application', was released in 2010 and has since then enormously gained in popularity among Dutch smartphone users. It was acquired by the Facebook company in 2014. Secondly, we have started collecting status updates, both public and non-public, posted on Facebook timelines. This social networking service was created in 2004. Its name comes from the 'face book' directories that are often given to university students in the United States, who were the initial members of this social network. The personal Facebook timeline was introduced in 2011, when the format of users' individual profile pages was changed. In this paper, we describe the collection of these two datasets. To the best of our knowledge, this is the first social media corpus with Dutch WhatsApp and Facebook messages.

2   Related Work

The corpus compiled for this project is an addition to existing corpora of computer-mediated communication, in particular SoNaR ('STEVIN Nederlandstalig Referentiecorpus'), a freely available reference corpus of written Dutch containing some 500 million words of text that was built by Dutch and Belgian computational linguists [3–6]. SoNaR contains a variety of text sources, including some social media genres, namely online chats, tweets, internet forums, blogs, and text messages. However, two media that are currently very popular in the Netherlands are lacking, that is, Facebook and WhatsApp. As such, there is a great need for the texts collected in the present project.

The creation and analysis of CMC corpora is currently an active research area. Yet, most projects explore language data that are publicly available, which are relatively easy to obtain, such as from Twitter, Wikipedia, discussion boards, or public social networking profiles. CMC corpora with non-public language data are still sparse: they are more time-consuming and difficult to obtain, because they require active participation of contributors. The following pioneering projects are in the vanguard of private social media message collection.

A notable project similar to our WhatsApp data collection is the 'What's up, Switzerland?' project [7,8], a follow-up of the 'Sms4science' project [9]. Researchers from four universities study the language used in Swiss WhatsApp chats. For this non-commercial large-scale project, over 838,000 WhatsApp messages (about 5 million tokens) by 419 users were collected in 2014. Contact with the project's coordinators provided us with information about the set-up of their data collection; this served as an inspiration for our own collection. A related project is 'What's up, Deutschland?' [10], conducted by researchers from seven German universities. They collected over 376,000 WhatsApp messages by 238 users in 2014 and 2015. Similar to our current project, the 'What's up' projects compare WhatsApp chats to SMS text messages, and several features are investigated, e.g. linguistic structures, spelling, and emoticons/emoji.

Our Facebook data collection is comparable to that of the DiDi project [11,12]. The DiDi corpus comprises German Facebook writings by 136 voluntary participants from the Italian province of South Tyrol (around 650,000 tokens). The corpus was collected in 2013. It contains not just status updates, but also comments on timeline posts, private messages, and chat conversations. The data and corresponding metadata were acquired by means of a Facebook web application. Their linguistic analysis focuses, among other things, on the use of dialects and age-related differences in language on social network sites.

3   Creation of Websites

We created two websites to gather WhatsApp chats and Facebook posts (see http://cls.ru.nl/whatsapptaal and http://cls.ru.nl/facebooktaal), where youths could donate their own WhatsApp and Facebook messages to science. The data thus represent authentic, original, unmodified messages that were composed in completely natural, non-experimental conditions. Besides the home page, the websites contain the tabs 'Prizes', 'Instructions', 'Consent', 'FAQ', 'About us', and 'Contact'. These pages present, respectively, information on the prizes youths can win by contributing their social media messages to the research project, instructions on how they can submit their messages, consent forms that they should sign for us to be allowed to use the data, frequently asked questions, brief info about ourselves (the researchers), and a contact form.

The main difference between the two websites for gathering social media data is that the WhatsApp collection website includes an 'Instructions' page with extensive explanations on how to submit chats depending on one's mobile phone type (Android, iPhone, or Windows Phone), whereas the Facebook collection website prominently features a button for donating messages. This difference stems from the technical possibilities of submitting messages: while WhatsApp chats can be sent via email from a mobile phone (to an email address created specifically for the purposes of this data collection, whatsapp-taal@let.ru.nl), Facebook posts cannot easily be submitted by users themselves, so we retrieved them by means of a self-built application.

4   Creation of Application

To automatically retrieve posts from volunteering youths' Facebook timelines, we created a Facebook app - a piece of software that has access to data stored by Facebook via the Facebook Graph API (application programming interface, https://developers.facebook.com/docs/graph-api). In practice, this means that users only have to click on a button on our website, telling the app to make a connection to Facebook, to collect their posts, and to save them in our database.

To protect the privacy of its users, Facebook has installed two layers of security with which the app needed to deal. The first layer entails that the volunteering user needs to allow the app access to every piece of information it collects. Facebook calls this allowance of access to personal data a 'permission'. Two permissions were required for our purpose, user_birthday (to make sure that we collected posts of youths of the intended ages) and user_posts. Users grant these permissions directly after they click the button on our website: a pop-up window appears which first asks them to log in to Facebook and then explains to what the app will have access if they proceed.

The second security layer entails that Facebook itself needs to allow the app to ask for permissions. During development, the app only worked for a predefined set of Facebook users for testing purposes; users that were not part of this set could not grant any permissions and thus donate their data with the app. To make the app available to all Facebook users, it had to be manually reviewed by a Facebook employee. Our app was accepted only after making clear that it is of value to Facebook users because it enabled volunteering users to effortlessly donate their posts without having to manually copy and paste these one by one. The source code of the app can be found at https://github.com/Woseseltops/ FB-data-donator. It can easily be adjusted to make another app that collects other user data in a similar way.

5   Promotion of Websites

The websites for collecting social media messages were promoted through free publicity in Dutch media. It attracted quite some media attention, which resulted in newspaper publications, both regional (*de Gelderlander*) and national (*AD.nl*), radio interviews on regional (*RTV Noord-Holland, Studio 040*) and national (*De Taalstaat, NPO Radio 1, 3FM, Radio FunX*) stations, and television interviews on regional (*NimmaTV*) and national (*Rtl4*) TV. University and student magazines reported on the data collection too (*Vox, ANS*). In addition, it was advertised in the digital newsletters of *Onze Taal* (the Dutch society for language buffs) - *Taalpost* for adults and *TLPST* for adolescents. The data collection was also promoted via the Radboud University's web pages and by researchers via social media channels, in particular Twitter and Facebook. We further promoted it during lectures and master classes for young audiences, i.e. students in secondary and tertiary education. Our aim was to promote the websites nationwide, in order to gather a representative sample of messages from youngsters throughout the country.

In order to stimulate youths to contribute their social media messages to our project, we decided to raffle off prizes - gift certificates at the value of 100, 50, and 20 euros. With respect to WhatsApp, individual contributors' odds of winning a prize increased as they sent in more chat conversations. We felt that this raffle was necessary to stimulate youths to donate their private messages to the corpus.

Importantly, it was emphasized on the websites that only those contributors who completely filled in the consent form stood a chance of winning the prizes. This was made explicit to motivate youths to give their informed consent.

6 Metadata

All WhatsApp chats and Facebook posts in our social media corpus are accompanied by a substantial amount of sociolinguistic information. Via the websites, the following metadata were obtained: name, place of residence, place and date of birth, age, gender, and educational level, as well as date and place of submission. These parameters are useful for sociolinguistic research, since they enable one to study the language use of different social groups in WhatsApp and Facebook.

7 IPR Issues and Ethical Approval

Intellectual property rights (IPR) were obtained by consent of both the Facebook company and individual contributors of Facebook and WhatsApp messages, since it is key to safeguard the authors' rights and interests [5] (p. 2270). For underage contributors, between 12 and 17 years old, written consent was also gained of one of their parents or guardians. By signing the consent web form, contributors declared the following:

- to have been informed about the purpose of the study;
- to have been able to ask questions about the study;
- to understand how the data from the study will be stored and to what ends they will be used;
- to have considered if they want to partake in the study;
- to voluntarily participate in the study.

Additionally, parents or guardians also declared:

- to be aware of the contents of their child's messages;
- to agree with their child's participation in the study.

Participants and their parents/guardians gave full permission for their (child's) submitted messages (i) to be used for scientific research and educational purposes; (ii) to be stored in a database, according to Radboud University's rules, and to be kept available for scientific research, provided they are anonymised and in no way traceable to the original authors; and (iii) to be used in scientific publications and meetings. If messages appear in publications or presentations, no parts that may harm the participants' interests will be made public.

Furthermore, ethical approval was obtained from our institution's Ethical Testing Committee (ETC). For the WhatsApp chats, it was crucial for the ETC that messages of conversation partners were deleted, since they have not given consent for the use of their messages. Accordingly, interlocutors' WhatsApp messages were immediately discarded. This procedure was explained on the FAQ page of the websites. In accordance with the ETC's further guidelines, we added downloadable information documents on the home pages.

8   Current Corpus Composition

The collection period of WhatsApp messages lasted from April until December 2015; the collection of Facebook messages started in December 2015. Up to the time of writing, over 332,000 word tokens of WhatsApp chats have been collected from youths between the ages of 12 and 23, which compares to the SoNaR subcorpora with texts by youths up to 20 years old from the Netherlands as follows - 44,012 word tokens in the SMS corpus (6.08 % of the total number of words of that corpus); 219,043 in the chat corpus (29.7 % of total); and 2,458,904 in the Twitter corpus (10.6 % of total). The scale of this corpus makes it suitable for fine-grained (manual) linguistic studies; it is not intended as a training data set for large-scale computational research.

We excluded chain messages from our corpus. Also not included were any visual or audio materials: since the study that prompted the data collection is completely linguistic in nature, images, videos, and sound files were not gathered, so the corpus is wholly textual rather than multimodal. Another deciding factor in asking contributors not to add media files when sending WhatsApp conversations from their smartphones is that adding them may prevent mails from arriving due to an exceeded data limit. More importantly, issues of copyright and privacy protection would make any inclusion of pictures, videos, or sounds highly problematic. The messages are stored as one WhatsApp chat conversation per file. Table 1 shows demographic details on the data collected so far, focusing on the age and gender distribution.

Table 1. Composition of WhatsApp dataset.

	Contributors		Conversations		Words	
	#	%	#	%	#	%
Adolescents	11	32.4	83	38.6	63,217	19.0
Young adults	23	67.6	132	61.4	269,440	81.0
Male	12	37.5	71	33.0	98,201	29.5
Female	22	68.8	144	67.0	234,456	70.5
Total	34	100	215	100	332,657	100

For the WhatsApp dataset, a relatively small number of youths (34) have contributed large quantities of data. At the time of writing, the number of contributors of Facebook posts was already considerably greater - 94, who together contributed 171,693 words. This difference may stem from the submission procedure: while users were asked to submit WhatsApp chats via separate emails, which required taking several steps on their mobile phones, they could easily submit all their Facebook posts with the click of a button. Young adults (18–23 years old, avg. age 20.1) submitted many more WhatsApp messages than adolescents (12–17, avg. age 14.4), not only in terms of number of contributors, but also in terms of number of conversations as well as words. The average age of all contributors was 18.3. In terms of gender, a higher percentage of WhatsApp chat contributors are female, with about two thirds girls versus one third boys (a distribution similar to that for donated text messages as reported in [4]). This corresponds to the percentages of words and conversations that were submitted by male versus female contributors.

9   Preliminary Data Analysis

This section presents the first findings of a linguistic corpus study of Dutch youths' WhatsApp chats. Their language use in social media often differs from Standard Dutch, in various dimensions of writing. A striking orthographic feature of written CMC are textisms: unconventional spellings of various kinds. We conducted a quantitative register analysis into the frequency of textisms, and investigated how the independent variable age group affects this linguistic feature by distinguishing between WhatsApp messages of adolescents and young adults. The following textism types were found (presented here with Dutch examples):

- textisms with letters:
    - initialism: first letters of each word/element in a compound word, phrase, sentence, or exclamation, e.g. *hvj (hou van je), omg (oh mijn God)*
    - contraction: omission of letters (mostly vowels) from middle of word, e.g. *vnv (vanavond), idd (inderdaad)*
    - clipping: omission of final letter of word, e.g. *lache (lachen), nie (niet)*
    - shortening: dropping of ending or occasionally beginning of word, e.g. *miss (misschien), wan (wanneer)*
    - phonetic respelling: substitution of letter(s) of word by (an)other letter(s), while applying accurate grapheme-phoneme patterns of the standard language, e.g. *ensow (enzo), boeiuh (boeien), okeej (oké), egt (echt)*
    - single letter/number homophone: substitution of word by phonologically resembling or identical letter/number, e.g. *n (een), t (het), 4 (for)*
    - alphanumeric homophone: substitution of part of word by phonologically resembling or identical letter(s)/number(s), e.g. *suc6 (succes), w88 (wachten)*
    - reduplication: repetition of letter(s), e.g. *neeee (nee), superrr (super)*
    - visual respelling: substitution of letter(s) by graphically resembling non-alphabetic symbol(s), e.g. *Juli@n (Julian), c00l (cool)*
    - accent stylisation: words from casual, colloquial, or accented speech spelled as they sound, e.g. *hoezut (hoe is het), lama (laat maar)*
    - inanity: other, e.g. *laterz (later)*
    - standard language abbreviations, e.g. *aug (augustus), bios (bioscoop)*
- textisms with diacritics:
    - missing, e.g. *carriere (carrière), ideeen (ideeën), enquete (enquête)*
- textisms with punctuation:
    - missing, e.g. *mn (m'n), maw (m.a.w.), ovkaart (ov-kaart)*
    - extra, e.g. *stilte-coupé (stiltecoupé)*

- reduplication, e.g. *!!!!!*, *??*, ...
- textisms with spacing:
  - missing (in between words), e.g. *hahaokeeedan (haha oké dan)*
  - extra (in between elements of compound words), e.g. *fel groen (felgroen)*
- textisms with capitalisation:
  - missing (of proper names, abbreviations), e.g. *tim (Tim)*, *ok (OK)*
  - extra, e.g. *WOW (wow)*

Figure 1 shows the results for the textisms, separating adolescents from young adults. The frequencies shown here have been standardised per 10,000 words, because the total number of words differs per age group in the WhatsApp dataset. The figure makes clear that textisms with letters were by far the most frequent in the WhatsApp chats. It also shows an age-based distinction: while textisms with diacritics, capitalisation, punctuation, and spacing occurred with more or less similar frequencies in the WhatsApp messages of the two age groups, those with letters were used much more by adolescents. Their greater use of orthographic deviations may be attributed to a desire to rebel against societal norms, including the standard language norms, and to play with language: the most non-conformist linguistic behaviour is said to occur around the ages of 15/16, when the 'adolescent peak' occurs. Young adults, on the other hand, may feel more social pressure to conform to norms set by society, also those about language.

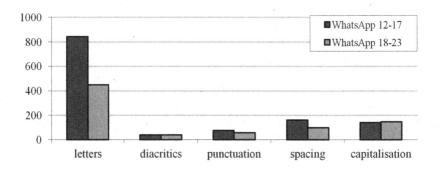

Fig. 1. Five types of textisms in WhatsApp dataset.

This preliminary analysis is part of a larger in-depth linguistic study of a broad range of linguistic features in WhatsApp chats. These focus on orthography (misspellings, typos, emoticons, symbols), syntax (omissions; complexity), and lexis (borrowings, interjections; diversity, density). Other lexical features that may be interesting for online youth communication are, for example, swearwords, intensifiers, and hyperbolic expressions. The WhatsApp data will be compared to the Facebook data, as well as to instant messages, text messages, and microblogs of the SoNaR corpus. This can reveal to what extent deviations from the standard language norms in CMC depend not just on individual user characteristics such as age, but also on genre characteristics.

10   Conclusions

The central role currently played by CMC in (especially) youths' lives makes social media corpora quite valuable for state-of-the-art sociolinguistic research. This paper discussed the compilation of such a corpus in the Netherlands. WhatsApp chats and Facebook posts were contributed by Dutch youths from 12 to 23 years old. This paper has made clear that a data collection method of voluntary donations, with the added incentive of a prize raffle, can yield a fair amount of data if sufficient public attention is obtained through e.g. media coverage. We have presented websites created for this purpose, and have explained how such websites can be promoted. The importance of collecting metadata and obtaining written consent and ethical approval have been stressed. Crucially, the application we created to gather Facebook posts, beside the process of gaining consent from the Facebook company, can serve as a model for future corpus builders.

11   Future Work

Eventually, if the WhatsApp and Facebook data are processed in a similar fashion as the rest of SoNaR, they can be incorporated into the corpus together with their metadata. This would require format conversion, tokenization, and anonymisation: the data should be (a) converted into the FoLiA xml-format, which was developed for linguistic resources, (b) tokenised by UCTO, a tokeniser adapted for social media, and (c) anonymised, if possible automatically, so that they contain no personal/place names, (email) addresses, telephone numbers, or bank accounts. Such additional processing was beyond the scope of the present project, but particularly data anonymisation is essential if the WhatsApp chats and Facebook messages are shared with the wider scientific community and become available for further research into social media texts. It would also be useful to apply part-of-speech tagging to this corpus. Moreover, we recognize the need for multimodal social media corpora: the next step in sociolinguistic social media research may be to focus on multimodality, given the increased options for incorporating visual materials (photographs, emoji, videos, etc.) and the use thereof in computer-mediated communication. The number of contributors so far suggest that youths remain hesitant to donate their private, often intimate, social media messages to science, despite significant gift certificates; perhaps a larger corpus could be obtained by even more publicity or even greater prizes. Nonetheless, albeit monomodal and of modest scale, the present corpus with its metadata can be a vital resource and an example of how social media texts can be collected for linguistic, sociological, or other research.

Acknowledgments. This research was funded by a grant of the Dutch Organisation for Scientific Research (NWO), under project number 322-70-006. Special thanks are due to Iris Monster, who constructed the WhatsApp website. Thanks also go to Wilbert Spooren and Ans van Kemenade, the supervisors of Lieke's PhD project. Finally, we thank all contributors of WhatsApp and Facebook messages to our corpus.

References

1. Thurlow, C.: From statistical panic to moral panic: the metadiscursive construction and popular exaggeration of new media language in the print media. J. Comput.-Mediated Commun. **11**(3), 667–701 (2006)
2. Postma, K.: Geen paniek! Een analyse van de beeldvorming van sms-taal in Nederland. Master thesis, VU University Amsterdam (2011)
3. Sanders, E.: Collecting and analysing chats and tweets in SoNaR. In: Proceedings LREC (Language Resources and Evaluation) 2012, pp. 2253–2256 (2012)
4. Treurniet, M., Sanders, E.: Chats, tweets and SMS in the SoNaR corpus: social media collection. In: Newman, D. (ed.) Proceedings of the 1st Annual International Conference Language, Literature & Linguistics, pp. 268–271. Global Science and Technology Forum, Singapore (2012)
5. Treurniet, M., De Clercq, O., van den Heuvel, H., Oostdijk, N.: Collecting a corpus of Dutch SMS. In: Proceedings LREC 2012, pp. 2268–2273 (2012)
6. Oostdijk, N., Reynaert, M., Hoste, V., Schuurman, I.: The construction of a 500-million-word reference corpus of contemporary written Dutch. In: Spyns, P., Odijk, J. (eds.) Essential Speech and Language Technology for Dutch: Results by the STEVIN Programme, pp. 219–247. Springer, Heidelberg (2013)
7. Dürscheid, C., Frick, K.: Keyboard-to-Screen-Kommunikation gestern und heute: SMS und WhatsApp. im Vergleich. In: Mathias, A., Runkehl, J., Siever, T. (eds.) Sprachen? Vielfalt! Sprache und Kommunikation in der Gesellschaft und den Medien. Eine Online-Festschrift zum Jubiläum von Peter Schlobinski, pp. 149–181. Networx 64, Hannover (2014)
8. Stark, E., Ueberwasser, S., Diémoz, F., Dürscheid, C., Natale, S., Thurlow, C., Siebenhaar, B.: What's up, Switzerland? Language, individuals and ideologies in mobile messaging. Universität Zürich, Universität Bern, Université de Neuchâtel, Universität Leipzig (2015). http://www.whatsup-switzerland.ch
9. Stark, E., Ueberwasser, S., Dürscheid, C., Béguelin, M.J., Moretti, B., Grünert, M., Gazin, A.-D., Pekarek Doehler, S., Siebenhaar, B.: Sms4science. Universität Zürich, Université de Neuchâtel, Universität Bern, Universität Leipzig (2015). http://www.sms4science.uzh.ch
10. Siebenhaar, B., et al.: What's up, Deutschland? WhatsApp-Nachrichten erforschen. Universität Leipzig, Technische Universität Dortmund, Technische Universität Dresden, Leibniz Universität Hannover, Universität Mannheim, Universität Koblenz-Landau, Universität Duisburg-Essen (2016). http://www.whatsup-deutschland.de
11. Frey, J.-C., Stemle, E.W., Glazniek, A.: Collecting language data of non-public social media profiles. In: Faaß, G., Ruppenhofer, J. (eds.) Workshop Proceedings of the 12th Edition of the KONVENS Conference, pp. 11–15. Universitätsverlag, Hildesheim (2014)
12. Frey, J.-C., Glaznieks, A., Stemle, E.W.: The DiDi corpus of South Tyrolean CMC data. In: Beißwenger, M., Zesch, T. (eds.) Proceedings of the 2nd Workshop of the Natural Language Processing for Computer-Mediated Communication/Social Media, pp. 1–6. University of Duisburg-Essen (2015)

A Dynamic Programming Approach
to Improving Translation Memory Matching
and Retrieval Using Paraphrases

Rohit Gupta[1(✉)], Constantin Orăsan[1], Qun Liu[2], and Ruslan Mitkov[1]

[1] University of Wolverhampton, Wolverhampton, UK
R.Gupta@wlv.ac.uk
[2] Dublin City University, Dublin, Ireland

Abstract. Translation memory tools lack semantic knowledge like paraphrasing when they perform matching and retrieval. As a result, paraphrased segments are often not retrieved. One of the primary reasons for this is the lack of a simple and efficient algorithm to incorporate paraphrasing in the TM matching process. Gupta and Orăsan [1] proposed an algorithm which incorporates paraphrasing based on greedy approximation and dynamic programming. However, because of greedy approximation, their approach does not make full use of the paraphrases available. In this paper we propose an efficient method for incorporating paraphrasing in matching and retrieval based on dynamic programming only. We tested our approach on English-German, English-Spanish and English-French language pairs and retrieved better results for all three language pairs compared to the earlier approach [1].

Keywords: Edit distance with paraphrasing · Translation memory · TM matching and retrieval · Computer aided translation · Paraphrasing

1 Introduction

Apart from retrieving exact matches, one of the core features of a TM system is the retrieval of previously translated similar segments for post-editing in order to avoid translation from scratch when an exact match is not available. However, this retrieval process is generally limited to edit-distance based measures operating on surface form (or sometimes stem) matching. Most commercial systems use edit distance [2] or some variation of it. Although these measures provide a strong baseline, they are not sufficient to capture semantic similarity between segments as judged by humans. For example, even though segments like *I would like to congratulate the rapporteur* and *I wish to congratulate the rapporteur* have the same meaning, current TM systems will consider them not similar enough to use one instead of the other. The two segments have only 71 % similarity based on word based Levenshtein edit-distance, even though one segment is a paraphrase of the other segment. To mitigate this limitation of TM, we propose an approach to incorporating paraphrasing in TM matching.

© Springer International Publishing Switzerland 2016
P. Sojka et al. (Eds.): TSD 2016, LNAI 9924, pp. 259–269, 2016.
DOI: 10.1007/978-3-319-45510-5_30

A trivial approach to implementing paraphrasing along with edit-distance is to generate all the paraphrased segments based on the paraphrases available and store these additional segments in the TM. This approach leads to a combinationtorial explosion and is highly inefficient both in terms of time necessary to process and space to store. For a TM segment which has n different phrases where each phrase can be paraphrased in m more possible ways, we get $(m + 1)^n - 1$ additional segments (still not considering that these phrases may contain paraphrases as well). To measure how many segments will be generated in reality, we randomly selected a sample of 10 segments from the TM and used the same paraphrase database as used in our experiments to generate additional segments. By limiting only to not more than three phrases to be paraphrased per segment we generated 1,622,115 different segments.

This paper presents a simple, novel and efficient approach to improve matching and retrieval in TM using paraphrasing based on dynamic programming. Our tool is available on Github.[1]

2   Related Work

Several researchers have pointed out the need for more advanced processing in TMs that go beyond surface form comparisons and proposed the use of semantic or syntactic information, but their methods are inefficient for large TMs [3–8].

Macklovitch and Russell [4] showed that using NLP techniques like named entity recognition and morphological processing can improve matching in TM, whilst Somers [5] highlighted the need for more sophisticated matching techniques that include linguistic knowledge like inflection paradigms, synonyms and grammatical alterations. Both Planas and Furuse [3] and Hodász and Pohl [6] proposed to use lemmas and parts of speech along with surface form comparison. Hodász and Pohl [6] extended the matching process to a sentence skeleton where noun phrases are either tagged by a translator or by a heuristic NP aligner developed for English-Hungarian translation. Planas and Furuse [3] tested a prototype model on 50 sentences from the software domain and 75 sentences from a journal with TM sizes of 7,192 and 31,526 segments respectively. A fuzzy match retrieved was considered usable if less than half of the words required editing to match the input sentence. The authors concluded that the approach gives more usable results compared to Trados Workbench, an industry standard commercial system, used as a baseline.

Pekar and Mitkov [7] presented an approach based on syntactic transformation rules. They evaluated the proposed method using a query sentence and found that the syntactic rules help in retrieving better segments. The use of alignments between source and target at word, phrase or character level was proposed in [9] as a way of improving matching.

Recent work focused on approaches which use paraphrasing in TM matching and retrieval [1,10,11]. Utiyama et al. [10] developed a method which relies on a finite state transducer limited to exact matches only. Gupta and Orăsan [1]

---

[1] https://github.com/rohitguptacs/TMAdvanced.

proposed a paraphrasing approach based on greedy approximation and dynamic programming. Gupta et al. [11] showed that post-editing time is reduced if paraphrases are used in the TM matching. Timonera and Mitkov [12] show clause splitting as a preprocessing stage improves the matching and retrieval.

The approach proposed in this paper is similar to the one described in [1] but instead of using greedy approximation we use dynamic programming which optimises edit-distance globally and makes the approach more simple and efficient.

3   Our Approach

In this section, we present our approach to include paraphrasing in the TM matching and retrieval process. In order to be able to compare our approach with the one proposed in [1], we use the same settings for the experiments. In the rest of the paper we will refer to the method proposed in [1] as DPGA (Dynamic Programming and Greedy Approximation) and the method proposed in this paper as DP (Dynamic Programming only).

3.1   Paraphrase Corpus and Classification

We use the PPDB 1.0 paraphrase database [13] for our work. This database contains lexical, phrasal and syntactic paraphrases automatically extracted using a large collection of parallel corpora. The paraphrases in this database are constructed using a bilingual pivoting method. The paraphrase database is available in six sizes (S, M, L, XL, XXL, XXXL) where S is the smallest and XXXL is the largest; we use size L (lexical and phrasal) in this research.

We use the four types proposed in [1] for classifying paraphrases on the basis of the number of words they contain (common part is shown in **bold** and can be removed after considering the context when computing edit-distance):

1. Paraphrases having one word in both the source and target sides, e.g. "period" ⇔ "duration"
2. Paraphrases having multiple words on both sides but differing in one word only, e.g. "in **the period**" ⇔ "during **the period**"
3. Paraphrases having multiple words, but the same number of words on both sides, e.g. "laid down **in article**" ⇔ "set forth **in article**"
4. Paraphrases in which the number of words in the source and target differ, e.g. "**a reasonable period** of time to" ⇔ "**a reasonable period** to"

The classification of paraphrases helps to implement the Type 1 and Type 2 paraphrases more efficiently (see Sect. 3.4 for further details).

3.2   Matching Steps

There are two options for incorporating paraphrasing in a typical TM matching pipeline: to paraphrase the input or to paraphrase the TM. For our approach we have chosen to paraphrase the TM on fly because it allows retrieval of matches in real time.

Our approach is also inspired by the one proposed in [1] and comprises of the following steps:

1. Read the Translation Memories available
2. Classify and store all paraphrases for each segment in the TM in their reduced forms according to the types presented in Sect. 3.1
3. Read the file that needs to be translated
4. For each segment in the input file
   (a) retrieve the potential segments for paraphrasing in the TM according to the filtering steps of Sect. 3.3
   (b) search for the most similar segment based on the approach described in Sect. 3.4
   (c) retrieve the most similar segment if it is above a predefined threshold

3.3   Filtering

Before processing begins, several filtering steps are applied to each input segment. The purpose of this filtering process is to remove unnecessary candidates and speed up the processing. These steps, based on [1], are as follows:

1. LENFILTER: Filter out segments based on length. If segments differ considerably in length, the edit-distance will also differ correspondingly. In our case, TM segments are discarded if the TM segments are shorter than 39 % of the input or vice-versa.
2. SIMFILTER: Filter out segments based on baseline edit-distance similarity. TM segments which have a similarity below a certain threshold will be removed. In our case, the threshold was set to 39 %.
3. MAXFILTER: Next, after filtering the candidates with the above two steps we sort the remaining segments in decreasing (non-increasing) order of baseline edit-distance similarity and pick the top 100 segments.
4. BEAMFILTER: Finally, segments within a certain range of similarity with the most similar segment are selected for paraphrasing. Here, the range adopted is 35 %. This means that if the most similar segment has 95 % similarity, segments with a similarity below 60 % will be discarded.

3.4   Edit-Distance with Paraphrasing Computation

For our implementation, we use the basic edit-distance procedure [2], which is word-based edit-distance with cost 1 for insertion, deletion and substitution. We obtain the similarity between two segments by normalising edit-distance with the length of the longer segment.

We have employed the basic edit-distance as a baseline and adapted it to incorporate paraphrasing. When edit-distance is calculated, the paraphrases of Types 1 and 2 can be implemented in a more efficient manner than for paraphrases of Types 3 and 4. They have the unique property that they can be reduced to single word paraphrases by removing the other matching words. The basic procedure works by comparing each token one by one in the input segment with each token in the TM segment. This procedure makes use of previous edit-distance computations to optimise the edit-distance globally (for the whole sentence). In our dynamic programming approach, at every step we consider the best matching path.

Table 1. Edit-distance calculation

$i \backslash j$	0	1	2	3	4	5	6	10	7	8	9	11
	#	the	period duration time	laid	down	referred	to	in	under	provided	for	by article
0 #	0	1	2	3	4	3	4	5	3	4	5	6
1 the	1	0	1	2	3	2	3	4	2	3	4	5
2 period	2	1	0	1	2	1	2	3	1	2	3	4
3 referred	3	2	1	1	2	0	1	2	1	2	3	3
4 to	4	3	2	2	2	1	0	1	2	2	3	2
5 in	5	4	3	3	3	2	1	0	3	3	3	1
6 article	6	5	4	4	4	3	2	1	4	4	4	0

Algorithm 1. Edit-Distance with Paraphrasing using Dynamic Programming

```
1:  procedure EDIT-DISTANCE(InputSegment, TMLattice)
2:      M ← length of TMLattice                        ▷ number of nodes in a TM lattice
3:      N ← length of InputSegment                     ▷ number of nodes in a input segment
4:      Initialise D, a two dimensional array of size M × N initialized with distance from null #  ▷
        See Table 1, all values at i = 0 and all values at j = 0
5:      for  j ← 1...M do
6:          for  i ← 1...N do
7:              cost ← 1                                ▷ substitution cost
8:              if TMLattice_j = InputSegment_i then
9:                  cost ← 0                            ▷ substitution cost if matches
10:             if cost = 1 and TMLattice_j is a TM token then▷  condition to avoid paraphrasing
        the paraphrase
11:                 OneWordPP ← get Type 1 and Type 2 paraphrases associated with TMToken
12:                 if InputToken ∈ OneWordPP then ▷ applying type 1 and type 2 paraphrasing
13:                     cost ← 0
14:             P ← previous indices of paths at TMLattice_j        ▷ get indices of previous nodes
        connecting to the present node TMLattice_j
15:                 D[i, j] ← minimum(D[i − 1, j] + 1, D[i, k] + 1, D[i − 1, k] + cost  for all k ∈ P)    ▷
        store minimum edit distance by considering insertion, deletion or substitution for all paths
16:             k ← index of minimum edit distance path at last node
17:             N ← length of InputSegment                 ▷ number of tokens in a input segment
18:             Return D[k, N]                             ▷ Return minimum edit-distance
```

Table 1 illustrates the edit-distance calculation of the first five tokens of the Input and TM segment with paraphrasing of the example given in Fig. 1. The second column represents the input segment (*the period referred to in article*) and

264   R. Gupta et al.

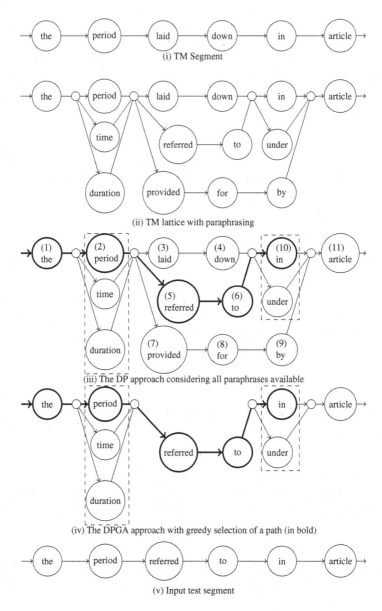

Fig. 1. (i) TM segment, (ii) TM lattice after considering paraphrases, (iii) The DP approach considering all paraphrases available (numbers show values of j as given in Table 1, dashed area indicates Types 1 and 2 paraphrasing), (iv) DPGA with greedy selection, (v) Input test segment

the second row represents the TM segment (*the period laid down in article*) along with the paraphrases. Figure 1(v) shows the input test segment and Fig. 1(ii) shows the TM lattice. In Algorithm 1, *InputSegment* is the segment that to be translated and *TMLattice* is the TM lattice (TM segment with paraphrasing). Table 1 shows the edit-distance calculation of the first five tokens of the input segment and TM lattice. In Table 1, if a word from the input segment matches any of the words "period", "time" or "duration", the cost of substitution will be 0. In Algorithm 1, *line 14* gets the previous indices at every step.

For example, in Table 1, when executing the token 'in' the algorithm will consider previous indices of 'down' and 'to' (see lattice in Fig. 1 (ii)) and store the minimum edit-distance path in *line 15*. As we can see in Table 1, the column 'in' is updated after considering both column 'down' and 'to' and column 'article' is updated after considering column 'in' and 'by'. In contrast to the DP approach, the DPGA approach makes a greedy selection and will update only on the bases of selected paraphrases and not all paraphrase paths as shown in Fig. 1.

3.5   Computational Considerations

The time complexity of the basic edit-distance procedure is $O(m\,n)$ where m and n are lengths of source and target segments, respectively. After employing paraphrasing of Type 1 and Type 2, the complexity increases to $O(m\,n\,log(p))$, where $p-1$ is the number of additional paraphrases of Type 1 and Type 2 per token of TM segment. Employing paraphrasing of Type 3 and Type 4 further increases the edit-distance complexity to $O(m\,n(log(p)+q\,l))$, where q is the number of Type 3 and Type 4 paraphrases stored per token and l is the average length of a Type 3 and Type 4 paraphrase. The time complexity of the DPGA approach is $O(l\,m\,n(log(p)+q))$, which is slightly more compared to $O(m\,n(log(p)+q\,l))$ complexity the DP approach. However, in practice we have not observed much difference in the speed of both approaches.

If we consider, Type 1 and Type 2 paraphrases in the same manner (comparing sequentially instead of searching in a list) as Type 3 and Type 4, the time complexity of the DP approach can be simply written as $O(k\,n)$, where k is the number of nodes in the TM lattice and n is the number of nodes in the input segment.

4   Experiments and Results

In this section, we present our experiments and the results obtained. For our experiments we used the English-French, English-German and English-Spanish pairs of the 2014 release of the DGT-TM corpus [14]. From this corpus, we filtered out segments of fewer than seven words and more than 40 words; the remaining pairs were used to create the TM and Test dataset. The test sets for all language pairs contain 20,000 randomly selected unique segments and the rest are used as the TM. The statistics of the datasets are given in Table 2.

When we use paraphrasing in the matching and retrieval process, the fuzzy match score of a paraphrased segment is increased, which results in the retrieval of more segments at a particular threshold. This increment in retrieval can be classified into two types: without changing the top rank and by changing the

Table 2. DGT-TM corpus statistics

	English-German		English-French		English-Spanish	
	TM	Test set	TM	Test set	TM	Test set
Segments	204, 776	20, 000	204, 713	20, 000	202, 700	20, 000
Source words	4, 179, 007	382, 793	4, 177, 332	382, 358	4, 140, 473	383, 694
Target words	3, 833, 088	343, 274	4, 666, 196	407, 495	4, 783, 178	433, 450

top rank. For example, for a particular input segment, we have two segments: A and B in the TM. Using simple edit-distance, A has a 65 % and B has a 60 % fuzzy score; the fuzzy score of A is better than that of B. As a result of using paraphrasing, we observe two types of score changes:

1. the score of A is still better than or equal to that of B, for example, A has 85 % and B has 70 % fuzzy score;
2. the score of A is less than that of B, for example, A has 75 % and B has 80 % fuzzy score.

In the first case, paraphrasing does not supersede the existing model and just facilitates it by improving the fuzzy score so that the top segment ranked using edit distance gets retrieved. However, in the second case, paraphrasing changes the ranking and now the top-ranked segment is different. In this case, the paraphrasing model supersedes the existing simple edit distance model. This second case also gives a different reference with which to compare. In the experiments reported below, we take the top segment retrieved using simple edit distance as a reference against the top segment retrieved using paraphrasing and compare to determine which one is better.

We performed experiments using both approaches (DPGA and DP). For both of them, we conducted experiments in two different preprocessing settings: In setting 1 (S1), we do not remove any punctuation marks and in the preprocessing step we perform only tokenization; In setting 2 (S2), along with tokenization, we also remove punctuation marks in the preprocessing stage. The DGT-TM dataset is of legal genre and contains many punctuation marks. We deleted the punctuation marks to see how it affects the baseline edit-distance and the edit-distance with paraphrasing. In S2, punctuation marks are also removed from the target side.[2]

Table 3 presents our results. We measure an increase in the number of segments in an interval when using paraphrasing as well as quality of those segments against simple edit-distance. Table 3 shows similarity threshold intervals (TH) for TM (the threshold intervals shown are on the basis of edit-distance with paraphrasing similarity), the total number of segments retrieved

---

[2] We have used Stanford tokenizer on the English side and tokenizer provided with Moses [15] on the target side. The source (English) tokenization is used for matching and target language tokenization is used when calculating BLEU score.

Table 3. Results on English-German (DE), English-French (FR) and English-Spanish (ES)

		TH	DP Approach				DPGA Approach			
			100	[85, 100)	[70, 85)	[55, 70)	100	[85, 100)	[70, 85)	[55, 70)
DE	S1	EditRetrieved	6629	1193	1029	1259	6629	1193	1029	1259
		+ParaRetrieved	54	114	146	445	44	88	89	244
		%Improve	0.81	9.56	14.19	35.35	0.66	7.38	8.65	19.38
		RankCh	7	8	27	217	4	5	21	115
		METEOR-EditRankCh	83.82	57.42	24.05	26.47	75.69	63.78	25.11	27.74
		METEOR-ParaRankCh	**90.76**	**64.15**	**33.88**	26.89	**89.93**	49.97	**28.53**	27.58
		BLEU-EditRankCh	72.79	38.21	12.13	15.38	65.26	50.27	6.67	17.69
		BLEU-ParaRankCh	**84.15**	**49.11**	**14.44**	14.40	**73.79**	23.68	**8.08**	15.61
	S2	EditRetrieved	6767	1078	889	1070	6767	1078	889	1070
		+ParaRetrieved	60	123	150	380	49	98	99	224
		%Improve	0.89	11.41	16.87	35.51	0.72	9.09	11.14	20.93
		RankCh	8	15	36	176	5	9	30	118
		METEOR-EditRankCh	80.01	57.79	37.99	24.90	69.20	49.19	38.58	25.00
		METEOR-ParaRankCh	**90.27**	**67.34**	**43.95**	27.59	**88.21**	54.42	**38.68**	28.23
		BLEU-EditRankCh	64.97	50.22	25.20	13.37	49.34	36.06	23.06	13.02
		BLEU-ParaRankCh	**84.90**	**54.43**	**29.67**	14.51	**77.85**	36.38	**23.48**	14.99
FR	S1	EditRetrieved	6611	1197	1040	1257	6611	1197	1040	1257
		+ParaRetrieved	54	116	141	443	44	90	87	241
		%Improve	0.82	9.69	13.56	35.24	0.67	7.52	8.37	19.17
		RankCh	7	8	26	217	4	5	20	115
		METEOR-EditRankCh	94.88	45.92	34.91	32.26	92.26	54.38	39.96	33.47
		METEOR-ParaRankCh	**96.17**	**70.23**	**42.92**	32.70	91.44	**61.28**	39.37	33.35
		BLEU-EditRankCh	78.21	17.81	20.06	19.75	79.60	28.87	18.76	20.87
		BLEU-ParaRankCh	**87.56**	**46.58**	**24.67**	18.15	69.32	27.42	17.19	17.50
	S2	EditRetrieved	6750	1082	894	1078	6750	1082	894	1078
		+ParaRetrieved	60	125	141	377	49	100	91	223
		%Improve	0.89	11.55	15.77	34.97	0.73	9.24	10.18	20.69
		RankCh	8	15	33	177	5	9	28	118
		METEOR-EditRankCh	81.73	50.35	49.02	31.45	71.18	41.02	50.27	31.17
		METEOR-ParaRankCh	**92.87**	**66.37**	**58.60**	33.99	85.27	52.54	**58.60**	35.77
		BLEU-EditRankCh	63.44	33.50	32.84	19.56	48.88	15.76	32.25	18.18
		BLEU-ParaRankCh	**84.80**	**47.38**	**39.70**	20.02	68.68	28.31	**37.85**	18.77
ES	S1	EditRetrieved	6620	1187	1028	1233	6620	1187	1028	1233
		+ParaRetrieved	54	115	141	433	44	89	86	234
		%Improve	0.82	9.69	13.72	35.12	0.66	7.50	8.37	18.98
		RankCh	7	8	29	209	4	5	21	110
		METEOR-EditRankCh	85.85	52.10	36.62	33.80	73.62	51.61	35.72	34.85
		METEOR-ParaRankCh	**88.45**	**64.83**	**44.48**	34.58	**83.36**	49.02	**44.94**	35.93
		BLEU-EditRankCh	72.79	24.74	17.49	19.20	55.75	38.20	14.49	20.11
		BLEU-ParaRankCh	**78.09**	**44.48**	**24.22**	19.09	**75.49**	18.61	**24.13**	19.48
	S2	EditRetrieved	6757	1076	885	1050	6757	1076	885	1050
		+ParaRetrieved	60	125	142	374	49	100	92	221
		%Improve	0.89	11.62	16.05	35.62	0.73	9.29	10.40	21.05
		RankCh	8	15	34	179	5	9	28	118
		METEOR-EditRankCh	82.11	58.01	49.32	35.13	69.89	51.35	49.58	35.46
		METEOR-ParaRankCh	**87.12**	**67.94**	**60.42**	36.07	82.52	**66.11**	60.39	37.14
		BLEU-EditRankCh	61.84	37.02	33.42	18.84	37.04	29.67	29.29	18.37
		BLEU-ParaRankCh	**73.83**	**47.07**	**44.08**	19.82	69.89	44.80	42.02	19.83

using the baseline approach (EditRetrieved), the additional number of segments retrieved using the paraphrasing approaches (+ParaRetrieved), the percentage increase in retrieval obtained over the baseline (%Improve), and the number of segments that changed their ranking and rose to the top because of paraphrasing (RankCh). BLEU-ParaRankCh and METEOR-ParaRankCh represent the BLEU score [16] and METEOR [17] score over translations retrieved by the DP approach for segments which changed their ranking and come up in the threshold interval because of paraphrasing and BLEU-EditRankCh and METEOR-EditRankCh represent the BLEU score and METEOR score on corresponding top translations retrieved by the baseline approach. Table 3 also shows results obtained by the DPGA approach.

The DP approach presented in this paper retrieves more matches than the DPGA approach for all language pairs. We see that the DP approach retrieves better results compared to using simple edit-distance for all language pairs in both settings.

Table 3 shows that for S1 (English-German), the DPGA approach does not retrieve better results compared to simple edit-distance for threshold interval [85, 100) and [55, 70). We have observed that removing punctuation marks in the preprocessing stage not only increases the retrieval but also increases the improvement in retrieval using paraphrases. Table 3 shows that there is an improvement around 9.56 % improvement for S1 and 11.41 % for S2 in the interval [85, 100) for the DP approach on English-German. Table 3 also suggests that the DPGA approach is more sensitive to preprocessing performed which can be seen in the difference between S1 and S2 results for all language pairs. The quality of the retrieved segments is influenced more for the DPGA approach compared to the DP approach.

5   Conclusion

In this paper, we presented our new dynamic programming based approach to include paraphrasing in the process of translation memory matching and retrieval. Using the DP approach, depending on the preprocessing settings, we observed an increase in retrieval around 9 % to 16 % for threshold intervals [100, 85) or [85, 70). We observe that the number of matches increased when using paraphrasing in every interval and also have better quality compared to those retrieved by simple edit-distance. The DP approach also yields better results compared to the DPGA approach for all three language pairs.

Acknowledgement. The research leading to these results has received funding from the People Programme (Marie Curie Actions) of the European Union's Seventh Framework Programme FP7/2007-2013/ under REA grant agreement No. 317471.

References

1. Gupta, R., Orăsan, C.: Incorporating paraphrasing in translation memory matching and retrieval. In: Proceedings of the European Association of Machine Translation (EAMT-2014) (2014)
2. Levenshtein, V.I.: Binary codes capable of correcting deletions, insertions, and reversals. Sov. Phys. Dokl. **10**, 707–710 (1966)
3. Planas, E., Furuse, O.: Formalizing translation memories. In: Proceedings of the 7th Machine Translation Summit, pp. 331–339 (1999)

4. Macklovitch, E., Russell, G.: What's been forgotten in translation memory. In: White, J.S. (ed.) AMTA 2000. LNCS (LNAI), vol. 1934, pp. 137–146. Springer, Heidelberg (2000)
5. Somers, H.: Translation memory systems. Comput. Transl.: Transl. Guide **35**, 31–48 (2003)
6. Hodász, G., Pohl, G.: MetaMorpho TM: a linguistically enriched translation memory. In: International Workshop, Modern Approaches in Translation Technologies (2005)
7. Pekar, V., Mitkov, R.: New generation translation memory: content-sensivite matching. In: Proceedings of the 40th Anniversary Congress of the Swiss Association of Translators, Terminologists and Interpreters (2007)
8. Mitkov, R.: Improving third generation translation memory systems through identification of rhetorical predicates. In: Proceedings of LangTech 2008 (2008)
9. Clark, J.P.: System, method, and product for dynamically aligning translations in a translation-memory system, 5 February 2002. US Patent 6,345,244
10. Utiyama, M., Neubig, G., Onishi, T., Sumita, E.: Searching translation memories for paraphrases. In: Machine Translation Summit XIII, pp. 325–331 (2011)
11. Gupta, R., Orăsan, C., Zampieri, M., Vela, M., Van Genabith, J.: Can translation memories afford not to use paraphrasing? In: Proceedings of EAMT (2015)
12. Timonera, K., Mitkov, R.: Improving translation memory matching through clause splitting. In: Proceedings of the Workshop on Natural Language Processing for Translation Memories (NLP4TM), Hissar, Bulgaria, pp. 17–23 (2015)
13. Ganitkevitch, J., Benjamin, V.D., Callison-Burch, C.: PPDB: the paraphrase database. In: Proceedings of NAACL-HLT, Atlanta, Georgia, pp. 758–764 (2013)
14. Steinberger, R., Eisele, A., Klocek, S., Pilos, S., Schlüter, P.: DGT-TM: a freely available translation memory in 22 languages. In: LREC, pp. 454–459 (2012)
15. Koehn, P., Hoang, H., Birch, A., Callison-Burch, C., Federico, M., Bertoldi, N., Cowan, B., Shen, W., Moran, C., Zens, R., et al.: Moses: open source toolkit for statistical machine translation. In: Proceedings of the 45th Annual Meeting of the ACL on Interactive Poster and Demonstration Sessions, pp. 177–180. Association for Computational Linguistics (2007)
16. Papineni, K., Roukos, S., Ward, T., Zhu, W.J.: BLEU: a method for automatic evaluation of machine translation. In: Proceedings of the ACL, pp. 311–318 (2002)
17. Denkowski, M., Lavie, A.: Meteor universal: language specific translation evaluation for any target language. In: Proceedings of the EACL 2014 Workshop on Statistical Machine Translation (2014)
18. Gupta, R.: Use of language technology to imporve matching and retrieval in translation memory. Ph.D. thesis, University of Wolverhampton (2016)

AQA: Automatic Question Answering System for Czech

Marek Medveď'[(⊠)] and Aleš Horák

Faculty of Informatics, Natural Language Processing Centre,
Masaryk University, Botanická 68a, 602 00 Brno, Czech Republic
{xmedved1,hales}@fi.muni.cz

Abstract. Question answering (QA) systems have become popular nowadays, however, a majority of them concentrates on the English language and most of them are oriented to a specific limited problem domain.

In this paper, we present a new question answering system called AQA (Automatic Question Answering). AQA is an open-domain QA system which allows users to ask all common questions related to a selected text collection. The first version of the AQA system is developed and tested for the Czech language, but we also plan to include more languages in future versions.

The AQA strategy consists of three main parts: question processing, answer selection and answer extraction. All modules are syntax-based with advanced scoring obtained by a combination of TF-IDF, tree distance between the question and candidate answers and other selected criteria. The answer extraction module utilizes named entity recognizer which allows the system to catch entities that are most likely to answer the question.

Evaluation of the AQA system is performed on a previously published Simple Question-Answering Database, or SQAD, with more than 3,000 question-answer pairs.

Keywords: Question Answering · AQA · Simple Question Answering Database · SQAD · Named entity recognition

1 Introduction

The number of searchable pieces of new information increases every day. Looking up a concrete answer to an asked question simply by information retrieval techniques thus becomes difficult and extremely time consuming. That is why new systems devoted to the Question Answering (QA) task are being developed nowadays [2,9,10]. The majority of them concentrates on the English language and/or relay on a specific limited problem domain or knowledge base.

This leads us to two main drawbacks of such QA systems: firstly, the specific limited problem domain and knowledge base are always pre-limiting the answers to what knowledge is stored in the system being thus possibly useful only for people working in this specific domain but useless for common people in daily

© Springer International Publishing Switzerland 2016
P. Sojka et al. (Eds.): TSD 2016, LNAI 9924, pp. 270–278, 2016.
DOI: 10.1007/978-3-319-45510-5_31

usage. The second problem arises when going multilingual, since in most cases the system functionality is not directly transferable to other languages without a decrease in accuracy.

In this paper, we present a new open domain syntax based QA system, named AQA. The first version of the system is developed and tested on the Czech language and uses the latest approaches from QA field. The evaluation was performed on Simple Question Answering Database (SQAD) [4].

2   The AQA System

In the following text we briefly describe the input format, the database format and the main AQA modules that are necessary to extract a concrete answer to a given question.

2.1   Input

Since the AQA system is aimed at processing texts in morphologically rich languages, it supports two input formats of the user question. The first is in a form of a plain text question without any structure or additional information; and the second one is in a vertical format[1] with words enriched by lexical and morphological information, see Fig. 1.

```
word/token      lemma       tag
<s>
Kdo             kdo         k3yQnSc1
je              být         k5eAaImIp3nS
autorem         autor       k1gMnSc7
novely          novela      k1gFnSc2
Létající        létající    k2eAgMnSc1d1
jaguár          jaguár      k1gMnSc1
<g/>
?               ?           kIx.
</s>
```

Fig. 1. Question *"Kdo je autorem novely Létající jaguár?"* (Who wrote the story called Flying jaguar?) in the vertical format.

Internally the syntax-based AQA system works with syntactic trees, which means that the plain text input form is automatically processed by a morphological tagger[2] and transformed into the (internal representation of) the vertical format.

---

[1] One text token per line with multiple attributes separated by tabs.
[2] In the current version, we use the morphological analyser Majka [5, 11] disambiguated by the DESAMB [7] tagger.

The morphologically annotated question input (obtained from the user or created by the tagger) is sent to the Question processor module that extracts all important information about the question. The detailed description of Question processor module is described in Sect. 2.3.

2.2 Knowledge Base

To find an appropriate answer to the given question the AQA system uses a knowledge database that can be created form any natural language text (in Czech, for now). The text can be again in a form of plain text or vertical text, the same as the input question.

To create a new knowledge database, the user enters just the source of the input texts (file, directory, ...) and the AQA system automatically processes the texts and gathers information to be stored in the knowledge base. The texts are processed in several steps and the following features are extracted:

- *Syntactic tree*: each sentence in the input text is parsed by a syntactic analyser and the resulting syntactic tree is stored in the database. We use the SET (Syntactic Engineering Tool) parser [6] in the current version. The SET parsing process is based on pattern matching rules for link extraction with probabilistic tree construction. For the purposes of AQA, the SET grammar was enriched by answer types that serve as hints to the Answer selection module (see Sect. 2.3) for matching among the candidate answers.
- *Tree distance*: for each syntactic tree, the system computes tree distance mappings between each words pair in a noun phrase. This information is used in the Answer selection module (see Sect. 2.3) to pick up the correct answer from multiple candidate answers.
- *Birth/death date, birth/death place*: this feature is specific for sentences where the birth/death date and birth/death place are present in the text just after a personal name, usually in parentheses (see example in Fig. 2).
- *Phrase extraction*: within the parsing process, the SET parser can provide a list of (noun, prepositional, ...) phrases present in the sentence. This information is used in in Answer extraction module (see Sect. 2.3) when the system searches for a match between the question and the (parts of the) knowledge base sentences.
- *Named entity extraction*: as mentioned above, the general SET grammar was modified to include, within the resulting syntactic tree, also the information about what kind of questions can be answered by each particular sentence. This information is also supplemented with a list of named entities found in the sentence. The AQA system recognizes three named entity types: a *place*, an *agent* and an *art work*. In the current version, AQA uses the Stanford Named Entity Recognizer [3] with a model trained on the Czech Named Entity Corpus 1.1 [8] and the Czech DBpedia [1] data.
- *Inverted index*: an inverted word index is created for fast search purposes, this serves mainly as a technical solution.

Sir Isaac Newton (January 4 1643 in Woolsthorpe – March 31 1727 in London) was a English physicist, mathematician ...

Fig. 2. Example sentence for birth/death date and place

For development and evaluation purposes, the AQA knowledge base was trained on the SQAD 1.0 database [4] that was manually created from Czech Wikipedia web pages and contains 3301 question-answer pairs. The answer is stated both in a form of the direct answer (noun, number, noun phrase, ...) and in the form of the answer context (several sentences) containing the direct answer. The results of an evaluation of AQA with the SQAD database are explicated in Sect. 3.

2.3   The AQA Modules

The AQA systems consists of tree main modules: the *Question processor*, the *Answer selection module* and the *Answer extraction module*. Figure 3 presents a schematic graphical representation of the AQA system parts.

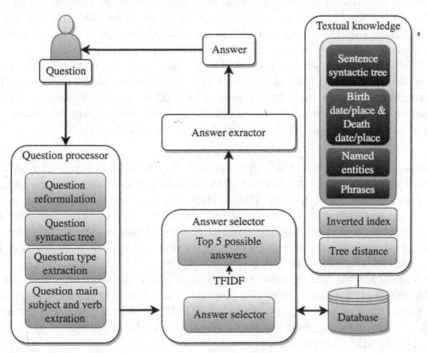

Fig. 3. The AQA system modules.

The Question Processor. The first step after the system receives the input question in the form of morphologically annotated tokens (either created by the tagger or obtained in the vertical format from the user) consists in extracting the following features:

– *Question reformulation*: the original question is reformulated (if possible) into a "normalized" question form, which allows to simplify the answer matching process later. For example, if the user asks a question starting with "*Jak se jmenuje ... osoba ...*" (What is the name of ... a person ...), the system reformulates the question to "*Kdo je ...*" (Who is ...). The meaning of the sentence remains the same but for further processing the reformulation directly corresponds to searching for a personal name or denotation among possible answers.
– *Question syntactic tree*: the same as the AQA system uses syntactic trees in the knowledge base creation process (Answer selection and extraction modules), the system automatically creates a syntactic tree for each question. The question syntactic tree is provided by the same parser as used for the knowledge base processing.
– *Question type extraction*: due to the enriched parsing grammar, the question type can be extracted during the syntactic tree creation. The question type itself is determined by the sentence structure and the specific pronouns present in the question. For example, whe user asks a question such as "*Kdo byl ...*" (Who was ...), the system assigns this question the WHO question type. The answer selection process then uses this information to filter the matching question and answer types.
– *Question main subject and main verb extraction*: the AQA system tries to find in the syntactic tree the main subject and the main verb of the question. This information is important for the answer extraction process. According to this information, the system knows which subject and verb should be present in the candidate answer (if the system picks up more than one).

Answer Selection Module. After extracting all the necessary information from the question, the Answer selection module is responsible for recognizing all possible candidate answers. This module communicates with the knowledge database and collects all pieces of text (sentence or paragraph) that contain a possible answer. The selection decision rules are based on the question type, the question main subject, the question main verb and specific words present in the question syntactic tree.

Each of the candidate answer texts are then assigned a score and the Answer selection module picks up first five top rated answers. The ranking score for each candidate answer is a combination of TF-IDF[3] score and the tree distance between the words in noun phrases in the answer and the question.

The TF-IDF score consists of two parts:

– TF-IDF match: for each word (nouns, adjectives, numerals, verbs, adverbs) that matches in the question-answer pair, the TF-IDF score is computed

$$tf\_idf_{match} = (1 + \log(tf)) * \log(idf)$$

– TF-IDF mismatch: TF-IDF for the rest of words that did not match in the question-answer pair.

---

[3] TF-IDF stands for Term Frequency-Inverse Document Frequency.

Then the resulting TF-IDF score is determined as:

$$tf\_idf_{res} = tf\_idf_{match} - tf\_idf_{mismatch}$$

The final score of a candidate answer according to the question is finally supplemented by the tree distance calculated between the words in the question and answer noun phrases. If the tree distance in the question-answer pair noun phrase is equal it does not influence the final score. But when the tree distance is not equal the final score is modified as follows:

$$final\_score = tf\_idf_{res} - |TreeDistance_q - TreeDistance_a|$$

The five best scored candidate answers are then send to the Answer extraction module that extracts particular parts of each sentence that will be considered as the required answer.

The Answer Extraction Module. The final step of the AQA processing is accomplished by the Answer extraction module where a particular part of each sentence is extracted and declared for the final answer. This part of the system works only with the five best scored candidate answers that were picked up by the Answer selector module.

The final answer for the asked question is extracted according to the following three factors:

- *Question focus*: as mentioned in Sect. 2.3, each question receives the question type according to the sentence structure and the words present in the question. The question type is then mapped to one or more named entity types.
- *Answer named entities*: within the knowledge base creation process, AQA extracts the supported named entities of three types, i.e. Place, Agent and ArtWork. The Answer extraction module then maps the question focus to the extracted answer named entities. In this process, AQA also excludes named entities that are present in the question to avoid an incorrect answer. This is the first attempt to get a concrete answer.
- *Answer noun phrases*: in case the previous step fails to find an answer, the system selects one phrase from the phrase list as the answer. The failure can be caused because of two reasons. First, the question focus can be a date or a number. In this case the Answer extraction module returns a noun phrase where the question main subject related to a number is present or it returns a birth/death date if the question asks about birth or death of some person (birth/death date is stored in database alongside the sentence tree, named entities and phrases), see Sect. 2.2.
  The second reason why the question focus mapping to answer named entities can fail is because of missing named entity in the candidate answer. In this case, the system checks the remaining candidate answers and if it does not succeed the Answer extraction module returns the noun phrase that contains the main question subject.

3   Evaluation

Within the evaluation process, we have used the SQAD v1.0 database containing 3,301 entries of a question, an answer and an answer text. The AQA knowledge base was build from all the answer texts from SQAD, and all 3,301 questions were answered by the AQA system. The results were evaluated in two levels (see Table 1). The first level evaluates the answer selection module and the second level evaluates the Answer extraction module. There are three possible types of a match between the AQA answer and the expected answer:

- a (full) *Match*:
  - for the Answer selection module: the module picked up the correct sentence/paragraph,
  - for the Answer extraction module: the first provided answer is equal to the expected answer;
- a *Partial match*:
  - the Answer selection module: a correct sentence/paragraph appears in the Top 5 best scored answers that were selected,
  - the Answer extraction module: the provided answer is not at the first position or it is not an exact phrase match;
- a *Mismatch*:
  - in case of the Answer selection module: the correct sentence/paragraph is not present in the Top 5 best scored answers,
  - for the Answer extraction module: incorrect or no answer produced.

Table 1. Evaluation of the AQA system on the SQAD v1.0 database

	Answer selection		Answer extraction	
		%		%
Match	2,645	80.1 %	1,326	40.2 %
Partial match	66	1.9 %	443	13.4 %
Mismatch	590	18 %	1,532	46.4 %

4   Error Analysis

Within a detailed analysis of the errors (mismatches and partial matches) in the evaluation, we have identified the causes and outlined the directions and tasks of next development steps of the AQA modules.

The Question processor module needs to use a fine grained grammar for the question type assignment task. In the evaluation process, 18 % of errors have been assigned an incorrect question type. The following modules then could not find the required answer.

The Answer selection module is in some cases too biased to preferring answers that contain a question word multiple times. This was a cause of 20 % of erroneous answers in the analysis.

The Answer extraction module uses a question subject word, named entities and phrase extraction in the extraction process. The evaluation showed that the module needs to include other phrase and tree matching techniques – 21 % of errors were caused by extracting a different phrase from a correctly selected answer sentence. The current AQA version also suffers from not applying anaphora resolution techniques, which are necessary when the correct answer needs to be extracted by a reference between two sentences.

5   Conclusions

The paper presented details about the architecture of a new open question-answering system named AQA. The system is aimed at morphologically rich languages (the first version is developed and tested with the Czech language).

We have described the AQA knowledge base creation from free natural language texts and the step-by-step syntax-based processing of the input question and processing and scoring the candidate answers to obtain the best specific answer. The AQA system has been evaluated with the Simple Question Answering Database (SQAD), where it has achieved an accuracy of 40 % of correct answers and 53 % of partially correct answers.

We have also identified prevailing causes of errors in the answer selection and answer extraction phases of the system and we are heading to amend them in the next version of the system.

Acknowledgments. This work has been partly supported by the Grant Agency of CR within the project 15-13277S. The research leading to these results has received funding from the Norwegian Financial Mechanism 2009–2014 and the Ministry of Education, Youth and Sports under Project Contract no. MSMT-28477/2014 within the HaBiT Project 7F14047.

References

1. Auer, S., Bizer, C., Kobilarov, G., Lehmann, J., Cyganiak, R., Ives, Z.G.: DBpedia: a nucleus for a web of open data. In: Aberer, K., et al. (eds.) ASWC 2007 and ISWC 2007. LNCS, vol. 4825, pp. 722–735. Springer, Heidelberg (2007)
2. Fader, A., Zettlemoyer, L., Etzioni, O.: Open question answering over curated and extracted knowledge bases. In: Proceedings of the 20th ACM SIGKDD International Conference on Knowledge Discovery and Data Mining, pp. 1156–1165. ACM (2014)
3. Finkel, J.R., Grenager, T., Manning, C.: Incorporating non-local information into information extraction systems by Gibbs sampling. In: Proceedings of the 43rd Annual Meeting on Association for Computational Linguistics, ACL 2005, pp. 363–370. Association for Computational Linguistics, Stroudsburg (2005). http://dx.doi.org/10.3115/1219840.1219885

4. Horák, A., Medved', M.: SQAD: simple question answering database. In: Eighth Workshop on Recent Advances in Slavonic Natural Language Processing, pp. 121–128. Tribun EU, Brno (2014)
5. Jakubíček, M., Kovář, V., Šmerk, P.: Czech morphological tagset revisited. In: Proceedings of Recent Advances in Slavonic Natural Language Processing 2011, pp. 29–42 (2011)
6. Kovář, V., Horák, A., Jakubíček, M.: Syntactic analysis using finite patterns: a new parsing system for Czech. In: Vetulani, Z. (ed.) LTC 2009. LNCS, vol. 6562, pp. 161–171. Springer, Heidelberg (2011)
7. Šmerk, P.: Towards morphological disambiguation of Czech (2007)
8. Ševčíková, M., Žabokrtský, Z., Straková, J., Straka, M.: Czech named entity corpus 1.1 (2014). http://hdl.handle.net/11858/00-097C-0000-0023-1B04-C, LINDAT/CLARIN digital library at Institute of Formal and Applied Linguistics, Charles University in Prague
9. Shtok, A., Dror, G., Maarek, Y., Szpektor, I.: Learning from the past: answering new questions with past answers. In: Proceedings of the 21st International Conference on World Wide Web, pp. 759–768. ACM (2012)
10. Yih, W.T., He, X., Meek, C.: Semantic parsing for single-relation question answering. In: Proceedings of ACL 2014, vol. 2, pp. 643–648. Citeseer (2014)
11. Šmerk, P.: Fast morphological analysis of Czech. In: Proceedings of the RASLAN Workshop 2009, Brno (2009)

Annotation of Czech Texts with Language Mixing

Zuzana Nevěřilová(✉)

Faculty of Informatics, NLP Centre, Masaryk University,
Botanická 68a, 602 00 Brno, Czech Republic
xpopelk@fi.muni.cz

Abstract. Language mixing (using chunks of foreign language in a native language utterance) occurs frequently. Foreign language chunks have to be detected because their annotation is often incorrect. In the standard pipelines of Czech texts annotation, no such detection exists. Before morphological disambiguation, unrecognized words are processed by Czech guesser which is successful on Czech words (e.g. neologisms, typos) but its usage makes no sense on foreign words.

We propose a new pipeline that adds foreign language chunk and multi-word expression (MWE) detection. We experimented with a small corpus where we compared the original (semi-automatic) annotation (including foreign words and MWEs) with the results of the new pipelines.

As a result, we reduced the number of incorrect annotations of interlingual homographs and foreign language chunks in the new pipeline compared to the standard one. We also reduced the number of tokens that have to be processed by the guesser. The aim was to use the guesser solely on potentially Czech words.

1 Introduction

Natural languages do not exist in isolation and their influence is one of their forming factors. The terms *language mixing* (*code mixing*, *code switching*, *code alteration* or *code shifting*) appear in different contexts: bilingual children language acquisition, social media content, pronunciation modeling in speech recognition, or group identity. A typology of language mixing can be found e.g. in [1].

1.1 Language Mixing and Automatic Language Analysis

Language mixing is also interesting from the viewpoint of automatic language analysis which typically starts with sentence segmentation, word segmentation (tokenization), morphological analysis, and morphological disambiguation (tagging).

Morphological analysis and tagging are language dependent and therefore it is very important to recognize chunks of foreign language *before* these steps or to have a possibility to review the results of morphological analysis and tagging.

© Springer International Publishing Switzerland 2016
P. Sojka et al. (Eds.): TSD 2016, LNAI 9924, pp. 279–286, 2016.
DOI: 10.1007/978-3-319-45510-5_32

Apparently, this approach is novel in morphological analysis and tagging of Czech. The standard pipeline tries to search individual tokens in a morphological dictionary and afterwards it tackles with the remaining tokens unknown to the morphological analyser. This procedure ignores completely the fact that in an utterance, language mixing can occur.

A new pipeline that detects automatically foreign language chunks and prevents incorrect assignment of lemma and morphological tag is proposed in this work. We use several approaches: collections of multi-word expressions (MWE) and corpus-based decision procedures.

We annotated a small Czech corpus (one million tokens) with the new pipeline and evaluated the results. We reduced the number of incorrectly annotated interlingual homographs either by including them into MWEs, or by identifying them as foreign language chunks.

2   Related Work

The field of language mixing is well covered from the linguistic viewpoint (see e.g. [2]). However, not much attention has been paid to the computation linguistics perspective. In this overview, we present works related to the latter domain.

2.1   Language Mixing and Adaptation

Sometimes, foreign expressions are easily included in the native language and they can be found in dictionaries of the formal language. On the other hand, many words or word expressions are used frequently but not standardized in the formal language (e.g. *business process*, *pole position*). The border between foreign and adapted expressions is unclear and it is a matter of linguistic debate. From the language analysis point of view, the border becomes clearer when language users *inflect* the words. For example, in Czech, the word *chatbot* (or *chat bot*) can be either English or adapted into Czech when mentioned in singular nominative. However, if language users decline the word according to Czech grammar (e.g. into plural *chatboty*), it becomes a Czech masculine inanimate word.

2.2   Language Mixing and Automatic Language Analysis

In Sect. 1, we stated that language mixing can cause serious errors in language analysis, especially when concerning interlingual homographs. It is difficult to estimate how extensive the problem is but the phenomenon seems to be ubiquitous. In [3, Sect. 4.3.1], the authors state that in case of web corpora, 6.8 % of the noise originates in foreign-language material. [4] states that "in specific domains, up to 6.4 % of the tokens of a German text can be English inclusions" and in the TIGER corpus, 7.4 % of sentences contain English inclusions.

To our knowledge, no estimate of foreign inclusions exists for Czech corpora but we can easily find examples of incorrect annotation caused by *non*-treating the interlingual homographs. For a preliminary insight, we examined two different corpora of Czech: SYN2015 from Czech National Corpus [5] and czTenTen12 [6]. Examples of incorrect tags can be seen in Table 1. English is not the only language that enters in Czech utterances, it concerns Latin but also Slovak.

Table 1. Examples of incorrect annotation in two different Czech corpora

Chunk	Explanation	Tags in SYN2015	Tags in czTenTen12
On the road	Named entity	Personal pronoun + unknown + unknown	Pers. pronoun + unknown + noun
A priori	Latin	Conjunction + adverb	Conjunction + adverb
Set top box	Set-top unit device	Masculine inanimate noun + adjective + noun	Noun + verb + noun
korzetový top	Sleeveless shirt	Adjective + adjective	Adjective + verb
Pole position	Motorsport term	Neuter noun + unknown	Neut. noun + masc. noun
Hot dog	Fast food item	Unknown, adjective or noun + feminine noun	Interjection + noun in genitive

2.3 Language Identification

Language identification is considered easy when the pattern length is sufficient [7,8] which is typically not the case of language mixing. Several works focus on identifying foreign words in native text with respect to social media communication [9] or further language analysis [4,10].

2.4 Language Analysis in Czech

To our knowledge, no language analysis of Czech texts takes foreign inclusions into account. The standard pipeline consists of tokenization, morphological analysis, tagging, and guessing unknown words. In Sect. 2.2, we mentioned two Czech corpora that were created by distinct subjects. From the examples in Table 1, it can be seen that both corpora contain annotation errors on foreign tokens. Guessing cannot handle foreign named entities and interlingual homographs.

The corpus SYN2015 is tagged with no foreign inclusion detection. [11] describes a guesser for unknown words but the guessing is limited to undoubtedly Czech words (e.g. typos, neologisms, language game). In [12], foreign words annotation is described but no method how to automatically detect them is proposed.

The czTenTen12 web corpus [6] was processed automatically without detecting language mixing. Unknown words are processed by the guesser [13].

3 Methods

In view of the facts mentioned in Sect. 2.2, we propose a new pipeline that retokenizes the input according to a collection of MWEs and detects foreign language chunks.

Multi-word Expressions Processing. We collected over 500,000 MWEs from different sources:

1. list of first names and surnames with their inflected forms[1]: from 1,000,154 first names, 648,938 were multi-word, from 3,626,826 surnames, 224,738 were multi-word.
2. Czech place names with their inflected forms[2]: from 32,185 unique place names, 6,837 were multi-word
3. Latin quotes and sayings from Wikipedia: from 3,956 Latin expressions, 3,827 were multi-word
4. Czech company names from Wikipedia[3]: from 676 Czech company names, 468 were multi-word
5. MWE collection from web corpus czTenTen12: 23,101 MWEs including inflected forms

The first three types of MWEs can be considered fixed (the number of different surnames used in Czech texts slowly increases), the fourth can be easily obtained after it updates.

The corpus-based collection was created from web corpus (therefore mostly unedited) texts. The assumption was that in case of fixed MWEs, language users are unsure about the correct orthography and several variants of the same expression can be found: several words (e.g. *a priori*), one word (e.g. *apriori*), one word with dashes (e.g. *a-priori*). More details about this study can be found in [14]. Using this method, we obtained over 28,000 MWE candidates that were annotated manually by four annotators. The annotation categories were: non-MWE (a random tuple of words), MWE with function of a noun, MWE with function of an adverb, English words, and other foreign words.

The agreement among four annotators is shown in Table 2. Since one of the annotations differed a lot from others, we tried to exclude it and to compute the agreements among the three remaining annotators. Finally, we included all annotations with majority agreement from both cases. Only 33 candidates left undecided. We are aware of the possible random agreement, however, the large number of 4-annotator agreements indicated that the result may be plausible for our purpose even without considering random agreement. Finally, we obtained a collection of MWEs with 3,219 nouns, 80 adverbs, 2,325 English and 140 non-English foreign items. The majority of the candidates were Czech words.

The unclear border between foreign and adapted words (discussed in Sect. 2.1) was apparent from the annotations: from 2,325 MWEs annotated as foreign, 144 had at least one variant annotated as Czech MWEs. The indistinctness is evident in case of the base form, e.g. *web hosting* was annotated as foreign, while the inflected form *web hostingu* was annotated as Czech noun.

---

[1] Collected over years from different sources, e.g. http://www.mvcr.cz/clanek/cetnost-jmen-a-prijmeni-722752.aspx.

[2] Source: Cadastre of Real Estate (State Administration of Land Surveying and Cadastre) available from CKAN site http://linked.opendata.cz/en/dataset/cz-ruian.

[3] https://cs.wikipedia.org/wiki/Kategorie:Firmy_podle_okres%C5%AF_%C4%8Ceska.

Table 2. Number of annotation the annotators agreed on. The top part shows agreement among four annotation collections, the bottom part shows agreement among three most similar annotation collections.

Agreement type	Num. of annots
4 annotator agreement	15,586
Majority agreement (3 vs. 1 or 2 vs. 1 and 1)	11,195
Non-agreement (2 vs. 2 or completely different annotations)	1,830
3 annotator agreement	19,967
Majority agreement (2 vs. 1)	7,887
Non-agreement (completely different annotations)	757
Total	28,611

We generated automatically the inflected forms for MWEs that contain an inflective word. For example, from the base form *web server*, we generated 12 inflected forms (6 cases, singular and plural) such as *web serveru, web serverem*. Using this technique, we increased the number of MWEs to 23,063. However, it is clear that many of the generated forms hardly ever occur in texts.

Foreign Chunks Detection. Next step in the analysis is the detection of foreign language chunks. These concern quotations, named entities such as artworks, and code-switching. In the decision procedure, we used unigram and bigram frequencies in large corpora, the occurrence in the morphological analyser dictionary, and an assumption that an interlingual homonym never occurs in isolation.

The decision procedure is based on two large web corpora: czTenTen12 and enTenTen13 [15] with normalized frequencies. The decision procedure is language independent and we can employ corpus of another foreign language for recognizing other than English foreign chunks.

It is difficult to guess the boundary of an foreign chunk, especially in case of interlingual homographs, non-words (e.g. numbers, punctuation). Some chunks still remain that can be recognized as Czech or foreign only by syntactic or semantic analysis (e.g. English *50 Cent, Jeopardy!*, Latin *per se*).

3.1 Morphological Analysis, Tagging, and Guessing

For morphological analysis and tagging, we use the analyser `majka` [16] and the tagger `desamb` [17]. Note that in this phase, MWEs recognized in the first phase have already assigned a base form and all possible tags. Unlike single words, we assume that MWEs can have only one base form.

In case of capitalized tokens, our pipeline first tries to find them in gazetteers of named entities. This is not a named entity recognition task since in this phase, a named entity is only one possibility. The decision whether a token is or is not a named entity is left on the tagger. With this improvement, the number of unknown words decreases. Such words are processed by the guesser.

4   Evaluation

We evaluated the new processing pipeline on a small, "clean" (i.e. containing mostly edited texts) corpus – DESAM [18]. Originally, DESAM contained 1,173,835 tokens, including 722 multi-word tokens. We selected DESAM for two reasons: it has annotation of MWEs and it was the only resource with marked foreign words. The disadvantage of DESAM is its cleanness – we expected that English mixing occurs less frequently in DESAM than in web corpora.

4.1   Multi-word Expressions

We retokenized DESAM data with the new pipeline (tokenization including MWEs, foreign chunks detection, analysis, tagging, guessing). We reduced the number of tokens processed by the guesser from 28,120 to 10,649 (4,094 tokens were found in gazetteers, 12,345 were detected as foreign, 1,032 were annotated as parts of MWEs).

We compared MWEs from the original (semi-automatically annotated) DESAM (409 unique MWEs) with the new one (410 unique MWEs): 50 MWEs were the same. The disagreement is caused by the unclear definition of a MWE. In semi-automatically annotated DESAM, MWEs were law numbers, foreign phrases, and multi-word named entities. In newly annotated DESAM, most MWEs were multi-word named entities.

4.2   Interlingual Homographs

In the next stage, we focused on frequent English-Czech homographs. In Table 3, it can be seen that our approach sometimes identifies a Czech word as foreign but in many cases (in bold in the table), it can identify English homographs with 100 % accuracy. The number of tokens in DESAM processed with the standard pipeline is sometimes higher than the number of tokens in DESAM processed with our new pipeline because some tokens became part of a MWE (e.g. *a priori*) and are no longer annotated separately. We evaluated manually how many tokens annotated as foreign were really foreign and for some homographs, we also evaluated how many tokens annotated as native were really native.

5   Conclusion and Future Work

We used a complex approach to corpus annotation of Czech texts with language mixing. We modified the standard processing pipeline in order to detect MWEs and foreign language chunks. We used collections of MWEs and gazetteers to reduce dramatically the number of unknown tokens. This reduction can lead to

Table 3. Number of interlingual homographs in DESAM annotated by the standard pipeline (2nd column), new pipeline (3rd column), number of detected foreign tokens, number of correct annotation of foreign tokens (5th column) and native tokens (7th column). The homographs are ordered by their length (2 letters, 3 letters, 4 letters). Dash means that correct annotation was not evaluated.

Token	Std DESAM	New DESAM	Annotated as foreign	Foreign correct	Annotated as native	Native correct
a	23,057	22,992	47	2	22,945	–
I	427	426	47	42	379	379
do	5,012	5,001	10	0	4,991	–
to	4,290	4,282	23	10	4,259	–
on	84	82	9	7	73	73
let	898	898	0	0	898	898
her	90	90	2	0	88	88
set	87	87	8	8	79	79
top	14	4	4	4	0	0
for	14	14	13	13	1	0
not	3	3	3	0	0	0
had	3	3	2	0	1	1
list	73	73	6	1	67	67
post	19	18	18	0	0	0
most	7	7	1	0	6	6

more appropriate use of the guesser. To our knowledge, this work is the first attempt to annotate Czech texts with language mixing. We plan to use our new pipeline for annotation of web corpus of Czech where we expect the phenomenon of language mixing to be more significant than in DESAM.

Acknowledgments. The research leading to these results has received funding from the Norwegian Financial Mechanism 2009–2014 and the Ministry of Education, Youth and Sports under Project Contract no. MSMT-28477/2014 within the HaBiT Project 7F14047.

This work has been partly supported by the Ministry of Education of CR within the national COST-CZ project LD15066.

References

1. Auer, P.: From codeswitching via language mixing to fused lects. Int. J. Bilingualism **3**(4), 309–332 (1999)
2. Alex, B.: Automatic detection of english inclusions in mixed-lingual data with an application to parsing. Ph.D. thesis. School of Informatics, University of Edinburgh, Edinburgh (2008)

3. Schäfer, R., Bildhauer, F.: Web Corpus Construction. Synthesis Lectures on Human Language Technologies. Morgan and Claypool, San Francisco etc. (2013). http://dx.doi.org/10.2200/S00508ED1V01Y201305HLT022

4. Alex, B., Dubey, A., Keller, F.: Using foreign inclusion detection to improve parsing performance. In: Proceedings of the 2007 Joint Conference on Empirical Methods in Natural Language Processing and Computational Natural Language Learning (EMNLP-CoNLL), pp. 151–160 (2007)

5. Křen, M., Cvrček, V., Čapka, T., Čermáková, A., Hnárková, M., Chlumská, L., Jelínek, M., Kováříková, D., Petkevič, V., Procházka, P., Skoumalová, H., Škrabal, M., Truneček, P., Vondřička, P., Zasina, A.: SYN2015: reprezentativní korpus psané češtiny [SYN2015: Representative Corpus of Written Czech] (2015)

6. Suchomel, V., Pomikálek, J.: Efficient web crawling for large text corpora. In: Adam Kilgarriff, S.S. (ed.) Proceedings of the Seventh Web as Corpus Workshop (WAC7), Lyon, pp. 39–43 (2012)

7. Baldwin, T., Lui, M.: Language identification: the long and the short of the matter. In: Human Language Technologies: The 2010 Annual Conference of the North American Chapter of the Association for Computational Linguistics, HLT 2010, pp. 229–237. Association for Computational Linguistics, Stroudsburg (2010)

8. Lui, M., Baldwin, T.: Langid.Py: an off-the-shelf language identification tool. In: Proceedings of the ACL 2012 System Demonstrations, ACL 2012, pp. 25–30. Association for Computational Linguistics, Stroudsburg (2012)

9. Eskander, R., Al-Badrashiny, M., Habash, N., Rambow, O.: Foreign words and the automatic processing of Arabic social media text written in Roman script. In: Proceedings of the First Workshop on Computational Approaches to Code Switching, pp. 1–12. Association for Computational Linguistics, Doha, October 2014

10. Ahmed, B.U.: Detection of foreign words and names in written text. Ph.D. thesis. Pace University, New York, NY, USA (2005). AAI3172339

11. Hlaváčová, J.: Morphological guesser of Czech words. In: Matoušek, V., Mautner, P., Mouček, R., Tauser, K. (eds.) TSD 2001. LNCS (LNAI), vol. 2166, pp. 70–75. Springer, Heidelberg (2001)

12. Hana, J., Zeman, D., Hajič, J., Hanová, H., Hladká, B., Jeřábek, E.: Manual for morphological annotation, revision for the Prague dependency treebank 2.0. Technical report TR-2005-27, ÚFAL MFF UK, Prague, Czech Rep. (2005)

13. Šmerk, P., Sojka, P., Horák, A.: Towards Czech morphological guesser. In: Proceedings of Recent Advances in Slavonic Natural Language Processing, RASLAN 2008, pp. 1–4. Masarykova univerzita, Brno (2008)

14. Nevěřilová, Z.: Annotation of multi-word expressions in Czech texts. In: Horák, A., Rychlý, P. (eds.) Ninth Workshop on Recent Advances in Slavonic Natural Language Processing, pp. 103–112. Tribun EU, Brno (2015)

15. Jakubíček, M., Kilgarriff, A., Kovář, V., Rychlý, P., Suchomel, V.: The TenTen corpus family. In: 7th International Corpus Linguistics Conference, CL 2013, pp. 125–127. Lancaster (2013)

16. Šmerk, P.: Fast morphological analysis of Czech. In: Proceedings of the Raslan Workshop 2009. Masarykovauniverzita (2009)

17. Šmerk, P.: K morfologické desambiguaci češtiny [Towards morphologicaldisambiguation of Czech]. Thesis proposal. Masaryk University (2008)

18. Rychlý, P., Šmerk, P., Pala, K.: DESAM morfologicky označkovaný korpus českých textů. Technical report. Masaryk University (2010)

Evaluation and Improvements in Punctuation Detection for Czech

Vojtěch Kovář[1]([✉]), Jakub Machura[1], Kristýna Zemková[1], and Michal Rott[2]

[1] NLP Centre, Faculty of Informatics, Masaryk University,
Botanická 68a, 602 00 Brno, Czech Republic
xkovar3@fi.muni.cz,{382567,415795}@mail.muni.cz
[2] Institute of Information Technology and Electronics,
Technical University of Liberec, Studentská 2, 461 17 Liberec, Czech Republic
michal.rott@tul.cz

Abstract. Punctuation detection and correction belongs to the hardest automatic grammar checking tasks for the Czech language. The paper compares available grammar and punctuation correction programs on several data sets. It also describes a set of improvements of one of the available tools, leading to significantly better recall, as well as precision.

1 Introduction

Punctuation detection belongs to important tasks in automatic checking of grammar, especially for Czech language. However, it is one of the most difficult tasks as well, unlike e.g. correcting simple spelling errors.

Czech, a free-word-order language with a rich morphology, has a complex system of writing commas in sentences – partly because the language norm defines it in very complicated and unintuitive way, partly because commas often affect the semantics of the utterances. It is based on linguistic structure of the sentences, and even native speakers of Czech, including educated people, have often problems with correct placement of punctuation.

There are several automatic tools that partly solve the problem: Two commercial grammar checkers for Czech (which also try to correct different types of errors, but here we exploit their punctuation correction features only) and some academic projects mainly focused on correcting punctuation. We list and briefly describe them in the next section.

This paper contains two important results: The first one is a significant accuracy improvement of one of the open-source academic tools, the SET system [6,7]; the other is a comprehensive comparison and evaluation of all the available tools that was missing for this task so far.

The structure of the paper is as follows: The next section briefly describes the past work done on the punctuation detection problem. Then we describe the punctuation detection in the SET tool and our improvements to it. Section 5 presents comparison and evaluation of all the available tools.

© Springer International Publishing Switzerland 2016
P. Sojka et al. (Eds.): TSD 2016, LNAI 9924, pp. 287–294, 2016.
DOI: 10.1007/978-3-319-45510-5_33

2   Related Work

The two mentioned commercial grammar checking systems are:

- **Grammar checker** built into the Microsoft Office, developed by the Institute of the Czech language [9,11],
- **Grammaticon** checker created by the Lingea company [8].

Both systems aim at manual description of erroneous constructions, and minimizing the number of false alerts. From the available comparisons [1,10,11] it seems that the Grammar checker generally outperforms Grammaticon; however, the testing data are rather small and present only the general results, whereas we are interested purely in punctuation corrections.

The following systems emerged from the academic world:

- Holan et al. [3] proposed using automatic dependency parsing for punctuation detection, but the result is only a prototype and not usable in practice.
- Jakubíček and Horák [5] exploit the grammar of the Synt parser [4] to detect punctuation, with both precision and recall slightly over 80 percent. The tool is unfortunately not operational, otherwise it would be very interesting to include it into our comparison.
- Boháč et al. [2] used punctuation detection for post-processing of automatic speech recognition. In their approach, a set of N-gram rules (with N up to 4, including the commas) was statistically induced from the training Czech corpora of news texts and books. Application of these rules is implemented via weighted final-state transducers[1] which enable both inserting a comma and suppressing the insertion by a more specific rule. We have included 2 versions of this system into our comparison – the original one, as referenced above, and a recent one – they are further referred to as FST 1 and FST 2.
- Kovář [6] reported on using the open-source SET parser [7] for punctuation detection, with promising results. Building on the existing grammar, we have significantly improved it. In the next sections, we describe the principle of punctuation detection within the SET parser, and some of our important improvements.

As for comparison presented in Sect. 5, there is no similar work, to our best knowledge.

3   Punctuation Detection with SET – Initial State

3.1   The SET Parser

The SET system[2] is an open-source pattern matching parser designed for natural languages [7]. It contains manually created sets of patterns ("grammars")

---

[1] http://openfst.org/.
[2] SET is an abbreviation of "syntactic engineering tool".

for Czech, Slovak and English – in the process of analysis, these patterns are searched within the sentence and compete with each other to build a syntactic tree. Primary output of the parser is a so-called "hybrid tree" that combines dependency and phrase structure features.

Before the syntactic analysis, the text has to be tokenized and tagged – for this purpose, we used the `unitok` tokenizer [13] and the `desamb` tagger for Czech [14] in all our experiments.

3.2 Punctuation Detection with SET

For the purpose of punctuation detection, a special grammar has been developed by Kovář [6], producing pseudo-syntactic trees that only mark where a comma should, or should not be placed (by hanging the particular word under a `<c>` or `<n>` node), as illustrated in Fig. 1.

```
TMPL: $NEG $PREP $REL   MARK 1 <c>   HEAD 1
      $REL(tag): k3.*y[RQ] k6.*y[RQ]
      $PREP(tag): k7
      $NEG(tag not): k7 k3.*y[RQ] k6.*y[RQ] k8
      $NEG(word not): a * " tak přitom
```

Input: Neví na jaký úřad má jít.

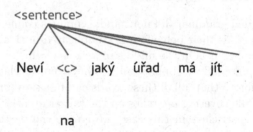

Output: Neví, na jaký úřad má jít.

Fig. 1. One of the existing punctuation detection rules in SET, and its realization on a sample Czech sentence: *"Neví na jaký úřad má jít"*. (missing comma before *"na"* – *"(He) does not know what bureau to go in."*) The rule matches preposition (k7) followed by relative pronoun (k3.*y[RQ]) or adverb (k6.*y[RQ]), not preceded by preposition or conjunction (k8) or relative pronoun/adverb and few other selected words (the `tag` not and `word` not lines express negative condition – token must not match any of the listed items). Ajka morphological tagset is used [12]; the example is borrowed from [6].

4   Punctuation Detection with SET – Improvements

The original grammar was a rather simplistic one – it contained only about 10 rules and despite relatively good results [6] it was clear that there is room for improvements. In this section, we describe our main changes to the original grammar, and motivation behind them.

The original set of rules is still available in the SET project repository for reference.[3]

All of the following changes have been tested on a development set of texts before including into the final grammar, and all of them proved a positive effect on precision and/or recall on this development set. Thanks to their good results also on the evaluation test sets (see Sect. 5), they were accepted into the SET parser repository and are also available online.[4] For this reason, we do not include the particular rules here in the paper – some of them are rather extensive and would unnecessarily fill the place available.

4.1 Adjusting to the Tagger Imperfections

As every tagger, `desamb` makes mistakes. For instance, relative pronouns *co* and *což* are sometimes tagged as particles and therefore not covered by the original grammar. We have added a special rule covering these particular words, independently on morphological tags. Also, several inconsistencies in the tagset were detected and fixed, e.g. the tag `k3.*y[RQ]` from Fig. 1 was sometimes recorded as `k3.*x[RQ]`, so we added these variants.

4.2 Special Conjunctions

The original grammar did not handle commas connected to conjuctions like *ale*, *jako* or *než*, as their behaviour is complicated in this respect: *ale* is preceded by comma unless it is in the middle of a clause (which is hard to detect). *Jako* and *než* are preceded by comma only if they connect clauses (rather than phrases). In addition to that, all of these words can function (and be tagged as) a particle.

The rule covering *ale* relies on the tagger abilities to distinguish conjunction from the particle – in this case, we ignore the particles. We also approximate the middle-clause position by listing a few stop words that often co-occur with *ale* in this position.

In case of *jako* and *než*, we have extended the rule for general conjunctions and place comma before them only if there is a finite verb later in the sentence.

4.3 Commas After Inserted Clauses

One of the punctuation rules in Czech orders delimiting insterted clauses by commas from both sides. The left side is usually easier, as it contains a conjunction or a relative pronoun. We implemented an approximate detection of the right side by two groups of rules:

---

[3] http://nlp.fi.muni.cz/trac/set/browser/punct.set.
[4] http://nlp.fi.muni.cz/trac/set/browser/punct2.set.

- There are two finite verbs close to each other – then the comma is placed before the second one.
- There is a clitic in a Wackernagel's position – that means that the inserted clause is the first constituent in the sentence, and the comma belongs right before the clitic.

We were experimenting also with other options (including detecting clauses by the full SET grammar for Czech) but they were not improving the results – in more complicated cases, placement of commas usually depends on semantics and pattern-based solution was not able to describe it sufficiently.

We have also introduced a number of rather small, or technical improvements. We do not include the full list here but the resulting grammar can be easily accessed in the SET repository (as referenced above).

5   Evaluation

In this section we present a thorough evaluation of the systems mentioned in the paper so far, that are currently operational. Namely, Grammaticon, Grammar checker, FST 1, FST 2, original SET punctuation grammar (SET:orig) and our improved set of rules (SET:new).

5.1   Data

Our aim was to conduct a rigorous evaluation of the available tools, so we put quite a lot of effort into selecting and processing the testing data. No part of the data was used for any part of the development.

We have collected 8 sets of Czech texts with different text types, as summarized in Table 1. The text types are:

- blog texts
- examples on punctuation from an official internet education portal
- horoscopes
- original Czech fiction (2 authors, one classic, one contemporary)
- Czech translations of English fiction (2 authors)
- basic and high school dictation transcriptions (focused on punctuation, with real mistakes recorded)

Blogs and horoscopes were manually corrected by 3 independent correctors so that we can be very certain that there are no mistakes in the data sets. The agreement rate among the correctors was very high, all 3 correctors agreed in 96.3 % (blogs) and 98.2 % (horoscopes) of cases. The other texts come from high-quality sources, so they were probably carefully corrected before publication and do not contain significant amount of errors.

Unfortunately, Grammaticon and Grammar checker do not offer any type of batch-processing mode, and each correction must be manually accepted. This feature makes it impossible to test them using large texts; for this reason, only 3 data sets were used for comparison of all the available programs: Blogs, language reference examples and dictations. The rest could be used only for evaluation of the newly introduced changes to the SET punctuation grammar and FSTs.

Table 1. Data sets used for testing.

Testing set	# words	# commas
Selected blogs	20, 883	1, 805
Internet Language Reference Book	3, 039	417
Horoscopes 2015	57, 101	5, 101
Karel Čapek – selected novels	46, 489	5, 498
Simona Monyová – Ženu ani květinou	33, 112	3, 156
J.K. Rowling – Harry Potter 1 (translation)	74, 783	7, 461
Neil Gaiman – The Graveyard Book (translation)	55, 444	5, 573
Dictation transcriptions	17, 520	2, 092

5.2 Method

Different people make different errors and it is not clear how a grammar checker should be properly evaluated when we do not have a huge corpus of target user errors. At this state, the most fair-play method is probably the one used already by Horák and Jakubíček [5]: To remove all the commas and measure how many can be properly placed by a checker. We used this method with all the data sets, except the dictations.

The dictation transcriptions do offer a small corpus of real people mistakes, so we are able to measure the real power of correcting commas, at least to some extent. Of course, the data is rather small and such evaluation is very specific to the text type and to the user type (basic school students), but it is an interesting number for comparison. We performed such measure (i.e. kept the original commas and let the tools correct the text as it is) on the dictations testing set.

We use standard precision and recall on detected commas or fixed errors, respectively.

5.3 Results

The results of the comparison are presented in Table 2. We can clearly see that Grammar checker outperforms Grammaticon, namely in recall (precisions are comparable). Similarly, our new SET grammar is always better than the original one, in both precision and recall.

Grammar checker compared to SET:new shows that SET has significantly higher recall whereas Grammar checker wins in precision. If we measured F-score, SET would be better – on the other hand, precision is more important in case of grammar checkers, as we want to minimize the number of false alerts; it is difficult to pick a winner from these two. Similarly, SET outperforms both final-state transducers in terms of F-score, although they reached higher precision in several cases.

Table 2. Results of the comparison. P, R stands for Precision, Recall (in percent). "Gr..con" stands for Grammaticon, "GCheck" is Grammar Checker. Note that the dictations testing set uses different method (fixed errors, rather than detected commas), as explained in Sect. 5.2. Best precision and recall result for each testing set is highlighted.

Testing set	Gr..con		GCheck		FST 1		FST 2		SET:orig		SET:new	
	P	R	P	R	P	R	P	R	P	R	P	R
Blogs	**97.5**	10.8	97.3	28.3	89.0	48.8	88.8	49.4	86.8	42.7	89.7	**58.2**
Language ref.	89.1	9.8	**92.0**	19.4	78.1	27.3	78.1	28.3	75.8	22.5	87.3	**36.2**
Horoscopes	-	-	-	-	93.5	53.8	92.2	54.3	89.5	46.7	**94.1**	**64.8**
K. Čapek	-	-	-	-	84.1	30.0	75.1	32.4	85.6	32.3	**87.2**	**34.6**
S. Monyová	-	-	-	-	**86.8**	47.8	86.2	49.2	82.1	51.0	84.0	**53.8**
J.K. Rowling	-	-	-	-	**90.0**	47.7	82.8	49.0	87.8	50.2	89.7	**53.4**
N. Gaiman	-	-	-	-	**91.5**	41.3	80.5	42.0	88.0	47.8	89.4	**48.4**
Dictations	**96.4**	5.7	93.3	14.6	60.9	18.2	51.0	17.4	68.2	14.7	78.7	**27.3**

Clearly some text types are easier than others – common internet texts and translations seem generally easier than original Czech fiction. Language reference examples are hard as well – but that was expected, as it contains examples for all the borderline cases that the automatic procedures cannot cover (yet) for various reasons.

In general, the results do not indicate that the tools will soon be able to correct all mistakes in a purely automatic way, the error rates are still too high and the recalls too low. But probably both Grammar checker, FSTs and the new SET grammar are able to productively assist people when correcting their texts.

6   Conclusions

In the paper, we described an improvement to an existing punctuation correction tool for Czech, and performed a thorough evaluation of all the available tools that are able to detect punctuation in Czech sentences automatically. The evaluation shows that our changes significantly improved the accuracy of the tool. Currently it has lower precision and significantly higher recall than state-of-the-art commercial grammar checker.

However, the figures also indicate that the results are still not good enough for purely automatic corrections of punctuation. Further research will be needed, probably aimed at exploiting more complicated syntactic and semantic features of the language.

Acknowledgments. This work has been partly supported by the Grant Agency of CR within the project 15-13277S. The research leading to these results has received funding from the Norwegian Financial Mechanism 2009–2014 and the Ministry of Education,

Youth and Sports under Project Contract no. MSMT-28477/2014 within the HaBiT Project 7F14047. This work was also partly supported by Student Grant Scheme 2016 of Technical University of Liberec.

References

1. Behún, D.: Kontrola české gramatiky pro MS Office - konec korektorů v Čechách? (2005). https://interval.cz/clanky/kontrola-ceske-gramatiky-pro-ms-office-konec-korektoru-v-cechach
2. Boháč, M., Blavka, K., Kuchařová, M., Škodová, S.: Post-processing of the recognized speech for web presentation of large audio archive. In: 2012 35th International Conference on Telecommunications and Signal Processing (TSP), pp. 441–445 (2012)
3. Holan, T., Kuboň, V., Plátek, M.: A prototype of a grammar checker for Czech. In: Proceedings of the 5th Conference on Applied Natural Language Processing, pp. 147–154. Association for Computational Linguistics (1997)
4. Horák, A.: Computer Processing of Czech Syntax and Semantics. Librix.eu, Brno (2008)
5. Jakubíček, M., Horák, A.: Punctuation detection with full syntactic parsing. Res. Comput. Sci. Spec. issue: Nat. Lang. Process. Appl. **46**, 335–343 (2010)
6. Kovář, V.: Partial grammar checking for Czech using the set parser. In: Sojka, P., Horák, A., Kopeček, I., Pala, K. (eds.) TSD 2014. LNCS, vol. 8655, pp. 308–314. Springer, Heidelberg (2014)
7. Kovář, V., Horák, A., Jakubíček, M.: Syntactic analysis using finite patterns: a new parsing system for Czech. In: Vetulani, Z. (ed.) LTC 2009. LNCS, vol. 6562, pp. 161–171. Springer, Heidelberg (2011)
8. Lingea s.r.o.: Grammaticon (2003). www.lingea.cz/grammaticon.htm
9. Oliva, K., Petkevič, V., Microsoft s.r.o.: Czech Grammar Checker (2005). http://office.microsoft.com/word
10. Pala, K.: Pište dopisy konečně bez chyb – Český gramatický korektor pro Microsoft Office. Computer, 13–14 (2005)
11. Petkevič, V.: Kontrola české gramatiky (český grammar checker). Studie z aplikované lingvistiky-Stud. Appl. Linguist. **5**(2), 48–66 (2014)
12. Sedláček, R., Smrž, P.: A new Czech morphological analyser ajka. In: Matoušek, V., Mautner, P., Mouček, R., Tauser, K. (eds.) TSD 2001. LNCS (LNAI), vol. 2166, pp. 100–107. Springer, Heidelberg (2001)
13. Suchomel, V., Michelfeit, J., Pomikálek, J.: Text tokenisation using unitok. In: Eighth Workshop on Recent Advances in Slavonic Natural Language Processing, pp. 71–75. Tribun EU, Brno (2014)
14. Šmerk, P.: Unsupervised learning of rules for morphological disambiguation. In: Sojka, P., Kopeček, I., Pala, K. (eds.) TSD 2004. LNCS (LNAI), vol. 3206, pp. 211–216. Springer, Heidelberg (2004)

Annotated Amharic Corpora

Pavel Rychlý and Vít Suchomel[⊠]

NLP Centre, Faculty of Informatics, Masaryk University,
Botanická 68a, 602 00 Brno, Czech Republic
{pary,xsuchom2}@fi.muni.cz

Abstract. Amharic is one of under-resourced languages. The paper presents two text corpora. The first one is a substantially cleaned version of existing morphologically annotated WIC Corpus (210,000 words). The second one is the largest Amharic text corpus (17 million words). It was created from Web pages automatically crawled in 2013, 2015 and 2016. It is part-of-speech annotated by a tagger trained and evaluated on the WIC Corpus.

1 Introduction

Annotated corpora are quite common even for under-resourced languages but there are languages with tens of million native speakers without high quality text corpora. Amharic is such a case.

Amharic is one of the official working languages of Ethiopia. It is the second most spoken Semitic language in the world with over 20 million native speakers. With so many speakers and being an official language it is hard to believe it counts as an under-resourced language. However, there are not many language resources for Amharic and most of those available are of poor quality, small sized and/or not easily accessible. That is also the case of text corpora. There are several text corpora available (see Sect. 2) but there is only one morphologically annotated corpus of small size and poor quality. One of the reasons for that situation is the special script used for writing Amharic: Ge'ez.

Ge'ez script, also called Fidel in Amharic, is a syllabic script. There are more than 300 characters, each representing a consonant – vowel pair. There are 26 consonant letters combined with 7 or more vowels. The Ge'ez script has also its own symbols for numbers and punctuation. See Table 1 for an example of the Ge'ez characters and Fig. 1 for an Amharic text written in Ge'ez. The Ge'ez script is used also for writing Tigrinya and several smaller languages of Ethiopia. Not all characters are used in all languages –there are characters used only in one language. Ge'ez script is supported by Unicode standard from version 3.0 (1999). Several rarely used characters were added into versions 4.1 (2005) and 6.0 (2010).

Because of the bad support for displaying and writing Ge'ez script on computers there were many attempts to use a transliteration of the script in other alphabets. There are at least ten different transliteration systems in Latin script. Not all of them define mapping of all Ge'ez characters but all are based on a

P. Sojka et al. (Eds.): TSD 2016, LNAI 9924, pp. 295–302, 2016.
DOI: 10.1007/978-3-319-45510-5_34

phonetic transcription, hence the differences are not big. The most complete and most different from others is SERA [2]. No accents are required, only ASCII characters (English alphabet) are used. Therefore, it is easy to type SERA on any keyboard. Transliteration of several Ge'ez characters is listed in Table 1. We are using SERA in all our corpora together with the original Ge'ez script.

Table 1. Transliteration of selected Ge'ez characters in SERA system.

ሰ=se	ሱ=su	ሲ=si	ሳ=sa	ሴ=sE	ስ=s	ሶ=so
ሸ=xe	ሹ=xu	ሺ=xi	ሻ=xa	ሼ=xE	ሽ=x	ሾ=xo
ኘ=Ne	ኙ=Nu	ኚ=Ni	ኛ=Na	ኜ=NE	ኝ=N	ኞ=No

2 Existing Corpora

2.1 WIC News Amharic Corpus

Amharic text corpora range from morphologically annotated to parallel corpora. Compared to similar corpora in other (even smaller) languages, all Amharic corpora are small. WIC Corpus [1] is the only manually morphologically annotated corpus. It consists of about 210,000 words in 1,065 documents. Texts were taken from the Web news published by the Walta Information Center (http://www.waltainfo.com) in 2001. A sample of the corpus in displayed in Fig. 1.

```
<document>
<filename>mes01a1.htm</filename>
<title>ልዩ <ADJ> ዞኑ <N> ያስገነባቸው <VREL> ፕሮጀክቶች <N> 50ሺ <NUMCR> ነዋሪዎችን
<N> ተጠቃሚ <ADJ> አደረጉ <V> ። <PUNC></title>
<dateline place="አዳማ" month="መስከረም" date="1/1994/(WIC)/" />
<body>
በአዳማ <NP> ልዩ <ADJ> ዞን <N> በተጠናቀቀው <VP> የበጀት <NP> አመት <N> በ6 ነጥብ
8ሚሊየን <NUMP> ብር <N> ከተጀመሩት <VP> 12 <NUMCR> ፕሮጀክቶች <N> መካከል
<PREP> አብዛኞቹ <ADJ> ተጠናቅቀው <V> 50ሺ <NUMCR> ነዋሪዎችን <N> ተጠቃሚ <N>
ማድረጋቸው <VN> ተገለጸ <V> ። <PUNC> በልዩ <ADJP> ዞኑ <N> የቴክኒክ <NP> አገልግሎትና
<NC> የከተማ <NP> ቦታ <N> አስተዳደር <N> ክፍል <N> ሃላፊ <N> አቶ <ADJ> ግዛው <N>
ድንቁ <N> ዛሬ <ADV> ለዋልታ <NP> እንፎርሜሽን <N> ማእከል <N> እንደገለጹት <VP> በልዩ
<ADJP> ዞኑ <N> 13 <NUMCR> ቀበሌዎች <N> ተገንብተው <V> ለአገልግሎት <NP> የበቁት
<VREL> የመንገድ <NP> ፣ <PUNC> የግብርናና <NPC> የመሰረታዊ <ADJP> ልማት <N>
ፕሮጀክቶች <N> ናቸው <AUX> ። <PUNC>
```

Fig. 1. Example of annotated WIC Corpus

2.2   Morphological Annotation

Amharic language has a rich morphology: Nouns and adjectives are inflected and there are complex rules for deriving verbs. Several part-of-speech tag systems were proposed earlier, all working with about 10 tags for basic part of speech. No existing tag-set includes any tags for annotating gender, number and other grammatical categories. In some cases, nouns, pronouns, adjectives, verbs and numerals have variants of words with attached prepositions and/or conjunctions. For example, there are N = noun, NP = noun with a preposition as a prefix, NC = noun with a conjunction as a suffix, NPC = noun with a preposition as a prefix and a conjunction as a suffix. In total, there are 30 different PoS tags in the WIC Corpus.

3   New Amharic Corpora

We have created two new corpora. The first one is a cleaned version of the WIC Corpus, the second one is a new big corpus from the Web. Both corpora are available for querying on the web page of the HaBiT project at https:// habit-project.eu/corpora.

3.1   Cleaned WIC Corpus

There were several attempts to use the WIC Corpus for training automatic part-of-speech taggers, for example [3,4,11]. All of them found that the corpus has many annotation inconsistencies: missing tags, misspelling of tags, multiword expressions and others. There were two separate versions of the corpus: one for original Ge'ez script and one with SERA transliteration. In several research papers, they report different number of tokens for each version. We have unified both versions and corrected non matching words either in Ge'ez or SERA depending on a native speaker decision. We have applied all cleaning procedures described in the above mentioned papers.

We have added more unifications of numbers and dates. For example, most of numbers containing decimal point were written as "6 ነጥብ 8" where "ነጥብ" means "point". It is the result of original transcription from hand-written "paper" annotation into computer. Sometimes such string formed one token while there were three tokens in other cases. We have normalised all such occurrences into the correct form (6.8 in this case) with the respective PoS tag. The size of the cleaned corpus is 200,561 tokens. Each token is represented by a word in Ge'ez, its transliteration in SERA and the respective PoS tag.

The cleaned WIC corpus was used to train a PoS tagger. Because of the small number of tags in the tag-set we chose TreeTagger [9], it works very well in such conditions. To evaluate an accuracy of created tagging model we have divided the corpus into 10 parts each containing 20,000 tokens. For each part, we trained a TreeTagger model on nine remaining parts, ran TreeTagger on that part, and compared the result with the manual annotation. The whole evaluation task was done separately on the Fidel part of the corpus and the SERA part, and for both on data before and after the final cleaning procedure. The results are summarised in Table 2, the average accuracy is 87.4 %. We can see that the final cleaning has not influenced the results much and the performance of TreeTagger is a bit better on the Fidel script than on the SERA transliteration.

Table 2. Accuracy of TreeTager on ten parts of the WIC corpus

Part	Fidel, before	SERA, before	Fidel, after	SERA, after
1	85.1	85.1	85.1	85.2
2	85.4	85.2	85.4	85.1
3	85.7	85.7	85.7	85.7
4	88.2	88.1	88.2	88.1
5	89.1	89.0	89.2	89.1
6	86.6	86.5	86.8	86.6
7	89.9	89.8	89.9	89.9
8	91.5	91.6	91.6	91.7
9	89.7	89.8	89.8	89.9
10	82.3	82.3	82.3	82.3
Average	**87.36**	**87.30**	**87.41**	**87.35**

3.2   Building an Amharic Web Corpus

We have used the following steps to create a big Web corpus: First, adopting the Corpus factory method [6] bigrams of Amharic words from the Crúbadán database[1] [8] were used to query Bing search engine for documents in Amharic. 354 queries yielded 6,453 URLs. URLs of 3,145 successfully downloaded documents were used as starting points for web crawler SpiderLing [10]. URLs of documents crawled in 2013 using a similar approach[2] were added to the set of starting points.

The following language models were created:

- Character trigram model for language detection.[3] 5.2 MB of text from the WIC Corpus and Amharic Wikipedia was used to train the model.
- Byte trigram model for character encoding detection. The model was trained using web pages obtained by the Corpus factory method.
- The most frequent Amharic words from the WIC Corpus wordlist were used as a resource for boilerplate removal tool jusText [7].

The crawler was set to harvest web domains in the Ethiopian national top level domain et and other general TLDs: com, org, info, net, edu. 3.6 GB of

---

[1] http://crubadan.org/languages/am, by K. Scannell.

[2] We made an unpublished attempt to crawl the Amharic web in 2013.

[3] http://code.activestate.com/recipes/326576-language-detection-using-character-trigrams/, by D. Bagnall.

http responses was gathered in the process. HTML tags and boilerplate paragraphs were removed from the raw data. 42 % of paragraphs were identified as duplicate or near duplicate and removed using tool onion [7]. 66 MB of deduplicated text obtained by the same process in 2013 was added to the data. Sentence boundaries were marked at positions with Amharic end of sentence characters ∷ and ፧. The final size of the corpus (containing data from years 2013, 2015 and 2016) is 461 MB or more than 17 million words. Finally, the corpus was tagged by TreeTagger with a model trained on the cleaned version of the WIC Corpus. The corpus is called amWaC 16.[4]

3.3   Corpus Properties

Basic properties of corpus sources are summarised in Tables 3 and 4.[5]

We observe the content of news/politic and religious portals has a significant presence in the corpus sources. Since there are only 138 domains with more than 10 documents represented in the corpus, we admit the result collection would benefit from a greater variety of sources.

The most frequent parts of speech in both corpora are nouns and verbs. For details see Fig. 2.

Table 3. The size of corpus structures.

Document count	33,542
Paragraph count	341,327
Sentence count	1,208,926
Word count	17,320,000
Ge'ez lexicon size	955,628
Sera lexicon size	948,553

Table 4. Document count – the most frequent web domains and domain size distribution.

Top level domains		Web domains		Domain size distribution	
org	14,582	*.jw.org	6,717	At least 1000 documents	7
com	11,927	*.gov.et	4,599	At least 500 documents	15
et	5,090	waltainfo.com	2,818	At least 100 documents	42
net	1,084	ginbot7.org	2,666	At least 50 documents	63
cz	724	eotcmk.org	1,141	At least 10 documents	149
info	85	ethsat.com	894	At least 1 document	573

---

[4] Amharic 'Web as Corpus' corpus, year 2016.
[5] TLD cz in Table 4 was set by the host server according to the location of the requesting IP address when downloading the data.

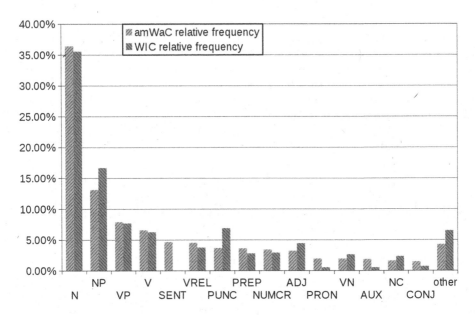

Fig. 2. Relative frequency of tags in both Amharic corpora. (End of sentence token is marked by a PUNCT tag in WIC.)

Table 5. Keyword comparison of amWaC 16 to WIC: words most characteristic for the web corpus, sorted by keyword score.

Word, translation		amWaC 16		WIC		KS
		Count	per million	Count	per million	
ነገር	thing	57927	2855.3	16	79.8	16.4
ነበር	was/where	69074	3404.8	28	139.6	14.6
እንዲህ	like this	27678	1364.3	1	5	13.9
ሥራ	job task	24372	1201.3	0	0	13.0
በ	by/on/at-	32125	1583.5	9	44.9	11.6
ቅዱስ	holy/saint	31392	1547.4	10	49.9	11.0
ዓመት	year	19295	951.1	0	0	10.5
መጽሐፍ	book	17938	884.2	0	0	9.8
አምላክ	God	17748	874.8	0	0	9.7
ማለት	it means	24235	1194.6	7	34.9	9.6
ቤት	house-	17902	882.4	1	5	9.4
ሰው	human/man	41861	2063.4	27	134.6	9.2
መንግሥት	government	16300	803.5	0	0	9.0
ክርስቲያን	Christian	16297	803.3	0	0	9.0
ኢየሱስ	Jesus	17968	885.7	2	10	9.0
ዓለም	world	15864	782	0	0	8.8
አበባ	flower	18109	892.6	4	19.9	8.3

Table 6. Keyword comparison of WIC to amWaC 16: words most characteristic for the news corpus, sorted by keyword score.

Word, translation		amWaC 16		WIC		KS
		Count	per million	Count	per million	
ማእከል	centre	455	22.4	1084	5404.8	45.0
አንፎርሜሽን	information	0	0	667	3325.7	34.3
ለዋልታ	to/for Walta	377	18.6	479	2388.3	21.0
ዋልታ	Walta	399	19.7	479	2388.3	20.8
ሚሊየን	million	952	46.9	501	2498.0	17.7
አስታወቀ	he announced	1578	77.8	565	2817.1	16.4
ሃላፊው	the head	177	8.7	315	1570.6	15.4
አንደገለጹት	as they stated	1721	84.8	522	2602.7	14.6
ዘግቧል	he has reported	1688	83.2	470	2343.4	13.3
ተቁመዋል	they pointed out	4	0.2	235	1171.7	12.7
አስታወቀዋል	they announced	1793	88.4	458	2283.6	12.7
ሃላፊ	head	842	41.5	325	1620.5	12.2
ኢንፎርሜሽን	information	754	37.2	292	1455.9	11.3
ወረዳዎች	districts	771	38	274	1366.2	10.6
ገለጻ	briefing	1008	49.7	292	1455.9	10.4
ኢትዮጵያ	Ethiopia	5	0.2	186	927.4	10.2
ምክርቤት	council	0	0	176	877.5	9.8

Tables 5 and 6 show main differences of corpora using keyword comparison: The language is much more formal and the main topic is politics in the news only corpus as expected. Religion related words are noticeable in the WaC corpus. Differences in tokenisation can be observed too, e.g. morpheme N is represented as a separate token in the WaC corpus.

The Keyword Score KS of a word is calculated according to [5] as

$$KS = \frac{fpm_{foc} + n}{fpm_{ref} + n}$$

where fpm_{foc} is the normalised (per million words) count of the word in the focus corpus, fpm_{ref} is the normalised count of the word in the reference corpus and $n = 100$ is the Simple Maths smoothing parameter.[6]

4 Conclusion

We have built a web corpus of Amharic texts comprising of more than 15 million words. To our knowledge it is the largest Amharic corpus for language technology use currently available. We expect the corpus linguistics, lexicography and language teaching in Ethiopia will greatly benefit from such a resource.

We have also cleaned the WIC corpus and unified its Fidel and SERA versions. This resource could be used for building language models (like the TreeTagger model) and for other natural language processing applications for Amharic.

---

[6] We selected $n = 100$ rather than $n = 1$ to prefer common words over rare words.

A similar approach is being applied to obtain web corpora in other East African languages: Afaan Oromo, Tigrinya and Somali. All corpora compiled within the project are available for browsing and querying by corpus manager Sketch Engine at https://habit-project.eu/corpora. The full source text was not made public because of possible copyright issues.

Acknowledgements. We would like to thank Dr. Derib Ado Jekale from Department of Linguistics, Addis Ababa University for checking seed bigrams of Amharic words, translating key words of the corpus comparison and answering questions about Amharic.

This work has been partly supported by the Grant Agency of CR within the project 15-13277S. The research leading to these results has received funding from the Norwegian Financial Mechanism 2009–2014 and the Ministry of Education, Youth and Sports under Project Contract no. MSMT-28477/2014 within the HaBiT Project 7F14047.

References

1. Demeke, G.A., Getachew, M.: Manual annotation of amharic news items with part-of-speech tags and its challenges. In: Ethiopian Languages Research Center Working Papers 2, pp. 1–16 (2006)
2. Firdyiwek, Y., Yaqob, D.: The system for Ethiopic representation in ASCII. J. EthioSci. (1997)
3. Gambäck, B., Olsson, F., Argaw, A.A., Asker, L.: Methods for amharic part-of-speech tagging. In: Proceedings of the First Workshop on Language Technologies for African Languages, pp. 104–111. Association for Computational Linguistics (2009)
4. Gebre, B.G.: Part of speech tagging for Amharic. Ph.D. thesis, University of Wolverhampton, Wolverhampton (2010)
5. Kilgarriff, A.: Getting to know your corpus. In: Sojka, P., Horák, A., Kopeček, I., Pala, K. (eds.) TSD 2012. LNCS, vol. 7499, pp. 3–15. Springer, Heidelberg (2012)
6. Kilgarriff, A., Reddy, S., Pomikálek, J., Avinesh, P.: A corpus factory for many languages. In: LREC (2010)
7. Pomikálek, J.: Removing boilerplate and duplicate content from web corpora. Ph.D. thesis, Masaryk University, Faculty of Informatics (2011)
8. Scannell, K.P.: The crúbadán project: corpus building for under-resourced languages. In: Building and Exploring Web Corpora: Proceedings of the 3rd Web as Corpus Workshop, vol. 4, pp. 5–15 (2007)
9. Schmid, H.: Treetagger: a language independent part-of-speech tagger. Institut für Maschinelle Sprachverarbeitung, Universität Stuttgart **43**, 28 (1995)
10. Suchomel, V., Pomikálek, J., et al.: Efficient web crawling for large text corpora. In: Proceedings of the Seventh Web as Corpus Workshop (WAC7), pp. 39–43 (2012)
11. Tachbelie, M.Y., Menzel, W.: Morpheme-based language modeling for inflectional language–Amharic. John Benjamin's Publishing, Amsterdam and Philadelphia (2009)

Speech

Evaluation of TTS Personification by GMM-Based Speaker Gender and Age Classifier

Jiří Přibil[1,2(✉)], Anna Přibilová[3], and Jindřich Matoušek[1]

[1] Faculty of Applied Sciences, Department of Cybernetics,
University of West Bohemia, Univerzitní 8, 306 14 Plzeň, Czech Republic
Jiri.Pribil@savba.sk, jmatouse@kky.zcu.cz
[2] SAS, Institute of Measurement Science,
Dúbravská cesta 9, 841 04 Bratislava, Slovakia
[3] Faculty of Electrical Engineering and Information Technology,
Slovak University of Technology in Bratislava,
Ilkovičova 3, 812 19 Bratislava, Slovakia
jmatouse@kky.zcu.cz

Abstract. This paper describes an experiment using the Gaussian mixture models (GMM)-based speaker gender and age classification for automatic evaluation of the achieved success in text-to-speech (TTS) system personification. The proposed two-level GMM classifier detects four age categories (child, young, adult, senior) as well as it discriminates gender for adult voices. This classifier is applied for gender/age estimation of the synthetic speech in Czech and Slovak languages produced by different TTS systems with several voices, using different speech inventories and speech modelling methods. The obtained results confirm the hypothesis that this type of classifier can be utilized as an alternative approach instead of the conventional listening test in the area of speech evaluation.

Keywords: GMM classifier · Gender and age classification · Text-to-speech system · Synthetic speech

1 Introduction

Automatic speaker gender and age identification can be used to verify the quality of the produced synthetic speech after the speaker identity transformation (gender and/or age) in the text-to-speech systems (TTS) which are usually the output part of the voice communication system by phone [1]. The objective evaluation results of such an identifier may be matched with the subjective results achieved by the standard listening test method. In addition, this approach can

The work has been done in the framework of the COST Action IC 1206 (A. Přibilová), and was supported by the Czech Science Foundation GA16-04420S (J. Matoušek, J. Přibil), and the Grant Agency of the Slovak Academy of Sciences VEGA 2/0013/14 (J. Přibil).

© Springer International Publishing Switzerland 2016
P. Sojka et al. (Eds.): TSD 2016, LNAI 9924, pp. 305–313, 2016.
DOI: 10.1007/978-3-319-45510-5_35

also be successfully applied for detection and localization of artefacts in the synthetic speech [2]. Speech processing research of recent years is accompanied by growing importance of the automatic identification of age and gender of a person from his/her speech in various forensic (age estimation in identifying suspected criminals [3], or in voice biometry [4]) as well as commercial (telephone dialogs in call centres), or many other applications. Mel frequency cepstral coefficient (MFCC) features are often used in the automatic speaker age estimation area. For seven speaker classes (children, young males/females, adult males/females, and senior males/females) the overall precision of about 75 % [1] or the classification accuracy over 60 % was reported [5]. Fusion of several subsystems using acoustic and prosodic features improved age classification by 4 % and gender classification by 6 % [6]. A similar approach including also voice quality features was presented in [7] with similar improvement. In spite of different age boundaries between the classes, the majority of researchers use seven classes consisting of three classes for each gender and a separate class of a child's voice as the gender is practically indistinguishable for the speech of children younger than 12 years [8].

2   GMM-Based Gender and Age Classification Method

The Gaussian mixture models represent a linear combination of multiple Gaussian probability distribution functions of the input data vector. For GMM creation, it is necessary to determine the covariance matrix, the vector of mean values, and the weighting parameters from the input training data. The maximum likelihood function of GMM is found using the expectation-maximization iteration algorithm [9] controlled by the number of mixtures N_{GMIX}. Practical realization of the proposed GMM-based gender and age classifier has two-level architecture as shown in a block diagram in Fig. 1. This classifier works in the following way: in the first step, the speech feature vectors (different types for each of the two levels) are extracted from the input sentence. The speaker gender recognition block (the first level) uses the GMM models that were created and trained on the data of the feature vectors obtained from the sentences of the speech corpora structured in dependence on the speaker gender (child/female/male). In the classification phase, the input feature vectors from the tested sentences are used to obtain the $scores(T, n)$ that are then sorted by the absolute size and quantized to N levels corresponding to N output classes. The second level consists of GMM age classification block working with the models created and trained on the data of each age class (young/adult/senior) separately for the previously determined speaker gender. The obtained individual values of the $score(T, m)$ are further used for calculation of the accumulated score m_{ACC}. Finally, the resulting age class is determined depending on the used M-level discrimination.

Unlike the other works on GMM-based speech classification using mostly the MFFC features with the energy and the prosodic parameters [1,5,6], we use the basic and supplementary spectral features determined from the smoothed magnitude or power spectrum envelope together with the prosodic parameters

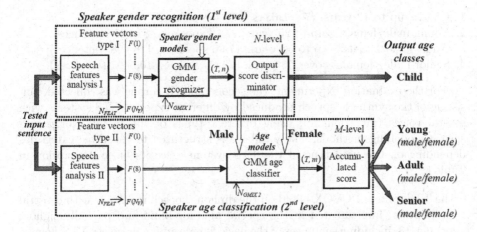

Fig. 1. Block diagram of the two-level architecture of the GMM age classifier.

determined from F0 contour [10]. The basic spectral properties are expressed by the additional statistical parameters as centroid (S_{centr}), flatness (S_{flat}), spread, etc. As the supplementary spectral features the following parameters are used: spectral decrease (tilt – S_{tilt}), harmonics-to-noise ratio (HNR), spectral entropy (SE), etc. The cepstral coefficients $\{c_n\}$ obtained during the process of cepstral analysis giving information about spectral properties of the human vocal tract [11] can also be successfully used in the feature vectors. The prosodic parameters consist of differential microintonation signal $F0_{DIFF}$, absolute jitter as vocal frequency perturbation (J_{abs}), and shimmer as relative amplitude perturbation (AP_{rel}). As the representative values in the feature vectors, the basic statistical parameters – mean values and standard deviations (std), and then also median values, ranges of values, and/or relative maximum and minimum – were used. In the case of the cepstral coefficients, the histograms of distribution were also calculated and the extended statistical parameters (skewness and kurtosis) were subsequently determined from these histograms. The detailed description of the tuning process of this type of the GMM-based speech classifier is mentioned in [12].

3 Material, Experiments, and Results

The main speech database consists of short sentences originating from stories and audio books uttered by professional speakers (actors) in a neutral (declarative) state (excluding the child category) in Czech and Slovak languages. For our purpose, the original MP3 audio records were resampled to 16 kHz and converted to 16-bit PCM format, the mean duration of the sentences was about 3 s. This speech corpus is divided into three basic speaker gender categories: male, female, and child. Adult voices are subdivided into another three age categories so that the resulting seven age classes are stored:

1. Child – up to 12 years, (7 speakers, 177 sentences);
2. Young male/female – up to 25 years, (7+6 speakers, 81+91 sentences);
3. Adult male/female – up to 50 years, (7+6 speakers, 118+98 sentences);
4. Senior male/ female – over 55 years, (7+6 speakers, 109+97 sentences).

For the evaluation experiment, the second speech corpus was collected consisting of the synthetic speech produced by different types of TTS systems with several voices (male/female), using different types of speech inventories and speech modelling methods. The database is structured to the three categories depending on the used type of the TTS system – see its basic description in Table 1:

- the TTS system PCVOX based on the diphone speech inventory with cepstral description, implemented in the special aids for blind and partially sighted people [13] including sentences of the basic female and male voices TTS-Female ("Ellen")/TTS-Male ("Kubec") and the transformed voices corresponding to a young male $(VT-Young)$, a female $(VT-Female)$, and a child $(VT-Child)$ [14],
- the TTS system Epos [15] based on the triphone speech inventory and using the PSOLA synthesis method (the male voices "Machac/Bob" and the female one "Violka") together with the female voice "Eliska" of the professional TTS system by the company Acapela (using the MBROLA synthesis approach) [16],
- the TTS system ARTIC using three speech synthesis methods: the TD-PSOLA, the unit selection (USEL) [17], and the Hidden Markov models (HMM) [18] with three male ("Jan/Jiri/Stanislav") and three female ("Melanie/Radka/Iva") voices where the voice "Melanie" is Slovak and the remaining ones are all Czech [19].

The corpus includes 40 sentences for each of the synthetic voices (1080 sentences in total) with the mean duration of about 1.5 s and the sampling frequency of 16 kHz.

In accordance with the previous related works [2, 10, 12] the feature sets with the length of $N_{FEAT} = 16$ were used separately for each of the functional levels – see their internal structures in Table 2. A simple diagonal covariance matrix of the GMM was applied in this classification experiment due to its lower computational complexity. The basic functions from the Ian T. Nabney "Netlab"

Table 1. Summary specification of the tested synthetic speech produced by TTS systems.

TTS type	Inventory (speech model)	TTS voices name
PC Vox	diphone (source-filter)	{Kubec,Ellen} / {VT-Young,VT-Female,VT-Child}
Epos/Acapela	triphone (PSOLA/MBROLA)	{Machac, Violka, Bob} / Eliska
ARTIC	triphone (TD-PSOLA), v. 2.06 unit selection (v. 2.10) HMM-based	{Melanie, Radka, Iva, Jan, Jiri, Stanislav} {Radka, Iva, Jan, Jiri, Stanislav}

Table 2. Setting of initial parameters and detailed structure of the used feature sets for the GMM gender/age classification in dependence on the functional level.

Feature set	Feature type	Statistical value	N_{FEAT}	N_{GMIX}
I. for the 1st level	$\{$HNR, S_{tilt}, S_{centr}, S_{flat}, SE, FO$_{DIFF}$, J_{abs}, $AP_{rel}\}$	$\{$min, rel. max, mean, median, std$\}$	16	64
II. for the 2nd level	$\{c_1$-c_4, S_{tilt}, S_{centr}, S_{flat}, SE, FO$_{DIFF}$, J_{abs}, $AP_{rel}\}$	$\{$skewness, kurtosis mean, median, std$\}$	16	128

pattern analysis toolbox [20] were used for creation of the GMM models, data training, and classification. The gender/age classification accuracy was calculated from X_A sentences with correctly identified the speaker gender/age class and the total number N_U of the tested sentences as $(X_A/N_U) * 100$ [%].

The performed basic verification experiment represents an evaluation of the gender/age classification accuracy of the automatic GMM-based classifier using the original Czech and Slovak speech database. The achieved accuracy values are subsequently compared with the results carried out manually by the conventional listening test called "$Determination of the speaker age category$" located on the web page http://www.lef.um.savba.sk/scripts/itstposl2.dll during the period from September 25 to December 20, 2015. Twenty two listeners (6 women and 16 men) took part in this listening test experiment. For each of the evaluated sentences there was a choice from the seven possibilities: "$Child$", "$Young$" Male/Female, "$Adult$" Male/Female, and "$Senior$" Male/Female. These choices comprise four age categories (child, young, adult, senior) with both genders for all but children's voices. The obtained results in the form of the 3D confusion matrix together with the bar-graph comparison for both tested approaches are presented in Fig. 2.

Secondary comparisons were aimed at investigation how the GMM gender/age classifier evaluates the voice type for the three mentioned types of the

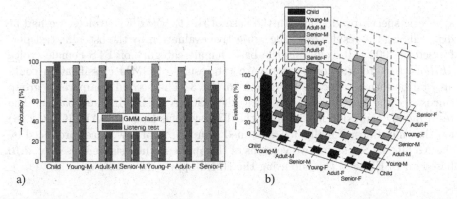

Fig. 2. The bar-graph of the gender/age recognition accuracy – comparison of the best obtained GMM results together with the results of the listening tests (a), confusion matrix for the GMM-based classification approach (b).

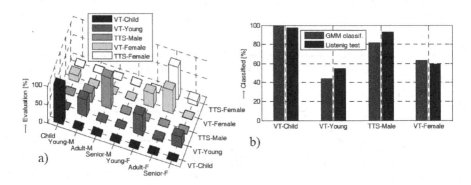

Fig. 3. 3D confusion matrix for the GMM classification of the voices produced by the PC Vox system for basic male and female voices ($TTS - Male/TTS - Female$) and its conversion to young male ($VT - Young$), female ($VT - Female$), and child ($VT - Child$) voices (a), a bar-graph comparison of the GMM classification accuracy together with the results of the listening test evaluation [14] excluding the $TTS - Female$ category (b).

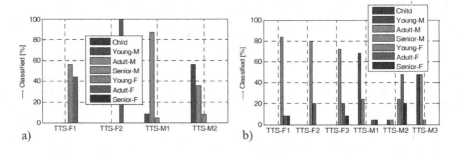

Fig. 4. 2D confusion matrices of the GMM gender/age classification for: Epos/Acapela TTS systems (a): TTS-F1=$Eliska$ (Acapela), TTS-F2=$Violka$, TTS-M1=$Machac$, and TTS-M2=Bob; ARTIC TTS system v. 2.06 (b); TTS-F1=$Melanie$, TTS-F2=$Radka$, TTS-F3=Iva, TTS-M1=Jan, TTS-M2=$Jiri$, and TTS-M3=$Stanislav$.

synthetic speech production. In the case of the PC Vox TTS voices, we had at disposal also the results from the subjective evaluation by the listening test performed in 2006 [14] excluding the basic female category of TTS-Female (voice "*Ellen*"). Therefore, comparison with the results of GMM classification could also be performed – see a bar-graph in Fig. 3. The second comparison group consists of the speech synthesis using the PSOLA/MBROLA approaches. The resulting 2D confusion matrices for the TTS systems Epos and ARTIC (v. 2.06) are shown in Fig. 4. Finally, for a detailed comparison of new voices produced by the ARTIC TTS system, two versions of the applied synthesis were used (v. 2.10 based on USEL, and the one using the HMM approach) – see Fig. 5.

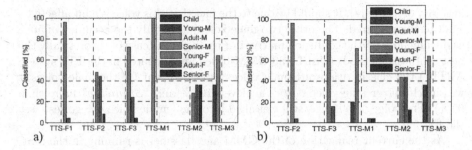

Fig. 5. 2D confusion matrices comparison of the GMM gender/age classification for the ARTIC TTS system: version 2.10-USEL (a), HMM-based synthesis (b); TTS-F1=*Melanie* (not included in the HMM synthesis corpus), TTS-F2=*Radka*, TTS-F3=*Iva*, TTS-M1=*Jan*, TTS-M2=*Jiri*, and TTS-M3=*Stanislav*.

4   Discussion and Conclusion

The performed experiments have successfully confirmed that the proposed two-level architecture of the speaker gender/age GMM classifier is functional and the achieved results of the classification accuracy are fully comparable with the manual evaluation by the standard listening test method – see the bar-graphs in Figs. 2a and 3b. A critical issue is a proper function of the first block (the gender type recognizer) as the age class determination block operates with different models trained for the male and the female voices. In the case of wrong gender class determination, the resulting age classification accuracy is decreased. According to the detailed analysis of the classification accuracy, the mean obtained value of 94.3 % (std=2.55) for all 7 classes was significantly higher than in the case of the used listening test evaluation (with the mean accuracy of 74.6 %, std=12.7). Therefore, it may be justified that the main goal of this work – to find an alternative approach to the standard listening tests – was successfully attained.

From the classification experiment based on the synthetic speech produced by the PC Vox TTS system follows that the highest speaker gender/age accuracy was achieved for the basic male and female voices. For the transformed voices the best value corresponds to the child voice (100 %) and a relatively bad accuracy was obtained for transformation to a young voice. On the other hand, this result was theoretically correct – the classification was divided into the "*Young – M*" and "*Young – F*" classes (see the 3-D confusion matrix in Fig. 3a). In summary, the received results are in a very good correspondence with the subjective evaluation by the listening test as documented by the bar-graph in Fig. 3b.

The comparison of TTS systems used for speech generation with the triphone-based inventory and PSOLA synthesis approach shows that the best result was obtained for the female voice "*Eliska*" (100 %) and the male voice "*Machac*" (89 %). On the hand, in the TTS system ARTIC, the relatively high accuracy was achieved for all three female voices (more than 85 %) but for the male voice "*Jiri*" the result was absolutely the worst (classified as a female voice type by the majority of listeners). It should be due to the fact that this voice is a bit affected – emotionally coloured. The other two male voices have only the accuracy of 50 % as documented by the confusion matrix in Fig. 4b. In the case of theoretically the best synthesis methods (USEL and HMM-based implemented in the TTS system ARTIC) the classification of female voices worsened, but for the male voices it was much better (the voice "*Jan*" achieved 100 % and "*Stanislav*" more than 65 %). The voice "*Jiri*" was still classified into the female voice categories – see the Fig. 5a.

As the current realization of the GMM age classifier is running in the Matlab environment, the processing in the identification phase must be performed off-line. After the program optimization and rewriting it in any of the higher programming languages, the real-time processing will be available, which is important for practical implementations. In the future, we plan to test the GMM age evaluation using larger speech databases or the speech material spoken in other languages (English, German).

References

1. Bocklet, T., Maier, A., Bauer, J.G., Burkhardt, F., Noth, E.: Age and gender recognition for telephone applications based on GMM supervectors and support vector machines. In: IEEE International Conference on Acoustics, Speech, and Signal Processing, pp. 1605–1608, 31 March–4 April 2008. IEEE, Las Vegas (2008)
2. Přibil, J., Přibilová, A., Matoušek, J.: Experiment with GMM-based artefact localization in Czech synthetic speech. In: Král, P., et al. (eds.) TSD 2015. LNCS, vol. 9302, pp. 23–31. Springer, Heidelberg (2015). doi:10.1007/978-3-319-24033-6_3
3. Bahari, M.H., McLaren, M., van Hamme, H., van Leeuwen, D.A.: Speaker age estimation using i-vectors. Eng. Appl. Artif. Intell. **34**, 99–108 (2014)
4. Fairhurst, M., Erbilek, M., Da Costa-Abreu, M.: Selective review and analysis of aging effects in biometric system implementation. IEEE Trans. Hum. Mach. Syst. **45**(3), 294–303 (2015)
5. van Heerden, C., et al.: Combining regression and classification methods for improving automatic speaker age recognition. In: IEEE International Conference on Acoustics, Speech, and Signal Processing, 14–19 March 2010, pp. 5174–5177. IEEE, Dallas (2010)
6. Meinedo, H., Trancoso, I.: Age and gender classification using fusion of acoustic and prosodic features. In: Interspeech 2010, Makuhari, Japan, pp. 2822–2825, 26–30 September 2010
7. Li, M., Han, K.J., Narayanan, S.: Automatic speaker age and gender recognition using acoustic and prosodic level information fusion. Comput. Speech Lang. **27**, 151–167 (2013)
8. Assmann, P., Barreda, S., Nearey, T.: Perception of speaker age in children's voices. In: Proceedings of Meeting on Acoustics, vol. 19, 060059 (2013)
9. Reynolds, D.A., Rose, R.C.: Robust text-independent speaker identification using Gaussian mixture speaker models. IEEE Trans. Speech Audio Process. **3**, 72–83 (1995)
10. Přibil, J., Přibilová, A.: Evaluation of influence of spectral and prosodic features on GMM classification of Czech and Slovak emotional speech. Eurasip J. Audio Speech Music Process. **2013**(8), 1–22 (2013)

11. Vích, R., Přibil, J., Smékal, Z.: New cepstral zero-pole vocal tract models for TTS synthesis. In: 2001 Proceedings of IEEE Region 8 EUROCON 2001, vol. 2, pp. 458–462 (2001)

12. Přibil, J., Přibilová, A.: Comparison of text-independent original speaker recognition from emotionally converted speech. In: Esposito, A., et al. (eds.) Recent Advances in Nonlinear Speech Processing. Smart Innovation, Systems and Technologies, pp. 137–149. Springer, Switzerland (2016)

13. Personal Computer Voices: PCVOX. Spektra v.d.n. http://www.pcvox.cz/pcvox/pcvox-index.html. Accessed 5 Feb 2014

14. Přibilová, A., Přibil, J.: Non-linear frequency scale mapping for voice conversion in text-to-speech system with cepstral description. Speech Commun. 48(12), 1691–1703 (2006)

15. The Epos Speech Synthesis System: Open Text-To-Speech Synthesis Platform. Text-to-speech synthesis demo. http://www.speech.cz/. Accessed 10 Feb 2014

16. Acapela Text to Speech Demo. Acapela Group Babel Technologies SA. http://www.acapela-group.com/. Accessed 15 Feb 2016

17. Matoušek, J., Tihelka, D., Romportl, J.: Current state of Czech text-to-speech system ARTIC. In: Sojka, P., Kopeček, I., Pala, K. (eds.) TSD 2006. LNCS (LNAI), vol. 4188, pp. 439–446. Springer, Heidelberg (2006)

18. Hanzlíček, Z.: Czech HMM-based speech synthesis. In: Sojka, P., Horák, A., Kopeček, I., Pala, K. (eds.) TSD 2010. LNCS, vol. 6231, pp. 291–298. Springer, Heidelberg (2010)

19. Interactive TTS Demo. SpeechTech, s.r.o. http://www.speechtech.cz/cz/demo-tts#Iva210. Accessed 17 Feb 2010

20. Nabney, I.T.: Netlab Pattern Analysis Toolbox. http://www.mathworks.com/matlabcentral/fileexchange/2654-netlab. Accessed 2 Oct 2013

Grapheme to Phoneme Translation Using Conditional Random Fields with Re-Ranking

Stephen Ash[1(✉)] and David Lin[2]

[1] University of Memphis, Memphis, TN, USA
sash@memphis.edu
[2] Baylor University, Waco, TX, USA
david_lin@baylor.edu

Abstract. Grapheme to phoneme (G2P) translation is an important part of many applications including text to speech, automatic speech recognition, and phonetic similarity matching. Although G2P models have been studied thoroughly in the literature, we propose a G2P system which is optimized for producing a high-quality top-k list of candidate pronunciations for an input grapheme string. Our pipeline approach uses Conditional Random Fields (CRF) to predict phonemes from graphemes and a discriminative re-ranker, which incorporates information from previous stages in the pipeline with a *graphone* language model to construct a high-quality ranked list of results. We evaluate our findings against the widely used CMUDict dataset and demonstrate competitive performance with state-of-the-art G2P methods. Additionally, using entries with multiple valid pronunciations, we show that our re-ranking approach out-performs ranking using only a smoothed *graphone* language model, a technique employed by many recent publications. Lastly, we released our system as an open-source G2P toolkit available at http://bit.ly/83yysKL.

Keywords: Grapheme-to-phoneme conversion · Conditional random fields · G2P

1 Introduction

Grapheme to phoneme (G2P) translation is the task of converting an input sequence of *graphemes*, the set of symbols in a writing system, to an output sequence of *phonemes*, the perceptually distinct units of sound that make up words in a spoken language. Few languages (e.g. Serbian) have a strict one-to-one correspondence between phonemes and graphemes, and therefore the difficulty of systematic conversion comes from the level of ambiguity and irregularity in the orthography. Typical G2P systems for English use 27 graphemes as input (26 Roman alphabet letters and the apostrophe) and between 40 and 45 phonemes. In this manuscript, graphemes will be written between angle brackets as ⟨aloud⟩ and phonemes will be written using the Arpabet phonetic alphabet [1] between slashes as /AH L AW D/.

© Springer International Publishing Switzerland 2016
P. Sojka et al. (Eds.): TSD 2016, LNAI 9924, pp. 314–325, 2016.
DOI: 10.1007/978-3-319-45510-5_36

There are a number of problems that make grapheme to phoneme translation difficult (examples from English): As previously mentioned, there is typically no one-to-one correspondence between graphemes and phonemes. For example, the c in ⟨cider⟩ sounds different from the c in ⟨cat⟩, which sounds similar to the k in ⟨kite⟩. The number of letters that produce a single phoneme is not always consistent. Usually one or a few graphemes produce a single phoneme, but there are exceptions where multiple phonemes are produced by a single grapheme: for example, the x in ⟨six⟩ produces two phonemes: /S I CK S/. Pronunciation can be affected by factors external to the word itself, such as its part of speech in the sentence. ⟨convict⟩ as a noun is /K AH N V IH K T/, but as a verb it is /K AA N V IH K T/. The output phoneme is not completely determined by local graphemes that are closest in sequence; there are some non-local effects. ⟨mad⟩ is pronounced /M AE D/, but adding a trailing e to make the word ⟨made⟩, changes the middle vowel. As language evolves, words are borrowed from other linguistic origins, and can be transliterated in a way that is inconsistent with other words in the target language. Existing dictionaries of words to pronunciations generally do not indicate which graphemes correspond to which phonemes. This *alignment* of individual graphemes to phonemes is generally treated as a latent factor in G2P systems, which complicates using dictionaries as training data for machine learning based solutions.

This paper describes a method of grapheme to phoneme translation that is optimized for use cases where the top-k predicted sequences are important and there is a high degree of complexity in how the input graphemes affect the spoken phonemes. For example, in the application of personal name matching based on phonetic similarity, producing multiple candidate pronunciations can increase recall of matching at little computational expense.

The contributions in this paper are: (1) We present a novel G2P pipeline using discriminative, probabilistic methods throughout, which are particularly well suited at incorporating expert domain knowledge. (2) We suggest an improvement to the many-to-many alignment formulation to increase alignment accuracy of training data. (3) We describe a re-ranking method which combines *graphone* language models and error information propagated from earlier stages to improve overall top-k prediction. (4) Our implementation is available as an open source G2P toolkit called **jG2P** available at http://bit.ly/83yysKL

2 Previous Work on G2P

The problem of grapheme to phoneme translation has been studied for many decades in different research communities. There are a number of different general approaches suggested in the literature including rules [2,3], local classification using neural networks [4] or decision trees [5], and a number of statistical formulations.

The statistical version of the G2P problem is as follows: let G be the set of graphemes in the orthography (e.g. a, b, etc.). G^* is the Kleene star of G, which is the set of all possible grapheme strings. Similarly, let P be the set of phonemes (e.g. the diphthong /OY/, the fricative /V/, etc.). P^* is the set of all possible phonetic pronunciations. G and P usually include ϵ as a dummy symbol which is used, for example, when graphemes are unvoiced. The probabilistic formulation of the G2P problem is the task of finding the best string of phonemes for the string of graphemes as in (1). g is a sequence of $g_{0..M}$ graphemes, and p is a sequence of $p_{0..N}$ phonemes.

$$\varphi(g) = \arg\max_{p \in P^*} P(g, p) \tag{1}$$

An *alignment* of graphemes to phonemes is a matching of substrings $g_{i:j}$ of g (or ϵ) to substrings $p_{k:l}$ of p (or ϵ). Typically the size of the substring of graphemes or phonemes is bounded. A n-m alignment limits the maximum size of the grapheme substring to length n and the maximum size of the phoneme substring to length m. For example, a 1-1 alignment of ⟨mixing⟩ to /M IH K S IH NG/ might be: m–/M/, i–/IH/, x–/K/, ϵ –/S/, i–/IH/, n–ϵ, g–/NG/. A 3-2 alignment might be: m–/M/, i–/IH/, x–/K S/, ⟨ing⟩– /IH NG/.

A pairing of a grapheme substring to a phoneme substring can be treated as a single unit, a *graphone*, such as the substring ⟨ing⟩ paired to substring /IH NG/ in the 3-2 alignment of ⟨mixing⟩. Therefore, the entire aligned grapheme and phoneme sequence can be represented by a sequence of graphones. [6] recognized that using graphones, one can factorize the probability in (1) similar to a language model, where the ith graphone is predicted by the n previous graphones as in (2).

$$\varphi(g) = \arg\max_{p \in P} \left(\prod_{q_i \in q} P(q_i | q_{i-1}, q_{i-2}, ...) \right) \tag{2}$$

[6] used a 9-gram language model with the modified version of Kneser-Ney smoothing. They evaluate this approach against a number of datasets and achieve a 24 % word error rate (WER) on the CMUDict dataset, which remains one of the top WERs reported for this dataset.

The first use of Conditional Random Fields (CRF) [7] for the problem of grapheme to phoneme translation comes from [8]. They use a 1-1 alignment model with ϵ only allowed on the grapheme side. Wang suggests that CRFs are particularly well-suited to the task because they perform global inference over the entire sequence unlike a Hidden Markov Model.

[9] created an open source framework building on the results from [6], by encoding graphone language models into Weighted Finite State Transducers (WFST). Novak also applied a recurrent neural network (RNN) based re-ranking technique to improve top-k sequence prediction. RNN based re-scoring improves the WER on CMUDict by <1 % [10].

A number of hybrid methods have been proposed recently. [11] uses alignments produced by [10] to train an *insertion predictor* to predict where epsilons should be inserted into the grapheme sequence. Then they train a CRF to predict a phoneme using the previously predicted phoneme and a local window of graphemes. The best published WER on CMUDict at present is demonstrated in

[12] using Long Short-term Memory Recurrent Neural Networks (LSTM) combined with a 5-gram graphone language model.

3 Discriminative Pipeline for Grapheme to Phoneme Translation

Our grapheme-to-phoneme system's use case is translating personal names into phonetic representations for the purpose of similarity matching. Personal names are particularly prone to phonetic irregularities and a high rate of being out of vocabulary (OOV). The following considerations influenced our design: (1) Our use case includes words from many different linguistic origins, and we want to be able to easily encode orthographic domain knowledge in order to improve translation accuracy. (2) Our use case prefers a strong top-k list of resulting candidate pronunciations over only optimizing for the top result.

We created **jG2P**, a G2P pipeline with three distinct components: an aligner, a pronouncer, and a re-ranker as illustrated in Fig. 1. The *aligner* is a Conditional Random Field (CRF) trained to predict alignment boundaries to partition the input grapheme string into a contiguous sequence of grapheme substrings. Then, then top-k_a alignments are sent to the pronouncer. The *pronouncer* is a first-order CRF trained to predict phoneme substrings from grapheme substrings and other features. The top-k_p pronunciations for each alignment are then evaluated by an 8-gram graphone language model built with Kneser-Ney smoothing. Finally, the pronunciation candidates are ordered by the *re-ranker*, which uses a discriminative model trained to minimize cross-entropy of output labels using information from each of the previous stages. Each component will be described in detail in the following sections.

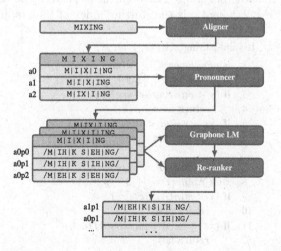

Fig. 1. The jG2P grapheme to phoneme translation pipeline

3.1   Grapheme to Phoneme Alignment

One challenging aspect of the G2P problem is the unequal number of symbols in G and P and the frequent many-to-many correspondence of graphemes to phonemes. When using a statistical formulation of the G2P problem, there are two aspects to alignment that need to be addressed: (1) at test-time when only presented with g, the system must reconstruct the alignment of g in order to apply structured prediction; (2) training data is often recorded as a word and its pronunciation, $\langle g, p \rangle$. Thus, the training data does not have a delineated correspondence to use in learning.

To address concern (1), at test-time jG2P uses a separate component for identifying grapheme sub-sequences, identified in Fig. 1 as the *aligner*. We treat alignment prediction at test-time as a sequential labeling task, assigning a binary label to each input grapheme indicating whether that letter is the *last* in the current grapheme substring.

To train jG2P, we created alignments of each of the training pairs using an approach similar to that described in [9]. We model the joint probability distribution of $\gamma(g_{i:j}, p_{k:l})$ and use Expectation Maximization (EM) to estimate each of the model parameters. Traditionally, in each step of the EM formulation, all graphones are enumerated by considering all possible pairings of substrings of $g \leq n$ with all substrings of $p \leq m$. We note that as a consequence of considering every combination, the probability mass is being spread *thinly* over many nonsensical graphones. Since we model many-to-many alignments, epsilons should be rare, and we observe that alignments with multiple adjacent epsilons are overwhelmingly invalid. Thus, in order to eliminate nonsensical graphones from the training data and preserve more probability mass for more likely pairings, we introduce a simple change to the algorithm: we only count pairings that require *at most* one epsilon transition at the beginning or end of the sequence. The pseudo-code for this is shown in Fig. 2, where x is the starting index of the grapheme substring

is-valid-pairing(x, n, xs, y, m, ys)

1 $minPhoBeforeX = \lceil x/\text{max\_n} \rceil$
2 $minPhoAfterX = \lceil (xs - x - n)/\text{max\_n} \rceil$
3 **if** ($y < minPhoBeforeX$)
4 **return** false
5 **if** ($ys - y - m < minPhoAfterX$)
6 **return** false

7 $minGraBeforeY = \lceil y/\text{max\_m} \rceil$
8 $minGraAfterY = \lceil (ys - y - m)/\text{max\_m} \rceil$
9 **if** ($x < minGraBeforeY$)
10 **return** false
11 **if** ($xs - x - n < minGraAfterY$)
12 **return** false

13 **return** true

Fig. 2. Constraint for allowing valid windows under the assumption that there should be few epsilons (base 0), n is the length of the grapheme substring, xs is the total length of the grapheme string; y, m, and ys are analogous for the phoneme string.

In jG2P, we use a maximum grapheme substring size of $n = 2$, a maximum phoneme substring size of $m = 2$, and we allow graphemes to match to ϵ in the aligned training examples, but we disallow phonemes from matching with ϵ. This last restriction is a consequence of the test-time *aligner* (Sect. 3.1) only representing grapheme boundaries.

3.2 Predicting Phoneme M-grams from Grapheme N-grams

The second stage in jG2P's pipeline is the structured prediction of phoneme substrings from grapheme substrings. A first-order CRF is trained using the aligned $\langle g, p \rangle$ examples. Since we trained with 2-2 alignments, the output labels of the pronouncer include both individual phonemes and some phoneme pairs. We take advantage of the feature engineering flexibility of CRFs and include a number of useful non-local features that would be more difficult to incorporate in other structured prediction methods. For example, we include a binary feature that indicates whether this vowel is the last vowel *before* a trailing e or y, because in English, these trailing letters often modify the proceeding vowel phoneme. The features used in the pronouncer are: (1) Letter windows: current letter, 4 individual letters before and after, 3 character substrings before and after. (2) Substring shapes: substrings transformed such that consonant characters are mapped to c and vowel characters to v. (3) Indicator true if current letter is the closest vowel before a trailing e or y. (4) Leading character shape, trailing character shape.

3.3 Reranking Candidate Phoneme Sequences

The *re-ranker* calculates a total ordering over all candidate phoneme sequences, Q_i. At test time, the aligner produces the top-k_a alignments from the best Viterbi paths in the Aligner's CRF lattice. Each path has a probability score based on the input sequence. Each candidate alignment is run through the *pronouncer* which produces the top-k_p phoneme sequences and probability scores (again from best Viterbi paths). In our system, we use $k_a = k_p = 5$, and therefore at the end of the second stage there can be at most 25 candidates, Q_i, each with alignment scores and pronouncer scores. In the typical way, we train a discriminative re-ranker to assign a binary *relavence* indicator for each output sequence. We train the re-ranker to minimize the cross-entropy of each candidate list.

The re-ranking model uses a number of features for inputs Q_i that incorporates the output from previous stages: (1) aligner output probability, (2) pronouncer output probability, (3) 8-gram graphone language model probability of Q_i, (4) normalized count of how many times Q_i's phoneme sequence appears in the candidate list Q, (5) a binary value indicating if Q_i's phoneme sequence matches the mode of Q (if there is a unique mode), (6) the original output rank of Q_i (normalized), (7) if the leading grapheme and leading phoneme are agreeing *simple* consonants, (8) the prefix of the *shape* of Q_i's grapheme sequence versus the *shape* of Q_i's phoneme sequence.

Feature 8 is a feature for the agreeing prefix shape of the grapheme and phoneme string. Prefixes size one through four are considered. For example, if the input word is ⟨steve⟩, which has shape CCvCv, and the candidate phoneme sequence is /S T IY V/, which has shape CCvC, then the binary indicator for feature tokens C, CC, CCv, and CCvC would be activated.

4   Experimental Setup and Results

To evaluate the results of jG2P, we use the CMUDict dataset of words to pronunciations [13]. This dataset contains ∼ 133,000 entries with a number of noisy examples (e.g. acronyms, pronounced punctuation symbols, transcription errors). Similar to [9], we test all unique grapheme input sequences and, in the case of multiple, acceptable pronunciations, count a win if the predicted phoneme sequence matches *any* of the accepted phoneme strings. We also use the exact same split of 90 % training data and 10 % test data as in [9], and thus our results are directly comparable to theirs.

jG2P uses the MaxEnt classifier and CRF implementations from the open source library: Mallet [14]. The complete system is trained using a Google Compute Cloud instance with 32 CPUs and 28 GB of RAM. Training the entire pipeline takes 5.5 h, which is currently a total cost under $2.00 USD. The test-time evaluation on that same hardware runs at ∼ 50 strings/second.

The two most frequently used metrics for measuring G2P performance are phoneme error rate (PER) and word error rate (WER). PER is the % of phonemes predicted incorrectly, and WER is the % of predicted phoneme strings with at least one incorrect phoneme. Therefore, for both WER and PER, lower scores are better. Table 1 compares the results of jG2P to a few recent publications. All of these publications use the CMUDict dataset, but in some cases (annotated with [†]) we cannot verify if the author used the exact same train/test split of data.

Table 1. Comparison of jG2P with recent state-of-the-art results

System	PER	WER
Galescu[†] [15]	7.0	28.5
Novak [9]	5.9	24.8
Kheang[†] [16]	5.6	29.4
We[†] [11]	**5.5**	23.4
Rao[†] [12]	9.1	**21.3**
jG2P Sect. 3	6.0	25.2

To evaluate the performance of the top-k produced by the pipeline, we present the results using three different ranking methods: (1) naïve, (2) graphone, (3) pipeline. The *naïve* method picks only the top scoring alignment and generates the top-k pronunciations from that single alignment. The results are ordered by the pronouncer's confidence. The *graphone* method picks the top-5 alignments, generates the top-5 phoneme sequences from each, and then sorts the results based on the normalized language model score from an 8-gram graphone language model. The *pipeline* method uses the re-ranking scheme described in Sect. 3.3. In the testing dataset (10 % of CMUDict), 673 of the graphemes contain more than one acceptable pronunciation in the dataset. We use these, treating each pronunciation as a *relevant* result in order to calculate the precision and recall of the returned list. We present precision and recall numbers considering top-2, top-3, and top-4 result lists. The overall precision and recall for each configuration is presented in Fig. 3. Some of the test query words do not have three or four acceptable pronunciations, and thus there cannot be a perfect precision score in the top-3 and top-4 cases (and similarly for the recall scores). The *max* score is shown for each configuration for comparison purposes.

Lastly, to evaluate the impact of constraining the pairs of grapheme and phoneme substrings in the training data as described in Sect. 3.1, we run a side-by-side comparison of the overall pipeline WER and PER with and without the alignment constraints. Running without constraints results in a PER of 6.3 % and WER of 26.8 %. Adding the constraints reduced the PER to 6.0 % and the WER to 25.2 %, a ≈ 6 % improvement.

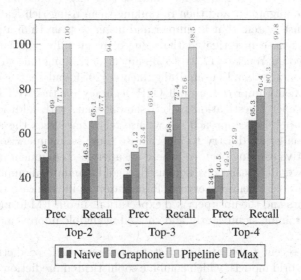

Fig. 3. Results of three different re-ranking strategies for G2P. *Max* is the theoretical maximum possible for that configuration

5 Discussion

The results in the previous section demonstrate that jG2P provides competitive results on the G2P task against the CMUDict dataset. This dataset is a good benchmark for our use case, because it is known to contain many irregularities and is viewed as 'hard' [17]. jG2P does not achieve the best PER or WER on CMUDict, but we believe that our pipeline is still compelling for a few reasons. First, it allows flexibility in incorporating expert domain knowledge for both alignments and pronunciation, even if the features are sparse and not linearly independent. This is not straightforward in a graphone language model approach, such as [9]. Second, jG2P is available as an open-source toolkit. To the best of our knowledge, the current state-of-the-art results from [11,12] are not available as open source toolkits.

jG2P with re-ranking outperforms the other re-ranking schemes described above in both precision and recall. In these experimental setups, no threshold analysis was done to improve precision when more results were returned. Therefore, precision generally decreases from the top-2 to the top-4 scenario. Each re-ranking approach suffered this handicap, but despite this, the jG2P method outperforms the others.

Previous research attempted to integrate alignment and pronunciation into the same probabilistic formulation [18,19]. The improvements in accuracy were minor or negligible, and (in one case) incorporating the two lead to a likelihood function that was not convex. In jG2P we decided to keep them as separate steps. We believe that by generating multiple candidate phoneme sequences for each candidate alignment and then re-ranking them using rich features, we end up with a robust system that is simpler and faster to train than a joint version.

Researchers have investigated the value of using fully supervised [20] and semi-supervised approaches [17,21] to alignment of training data. The NETTalk dataset [22] contains hand curated alignments. [20] found that using the hand curated alignments improves overall G2P accuracy on the NETTalk dataset. However, a deeper investigation in [17] showed that, with additional training data, the unsupervised approach (as we use in jG2P) closes the accuracy gap quickly. Originally, we did try to incorporate some semi-supervised techniques using an iterative training protocol where unsupervised alignment decisions that lead to good overall translations were used as gold standard alignments in future iterations. We used a semi-supervised formulation to incorporate the gold standard alignments and the unsupervised expectations in our EM formulation. This complicated training significantly and showed negligible improvement to overall G2P accuracy.

The most recent publications on state-of-the-art G2P prediction highlight the value of hybrid models, which combine sophisticated prediction of individual phonemes with accurate global prediction through the use of smoothed, high-order language models [11,12,23]. We incorporate the rich global sequence knowledge through an 8-gram graphone language model, and include the sequence probability as a feature to the re-ranker. An analysis of the feature weights learned by the re-ranking model shows that the graphone model score is the most important feature, followed by the aligner score.

An error analysis of bad cases shows that, as expected, the most frequent errors are incorrect vowels. Of the $\sim 3,000$ wrong sequences from the test data, 56 % have the correct consonant *structure*, meaning that the predicted sequence matches the expected in consonants exactly and in the *placement* of vowels, but at least one of the vowels is predicted incorrectly. Of those that match structurally, 82 % differ only in one vowel phoneme and in 57 % of those cases it is the first vowel that is incorrect. This is partially attributed to the fact that some words only have one vowel, but for other cases, we believe this is exacerbated by the fact that the pronouncer is a low-order linear chain CRF. If the CRF were not a linear chain and subsequent phoneme predictions could influence previous ones, perhaps this error could be improved.

There are a number of opportunities for future research. First, we would like to incorporate syllable information as input features as that has been shown to improve G2P accuracy [24]. We wish to enhance the aligner to predict a label sequence that combines syllable structure and grapheme alignments. This should improve alignments as adjacent graphemes that straddle a syllable boundary would not contribute to the same phoneme, and the syllable structure (onset, nucleus, coda) will be input to the pronouncer, which might improve phoneme labeling Second, we would like to investigate the use of general CRFs (in contrast to linear chain CRFs as we have used here) to improve performance by allowing all phoneme predictions to affect each other. Lastly, we would like to investigate the quality of the ranked output in a more thorough way by using multiple adjudicators to rank output pronunciations.

6   Conclusion

In this paper, we describe jG2P, a sophisticated pipeline solution to the problem of grapheme to phoneme translation. It is available as an open source toolkit and has the advantage of being able to incorporate domain knowledge in a flexible way. In addition, jG2P includes a re-ranking method which incorporates the error information from previous stages, predicted sequence information, and a high-order *graphone* language model. Using this approach, we produce a high quality top-k list of candidate phonemes and exhibit competitive performance compared to the current state of the art.

References

1. Klatt, D.H.: Review of the ARPA speech understanding project. J. Acoust. Soc. Am. **62**(6), 1345–1366 (1977)
2. Kaplan, R.M., Kay, M.: Regular models of phonological rule systems. Comput. Linguist. **20**(3), 331–378 (1994)
3. Black, A.W., Lenzo, K., Pagel, V.: Issues in building general letter to sound rules. In: ESCA Synthesis Workshop, Australia, pp. 77–80 (1998)

4. McCulloch, N., Bedworth, M., Bridle, J.: NETspeak a re-implementation of NETtalk. Comput. Speech Lang. **2**(3), 289–302 (1987)

5. Torkkola, K.: An efficient way to learn english grapheme-to-phoneme rules automatically. In: IEEE International Conference on Acoustics, Speech, and Signal Processing (ICASSP), vol. 2, pp. 199–202. IEEE (1993)

6. Bisani, M., Ney, H.: Joint-sequence models for grapheme-to-phoneme conversion. Speech Commun. **50**(5), 434–451 (2008)

7. Lafferty, J., McCallum, A., Pereira, F.: Conditional random fields: probabilistic models for segmenting and labeling sequence data. In: Proceedings of ICML, pp. 282–289 (2001)

8. Wang, D., King, S.: Letter-to-sound pronunciation prediction using conditional random fields. IEEE Signal Process. Lett. **18**(2), 122–125 (2011)

9. Novak, J.R., Minematsu, N., Hirose, K.: WFST-based grapheme-to-phoneme conversion: open source tools for alignment, model-building and decoding. In: 10th International Workshop on Finite State Methods and Natural Language Processing, p. 45 (2012)

10. Novak, J.R., Minematsu, N., Hirose, K., Hori, C., Kashioka, H., Dixon, P.R.: Improving WFST-based G2P conversion with alignment constraints and RNNLM n-best rescoring. In: Interspeech (2012)

11. Wu, K., Allauzen, C., Hall, K., Riley, M., Roark, B.: Encoding linear models as weighted finite-state transducers. In: Interspeech (2014)

12. Rao, K., Peng, F., Sak, H., Beaufays, F.: Grapheme-to-phoneme conversion using long short-term memory recurrent neural networks. In: IEEE International Conference on Acoustics, Speech, and Signal Processing (ICASSP) (2015)

13. Weide, R.: The CMU pronunciation dictionary, release 0.7a (2014). http://www.speech.cs.cmu.edu/cgi-bin/cmudict

14. McCallum, A.K.: Mallet: a machine learning for language toolkit (2002). http://mallet.cs.umass.edu

15. Galescu, L., Allen, J.F.: Pronunciation of proper names with a joint n-gram model for bi-directional grapheme-to-phoneme conversion. In: 7th International Conference on Spoken Language Processing, pp. 109–112 (2002)

16. Kheang, S., Katsurada, K., Iribe, Y., Nitta, T.: Solving the phoneme conflict in grapheme-to-phoneme conversion using a two-stage neural network-based approach. IEICE Trans. Inf. Syst. **97**(4), 901–910 (2014)

17. Eger, S.: Do we need bigram alignment models? On the effect of alignment quality on transduction accuracy in G2P. Proc. EMNLP **18**, 127–136 (2015)

18. Jiampojamarn, S., Kondrak, G.: Online discriminative training for grapheme-to-phoneme conversion. In: Interspeech, pp. 1303–1306 (2009)

19. Lehnen, P., Allauzen, A., Lavergne, T., Yvon, F., Hahn, S., Ney, H.: Structure learning in hidden conditional random fields for grapheme-to-phoneme conversion. In: Interspeech, pp. 2326–2330 (2013)

20. Lehnen, P., Hahn, S., Guta, A., Ney, H.: Incorporating alignments into conditional random fields for grapheme to phoneme conversion. In: IEEE International Conference on Acoustics, Speech, and Signal Processing (ICASSP), pp. 4916–4919. IEEE (2011)

21. Jiampojamarn, S., Kondrak, G.: Letter-phoneme alignment: an exploration. In: Proceedings of the 48th Annual Meeting of the Association for Computational Linguistics (ACL), pp. 780–788 (2010)

22. Sejnowski, T.J., Rosenberg, C.R.: Parallel networks that learn to pronounce English text. J. Complex Syst. **1**(1), 145–168 (1987)

23. Wang, X., Sim, K.C.: Integrating conditional random fields and joint multi-gram model with syllabic features for grapheme-to-phone conversion. In: Interspeech, pp. 2321–2325 (2013)
24. Bartlett, S., Kondrak, G., Cherry, C.: On the syllabification of phonemes. In: Proceedings of NAACL-HLT, pp. 308–316 (2009)

On the Influence of the Number of Anomalous and Normal Examples in Anomaly-Based Annotation Errors Detection

Jindřich Matoušek[✉] and Daniel Tihelka

Department of Cybernetics, New Technology for the Information Society (NTIS),
Faculty of Applied Sciences, University of West Bohemia, Plzeň, Czech Republic
jmatouse@kky.zcu.cz, dtihelka@ntis.zcu.cz

Abstract. Anomaly detection techniques were shown to help in detecting word-level annotation errors in read-speech corpora for text-to-speech synthesis. In this framework, correctly annotated words are considered as normal examples on which the detection methods are trained. Mis-annotated words are then taken as anomalous examples which do not conform to normal patterns of the trained detection models. As it could be hard to collect a sufficient number of examples to train and optimize an anomaly detector, in this paper we investigate the influence of the number of anomalous and normal examples on the detection accuracy of several anomaly detection models: Gaussian distribution based models, one-class support vector machines, and Grubbs' test based model. Our experiments show that the number of examples can be significantly reduced without a large drop in detection accuracy.

Keywords: Annotation error detection · Anomaly detection · Read speech corpora · Speech synthesis

1 Introduction

Word-level annotation of speech data is still one of the most important processes for many speech-processing tasks. Concretely, concatenative speech synthesis methods including very popular unit selection assume the word-level (textual) annotation to be correct, i.e. that textual annotation literally matches the corresponding speech signal. Such an assumption could hardly be guaranteed for corpus-based speech synthesis in which large speech corpora are typically exploited. Manual annotation of the corpora is time-consuming and costly, but, given the large amount of data, still not errorless process [1]. Automatic or semi-automatic annotation approaches could be a solution but they are still far from perfect, see, e.g. [2–7]. If not detected, any mismatch between speech data and its annotation may inherently result in audible glitches in synthetic speech [8].

This research was supported by the Czech Science Foundation (GA CR), project No. GA16-04420S. The access to the MetaCentrum clusters provided under the programme LM2010005 is highly appreciated.

© Springer International Publishing Switzerland 2016
P. Sojka et al. (Eds.): TSD 2016, LNAI 9924, pp. 326–334, 2016.
DOI: 10.1007/978-3-319-45510-5_37

Research conducted by Matoušek et al. [9,10] showed that annotation errors in read-speech corpora for text-to-speech (TTS) synthesis could be detected automatically using anomaly detection techniques. In this framework, the problem of the automatic detection of misannotated words could be viewed as a problem of *anomaly detection* (also called *novelty detection, one-class classification,* or *outlier detection*), an unsupervised detection technique under the assumption that the majority of the examples in the unlabeled data set are normal. By just providing the normal training data, an algorithm creates a representational model of this data. If newly encountered data is too different from this model, it is labeled as anomalous. This could be perceived as an advantage over a standard classification approach in which substantial number of both negative (normal) and positive (anomalous) examples is needed [9]. Nevertheless, if some anomalous examples are given in the anomaly detection framework, they can be used to tune the detector and to evaluate its performance. In the annotation error detection framework, misannotated words are considered as *anomalous examples*, and correctly annotated words are taken as *normal examples*.

As it could be hard to collect enough data (especially the misannotated words) to set up an anomaly detector, we investigate the influence of both the number of correctly annotated and misannotated words on the anomaly detection accuracy in this paper.

2   Experimental Data

We used a Czech read-speech corpus of a single-speaker male voice, recorded for the purposes of unit-selection speech synthesis [11,12]. The voice talent was instructed to speak in a "news-broadcasting style" and to avoid any spontaneous expressions. The full corpus consisted of 12242 utterances (approx. 18.5 h of speech) segmented to phone-like units using HMM-based forced alignment (carried out by the HTK toolkit [13]) with acoustic models trained on the speaker's data [14]. From this corpus we selected $N_\mathrm{n} = 1124$ words, which were annotated correctly (i.e. *normal examples*), and $N_\mathrm{a} = 273$ words (213 of them being different), which contained some annotation error (i.e. *anomalous examples*). The decision whether the annotation was correct or not was made by a human expert who analyzed the phonetic alignment.

For the purposes of anomaly detection model training and selection, the normal examples were divided into training and validation examples using 10-fold cross validation with 60 % of the normal examples used for training and 20 % of the normal examples used for validation in each cross-validation fold (i.e., $N_\mathrm{nd} = 899$ normal examples were available for detector development). The remaining 20 % of the normal examples were used as test data for the final evaluation of the model. As for the anomalous examples, 50 % of them (i.e. $N_\mathrm{ad} = 136$) were used in cross validation when selecting the best model parameters, and the remaining 50 % of anomalous examples were used for the final evaluation.

Various word-level feature sets were tested with each anomaly detector. The sets incorporated various acoustic, spectral, phonetic, positional, durational, and other features. To emphasize anomalies in the feature values, histograms and deviations from their expected values were also used [10].

3   Anomaly Detection Models

Let us denote $\mathbf{x}^{(1)}, \ldots, \mathbf{x}^{(N_n)}$ the training set of normal (i.e. not anomalous) examples where N_n is the number of normal training examples with each example $\mathbf{x}^{(i)} \in \mathbb{R}^{N_f}$ and N_f being the number of features.

3.1   Gaussian Distribution Based Detectors

Gaussian distribution based detectors model normal examples using Gaussian distribution. In *univariate Gaussian distribution* (UGD), each feature x_j ($j = 1, \ldots, N_f$) is modeled separately with mean $\mu_j \in \mathbb{R}$ and variance $\sigma_j^2 \in \mathbb{R}$ under the assumption of feature independence, i.e. $x_j \sim \mathcal{N}(\mu_j, \sigma_j^2)$. The probability of x_j being generated by $\mathcal{N}(\mu_j, \sigma_j^2)$ can be then written as $p(x_j; \mu_j, \sigma_j^2)$.

Multivariate Gaussian distribution (MGD) is a generalization of the univariate Gaussian distribution. In this case, $p(\mathbf{x})$ is modeled in one go using mean vector $\boldsymbol{\mu} \in \mathbb{R}^{N_f}$ and covariance matrix $\boldsymbol{\Sigma} \in \mathbb{R}^{N_f \times N_f}$, i.e. $\mathbf{x} \sim \mathcal{N}_{N_f}(\boldsymbol{\mu}, \boldsymbol{\Sigma})$.

The training of the detectors can be described as follows

$$\text{UGD}: \quad \mu_j = \frac{1}{N_n} \sum_{i=1}^{N_n} x_j^{(i)}, \quad \sigma_j^2 = \frac{1}{N_n} \sum_{i=1}^{N_n} (x_j^{(i)} - \mu_j)^2 \tag{1}$$

$$\text{MGD}: \quad \boldsymbol{\mu} = \frac{1}{N_n} \sum_{i=1}^{N_n} \mathbf{x}^{(i)}, \quad \boldsymbol{\Sigma} = \frac{1}{N_n} \sum_{i=1}^{N_n} (\mathbf{x}^{(i)} - \boldsymbol{\mu})(\mathbf{x}^{(i)} - \boldsymbol{\mu})^{\mathsf{T}}. \tag{2}$$

Having the estimated parameters, probability of a new example $\mathbf{x}$ (either normal or anomalous) can be computed as

$$\text{UGD}: \quad p(\mathbf{x}) = \prod_{j=1}^{N_f} p(x_j; \mu_j, \sigma_j^2) = \prod_{j=1}^{N_f} \frac{1}{\sqrt{2\pi}\sigma_j} \exp(-\frac{(x_j - \mu_j)^2}{2\sigma_j^2}) \tag{3}$$

$$\text{MGD}: \quad p(\mathbf{x}) = \frac{1}{\sqrt{(2\pi)^{N_f} |\boldsymbol{\Sigma}|}} \exp\left(-\frac{1}{2}(\mathbf{x} - \boldsymbol{\mu})^{\mathsf{T}} \boldsymbol{\Sigma}^{-1} (\mathbf{x} - \boldsymbol{\mu})\right). \tag{4}$$

If $p(\mathbf{x})$ is very small, i.e. $p(\mathbf{x}) < \varepsilon$, then the example $\mathbf{x}$ does not conform to the normal examples distribution and can be denoted as anomalous.

3.2   One-Class SVM Based Detector

One-class SVM (OCSVM) algorithm maps input data into a high dimensional feature space via a *kernel function* and iteratively finds the maximal margin hyperplane which best separates the training data from the origin. This results in a binary decision function $f(\mathbf{x})$ which returns $+1$ in a "small" region capturing the (normal) training examples and -1 elsewhere (see Eq. 8) [15].

The hyperplane parameters $\mathbf{w}$ and ρ are determined by solving a quadratic programming problem

$$\min_{\mathbf{w},\xi,\rho} \frac{1}{2}||\mathbf{w}||^2 + \frac{1}{\nu N_{\mathrm{n}}} \sum_{i=1}^{N_{\mathrm{n}}} \xi_i - \rho \tag{5}$$

subject to

$$\mathbf{w} \cdot \mathbf{\Phi}(\mathbf{x}^{(i)}) \geq \rho - \xi_i, \qquad i = 1, 2, \ldots, N_{\mathrm{n}}, \quad \xi_i \geq 0, \tag{6}$$

where $\mathbf{\Phi}(\mathbf{x}^{(i)})$ is the mapping defining the kernel function, ξ_i are slack variables, and $\nu \in (0, 1]$ is an a priori fixed constant which represents an upper bound on the fraction of examples that may be anomalous. We used a Gaussian radial basis function kernel

$$K(\mathbf{x}, \mathbf{x}') = \exp(\gamma||\mathbf{x} - \mathbf{x}'||^2) \tag{7}$$

where γ is a kernel parameter and $||\mathbf{x} - \mathbf{x}'||$ is a dissimilarity measure between the examples $\mathbf{x}$ and $\mathbf{x}'$.

Solving the minimization problem (5) using Lagrange multipliers α_i and using the kernel function (7) for the dot-product calculations, the decision function for a new example $\mathbf{x}$ then becomes

$$f(\mathbf{x}) = \mathrm{sgn}(\mathbf{w} \cdot \mathbf{\Phi}(\mathbf{x}) - \rho) = \mathrm{sgn}(\sum_{i=1}^{N_{\mathrm{n}}} \alpha_i K(\mathbf{x}^{(i)}, \mathbf{x}) - \rho). \tag{8}$$

3.3 Grubbs' Test Based Detector

Grubbs's test is typically used to detect a single outlier (i.e. anomalous example in our case) in a univariate data set x of length N assumed to come from a normally distributed population [16].

The Grubbs' two-sided test statistic is defined as

$$G = \frac{\max\limits_{i=1,\ldots,N} |x_i - \mu|}{\sigma} \tag{9}$$

with x_i being i-th (one-dimensional) example, and μ and σ denoting sample mean and standard deviation, respectively.

The hypothesis of no outliers is rejected at significance level α if

$$G > \frac{N-1}{\sqrt{N}} \sqrt{\frac{t^2_{\alpha/(2N),N-2}}{N - 2 + t^2_{\alpha/(2N),N-2}}} \tag{10}$$

with $t^2_{\alpha/(2N),N-2}$ denoting the upper critical value of the t-distribution with $N-2$ degrees of freedom and a significance level of $\alpha/(2N)$.

For the purposes of anomaly detection we modified the underlying Grubbs' test in the following ways:

1. Since we have a training set $\mathbf{x}^{(1)}, \ldots, \mathbf{x}^{(N_n)}$ of normal (i.e. not outlying) examples, μ and σ were computed as sample mean and standard deviation of this training set, and the Grubbs' statistic (9) was calculated for each tested example $\mathbf{x}$.
2. Having multidimensional examples $\mathbf{x}^{(i)} \in \mathbb{R}^{N_f}$, the Grubbs' test (10) was carried out independently for each feature x_j $(j = 1, \ldots, N_f)$, and the tested example $\mathbf{x}$ was detected as outlier if at least n features were detected as outlying.

3.4   Model Training and Selection

The standard training procedure was utilized to train the models described in previous sections. Models' parameters were optimized during *model selection*, i.e. by selecting their values that yielded best results (in terms of $F1$ score, see Sect. 3.5) applying a grid search over relevant values of the parameters and various feature set combinations with 10-fold cross validation [10]. The optimal parameter values are shown in Table 1. *Scikit-learn* toolkit [17] was employed in our experiments.

Table 1. Optimal model parameter values as found by cross validation.

OCSVM	UGD	MGD	GT
$\nu = 0.005$	$\varepsilon = 0.005$	$\varepsilon = 2.5e{-}14$	$n = 1$
$\gamma = 0.03125$			$\alpha = 0.0375$

3.5   Model Evaluation

Due to the unbalanced number of normal and anomalous examples, $F1$ *score* was used to evaluate the performance of the proposed anomaly detection models

$$F1 = \frac{2 * P * R}{P + R}, \quad P = \frac{t_p}{p_p}, \quad R = \frac{t_p}{a_p} \tag{11}$$

where P is *precision*, the ability of a detector not to detect as misannotated a word that is annotated correctly, R is *recall*, the ability of a detector to detect all misannotated words, t_p means "true positives" (i.e., the number of words correctly detected as misannotated), p_p stands for "predicted positives" (i.e., the number of all words detected as misannotated), and a_p means "actual positives" (i.e., the number of actual misannotated words). $F1$ score was also used to optimize all parameters during model training and selection (see Sect. 3.4).

McNemar's statistical significance test at the significance level $\alpha = 0.05$ was used to see whether the achieved results were comparable in terms of statistical significance [10, 18].

The results in Fig. 1 suggest that MGD detector performs best but the differences among the individual detectors were not proved to be statistically significant.

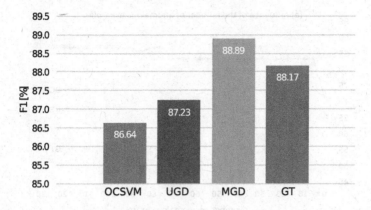

Fig. 1. Comparison of the anomaly detection models on test data.

4   Influence of the Number of Examples on Detection Accuracy

4.1   Influence of Anomalous Examples

While there are usually many correctly annotated words (i.e. normal examples) in a TTS corpus, it could be hard to collect a set of misannotated words (anomalous examples). The aim of the following experiment was to investigate the influence of the number of anomalous examples used during the anomaly-detector development on the resulting detection accuracy. The number of normal examples used for detector development and the number of normal and anomalous examples used for testing were kept the same as described in Sect. 2.

As can be seen in Fig. 2(a), reasonably good performance can be achieved with substantially less anomalous examples. The 1st row of Table 2 shows the minimum numbers of anomalous examples which lead to statistically same performance (McNemar's test, $\alpha = 0.05$) compared to the case when all 136 anomalous examples were used. It is also evident that in the case no anomalous examples were available at all, the detectors' performance dropped significantly.

4.2   Influence of Normal Examples

This experiment was aimed at investigating the influence of the number of normal examples used during anomaly-detector development on the detection accuracy. All 136 anomalous examples available for cross validation were used. Again, Fig. 2(b) suggests that much less normal examples could be used and the detection performance remains good. The minimum numbers of normal examples which yield statistically same performance (McNemar's test, $\alpha = 0.05$) compared to the case when all 899 normal examples were used are shown in the 2nd row of Table 2.

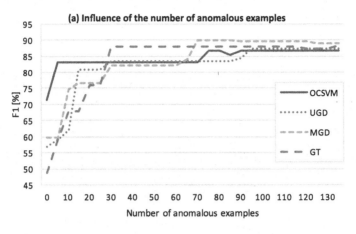

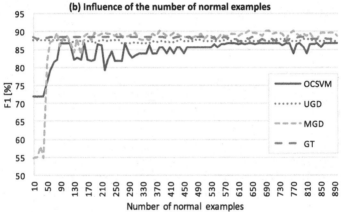

Fig. 2. The influence of the number of anomalous (a) and normal (b) examples used during anomaly-detector development on the detection accuracy.

Table 2. Minimum numbers of anomalous and normal examples that yield statistically same performance when compared to the corresponding "full" anomaly detector.

	OCSVM	UGD	MGD	GT
# anomalous examples	5	15	28	27
# normal examples	90	10	60	10

4.3   Influence of Both Anomalous and Normal Examples

Putting it all together, Fig. 3 shows detection accuracy of the individual anomaly detectors for combinations of various number of anomalous N_{ad} and normal N_{nd} examples used during detectors' development.

As can be seen, there are some areas where the detection accuracy remains very high. For MGD detector, $N_{ad} \gtrsim 30 \wedge N_{nd} \gtrsim 200$ yield top performance. In the case of GT, very good detection accuracy is guaranteed for $N_{ad} \gtrsim 20 \wedge$

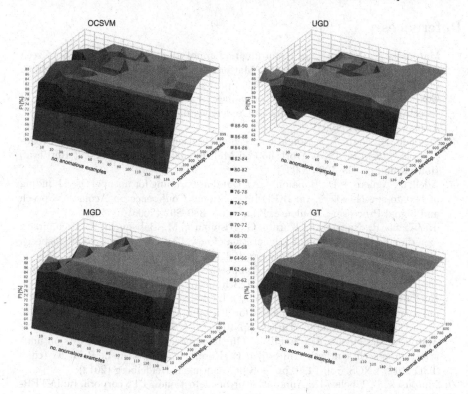

Fig. 3. The influence of both normal and anomalous development examples on the detection accuracy.

$N_{nd} \gtrsim 600$. For UGD, reasonable performance is achieved for $N_{ad} \gtrsim 20 \wedge 200 \lesssim N_{nd} \lesssim 400$ and for $N_{ad} \gtrsim 50 \wedge N_{nd} \gtrsim 400$. Values $N_{ad} \gtrsim 60 \wedge N_{nd} \gtrsim 700$ lead to acceptable detection accuracy in the case of OCSVM.

5   Conclusion

We experimented with different sized data sets used for detection of annotation errors in read-speech TTS corpora. We focused on the influence of the number of both anomalous and normal examples on the detection accuracy. Several anomaly detectors were taken into account: Gaussian distribution based models, one-class support vector machines, and Grubbs' test based model. The experiments show that the number of examples can be significantly reduced without a large drop in detection accuracy. When very few anomalous examples (or even none) are available for the development of an anomaly detection model, one-class SVM seems to perform best. As for the case very few normal examples are available, univariate Gaussian distribution and Grubb's test based models give best results. When there are both few anomalous and normal examples available, multivariate Gaussian distribution based detector performs best.

References

1. Matoušek, J., Romportl, J.: Recording and annotation of speech corpus for Czech unit selection speech synthesis. In: Matoušek, V., Mautner, P. (eds.) TSD 2007. LNCS (LNAI), vol. 4629, pp. 326–333. Springer, Heidelberg (2007)
2. Cox, S., Brady, R., Jackson, P.: Techniques for accurate automatic annotation of speech waveforms. In: International Conference on Spoken Language Processing, Sydney, Australia (1998)
3. Meinedo, H., Neto, J.: Automatic speech annotation and transcription in a broadcast news task. In: ISCA Workshop on Multilingual Spoken Document Retrieval, Hong Kong, pp. 95–100 (2003)
4. Adell, J., Agüero, P.D., Bonafonte, A.: Database pruning for unsupervised building of text-to-speech voices. In: IEEE International Conference on Acoustics Speech and Signal Processing, Toulouse, France, pp. 889–892 (2006)
5. Tachibana, R., Nagano, T., Kurata, G., Nishimura, M., Babaguchi, N.: Preliminary experiments toward automatic generation of new TTS voices from recorded speech alone. In: INTERSPEECH, Antwerp, Belgium, pp. 1917–1920 (2007)
6. Aylett, M.P., King, S., Yamagishi, J.: Speech synthesis without a phone inventory. In: INTERSPEECH, Brighton, Great Britain, pp. 2087–2090 (2009)
7. Boeffard, O., Charonnat, L., Maguer, S.L., Lolive, D., Vidal, G.: Towards fully automatic annotation of audiobooks for TTS. In: Language Resources and Evaluation Conference, Istanbul, Turkey, pp. 975–980 (2012)
8. Matoušek, J., Tihelka, D., Šmídl, L.: On the impact of annotation errors on unit-selection speech synthesis. In: Sojka, P., Horák, A., Kopeček, I., Pala, K. (eds.) TSD 2012. LNCS, vol. 7499, pp. 456–463. Springer, Heidelberg (2012)
9. Matoušek, J., Tihelka, D.: Annotation errors detection in TTS corpora. In: INTERSPEECH, Lyon, France, pp. 1511–1515 (2013)
10. Matoušek, J., Tihelka, D.: Anomaly-based annotation errors detection in TTS corpora. In: INTERSPEECH, Dresden, Germany, pp. 314–318 (2015)
11. Matoušek, J., Tihelka, D., Romportl, J.: Building of a speech corpus optimised for unit selection TTS synthesis. In: Language Resources and Evaluation Conference, Marrakech, Morocco, pp. 1296–1299 (2008)
12. Kala, J., Matoušek, J.: Very fast unit selection using Viterbi search with zero-concatenation-cost chains. In: IEEE International Conference on Acoustics Speech and Signal Processing, Florence, Italy, pp. 2569–2573 (2014)
13. Young, S., Evermann, G., Gales, M.J.F., Hain, T., Kershaw, D., Liu, X., Moore, G., Odell, J., Ollason, D., Povey, D., Valtchev, V., Woodland, P.: The HTK Book (for HTK Version 3.4). Cambridge University, Cambridge (2006)
14. Matoušek, J., Tihelka, D., Psutka, J.V.: Experiments with automatic segmentation for Czech speech synthesis. In: Matoušek, V., Mautner, P. (eds.) TSD 2003. LNCS (LNAI), vol. 2807, pp. 287–294. Springer, Heidelberg (2003)
15. Schölkopf, B., Platt, J.C., Shawe-Taylor, J.C., Smola, A.J., Williamson, R.C.: Estimating the support of a high-dimensional distribution. Neural Comput. **13**, 1443–1471 (2001)
16. Grubbs, F.E.: Procedures for detecting outlying observations in samples. Technometrics **11**, 1–21 (1969)
17. Pedregosa, F., Varoquaux, G., Gramfort, A., Thirion, V.M.B., Grisel, O., Blondel, M., Prettenhofer, P., Weiss, R., Dubourg, V., Vanderplas, J., Passos, A., Cournapeau, D., Brucher, M., Perror, M., Duchesnay, É.: Scikit-learn: machine learning in Python. J. Mach. Learn. Res. **12**, 2825–2830 (2011)
18. Dietterich, T.G.: Approximate statistical tests for comparing supervised classification learning algorithms. Neural Comput. **10**, 1895–1923 (1998)

Unit-Selection Speech Synthesis Adjustments for Audiobook-Based Voices

Jakub Vít[✉] and Jindřich Matoušek

Department of Cybernetics, University of West Bohemia in Pilsen,
Pilsen, Czech Republic
{jvit,jmatouse}@kky.zcu.cz

Abstract. This paper presents easy-to-use modifications to unit-selection speech-synthesis algorithm with voices built from audiobooks. Audiobooks are a very good source of large and high quality audio data for speech synthesis; however, they usually do not meet basic requirements for standard unit-selection synthesis: "neutral" speech properties with no expressive or spontaneous expressions, stable prosodic patterns, careful pronunciation, and consistent voice style during recording. However, if these conditions are taken into consideration, few modifications can be made to adjust the general unit-selection algorithm to make it more robust for synthesis from such audiobook data. Listening test shows that these adjustments increased perceived speech quality and acceptability against a baseline TTS system. Modifications presented here can also allow to exploit audio data variability to control pitch and tempo of synthesized speech.

Keywords: Speech synthesis · Audiobooks · Unit selection · Target cost modification

1 Introduction

Unit selection ranks among the most popular techniques for generating synthetic speech. It is widely used and it is known for its ability to produce high-quality speech. The unit-selection algorithm is based on a concatenation of units from a speech database. Each unit is represented by a set of features describing its prosodic, phonetic, and acoustic parameters. Target cost determines a distance of each unit candidate to its target unit using features such as various positional parameters, phonetic contexts, phrase type, etc. When the algorithm is searching for an optimal sequence of unit, it minimizes total cumulative cost which is composed of the target cost and join cost. Join cost measures the quality of adjacent unit concatenation using prosodic and acoustic features like $F0$, energy, duration and spectral parameters. More detailed explanation of this method can be found in [1].

The work has been supported by the grant of the University of West Bohemia, project No. SGS-2016-039, and by the Technology Agency of the Czech Republic, project No. TA01011264.

© Springer International Publishing Switzerland 2016
P. Sojka et al. (Eds.): TSD 2016, LNAI 9924, pp. 335–342, 2016.
DOI: 10.1007/978-3-319-45510-5_38

The speech database is usually recorded by professional speakers in a sound chamber. Sentences for the recording are selected to cover as many unit combinations in various prosodic contexts as possible. When recording the speech corpus, the speaker is instructed to keep a consistent speech rate, pitch and prosody style during the entire recording. This method produces a very high quality synthetic voice but it is very expensive and time consuming as the number of sentences to record is very high (usually more than ten thousands—approximately 15 h of speech).

Audiobooks offer an alternative data source for building synthetic voices. They are also recorded by professional speakers and they have good sound quality. Unfortunately, they do not meet basic requirements for standard unit-selection synthesis: "neutral"[1] speech properties with no expressive or spontaneous expressions, stable prosodic patterns, careful pronunciation, and consistent voice style during recording. This problem is not so significant for the HMM; however, it greatly reduces quality of unit-selection based synthetic speech.

Unlike [2–5] and [6] where various styles were exploited to build an HMM synthesizer with the capability of generating expressive speech, our primary goal is to build only neutral voice but with highest quality possible. Therefore, unit-selection algorithm was used to ensure naturalness of synthetic speech.

This paper presents adjustment to unit selection algorithm to better cope with non-neutral and inaccurate speech database. It introduces a statistical analysis step to synthesis algorithm which allows to penalize units which would drop quality of speech. This step also allows to partially modify speech prosody parameters. It also summarizes the process of creating synthetic voice from audiobook for unit selection speech synthesis.

2   Audio Corpus Annotation

Unit selection voice requires a speech corpus which is in fact a database of audio files containing sentences and text transcriptions [7]. These sentences have to be aligned on a unit level, i.e., usually on a phone level. Text representation is often available for audiobooks but only in a form of a formatted text, not of a unit-level alignment. Also, the text form is usually optimized for reading, not for the computer analysis, therefore there must be some text preprocessing which removes formatting and converts text to its plain form. It is also necessary to perform text normalization, replace abbreviations, numbers, dates, symbols and delete all texts, which do not correspond to the audio content of a book. Due to the large volume of data, this step is no longer possible to perform by hand; therefore, it must be done automatically or at least semi-automatically.

The normalized text is then ready to be aligned to phone levels. However, segmentation and alignment of large audio files is not a trivial task [8]. Standard forced-alignment techniques which are used for the alignment of single sentences cannot be used here primarily because of memory and complexity requirements.

---

[1] In this paper, neutral speech is meant as news broadcasting style, which is very often used by modern commercial TTS system.

Text must be either cut into smaller chunks with some heuristics or annotated with the help of a speech recognizer run in forced-alignment mode. This approach tends to produce much more annotation errors when compared to the alignment of single sentences.

This problem was already dealt with in [8,9] or [10] where new techniques to reduce the number of errors were proposed. However, it must be noted that these errors can still occur and that the corpus database could contain badly aligned or otherwise unsuitable units.

2.1 Automatic Annotation Filtering

For our experiment, a simple procedure was used to check whether text annotation and alignment matches audio data. This procedure helped to remove the worst annotated sentences, i.e. sentences where the speech recognizer was desynchronized or text did not match audio representation. For every sentence from the source text, a sentence with the same text was synthesized using an existing high-quality voice, which was selected to be similar to the voice of the audiobook speaker. These sentences were compared using dynamic time warping (DTW) using mel-frequency cepstrum coefficients (MFCC) and euclidean distance. Distance was then divided by number of phones in a sentence to ensure the final score to be independent on the sentence length. Ten percent of sentences with the worst score were then removed from the speech corpus. Manual inspection confirmed that textual transcription of these sentences did not match the corresponding audio signal. A more sophisticated algorithm for detecting wrong annotation was proposed e.g. in [11].

3 Unit Selection Modification

The following subsections describe various modifications that were made to the described baseline algorithm to achieve better synthesis quality when using voices built from audiobooks.

3.1 Weights Adjustments

The total cost is composed of a large number of features which are precisely tuned to select the best possible sequence of units given the input specification. Source data from audiobooks have different characteristics. To reflect that, features' weights must be adjusted. Due to the higher prosodic variability of audiobook speech data, it is suitable to increase the weight of prosodic features (intonation, tempo, etc.) to keep speech as neutral as possible. Also, having a big database of audio data, it is possible to incorporate more specific features like more distant phonetic contexts or stricter rules for comparing positional parameters.

However, some of the features tend to be problematic in audiobooks, for example, phrasing. Narrators usually do not follow phrasing rules typical for read speech. They adjust their phrasing style based on the current context and actual sentence meaning. They simply do not use the same prosodic sentence pattern. So, relying on positional parameters is not always useful and its contribution to the cost function should be reduced. It is better to focus more on "neutral" prosody to ensure the requested "neutral" speech.

Audiobooks also contain a lot of sentences with direct speech which are usually pronounced with different (more expressive) style. Moreover, this change of style can also affect neighboring sentences. Such problematic sentences can be either completely removed or penalized with another component of the target cost. If this component is tuned well, it could preserve those direct speech segments which do not have a different style than another parts of the book.

3.2   Target Cost Modification

A typical voice database created for speech synthesis contains precisely annotated units. All of them could be used during synthesis. Audiobooks contain much more "unwanted" data. Some units just do not fit into neutral style because of their dynamic prosodic parameters and some units might be wrongly annotated (see Sect. 2).

During the synthesis, each unit is assigned a set of possible candidates. Target cost is computed for each of the candidates. This cost is composed of many features which evaluate how well a candidate fits this unit. At this point, the algorithm is modified with an another step, in which all candidates are analyzed together. More concretely, their prosodic and spectral features (which are typically used also in join cost) are analyzed. For each individual feature ($F0$, energy, duration, MFCCs) a statistical model is built. The model is described by its mean and variance so that "expected" values for each feature can be predicted.

The target cost of every candidate is then modified with a value representing how much its features differ from the its statistical model. For each candidate, the *modified target cost* T_{modif} is then computed as the sum of original target costs T and the sum of features' *diversity penalty* $|d_i - 0.5|$

$$T_{modif} = T + \sum_{i=1}^{n_f} w_i \cdot |d_i - 0.5| \tag{1}$$

$$d_i = \frac{1}{2}\left[1 + \mathrm{erf}\left(\frac{f_i - \mu_i}{\sigma_i \sqrt{2}}\right)\right] \tag{2}$$

$$\mathrm{erf}(x) = \frac{2}{\sqrt{\pi}} \int_0^x e^{-t^2}\, dt \tag{3}$$

where f_i is the i-th feature value of the candidate, μ_i and σ_i are mean and standard deviation of i-th feature across all candidates, d_i is a value of cumulative distribution function of the i-th feature in its statistical model. The value $d_i = 0.5$ means that $f_i = \mu_i$, w_i is a weight for the i-th feature and n_f is the number of features. Function $\mathrm{erf}(x)$ stands for an error function. Scheme of the target cost modification is shown in Fig. 1.

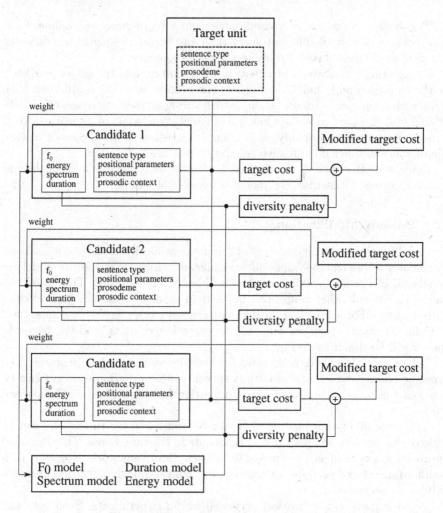

Fig. 1. Scheme of target cost modification.

Since prosodic feature values of candidates change a lot with different phonetic contexts, it would be unwise to build a model of the unit from all candidates with the same weight. Therefore, a weighted mean and variance were used with the weight being the inverted value of the target cost. If the candidate has a low target cost (meaning it fits unit nicely) its significance to mean and variance calculation is high. The weighted formula for mean value calculation is:

$$\mu_i = \frac{1}{c} \cdot \sum_{k=1}^{n_c} \frac{1}{1 + T_k} \cdot f_{k,i} \tag{4}$$

$$c = \sum_{k=1}^{n_c} \frac{1}{1 + T_k} \tag{5}$$

where T_k is the original target cost, n_c is the number of candidates, $f_{k,i}$ is the i-th feature of the k-th candidate.

This is also the reason why candidates cannot be preprocessed offline. Phonetic context of units is different every time and therefore weights are different resulting in different model parameters for each sentence.

By selecting candidates which feature values are close to the values predicted by the corresponding model, the outliers (i.e., candidates with a different voice style, with an unusual pronunciation or with wrong segmentation) are effectively filtered out. If a unit candidate has a bad annotation, some of its features (e.g. duration) would very probably differ from its expected values. Prosodic feature values will also differ if expressive voice style was used.

Being statistical based, this approach works only if there is a lot of data in a speech corpus. Otherwise, the model is not reliable enough.

3.3   Prosody Modification

The modification presented in Sect. 3.2 is used primarily to filter out outliers. Candidates whose feature values differ significantly from its statistical model are penalized. In Formula 1, 0.5 is used as an ideal reference value. This value is a logical choice but other values can be used to modify (prosodic) properties of output speech. For instance, if the duration reference is set to 0.6 (60 % quantile), the unit-selection algorithm will tend to select longer units, and the resulting speech will be slower in average.

Let us note that there is no need for absolute values of these features. The prosodic characteristics of the output speech (pitch, duration, energy) can be controlled using relative probability distribution function values on the interval <0,1>.

The amount of this modification can be controlled by tuning the weight ratio of this penalty against other components in the target cost. The described approach works even in the standard unit-selection framework. However, as a standard speech corpus does not contain so much variable data, the modification will be not so powerful.

Prosody modification worked very well in our experiments. Even very low weight w_i in Formula (1) was enough for prosody modification to work, especially for pitch and tempo. These parameters were changing on very large scale from very slow to very fast speech (or very low to very high pitch). On the other hand, energy modification was not so useful.

4   Evaluation

A three-point Comparison Category Rating (CCR) listening test was carried out to verify whether the modifications presented in this paper improved the quality of speech synthesized using audiobook-based voices. Ten listeners participated in the test. Half of them had experience with speech synthesis. Each participant got 70 pairs of synthesized sentences. In each pair, one sentence was synthesized by the baseline TTS system and the other one was synthesized by the modified TTS system. The set of possible answers was following: A is much better, A is slightly better, A and B are the same, B is slightly better and B is much better than A. Order of A and B was shuffled. Listeners were instructed to choose a sentence with better general quality and naturalness.

Listening test results are shown in Figs. 2 and 3. The results show that proposed modification helped to improve the quality of synthesized speech. Nearly 70 % of all answers preferred the new system with the proposed modifications. Average answer was 0.85 on interval $< -2.0, 2.0 >$. The results emphasize the importance of the modifications proposed in the paper when such variable speech data as the one from audiobooks are used. On other hand, the baseline unit-selection system does not take the speech data variability into account; the synthetic speech then suffers from more frequent occurrence of audible artifacts and inconsistent prosodic characteristics.

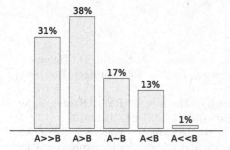

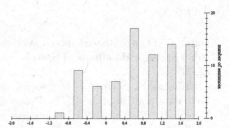

Fig. 2. Listening tests: distribution of answers. (A is the proposed system, B is the baseline system).

Fig. 3. Histogram of sentence average answers.

5   Conclusion

In this paper, we have described a modification of unit-selection algorithm for audiobook-based voices. As audiobooks do not always meet strict requirements for voices used by a standard unit-selection algorithm, our primary goal was to maximize quality and naturalness of speech synthesized from such variable speech sources. We presented modifications of the baseline unit-selection scheme which allow to lower these requirements by introducing a special step into the unit-selection procedure. In this step, candidates' prosodic features are analyzed and statistical models are built to describe expected feature values. Then, each candidate is penalized based on how much its feature values are different from its corresponding model. This approach leads to a penalization of units with different properties such as the ones frequent in audiobooks (e.g., non-neutral, expressive, and spontaneous speech properties, dynamic prosodic patterns, etc.). The proposed modification can also handle errors resulted from a mismatch between text and audio representation of an audiobook (i.e. annotation errors). Since this modification affects only target cost, which is far less computationally expensive than join cost, the computational complexity of the proposed algorithm is not significantly affected.

We also showed a way of taking advantage of this modification. The proposed algorithm enables to control output speech properties like pitch or tempo. According to a listening test, people preferred the proposed system much more than the baseline system. The test proved that this modification was beneficial to speech quality.

In future work, more detailed analysis could be done to identify benefits of individual adjustments to overall quality of speech. Also, expressive part of audiobooks could be used to build synthesis with expressive capability as proposed for instance by Zhao et al. [4]. Lastly, as more audiobooks narrated by the same speaker offer a possibility to use more data at the expense of introducing more inconsistencies, we plan to investigate how a mixture of recordings from more audiobooks will affect the quality of the resulting synthetic speech.

References

1. Dutoit, T.: Corpus-based speech synthesis. In: Benesty, J., Sondhi, M., Huang, Y. (eds.) Springer Handbook of Speech Processing, pp. 437–455. Springer, Dordrecht (2008)
2. Charfuelan, M., Steiner, I.: Expressive speech synthesis in MARY TTS using audiobook data and EmotionML. In: Proceedings of INTERSPEECH (2013)
3. Eyben, F., Buchholz, S., Braunschweiler, N., Latorre, J., Wan, V., Gales, M., Knill, K.: Unsupervised clustering of emotion and voice styles for expressive TTS. In: ICASSP, pp. 4009–4012 (2012)
4. Zhao, Y., Peng, D., Wang, L., Chu, M., Chen, Y., Yu, P., Guo, J.: Constructing stylistic synthesis databases from audio books. In: INTERSPEECH, Pittsburgh, PA, USA (2006)
5. Székely, E., Cabral, J.P., Cahill, P., Carson-Berndsen, J.: Clustering expressive speech styles in audiobooks using glottal source parameters. In: INTERSPEECH, pp. 2409–2412 (2011)
6. Székely, E., Cabral, J.P., Abou-Zleikha, M., Cahill, P., Carson-Berndsen, J.: Evaluating expressive speech synthesis from audiobook corpora for conversational phrases. In: Proceedings of LREC 2012 (2012)
7. Matoušek, J., Tihelka, D., Romportl, J.: Building of a speech corpus optimised for unit selection TTS synthesis. In: Proceedings of LREC 2008 (2008)
8. Prahallad, K., Toth, A.R., Black, A.W.: Automatic building of synthetic voices from large multi-paragraph speech databases. In: INTERSPEECH, pp. 2901–2904 (2007)
9. Braunschweiler, N., Buchholz, S.: Automatic sentence selection from speech corpora including diverse speech for improved HMM-TTS synthesis quality. In: INTERSPEECH, pp. 1821–1824 (2011)
10. Prahallad, K., Black, A.W.: Handling large audio files in audio books for building synthetic voices. In: The Seventh ISCA Tutorial and Research Workshop on Speech Synthesis, pp. 148–153, Japan, Kyoto (2010)
11. Matoušek, J., Tihelka, D.: Annotation errors detection in TTS corpora. In: Proceedings of INTERSPEECH, pp. 1511–1515, Lyon, France (2013)

The Custom Decay Language Model for Long Range Dependencies

Mittul Singh[1,2(✉)], Clayton Greenberg[1,2,3], and Dietrich Klakow[1,2,3]

[1] Spoken Language Systems (LSV), Saarland University, Saarbrücken, Germany
{mittul.singh,clayton.greenberg,dietrich.klakow}@lsv.uni-saarland.de
[2] Saarbrücken Graduate School of Computer Science, Saarland University,
Saarland Informatics Campus, Saarbrücken, Germany
[3] Collaborative Research Center on Information Density and Linguistic Encoding,
Saarland University, Saarbrücken, Germany

Abstract. Significant correlations between words can be observed over long distances, but contemporary language models like N-grams, Skip grams, and recurrent neural network language models (RNNLMs) require a large number of parameters to capture these dependencies, if the models can do so at all. In this paper, we propose the Custom Decay Language Model (CDLM), which captures long range correlations while maintaining sub-linear increase in parameters with vocabulary size. This model has a robust and stable training procedure (unlike RNNLMs), a more powerful modeling scheme than the Skip models, and a customizable representation. In perplexity experiments, CDLMs outperform the Skip models using fewer number of parameters. A CDLM also nominally outperformed a similar-sized RNNLM, meaning that it learned as much as the RNNLM but without recurrence.

Keywords: Reduction in number of parameters · Robust training · Long range context · Language model

1 Introduction

The central task of automatic speech recognition (ASR) is predicting the next word given sequential acoustic data. Language models (LMs), which predict words given some notion of context, inform ASR systems about which word choices fit well together, thus acting complementarily to acoustic models which directly assign probabilities to words given the acoustic input. Within ASR Systems, smoothed N-gram LMs have been very successful and also are very simple to build. These standard LMs work well on short context sizes because the model directly enumerates them. But enumerating dependencies of longer distances is unfeasible due to the exponential growth of parameters it would require.

D. Klakow—The work was supported by the Cluster of Excellence for Multimodal Computing and Interaction, the German Research Foundation (DFG) as part of SFB 1102 and the EU FP7 Metalogue project (grant agreement number: 611073).

© Springer International Publishing Switzerland 2016
P. Sojka et al. (Eds.): TSD 2016, LNAI 9924, pp. 343–351, 2016.
DOI: 10.1007/978-3-319-45510-5_39

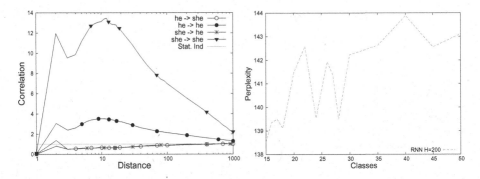

Fig. 1. Variation of word triggering correlations for pronouns over large distances

Fig. 2. Variation of perplexity against the number of classes for a RNNLM with 200 hidden nodes

To quantify information in dependencies of long distances, we use a variant of pointwise mutual information. Specifically, for a given pair of words (w_1, w_2) separated over a distance d, we examine the ratio of the actual co-occurrence rate to the statistically predicted co-occurrence rate: $c_d(w_1, w_2) = \frac{P_d(w_1, w_2)}{P(w_1)P(w_2)}$. A value greater than 1 shows it is more likely that the word w_2 follows w_1 at a distance d than otherwise expected according to the unigram frequencies of the two words. In Fig. 1, we show an example variation of this correlation for pronouns with the distance d on the English Gigawords corpus [5].

In this corpus, seeing another "she" about twenty words after seeing a first "she" is more than 13 times more likely than seeing a "she" in general. A similar, but interestingly weaker, observation can be made for the word "he". Note also that "she" somewhat suppresses "he" and vice versa, and these cross-correlations, although negative, are still informative for a prediction system. In summary, Fig. 1 demonstrates that plenty of word triggering information is spread out over long distance dependencies that is typically beyond the reach of N-gram LMs.

Several models, such as the cache-based LM [8], Skip models [6,11], and recurrent neural network language models (RNNLMs) [10] have been proposed to capture triggering in large contexts, but they usually only handle auto-triggering and/or have too many parameters to scale with vocabulary size. In this paper, we develop a novel modelling scheme, the Custom Decay Language Model (CDLM), which is specifically built to capture long range dependencies while growth in number of parameters remains sub-linear in vocabulary size. CDLMs outperform large-context-size Skip models, which are not constrained this way. Additionally, CDLMs show a more robust variation of performance metric against the variation of meta-parameters than RNNLMs, and they allow us to study the sparseness of word representations over different context sizes.

In the rest of the paper, we first briefly describe Skip models and RNNLMs and their limitations in Sect. 2, leading up to the detailed description of our new modelling technique in Sect. 2.3. We then set up experiments to analyze performance of these models in Sect. 3. Section 4 gives a robustness analysis of our model in addition to perplexity results for comparing the performance of various LM types and finally, Sect. 5 gives some concluding remarks.

2 Language Models

In this section, we first briefly describe and outline the numbers of parameters needed by Skip models and RNNLMs for handling long range dependencies. We then describe our novel CDLM which has been designed to overcome the limitations of skip models by reducing the number of parameters.

2.1 Skip Models

Skip models enumerate dependencies like N-grams, but allow wildcards (skips) at specific positions. This technique in combination with distance-specific smoothing methods spans larger context sizes and reduces the sparseness problem. However, the number of parameters still grow by $O(V^2)$ (where V is the vocabulary size) each time the context size is increased by one, making them computationally inefficient. In addition, the skip modeling framework lacks representational power when compared to neural network based LMs.

For our experiments, we build skip models by combining unified-smoothing trigrams and distance bigrams, which extend the range. Previously, such a combination has been shown to outperform state-of-the-art smoothed N-gram LMs [12].

2.2 RNNLMs

RNNLMs provide impressive performance gains when compared to other state-of-the-art language models. Through recurrence, the context size for these models is essentially infinite, or at least, formally unconstrained. This makes them especially suitable for long range dependencies. However, training RNNLMs can be slow, especially because the output must be normalized for each word in the vocabulary. Hierarchical softmax and related procedures that involve decomposing the output layer into classes can help with this normalization [4]. Unfortunately, using classes for normalization complicates the training process, since it creates a particularly volatile metaparameter. This can be observed in Fig. 2, where even for small variation in classes, RNNLMs show unstable variation in perplexity.

In our experiments, we employ a widely used class-based RNNLM implementation [10] builds networks that require $H^2 + 2HV + HC$ parameters, where H is the number of hidden units and C is the number of normalization classes. To produce better RNNLMs, we can increase the hidden layer size by one which in turn increases the number of parameters linearly in vocabulary size ($O(V + H + C)$).

2.3   Custom Decay Language Models

Our new modelling scheme was inspired by log-linear language models, which are characterized by sub-linear growth in the number of parameters with context size [7]. This model consists of two parts: a log-linear model and an N-gram model. For a history of size M, the N-gram part looks at the first $N - 1$ ($N < M$) predecessor words and the log-linear part captures the triggering information stored in distances d in the range $[N, M]$. Given the string of words $\{w_{i-M+2}, \cdots, w_{i-1}, w_i, w_{i+1}\}$ where $h = \{w_{i-M+2}, \cdots, w_i\}$, and supposing that $N = 3$, CDLM can be defined as:

$$P(w_{i+1}|h_i) = \frac{1}{Z(h_i)} \times P_{3\text{-}gram}(w_{i+1}|w_{i-1}, w_i)$$

$$\times e^{(E^{w_{i+1}} v_{w_{i-2}} + \sum_{k=i-N+2}^{i-3} E^{w_{i+1}} T_k v_{w_k})} \tag{1}$$

where i is the position in the document, $P_{3\text{-}gram}$ is a standard trigram LM and v_{w_k} is the vector representation of the word at a distance k from the word to be predicted in a C-dimensional, continuous, dense latent space ($C < V$). Here, the dimensions of C can be understood as "classes" capturing latent semantic information in the data.

E^{w_i} refers to a column of the emission matrix E, which weighs the word vectors v_{w_k} to predict the next word. Such a matrix can be thought of as an interpretation function for the current latent state of the model. These latent states exist in the same space as the word vectors. Presumably, some words are closer to this state than others. In this way, the latent states represent semantic concepts that the E matrix can translate into words.

The model also includes a distance specific transition matrix T_k to take word vectors from one distance-based latent space to another. More directly, the T_k matrices control the decay of each word within the latent state. Since the T_k are matrices, as opposed to scalars, which would provide a uniform decay, and as opposed to vectors which would provide a class-based decay, the shape of the decay function is *custom* to each word, which is why this model is named the Custom Decay Language Model.

This setup allows the model to constrain the number of parameters, as each time a word is added to the latent state, only the T_k matrix needs to be updated. Apart from the $O(V^3)$ parameters required to construct the trigram, it needs $O(VC)$ parameters to train the E matrix and the word vectors v_{w_k}, and it needs $O(C^2)$ parameters for training the T_k matrices. In all, CDLM parameters increase sub-linearly with V.

As shown in the last line of Eq. 1, the model log-linearly combines $T_k v_{w_k}$ at each context position to form a long-distance predictor of the next word. This approach, though inspired by skip models, is more customizable as it allows the exponent parameters to include matrix based formulations and not be constrained only to single values like skip models. Though the exponential element captures the latent/topical information well, the effects are too subtle to capture many simple short-distance dependencies (sparse sequential details). In order to make the model richer in sparse sequential details, we log-linearly combine the long-distance component with an N-gram LM.

In order to estimate the parameters E, v_{w_k} and T_k, we use the stochastic gradient descent algorithm and minimize the training perplexity of CDLM.

3 Language Modeling Experiments

3.1 Corpus

We trained and evaluated the LMs on the Penn Treebank as preprocessed in [1]. We used the traditional divisions of the corpus: sections 0–20 for training (925K tokens), sections 21–22 for parameter setting (development: 73K tokens), and sections 23–24 for testing (82K tokens). Despite its final vocabulary of 9,997 words and overall small size, this particular version has become a standard for evaluating perplexities of novel language models [2,9]. The small size makes training and testing faster, but also makes demonstrating differences in performance more difficult. We expect our results would scale for larger datasets.

3.2 Experimental Setup

In our experiments, we use perplexity as the performance metric to compare the language modelling techniques described in this paper.

In order to establish the most competitive baselines, the RNNLMs trained in our experiment were optimized for number of classes. Recall that these classes just aid the normalization process, as opposed to CDLM classes, which form a very integral part of the model. If classes were overhauled from the RNNLM altogether, training would take much longer, but the perplexity results would be slightly lower. We found that 15 classes optimized perplexity values for RNNLMs with 50 and 145 hidden nodes, and 18 classes optimized perplexity values for RNNLMs with 500 nodes. These models were trained using the freely available RNNLM toolkit, version 0.4b, with the `-rand-seed 1` and `-bptt 3` arguments.

The N-gram models used were trained with SRILM. They were a unified smoothing trigram (*UniSt*) and an interpolated modified Kneser-Ney 5-gram (*KN*). The *KN* model was trained with the following arguments: `-order 5 -gt2min 1 -gt4min 1 -gt3min 1 -kndiscount -interpolate`.

CDLM uses the unified-smoothing trigram as the short-distance dependency component of its model and the long-distance (exponential) element of the model considers up to five words after the trigram.

The learning rate (η) adaptation scheme is managed by the adaptive gradient methods [3]. After optimizing on the development set, η was fixed to 0.1 and the dimensionality of the latent space C was fixed at 45.

While building CDLMs, we first trained a CDLM $M = 4$ and reused its constituent parameters E and v to build CDLM $M = 5$, only updating T_k while training. This process iterated up to $M = 8$.

4   Results and Discussion

4.1   CDLM Robustness Analysis

CDLM shows robust variation of perplexity with changes in classes, as shown in Fig. 3. The perplexity values decrease monotonically with increasing classes, as expected since each increase in class creates more parameters that can be tuned. Note that moving from $M = 4$ to $M = 5$ doubles the number of T_k matrices, which caused the large perplexity drop.

Along with the robustness shown by CDLM, the log-linear formulation of CDLM allows us to study and analyze the sparseness of the transformed word space matrices represented by $T_k v_{w_k}$ for different distances. We measure sparseness by counting the matrix entries below a given threshold. By this measure, a more sparse matrix will have large number of entries below the threshold than a less sparse matrix. We plot the variation of the sparseness for $T_k v_{w_k}$ matrices for different thresholds against the context size of CDLM in Fig. 4. In most cases, we observe that as the context size increases the transition matrices have fewer number of entries below the threshold making them less sparse. Therefore, we believe that this matrix formulation alleviates the sparseness problem and also allows the exponent part to capture latent information.

4.2   Perplexity Results

Table 1 presents our comparison of CDLM with different language models on the basis of their total numbers of parameters and their perplexities. As shown, skip models (*Skip*) outperform the unified smoothing trigram (*UniSt3*) as they have more parameters and hence, they better encode information spread over larger distances.

CDLM outperforms *UniSt3* because of spanning larger context size and greater number of parameters at its disposal. *CDLM45* also outperforms the

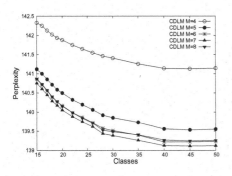
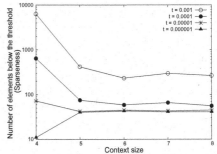

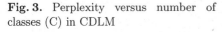

Fig. 3. Perplexity versus number of classes (C) in CDLM

Fig. 4. Sparseness of CDLM's transformed word space $(T_l v_{w_l})$ measured at different threshold (t) versus its context size

Table 1. Test set perplexity (PPL) and total number of parameters (PAR) for each language model (LM).

LM	Range	Hidden	PPL	PAR
UniSt	3	-	162.1	2.0M
	4		160.0	
	5		155.8	
Skip	6	-	154.4	4.1M
	7		153.6	
	8		153.2	
KN	5	-	141.8	3.2M
		50	156.5	1.0M
RNNLM	∞	145	139.3	2.9M
		500	136.6	10.3M

LM	Range	Hidden	PPL	PAR
	4		141.1	
	5		139.5	
CDLM	6	45	139.2	2.9M
	7		139.1	
	8		139.2	
	5 + 4		137.2	
	5 + 5		135.7	
KN+CDLM	5 + 6	45	135.2	6.1M
	5 + 7		134.9	
	5 + 8		134.9	
KN+RNNLM	5 + ∞	50	120.3	4.2M
CDLM+RNNLM	7 + ∞	45 + 50	120.2	3.9M

Skip models. In fact, increasing the context size of *Skip* to eight words obtains a perplexity of 153.2, which is still less than the CDLM perplexity of 141.1 for a context size of four words. Also, *Skip* requires 4.1 million parameters which is more than a third greater than those required to build the seven-word CDLM. Also, *CDLM* is able to perform better than *KN* with fewer number of parameters. When combining *CDLM* with *KN*, increasing the context size for CDLM obtains progressively-better performance than *KN*. This is due to more number of parameters in CDLM formulation.

We further compare CDLMs with RNNLMs. An RNNLM with 145 hidden nodes has about the same number of parameters as *CDLM* and performs 0.1 perplexity points worse than *CDLM*. Increasing the hidden units for RNNLM to 500, we obtain the best performing RNNLM. This comes at a cost of using a lot of parameters. To produce better performing LMs with fewer parameters we constructed an RNNLM with 50 hidden units, which when linearly combined with *CDLM* (*CDLM+RNNLM*) outperforms the best *RNNLM* using less than half as many parameters. It even nominally outperforms the combination of *KN* and *RNNLM* using fewer parameters, but this difference is likely not significant. Combinations of the three different LMs do not result in any large improvements, suggesting that there is redundancy in the information spread over these three types of LMs.

Finally, we note that the increase in parameters does not always lead to better performance. We observe this increase while comparing a *Skip* model with *CDLM* and this increase can be attributed to the richer formulation of *CDLM*. Increase in parameters for *CDLM+KN* does not also lead to a better performance against the fewer-parameters *CDLM+RNNLM*. This is also observed when we compare, *CDLM+KN* and *KN+RNNLM*. In this case, we suspect that the lower performance is due to CDLM's lack of recursive connections which form an integral part of RNNLMs. But note that CDLMs, which are not recurrent, can capture much of the long-distance information that the recurrent language models can.

5 Conclusion

In this paper, we proposed Custom Decay Language Model, inspired by Skip models' log-linear technique of dividing context into smaller bigrams and then recombining them. In contrast with Skip models, CDLM uses a richer formulation by employing a matrix based exponentiation method to capture long range dependencies. Additionally, CDLM model uses an N-gram model to capture the short range regularities.

Perplexity improvements are observed for CDLM even when compared to Skip models with larger range and Kneser Ney five-grams. This improvement is observed even though CDLM uses fewer parameters compared to larger range Skip model and $KN5$ with more parameters. In conclusion, CDLM provides a rich formulation for language modeling where the growth of number of parameters is constrained. We look forward to further enhancing CDLM with recurrent connections and analyzing its performance on other language datasets with a focus on ASR tasks.

References

1. Charniak, E.: Immediate-head parsing for language models. In: Proceedings of 39th Annual Meeting of the Association for Computational Linguistics, pp. 124–131, Toulouse, France, July 2001
2. Cheng, W.C., Kok, S., Pham, H.V., Chieu, H.L., Chai, K.M.A.: Language modeling with sum-product networks. In: 15th Annual Conference of the International Speech Communication Association (INTERSPEECH), pp. 2098–2102, Singapore, September 2014
3. Duchi, J., Hazan, E., Singer, Y.: Adaptive subgradient methods for online learning and stochastic optimization. J. Mach. Learn. Res. **12**, 2121–2159 (2011)
4. Goodman, J.T.: Classes for fast maximum entropy training. In: 2001 IEEE International Conference on Acoustics, Speech, and Signal Processing (ICASSP), vol. 1, pp. 561–564. IEEE, Salt Lake City, May 2001
5. Graff, D., Cieri, C.: English gigaword LDC2003t05. Web Download. Linguistic Data Consortium, Philadelphia (2003)
6. Guthrie, D., Allison, B., Liu, W., Guthrie, L., Wilks, Y.: A closer look at skip-gram modelling. In: Proceedings of the 5th International Conference on Language Resources and Evaluation (LREC), pp. 1222–1225 (2006)
7. Klakow, D.: Log-linear interpolation of language models. In: Fifth International Conference on Spoken Language Processing (ICSLP), pp. 1695–1698 (1998)
8. Kuhn, R., De Mori, R.: A cache-based natural language model for speech recognition. IEEE Trans. Pattern Anal. Mach. Intell. **12**(6), 570–583 (1990)
9. Mikolov, T., Karafiát, M., Burget, L., Černocký, J., Khudanpur, S.: Recurrent neural network based language model. In: 11th Annual Conference of the International Speech Communication Association (INTERSPEECH), pp. 1045–1048, Makuhari, Japan, September 2010

10. Mikolov, T., Kombrink, S., Burget, L., Černocký, J., Khudanpur, S.: Extensions of recurrent neural network language model. In: 2011 IEEE International Conference on Acoustics, Speech, and Signal Processing (ICASSP), pp. 5528–5531, Prague, Czech Republic (2011). http://dx.doi.org/10.1109/ICASSp.2011.5947611
11. Momtazi, S., Faubel, F., Klakow, D.: Within and across sentence boundary language model. In: 11th Annual Conference of the International Speech Communication Association (INTERSPEECH), pp. 1800–1803, Makuhari, Japan, September 2010
12. Singh, M., Klakow, D.: Comparing RNNs and log-linear interpolation of improved skip-model on four babel languages: Cantonese, Pashto, Tagalog, Turkish. In: 2013 IEEE International Conference on Acoustics, Speech and Signal Processing (ICASSP), pp. 8416–8420, May 2013

Voice Activity Detector (VAD) Based on Long-Term Mel Frequency Band Features

Sergey Salishev[1], Andrey Barabanov[1], Daniil Kocharov[1], Pavel Skrelin[1], and Mikhail Moiseev[2(✉)]

[1] Saint-Petersburg State University, Saint-Petersburg, Russia
[2] Intel Labs, Intel Corporation, Santa Clara, CA 95054-1549, USA
mikhail.moiseev@intel.com

Abstract. We propose a VAD using long-term 200 ms Mel frequency band statistics, auditory masking, and a pre-trained two level decision tree ensemble based classifier, which allows capturing syllable level structure of speech and discriminating it from common noises. Proposed algorithm demonstrates on the test dataset almost 100 % acceptance of clear voice for English, Chinese, Russian, and Polish speech and 100 % rejection of stationary noises independently of loudness. The algorithm is aimed to be used as a trigger for ASR. It reuses short-term FFT analysis (STFFT) from ASR frontend with additional 2 KB memory and 15 % complexity overhead.

Keywords: Voice Activity Detector · Classification · Decision tree ensemble · Auditory masking

1 Introduction

The problem of low complexity accurate VAD is important for many applications in Consumer Electronics, Wearables, Smart Home and other areas, where VAD serves as a low-power gatekeeper for a more complex and energy consuming Automatic Speech Recognition (ASR) system. Existing approaches vary by maximum allowed latency. For the telecom and VoIP applications, typical latency is 10–30 ms. Existing solutions, e.g. Sohn et al. [12], Graf et al. [8] exploit various statistical models, e.g. GMM, SVM, and multiple short-term features: energy, energy envelope, pitch, multi-band peak-factors, and formants. Small latency makes VAD highly sensitive to non-stationary noises. For ASR applications, VAD latency is chosen to allow 200–300 ms total recognition latency of command/sentence. Ramirez et al. [11] use latencies up to 100 ms, energy envelope divergence statistics and demonstrate substantial improvement over more complex VADs with shorter statistics on non-stationary noises.

Still 100 ms latency does not allow a VAD to capture the syllable level structure of speech, which leads to missing onset and coda of syllables especially in presence of fricatives and corruption of speech in low SNR noise.

© Springer International Publishing Switzerland 2016
P. Sojka et al. (Eds.): TSD 2016, LNAI 9924, pp. 352–358, 2016.
DOI: 10.1007/978-3-319-45510-5_40

We propose to use long-term 200 ms speech statistics in combination with a pre-trained complex non-linear classifier, which allows capturing syllable level structure of speech and distinguishing it from common noises. Proposed algorithm substantially outperforms competitive solutions in different non-stationary noises and demonstrates on the test dataset almost 100 % acceptance of clear voice and 100 % rejection of stationary noises at the cost of higher latency. The algorithm reuses short-term FFT analysis (STFFT) in ASR front-end; therefore, the complexity increase to MFCC ASR front-end is small.

2   VAD Algorithm Description

The algorithm (Fig. 1) consists of feature extraction, feature space dimensionality reduction and two level classifier. It uses Mel band spectral envelope and Mel band peak factor as features.

Spectral envelope is a standard ASR feature usually manifesting as Mel Frequency Cepstrum Coefficients (MFCC) or Linear Prediction Coefficients (LPC). According to the acoustic theory of speech production by Gunnar Fant [10], the harmonics of fundamental frequency (pitch) contain most speech energy, which makes them robust to noise due to high SNR, it distinguishes speech from most types of noise. Spectral envelope does not capture pitch; therefore, we added band peak factors, which are ratios of peak to mean. We do not estimate pitch due to complexity and sensitivity to noise. To improve noise robustness, tonal and temporal auditory maskings [6] are applied to spectral envelope. Features form a column per 20 ms frame, 50 % overlap. Columns are aggregated by sliding 19 frame window forming 26×19 feature-space. Then scores are calculated only for the most important PCA components.

Features are classified by a soft classifier using an ensemble of deep decision trees [13].

$$H_0(f) = \sum_{i=1}^{N} \omega_i h_i(f) \approx P(v^*|f),$$

where H_0 – output of the first-level classifier, N – number of trees in the ensemble, ω_i – soft voting classifier weights, h_i – deep decision tree soft classifier outputs, v^* – binary ground truth value, $v^* = 1$, if features f correspond to whole speech window.

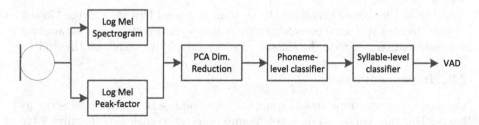

Fig. 1. VAD overview

As a window is shorter than a syllable, the first level classifier tends to generate false positives on noise signal attacks similar to the syllable onsets containing only unvoiced consonants and to skip unvoiced codas. In addition, onsets and codas are more sensitive to noise due to lower loudness, and therefore lower SNR. To improve this, the second-level binary classifier is used.

$$X = \alpha X^* + (1 - \alpha)H_0(f)$$
$$Y = \max(Y^* - \beta, X)$$
$$H_1 = \begin{cases} 1, Y \geq T \\ 0, Y < T \end{cases}$$

where H_1 - second classifier output, X^*, Y^* - values on previous frame. The second-level classifier design was influenced by Nuance SREC VAD [3] decision rule. X part is a smoothing IIR filter on first-classifier output. It handles fricatives in syllable onset allowing a higher threshold due to accumulation. Part of the fricative is skipped, but it is not important as long as transition between fricative and voiced phoneme is covered. Y part is a latch with hangover. It snaps on vowel and covers the typical length of syllable coda. Parameters α, T determine the minimal length of false alarm, β determines the length of the shadow cast by a vowel.

2.1   Classifier Training

For classifier training we used database of continuous English speech TIMIT [7]; noise databases Aurora 2 [9], ETSI [4], and SISEC10 [5].

For the algorithm to work, we needed to select PCA components, decision tree depth, training algorithm, number of decision trees. To evaluate PCA component importance, we performed a preliminary round of deep tree classifier training and arranged PCA components by predictor importance.

For compact decision tree implementation, we put additional constraint on memory. By resubmission loss criteria we chose 4 weight quantization with 32 PCA components meaning 9 bit per tree node.

To choose the number of trees in ensemble, we manually selected a representative subset of noise files not used in training. Then we evaluated all combinations of algorithms and tree numbers up to 64 trees. We used Ada boost, Logit boost, Gentle boost, and Bagging training algorithm. The Logit boost with 16 trees provides the best results.

To train the second classifier, all the training sound files were merged into a continuous file with guard pauses between sentences in speech. T, α, β parameters were selected to minimize the classifier loss by exhaustive search on the grid.

2.2   Implementation

Decision trees were represented using balanced binary trees packed in array as heaps. This representation does not require pointers so each node requires 9 bit of memory. As a result the total memory size of the first-level classifier is about

1KB. With the quantized features the memory size of feature 200 ms feature buffer is 0.5 KB. PCA table coefficients occupy additional 0.1 KB. So the total memory overhead is less than 2 KB.

Total computational complexity overhead of VAD compared to ASR front-end is about 15 %.

3   Comparison

For comparison, we used two state of the art VADs: Google WebRTC VAD [2] which has default latency of 10 ms and uses multiple frequency band features with pre-trained GMM classifier; Nuance SREC VAD [3] which has latency 70 ms and analyses energy envelope statistics. For testing, we used sound files completely unused in training.

We separately performed False Accept testing on noise database and False Reject testing on speech database with various noise levels.

3.1   False Accept Testing

For false accept test, we used DEMAND [1] database containing background noises for 18 environments. Noises 'dliving', 'npark', 'pcafeter', 'presto', 'scafe', 'spsquare', 'tbus', 'tmetro' contain minor amount of comprehensible speech and babble noise. Noise 'omeeting' consist of intelligible speech. Comparison is shown in Table 1. We conclude that proposed algorithm performs well for stationary and

Table 1. False accept rate comparison in % for various enviromental noises

Noise	SREC	WebRTC	Proposed
dkitchen	11.7	12.4	10.9
dliving	30.7	90.3	4.5
dwashing	20.9	84.5	5
nfield	0	74.1	0
npark	48.1	30	4.6
nriver	0.5	15.3	0
ohallway	23.4	15.4	2.7
omeeting	71.7	67.8	78.3
ooffice	28.6	20.8	0.3
pcafeter	77.4	80.9	38.3
presto	42.8	83.2	1.6
pstation	1	100	0
scafe	75.2	89.7	22.1
spsquare	31.4	73.3	11.8
straffic	10.6	82.9	0
tbus	67.1	77.2	30.4
tcar	1.2	95.5	0
tmetro	27.5	89.1	14.3

non-stationary wideband noises and on babble noise except for 'pcafeter', which contains distinguishable vowels in speech and clutter of dishes. The most complicated noises in the set are 'dkitchen' containing clutter of dishes and 'dwashing' containing washing machine noise and sound of clothes' buttons clinking the glass. On 'dkitchen' all VADs performed similarly. We conclude that the new VAD outperforms competitors.

We tested false accept rate on 3 tracks of Rock, Pop, and Classic music genres not used in training (Table 2). We conclude that new algorithm substantially outperforms competitors, still false accept on music is about 20 %. The algorithm falsely accepts solo instruments and specifically guitar riffs and piano chords.

Table 2. False accept rate comparison in % for music

Noise	SREC	WebRTC	Proposed
rock	91	97.6	11.1
pop	88.8	82.4	19.7
classic	91	90.7	18.1

3.2 False Reject Testing

For false reject testing, we used speech database of 5 min recordings in 4 languages (Chinese, English, Polish, Russian), male and female speakers for each language with manual VAD markup. Noise was synthetically added to with various SNRs calculated as total speech to total noise power after high-pass filter with 100 Hz cutoff. We conclude that the new VAD is highly accurate and language and speaker insensitive for high SNR (up to 10 dB). We tested with various noises (Table 3):

Table 3. Proposed VAD false reject rate in % for various enviromental noises and SNRs

Language	Sex	tcar			nriver			presto		
		20 dB	10 dB	0 dB	20 dB	10 dB	0 dB	20 dB	10 dB	0 dB
Chinese	f	0.1	0	0	0	1.4	5.1	0	27.6	40.8
Chinese	m	0	0	0	0	0.4	0.1	0	8.8	51.7
English	f	0	0	0	0	1.6	0.7	0	25.6	54.7
English	m	0.2	0.2	0.2	0.2	3.7	3.3	0.9	23.1	37.2
Polish	f	0.5	0.7	0.5	0.6	5.1	5.8	1.6	41.1	53
Polish	m	0.2	0.3	0.4	0.3	7.6	11.3	1.3	42.3	57.4
Russian	f	0.1	0.3	0	0.2	4	6.1	0.5	31.1	63.6
Russian	m	0.2	0.2	0.3	0.2	3.1	3.1	0.1	20.4	65.8
Mean		0.2	0.2	0.2	0.2	3.4	4.4	0.6	27.5	53

- 'tcar' - stationary wide-band noise;
- 'nriver' - non-stationary wide-band noise;
- 'presto' - non-stationary babble noise.

The new VAD algorithm is highly accurate in car noise with FAR about 1 %
at SNR 0 dB. For non-stationary noises, it demonstrates similar performance
up to SNR 10 dB and degrades for lower SNR on babble noise. This correlates
with subjective intelligibility of the speech.

We compared new VAD false reject rate with other algorithms (Table 4). The
new VAD outperforms the SREC VAD on all SNR at least by 50 %, except for
'presto' 5 dB and lower SNR. It also outperforms WebRTC at least by 50 %
for high SNR (up to 10 dB) and performs on pair for lower SNR, except for
'presto' 5 dB and lower SNR. I may seem that SREC and WebRTC outperform
on 'presto' at low SNR, but it comes at the cost of false accept rate of 43 % and
83 % (Table 1).

Table 4. False reject rate comparison in % for various enviromental noises and SNRs

Noise	VAD	Inf	20 dB	15 dB	10 dB	5 dB	0 dB
tcar	srec	0.9	1.3	1.6	1.9	2.2	2.5
	webrtc	1.1	1.3	1.3	1.2	0.8	0.4
	proposed	0.1	0.2	0.1	0.2	0.3	0.6
nriver	srec	0.9	3.5	5.8	11.3	23.4	54.2
	webrtc	1.1	3.1	4.7	7.1	12	22.8
	proposed	0.1	0.2	0.8	3.4	9.8	27.5
presto	srec	0.9	2	2.6	4.3	9.2	21.4
	webrtc	1.1	3.3	4.9	5.9	4.8	1.8
	proposed	0.1	0.2	0.9	4.4	19.8	53

4   Conclusion

The proposed algorithm substantially outperforms competitive solutions in various environments and demonstrates on the test dataset almost 100 % acceptance
of clear voice and 100 % rejection of stationary noises with 15 % complexity
increase compared to MFCC based ASR front-end. The algorithm has a latency
of 200 ms, which is not acceptable for some scenarios such as VoIP. Th algorithm
in some cases falsely accepts some noises as voice: clatter of dishes; sound of flow-
ing water; resonant strokes; tonal beeps; babble noise; bird songs. The algorithm
falsely rejects speech in the presence of high amplitude non-stationary noise
especially babble noise.

References

1. Demand: Diverse environments multichannel acoustic noise database. http://parole.loria.fr/DEMAND/. Accessed 20 Mar 2016
2. Google WebRTC. https://webrtc.org/. Accessed 20 Mar 2016
3. Nuance SREC. https://android.googlesource.com/platform/frameworks/base/+/android-4.4_r1/core/java/android/speech/srec/Recognizer.java. Accessed 20 Mar 2016
4. Tsi EG 202 396–1 speech, multimedia transmission quality (STQ); part 1: Background noise simulation technique and background noise database, March 2009
5. Source separation in the presence of real-world background noise: test database for 2 channels case (2010). http://www.irisa.fr/metiss/SiSEC10/noise/SiSEC2010_diffuse_noise_2ch.html. Accessed 20 Mar 2016
6. Fastl, H., Zwicker, E.: Psychoacoustics: Facts and Models, vol. 22. Springer, Heidelberg (2006)
7. Garofolo, J.S., Lamel, L.F., Fisher, W.M., Fiscus, J.G., Pallett, D.S.: DARPA TIMIT acoustic-phonetic continous speech corpus CD-ROM. NIST speech disc 1–1.1. NASA STI/Recon Technical Report N 93 (1993)
8. Graf, S., Herbig, T., Buck, M., Schmidt, G.: Features for voice activity detection: a comparative analysis. EURASIP J. Adv. Signal Process. **2015**(1), 1–15 (2015)
9. Hirsch, H.G., Pearce, D.: The Aurora experimental framework for the performance evaluation of speech recognition systems under noisy conditions. In: ASR2000-Automatic Speech Recognition: Challenges for the new Millenium ISCA Tutorial and Research Workshop (ITRW) (2000)
10. Fant, G.: Acoustic Theory of Speech Production: with Calculations based on X-Ray Studies of Russian Articulations. Description and Analysis of Contemporary Standard Russian. De Gruyter (1971). ISBN: 9783110873429. https://books.google.ru/books?id=UY0iAAAAQBAJ
11. Ramırez, J., Segura, J.C., Benıtez, C., De La Torre, A., Rubio, A.: Efficient voice activity detection algorithms using long-term speech information. Speech Commun. **42**(3), 271–287 (2004)
12. Sohn, J., Kim, N.S., Sung, W.: A statistical model-based voice activity detection. IEEE Signal Process. Lett. **6**(1), 1–3 (1999)
13. Zhou, Z.H.: Ensemble Methods: Foundations and Algorithms. CRC Press, Florida (2012)

Difficulties with Wh-Questions in Czech TTS System

Markéta Jůzová[⊠] and Daniel Tihelka

Departement of Cybernetics and New Technologies for the Information Society,
University of West Bohemia, Univerzitní 8, Plzeň, Czech Republic
juzova@kky.zcu.cz, dtihelka@ntis.zcu.cz

Abstract. The sentence intonation is very important for differentiation
of sentence types (declarative sentences, questions, etc.), especially in
languages without fixed word order. Thus, it is very important to deal
with that also in text-to-speech systems. This paper concerns the problem of wh-question, where its intonation differs from the intonation of
another basic question type – yes/no question. We discuss the possibility
to use wh-questions (recorded during the speech corpus preparation) in
speech synthesis. The inclusion and appropriate usage of these recordings
is tested in a real text-to-speech system and evaluated by listening tests.
Furthermore, we focus on the problem of the perception of wh-question
by listeners, with the aim to reveal whether listeners really prefer phonologically correct (falling) intonation in this type of questions.

Keywords: Speech synthesis · *wh*-question · Prosody · Prosodic clause ·
Prosodic phrase · Prosodic word · Intonation · Prosodeme

1 Introduction

In linguistics, intonation means variation of spoken pitch (also *fundamental frequency*, $f0$) and it is one of the basics of prosodic language, called supraseg-
mental phenomena. Besides intonation, the other prosodic variables are *tempo*,
stress (intensity) and *rhythm*. This phenomenon is not used to distinguish words,
except a few tonal languages, but at sentence level, it is used e.g. for differentiation between statements and questions, and between different types of questions,
indicating attitudes and emotions of the speaker, etc.

According to [1,2], there are two main and two secondary intonation patterns:

- *rising intonation* – the pitch of the voice rises over time,
- *falling intonation* – the pitch falls down,
- *dipping (fall-rise) intonation* – the pitch falls and then rises,
- *peaking (rise-fall) intonation* – the pitch rises and then falls.

M. Jůzová—This research was supported by Ministry of Education, Youth and
Sports of the Czech Republic, project No. LO1506, and by the grant of the University of West Bohemia, project No. SGS-2016-039.

© Springer International Publishing Switzerland 2016
P. Sojka et al. (Eds.): TSD 2016, LNAI 9924, pp. 359–366, 2016.
DOI: 10.1007/978-3-319-45510-5_41

The intonation patterns are applied on every *prosodic clause* (*PC*) in a sentence which is a linear unit of prosodic sentence delimited by pause. As defined in [3], prosodic clauses can consist of more *prosodic phrases* (*PP*), where a certain intonation scheme is realized continuously. *PP* is further divided to *prosodic words* (*PW*), a group of words subordinated to one word accent.

Since the Czech has not a fixed word order, the sentence intonation is an important factor which helps the listeners to distinguish the sentence type. It is sometimes the only factor, especially for the recognition of declarative sentences and *yes/no* questions ([2], Chap. 8). While the common intonation progress in declarative sentences (those ending with a full stop sign) is evident in neutral speech[1] – the falling intonation is used, there are more types of sentences ending with a question mark regarding the intonation. There are two main types of questions – yes/no questions and *wh*-questions. First of them, as the name indicates, expects the answer "yes" or "no". In Czech, the only difference from a declarative sentence[2] is a noticeable raising intonation in the last prosodic word of the sentence. On the other hand, *wh*-questions, which contain any of question words (what, who, why, when, etc.) are characterized by the peaking intonation at the beginning and the falling intonation at the end of the sentence. Figure 1 contains three different sentences represented by a schematic representation of intonation adopted from [4]. Sometimes, mainly in the heat of passion, there can be an intonation disruption and the listener (even native speakers) does not know what type of sentence is involved.

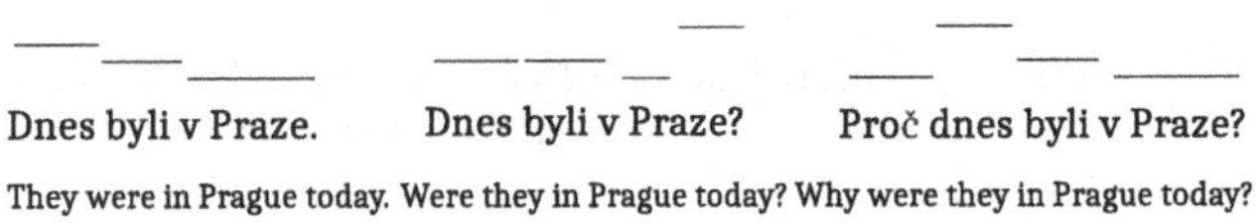

Dnes byli v Praze.   Dnes byli v Praze?   Proč dnes byli v Praze?

They were in Prague today. Were they in Prague today? Why were they in Prague today?

Fig. 1. Schematic representation of intonation comparing a declarative sentence, yes/no question and a *wh*-question.

2 Current State of Our TTS System

The current version of our text-to-speech (TTS) system *ARTIC* [5] uses all three widespread speech synthesis methods – *HMM, single unit instance* and *unit selection*. This paper will focus only on unit selection, a method of concatenative speech synthesis in which the optimal sequence of speech unit candidates is searched for the given input text. Generally, after decomposition of the text into

---

[1] Note that only neutral speech is taken into account in this paper since our TTS system does not currently involve emotions.

[2] The important communication function in (complex/compound) declarative sentences is manifested with falling intonation only in the last phrase. Therefore, we sometimes substitute a term *declarative sentence* with a term *declarative phrase*.

speech units, candidates of these units stored in the speech units database are used to build the graph being evaluated by *target cost* (nodes) and *concatenation cost* (edges), and the Viterbi algorithm is used for optimal path search, see [6].

One of the features used for the *target cost* computation, besides contextual and positional features (see [7]), is a symbolic prosodic feature, called *prosodeme* [3]. This abstract unit corresponds to an important prosodic phenomenon, a certain communication function within the language system. In [3], the authors distinguish 4 basic prosodeme types which are also used in our TTS system:

- *prosodeme terminating satisfactory* (no reply is expected) - *P1*
    - it corresponds mainly to declarative phrases
- *prosodeme terminating unsatisfactory* (a reply is expected) - *P2*
    - it refers to questions
- *nonterminating prosodeme* - *P3*
    - it corresponds to phrase not terminating a sentence (i.g. before a comma)
- *null prosodeme* - *P0*
    - it does not represent any distinguishable communication function

When assigning prosodemes to units in our speech corpus, it is performed from the text data after segmentation process:

- The last *PW* in final *PP* ending with a full stop is tied with *P1*.
- The last *PW* in final *PP* ending with a question mark is tied with *P2*.
- Last *PWs* in other *PPs* (before a pause) are tied with *P3*.
- Units in all other prosodic words are marked as *P0*.

The situation is illustrated on Fig. 2. The consequence of this labeling is that, on the surface level, *P2* is manifested with raising intonation, *P1* is manifested with falling intonation etc.

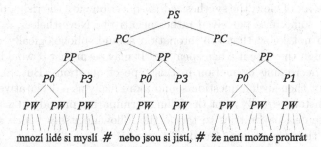

Fig. 2. The illustration of the tree build using the prosodic grammar for the Czech sentence "Many people think or are convinced that it is impossible to lose"; taken from [8] (symbol # means a pause).

Although the corpora are usually recorded by professional speakers, there can be some inconsistencies between aligned prosodeme type (taken from the text only!) and the real intonation pattern pronounced by the speaker during the recording. This issue is still investigated and we are trying to find some acoustic similarities which would enable to automatically control the corpora and change some prosodemes if needed [9].

When a text is synthesized with our TTS system, it is divided to prosodic phrases (using commas and other punctuations instead of pauses) and each PW is labeled with a prosodeme type in the same way as described above. The described approach is quite simple, nevertheless, it ensures the coverage of basic intonational phenomena in speech. However, since the system does not distinguish between various type of questions, the terminal intonation of synthesized sentence is not always phonologically correct.

3 Problem Definition

The corpora created for the purpose of TTS systems are intentionally built to contain both declarative sentences and questions of different types to ensure the coverage of all speech units in all prosodeme types (see [10]). The current version of a speech corpus contains about 12 thousand sentences, 700 of them are wh-questions. These sentences, together with several tens of choice questions, were excluded from the final TTS system till now. The main purposes of the exclusion were simply "security" reasons – these sentences have more complex pitch course which has not yet been satisfactory described by the used symbolic prosodic labeling. This paper focuses on finding a way how to avoid dropping almost 6 % of the data and how to use the wh-question reliably in the system.

In addition, the way of synthesizing wh-questions with our TTS system is not phonologically correct – they are synthesized equally to yes/no questions, i.e. with the raising intonation. Phonetic experts in [4] state that, in a neutral speaking style in Czech, the intonation at the end of wh-questions should be falling. Also several addressed phoneticians agreed on the fact that, from the perception point of view, the synthesized wh-question with the rising terminating intonation is sometimes perceived rather negatively. Nevertheless, many people explicitly do not know that the intonation should, phonologically correct, be falling although they use it when speaking. It may seem that it would be correct to ensure a decreasing intonation in this type of questions. But, on the other hand, will not then such a question sound more likely as a declarative sentence? And do the listeners really want the falling terminating prosody in wh-question?

The presented paper tries to tackle the following issues which are further discussed in the subsections *I1* and *I2* (page 353).

I1: Does the process of adding wh-questions in the corpus influence the quality of synthesized speech?

First, we synthesized a huge number (hundreds of thousands) of various sentences with two TTS system - the baseline version not using wh-questions (TTS) and the enhanced one, with the corpus containing the wh-questions (TTS^{wh}). Let us note that prosody of all added wh-questions was labeled with the symbolic prosody marking as declarative sentences ($P1$ prosodemes on the last PW in final *PP*, *P3* prosodemes before pauses and *P0* otherwise). We could afford to do that because we analyzed the recordings from the corpus and found out that the professional speakers who recorded that speech corpus were strictly using the falling intonation all the time in these sentences. However, since we have divided our experiment with adding and synthesizing *wh*-questions in 2 issues, for this experiment we did not change the way of synthesizing questions in TTS system. And so, when synthesizing, all *wh*-questions are described with the same symbolic prosodic labeling as yes/no questions – *P2* prosodeme at the last *PW* which corresponds, in our TTS system, to raising intonation.

Having generated all synthesized sentences, we adapted the way of selecting sentences for the listening tests presented in [11]. In this method, the criterion for selection of a sentence pair was the number of differences between unit sequences of synthesized sentences. From the set of synthesized pairs, we randomly selected those with a high number of differences of used units (taking into account a number of units from *wh*-questions in the second sentence) and divided them into 3 groups which were first evaluated separately, then together: 20 declarative phrases, 12 short *wh*-questions (max. 3 words) and 20 longer *wh*-questions. These 3 groups were selected intentionally. First, we wanted to know whether speech units from *wh*-questions did not worsen the quality of synthesizing "normal" sentences. Second, we inspect how these units influence the quality of wh-question. Furthermore, the intonation course of short and longer wh-question can differ [12], so we split the remaining sentences into 2 other groups.

All the pairs were evaluated by 7 people, all of them were speech synthesis experts. Pairs of sentences were, one by one, presented to the listeners, who did not know which sentence was synthesized by TTS^{wh}. Listeners could listen to them repeatedly and after that they had to choose which one sounds better.

For the evaluating, we used *CCR tests* with a simple 3–point scale:

- 1 (E) – the output of (enhanced) TTS^{wh} sounds better,
- 0 (S) – the outputs are about the same,
- −1 (B) – the output of baseline TTS sounds better.

The quality is then defined as $q = \frac{1 \cdot E + (-1) \cdot B}{E+S+B}$, where E, S, B are the numbers of listeners' answers in the given category. The positive value of q means that the use of TTS^{wh} (with *wh*-questions added) brings the improvement of the overall quality of synthesized speech.

To ensure the validity of results, we also carried out the *sign test* with the null hypothesis H_0 : *"The outputs of the both TTS systems are of the same quality."* compared against the alternative hypothesis H_1 : *"The output of one TTS system sounds better."*. The *p*-values were determined for the significance level $\alpha = 0.01$. If $p < \alpha$, H_0 is rejected and H_1 is valid. The numbers of listeners' answers and the computed quality value q, as well as the computed *p*-values, are shown in the Table 1.

It is evident, that the process of including *wh*-questions in the speech corpus influences positively the quality of synthesized speech. The quality value q is positive for all evaluated groups except one containing declarative phrases – there is

Table 1. Comparison of the baseline version of TTS system and the enhanced version TTS^{wh} with wh-questions in the speech corpus.

	Declarative	Short wh	Longer wh	All sentences
TTS^{wh} better (E)	30	44	78	152
Same quality (S)	72	23	43	138
TTS better (B)	38	17	19	74
Quality value q	-0.057	0.321	0.421	0.214
p-value for $\alpha = 0.01$	0.3961	0.0007	< 0.0001	< 0.0001
Conclusion	The same quality	TTS^{wh} better	TTS^{wh} better	TTS^{wh} better

q slightly negative. But as the p-value shows, this negative result is not significant which means that the outputs of the baseline TTS and the enhanced TTS^{wh} are approximately of the same quality. To summarize, wh-questions, recorded during the speech corpus building, can be used in the TTS system, having the same prosodic labeling as declarative phrases.

I2: Should be the wh-questions really synthesized with the falling terminating intonation?

Once having added the recorded wh-questions in the speech corpus, we tried to find out what terminal intonation (raising/falling) of synthesized sentences sounds better to the listeners. As written in Sect. 1, the falling intonation is the only expected final intonation in wh-questions in neutral speaking style. But, as people often use this type of question in many situations, the perception may differ. Therefore, we have created second listening test to check the theoretical statement. Since our TTS system is used by many different people, we decided to spread out the test not only between speech synthesis experts, but also between naive listeners – non-experts (i.e. users, not developers of TTS systems).

For this listening test, we randomly selected 35 wh-questions and synthesized them using TTS^{wh}. Every sentence was generated twice – first with the prosodeme $P2$ assigned to the last PW (in the surface level realized by raising intonation), second with the prosodeme $P1$. The listeners, just as in the first listening test (Sect. 3, $I1$), did not know the order of the presented sentences in the pair and had to choose one of 3 choices described above. In addition, there was an optional possibility to write a comment on the current sentence pair in a text-field. The test was evaluated by 22 listeners, 11 of them being speech synthesis experts. The second half of the listeners did not know the purpose of the listening test (that it concerns wh-questions) and they were not also informed in advance about the phonological theory. The results are presented in Table 2.

The numbers in the table show that the listeners rather prefer the falling intonation. These results are overall statistically significant, except the evaluating of the group of TTS experts which is almost on the significant boundary for the chosen α. We could not find a reasonable explanation why the experts

Table 2. Comparison of the TTS systems which use raising (baseline - B), respectively falling (enhanced mode - E) terminal intonation.

	TTS experts	Naive listeners	All listeners
Falling is better (E)	170	185	355
Same quality (S)	89	81	170
Rising is better (B)	123	119	245
Quality value q	0.114	0.171	0.143
p-value for $\alpha = 0.01$	0.0071	< 0.0001	< 0.0001
Rising vs. falling	Falling	Falling	Falling

evaluated the sentences with the falling intonation worse than naive listeners. Nevertheless, that problem is out of the scope of this paper.

We have also inspected listeners' comments. Many times, they wrote that the variant with the raising intonation sounded very affected and artificial in the presented questions, they also complained about unnatural stress at the end of the sentences. Nevertheless, in some comments, they admitted that they could imagine this raising intonation in some special contexts or situations (e.g. as a rhetorical question), nevertheless, TTS system is designed to generate non-expressive speech. On the other hand, the listeners sometimes criticized that the variant with the falling intonation sounded more likely as a declarative phrase than a question. This effect is caused by a lack of peaking intonation in the beginning of these *wh*-questions which was not explicitly modeled in our TTS system (see Fig. 3). However, in many sentences this "peak" was implicitly created because speech units used at the sentence's beginning usually originated from beginnings of *wh*-questions in the speech corpus – it was due to the presence of *wh*-words in the right position of the phrase.

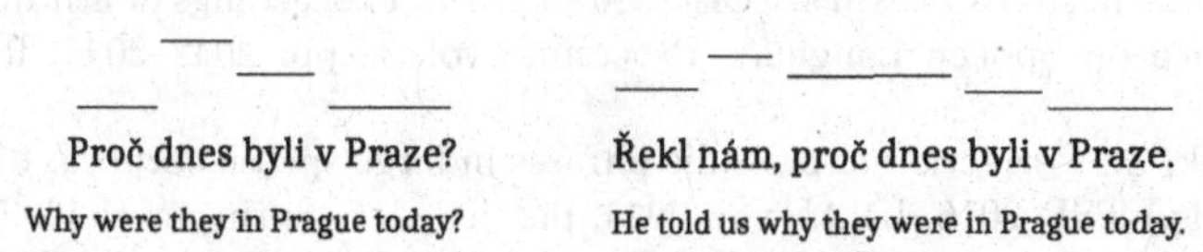

Proč dnes byli v Praze?   Řekl nám, proč dnes byli v Praze.

Why were they in Prague today?   He told us why they were in Prague today.

Fig. 3. Schematic representation of intonation comparing a *wh*-question and a complex declarative sentence.

4   Conclusion and Future Work

The presented paper focused on the problem of including *wh*-questions into the speech corpus used by TTS system and on the problem of synthesizing them correctly. The experiments confirmed that all recorded *wh*-questions in the speech corpus could be included in the TTS system and used for the synthesis without worsening of the overall quality of synthesized speech. Since these sentences have similar intonation course to declarative sentences, they are labeled in the same way with the abstract prosodic feature – prosodeme. The same prosodeme ($P1$, used in declarative phrases) is also better to use for synthesizing this type of questions as follows from the second listening test.

Although the presentend way of including and synthesizing *wh*-questions is quite simple, it brings the improvement of the speech synthesis from the perception point of view. Nevertheless, further experiments may be realized to describe better the special "peak" in these questions with, possibly, another prosodeme type representing this intonation pattern, which can not be expressed by any of the current prosodeme types.

References

1. Cruttenden, A.: Intonation. Cambridge University Press, Cambridge (1997)
2. Skarnitzl, R., Šturm, P., Volín, J.: Zvuková báze řečové komunikace: Fonetický a fonologický popis řeči. Univerzita Karlova, vydavatelství Karolinum, Praha (2016)
3. Romportl, J., Kala, J.: Prosody modelling in Czech text-to-speech synthesis. In: Proceedings of the 6th ISCA Workshop on Speech Synthesis, pp. 200–205. Rheinische Friedrich-Wilhelms-Universität Bonn, Bonn (2007)
4. Palková, Z.: Fonetika a fonologie češtiny: s obecným úvodem do problematiky oboru. Univerzita Karlova, vydavatelství Karolinum, Praha (1994)
5. Matoušek, J., Tihelka, D., Romportl, J.: Current state of Czech text-to-speech system ARTIC. In: Sojka, P., Kopeček, I., Pala, K. (eds.) TSD 2006. LNCS (LNAI), vol. 4188, pp. 439–446. Springer, Heidelberg (2006)
6. Tihelka, D., Kala, J., Matoušek, J.: Enhancements of viterbi search for fast unit selection synthesis. In: INTERSPEECH 2010, Proceedings of 11th Annual Conference of the International Speech Communication Association, pp. 174–177 (2010)
7. Matoušek, J., Legát, M.: Is unit selection aware of audible artifacts?. In: Proceedings of the 8th Speech Synthesis Workshop, Barcelona, Spain, pp. 267–271 (2013)
8. Tihelka, D., Matoušek, J.: Unit selection and its relation to symbolic prosody: a new approach. In: INTERSPEECH 2006 - ICSLP, Proceedings of 9th International Conference on Spoken Language Procesing, vol. 1, pp. 2042–2045. ISCA, Bonn (2006)
9. Hanzlíček, Z.: Correction of prosodic phrases in large speech corpora. In: Sojka, P., et al. (eds.) TSD 2016. LNAI, vol. 9924, pp. 408–417. Springer, Heidelberg (2016)
10. Matoušek, J., Romportl, J.: On building phonetically and prosodically rich speech corpus for text-to-speech synthesis. In: Proceedings of the 2nd IASTED International Conference on Computational Intelligence, pp. 442–447. ACTA Press, San Francisco (2006)
11. Tihelka, D., Grůber, M., Hanzlíček, Z.: Robust methodology for TTS enhancement evaluation. In: Habernal, I. (ed.) TSD 2013. LNCS, vol. 8082, pp. 442–449. Springer, Heidelberg (2013)
12. Volín, J., Bořil, T.: General and speaker-specific properties of F0 contours in short utterances. AUC Philologica 1/2014, Phonetica Pragensia XIII, pp. 101–112 (2013)

Tools rPraat and mPraat
Interfacing Phonetic Analyses with Signal Processing

Tomáš Bořil[✉] and Radek Skarnitzl

Faculty of Arts, Institute of Phonetics, Charles University in Prague,
nám. Jana Palacha 2, Praha 1, Czech Republic
{tomas.boril,radek.skarnitzl}@ff.cuni.cz
http://fu.ff.cuni.cz/

Abstract. The paper presents the rPraat package for R/mPraat toolbox for Matlab which constitutes an interface between the most popular software for phonetic analyses, Praat, and the two more general programmes. The package adds on to the functionality of Praat, it is shown to be superior in terms of processing speed to other tools, while maintaining the interconnection with the data structure of R and Matlab, which provides a wide range of subsequent processing possibilities. The use of the proposed tool is demonstrated on a comparison of real speech data with synthetic speech generated by means of dynamic unit selection.

Keywords: Matlab · R · Praat · Phonetics · Speech synthesis

1 Introduction

Speech sciences have always been an interdisciplinary field of study. With this cross-disciplinary focus, it is natural that scientists from diverse backgrounds – linguists, speech engineers, psychologists, sociologists, neurologists and others – have to be able to communicate their research to each other, and to cooperate on mutual research goals. Although arguments for increased collaboration within the speech sciences have been put forward [1], this is no easy task. One of the reasons may include differing terminology, divergent mindsets, and also tools used to process speech data.

This paper addresses the last of these issues. Linguistically trained phoneticians and also psycholinguists typically use software tools such as Praat [2] to perform phonetic analyses. On the other hand, speech engineers typically rely on more general programmes like Matlab [3] or R [4]. The goal of this paper is to at least partially bridge the gap and introduce a tool which constitutes an interface between Praat on the one hand and Matlab and R on the other; the tool is called *mPraat* in the Matlab environment and *rPraat* in the R environment.

Praat works with various types of objects, most typical of which is the Sound and a corresponding text object (TextGrid) which includes the desired labels on interval tiers (e.g., word and phone boundaries) and on point tiers (e.g., prosodic break locations or midpoints of syllable nuclei); see Fig. 1a for an example.

P. Sojka et al. (Eds.): TSD 2016, LNAI 9924, pp. 367–374, 2016.
DOI: 10.1007/978-3-319-45510-5_42

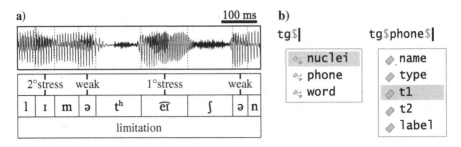

Fig. 1. (a) An illustration of a TextGrid with one point (*nuclei*) and two interval (*phone, word*) tiers. (b) Access to tier names and variables using RStudio [5] code completion.

The tool introduced here will be of assistance to phoneticians who want to make their analyses more accessible to technically oriented scientists, as well as to speech engineers who can use Praat TextGrids for segmenting sound files (freely available forced alignment tools often provide TextGrids as their output [6,7]). The general applicability of our tool allows for subsequent advanced DSP processing (e.g., computing power spectral density, spectral slope [8], or adaptive threshold algorithms working with the spectral envelope [9]), statistical analyses, or generating high-quality graphics in Matlab or R.

Other tools have already been introduced which make phonetic analyses more compatible with general programmes. PhonTools [10] is an R package for phonetic analyses: users can display spectra or spectrograms, create vowel formant plots and F0 tracks, or filter sound files. PraatR [11] is a package for R which operates on the principle of external calling of Praat from R. Our rPraat differs from PraatR in two main respects. First, and importantly, rPraat does not run Praat in the background but performs all operations directly on the TextGrid, which makes the analyses faster (see Sect. 3 for details). Also, it is possible to exploit the comfort of R's internal data structure, like the advantages of the effective vectorized access or a universal access to TextGrid components via tier names (see Fig. 1b) etc. All these benefits are lost in PraatR. Second, unlike in PraatR, the range of rPraat commands extends beyond those offered in Praat, making some measurements more accessible and intuitive. The package thus does not merely provide an interface between Praat and R, but builds upon Praat in the R environment. This includes more effective work with PitchTiers (objects which contain the modifiable fundamental frequency (F0) track); our tool can be used to cut PitchTiers according to information in TextGrids. Additionally, the same functionality is provided for Matlab, for which no similar interface with Praat is, to our best knowledge, available.

Our rPraat (or mPraat) will be introduced in more detail in the following section. Section 3 presents an analysis of the tools' performance. In Sect. 4, we will illustrate its use in a comparison of real speech data from actual conversations with synthetic speech; we will focus on several acoustic parameters of fricative sounds and point to possible problems of concatenative speech synthesis.

2   Introducing rPraat and mPraat

The current version of the tool is available on http://fu.ff.cuni.cz/praat/, along with detailed instructions and a demonstration. Table 1 provides an overview of rPraat/mPraat functions. These include reading and writing, querying and modifying TextGrids. As was mentioned in the Introduction, some of the functions, such as `tg.insertInterval`, are new (i.e., not featured in Praat). These functions make work with TextGrids considerably easier.

Table 1. Summary of rPraat functions. mPraat features similar names, but without fullstops and with the following letter capitalized, e.g., tg.getLabel() *vs.* tgGetLabel().

tg.read / tg.write / tg.plot / tg.createNewTextGrid / tg.repairContinuity
pt.read / pt.write *(PitchTiers containing F0 tracks)*
tg.getStartTime / tg.getEndTime / tg.getTotalDuration
tg.getNumberOfTiers
tg.getTierName / tg.setTierName / tg.isIntervalTier / tg.isPointTier
tg.insertNewIntervalTier / tg.insertNewPointTier / tg.duplicateTier / tg.removeTier
tg.getNumberOfIntervals / tg.getIntervalStartTime / tg.getIntervalEndTime
tg.getIntervalDuration / tg.getIntervalIndexAtTime
tg.insertBoundary / tg.insertInterval / tg.removeIntervalLeftBoundary
tg.removeIntervalRightBoundary / tg.removeIntervalBothBoundaries
tg.getNumberOfPoints / tg.getPointTime / tg.insertPoint / tg.removePoint
tg.getPointIndexLowerThanTime / tg.getPointIndexHigherThanTime /
tg.getPointIndexNearestTime
tg.getLabel / tg.setLabel / tg.countLabels

Many operations may be performed at a high level, using the `tg` functions which correspond to their equivalents in Praat with thorough checks and warnings, or directly using low-level commands. For instance, let us assume a TextGrid read in the TG variable. We may find out the number of descriptive tiers using `tg.getNumberOfTiers(TG)` or merely `length(TG)`. To find out the start time of the 7th interval on the phone tier, the command is

```
tg.getIntervalStartTime(TG, "phone", 7) or TG$phone$t1[7].
```

It is possible to use the pipeline operator `%>%` for sending output from one function to the following function, instead of its first parameter, for example:

```
tg.read("H.TextGrid") %>% tg.removeTier("word") %>%
    tg.write("out.TextGrid")
```

The tremendous advantage of vectorized operations can be illustrated on the calculation of the mean duration of all [e/eː] vowels on the phone tier:

```
condition <- TG$phone$label %in% c("e", "eː")
```

```
meanDur <- mean(TG$phone$t2[condition] - TG$phone$t1[condition])
```

A single change – `hist` for `mean` – would produce the histogram of the duration values. As opposed to this simple procedure, we would have to create a for-loop in PraatR and repeat the following code for every index of an interval:

```
lab <- praat("Get label of interval...", list(2, index),
                input = "file.TextGrid")
if (lab == "e" | lab == "e:") {
    t1 <- as.numeric(praat("Get start point...", list(2, index),
                input = "file.TextGrid", simplify = TRUE))
    t2 <- as.numeric(praat("Get end point...", list(2, index),
                input = "file.TextGrid", simplify = TRUE))
    dur <- t2 - t1; totalDur <- totalDur + dur; cnt <- cnt + 1
}
```

Praat would also require the use of a for-loop, but the notation is considerably more concise. Moreover, it is not necessary to read each file again when making a new query, for example: `t1 = Get start point: 2, index`.

3   Performance Analysis

The objective of the three experiments reported in this section was to compare the performance of our Matlab and R package with that of Praat and the PraatR package. For this purpose, we used recordings of 200 short dialogues, each composed of five speaking turns. The dialogues were recorded in a sound-treated studio of the Institute of Phonetics in Prague, as mono recordings with a sampling frequency of 32 kHz and 16-bit depth. The recordings were segmented manually; each recording contains, on average, 29 phones (mean recording duration 2.065 s) and 3.4 [e/e:] vowels (mean duration 216 ms). The duration of processing a given task was measured on all 1,000 files. Each task was repeated 12 times, and the results are based on the last 10 cycles: the first two cycles were ignored so as to eliminate the initial reading of the files from the hard drive into memory. In PraatR, only 10 files were tested with 5 repetitions (and the first two ignored), due to the extremely slow processing speed. The TextGrids contain seven tiers (see Sect. 1). The mean duration values, as well as the 95 % confidence intervals using bootstrap are reported below.

All TextGrid formats were used for the experiments: the Full format (full text with each line led by its name, such as 'xmin = '); the Short format (an abbreviated form of the Full format which includes only the values); and the Binary format (a separate native binary format for each software programme).

The experiments were run on a Samsung 530U laptop with 64-bit Windows 7 Home Premium, Intel(R) Core(TM) i5-2467M CPU @ 1.60 GHz processor, 4 GB DDR3 RAM, and a hybrid (500 GB HDD, 5400 rpm + 16 GB SSD) drive. The following software versions were used: Praat 6.0.14 64 bit, Matlab R2013b 64 bit, and R version 3.2.4 64 bit with RStudio 0.99.891, and the attached packages dplyr_0.4.3, dygraphs_0.8, PraatR_2.3, stringr_1.0.0, tidyr_0.4.1.

Experiment 1 consisted in calculating the mean average duration of all [e/e:] vowels. It was calculated as the difference between the End and Start time of those intervals which correspond to the label criterion (i.e., [e] or [e:]).

In *Experiment 2*, we calculated the mean energy of all [e/eː] vowels. The task thus involved the reading of the wav files, along with the TextGrids. The computations were slightly different in R and Matlab, as we always chose that procedure which yielded a faster result. In R, energy was computed as the sum of the squared amplitudes of the individual samples, `sum(segment^2)`. Matrix multiplication `segment'*segment` was used in Matlab and the `Get energy` function in Praat and PraatR.

Experiment 3 consisted in listing all the labels of the second (phone) tier from all TextGrids into one single file each label written on a separate line, yielding a table with one column. This can be useful if we are interested in calculating the occurrences of individual phones in all the files. This task is the simplest of the three, with labels being merely listed in one file.

At this point, we would like to mention one way of listing results in Praat, the Info window; less experienced Praat users often write results into the Praat Info window rather than into a file, but they should be warned that this is an extremely slow process. We carried out Experiment 3 with the binary format TextGrids: it took 17.11 s to list the results of 100 files, 80.76 s for 200 files, and 370.65 s for 500 files. In other words, listing into the Praat Window becomes slower with every new line.

The results of all the three performance experiments are shown in Fig. 2. It is clear that PraatR is by a magnitude slower: the processing of one file took,

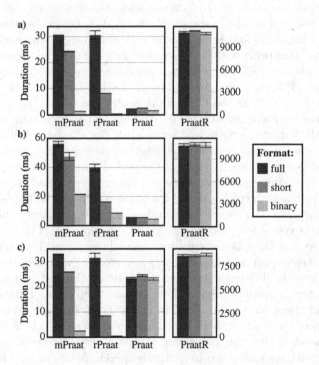

Fig. 2. Results of (a) experiment 1, (b) experiment 2, (c) experiment 3.

in all the experiments, more than 9 s. This is caused by the fact that for every single query, Praat has to be run in the background and the file has to be read into memory again.

The results of Experiment 1 (Fig. 2a) show that mPraat and rPraat are slower when reading the text-format TextGrids because the operations are performed in a scripting language, while Praat is programmed in C and compiled into machine code. Still, the processing is quite fast in mPraat and rPraat; when a repeated access to TextGrids is required, it is beneficial to convert them into the binary format, where the reading is ca. twice as fast as in Praat.

In Experiment 2 (Fig. 2b), the analyses which include access to sounds are fastest in Praat, due to its low-level implementation, but the performance of mPraat and rPraat is quite acceptable. The reading of the text formats of the TextGrid is, again, what takes longest: especially rPraat approaches Praat when binary files are used.

In Experiment 3 (Fig. 2c), where individual labels are written into a text file, the processing speed of mPraat and rPraat is similar as in Experiment 1, but Praat is considerably slower. The performance is best in rPraat with binary files, where processing time per file equals 0.4 s.

4   Comparison of Real and Synthetic Speech

The aim of this experiment was to compare selected acoustic properties of speech sounds from real and synthetic speech. We focused on the postalveolar voiceless fricative [ʃ]. This sound appeared twice in each of 31 target sentences. In type 1 sentences, one [ʃ] appears within the sentence and the other is (near-)final (e.g., "V naší vile občas straší" (*Our villa is sometimes haunted*), while in type 2 sentences both [ʃ] sounds are non-final (e.g., "Budeš jist, že dojdeš jistě k cíli." (*You'll know for sure that you'll reach the goal*). The target [ʃ] sounds always appeared in the same segmental context. That allowed us to compare the acoustic make-up of the target fricatives and to examine the effect of position within a sentence. In other words, we were interested whether real and synthetic speech would yield different results in terms of the declination of acoustic parameters.

Eight female students of the Faculty of Arts were asked to read the 31 sentences. They were recorded under the same conditions as those mentioned in Sect. 3. The same sentences were synthesized using dynamic unit selection in three female voices featured in the ARTIC system [12]. The target phones were identified and their boundaries set manually in Praat TextGrids, so that only their voiceless portions (and no carryover voicing) were included.

We analyzed the difference in the values between the first and second [ʃ] in three parameters: duration, power (intensity) and the centre of gravity (COG). The results of the analyses performed in R are shown in Fig. 3. When the box-plots are centred around 0, there is no significant trend which would point to differences between the first and second [ʃ]. An interesting trend found in real speech data, with no such trend in synthetic speech, points to possible deficiencies of speech synthesis.

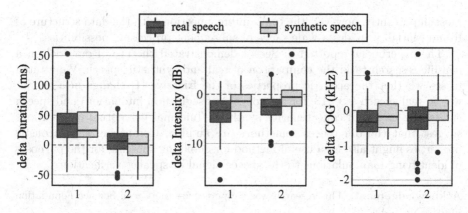

Fig. 3. Comparison of real and synthetic [ʃ] in type 1 and type 2 sentences (see text), showing the difference in duration (left), intensity (centre) and centre of gravity (right) between the second and first [ʃ]. A positive value thus means, e.g., longer duration of the second [ʃ].

As for duration, we can see in both real and synthetic speech higher values in the second [ʃ] in type 1 sentences (those where the second [ʃ] is located towards the sentence end). No such trend is apparent in type 2 sentences (the boxplots are more or less centred around zero). The agreement between real and synthetic speech is not surprising, since diphones are selected from the corpus according to the context.

Similarly, the data document the declination of intensity in type 1 sentences in both real and synthetic speech, while in type 2 sentences some degree of declination is apparent only in real speech; this may point to natural, gradual (but also inconspicuous) declination. Since non-final diphones are selected from various portions of the original sentences, on average there is no change between the first and second [ʃ] in synthetic speech.

Centre of gravity show a declination tendency in both sentence types in real speech: as the sentence end approaches and breath becomes weaker, the less "sharp" the sound of the fricative is and the COG drops. Such lower resonances are easily audible (*cf.* You don't wa**sh** where I wa**sh**). Some degree of COG drop is visible also in type 1 synthetic sentences but not in type 2, due to the random selection of segments from the original sentences.

5   Conclusions

The objective of this paper was to introduce a new tool which provides an interface of Praat with R and Matlab, to evaluate its performance with respect to other tools, and to demonstrate some applications on real data. The rPraat R package/mPraat Matlab toolbox proved to be computationally very effective especially when we use the binary format (see Sect. 3). The package offers more options for processing TextGrids and also PitchTiers which are absent in Praat.

Most importantly, however, the tool's interconnection with the data structure of R and Matlab provides a wide range of subsequent processing possibilities.

The experiment reported in Sect. 4 demonstrated the use of our tool on a real-life research task: the comparison of real and synthetic speech. We wanted to see whether the acoustic properties of the fricative [ʃ] change in a similar way in real speech (see [13] for the concept of segmental intonation) and speech created by means of concatenative synthesis. Informal perceptual observations suggest that the differences sometimes are audible. Comparing such acoustic properties might also be a useful method for detecting spoofing for the purposes of identifying manipulations to the speech signal or speaker verification.

Acknowledgement. This research was supported by the Czech Science Foundation project No. 16-04420S.

References

1. Barry, W.J., Van Dommelen, W.A. (eds.): The Integration of Phonetic Knowledge in Speech Technology. Springer, Dordrecht (2005)
2. Boersma, P., Weenink, D.: Praat: doing phonetics by computer (version 6.0.14) (2016). http://www.praat.org/
3. MathWorks: MATLAB Release 2013b. MathWorks Inc., Natick (2013)
4. R Core Team: R: a language and environment for statistical computing. R Foundation for Statistical Computing, Vienna (2016). http://www.R-project.org/
5. RStudio Team: RStudio: Integrated Development Environment for R. RStudio Inc., Boston (2015). http://www.rstudio.com/
6. Yuan, J., Liberman, M.: Speaker identification on the SCOTUS corpus. In: Proceedings of Acoustics 2008, pp. 5687–5690 (2008)
7. Pollák, P., Volín, J., Skarnitzl, R.: HMM-based phonetic segmentation in Praat environment. In: Proceedings of SPECOM 2007, pp. 537–541 (2007)
8. Volín, J., Zimmermann, J.: Spectral slope parameters and detection of word stress. In: Proceedings of Technical Computing Prague, pp. 125–129 (2011)
9. Šturm, P., Volín, J.: P-centres in natural disyllabic Czech words in a large-scale speech-metronome synchronization experiment. J. Phonetics **55**, 38–52 (2016)
10. Barreda, S.: phonTools, Functions for phonetics in R. R package (version 0.2-2.1) (2015). http://www.santiagobarreda.com/
11. Albin, A.: PraatR: an architecture for controlling the phonetics software "Praat" with the R programming language. J. Acoust. Soc. Am. **135**(4), 2198–2199 (2014)
12. Matoušek, J., Tihelka, D., Romportl, J.: Current state of Czech text-to-speech system ARTIC. In: Sojka, P., Kopeček, I., Pala, K. (eds.) TSD 2006. LNCS (LNAI), vol. 4188, pp. 439–446. Springer, Heidelberg (2006)
13. Niebuhr, O.: At the edge of intonation: the interplay of utterance-final F0 movements and voiceless fricative sounds. Phonetica **69**(1–2), 7–27 (2012)

A Composition Algorithm of Compact Finite-State Super Transducers for Grapheme-to-Phoneme Conversion

Žiga Golob[1], Jerneja Žganec Gros[1], Vitomir Štruc[2], France Mihelič[2], and Simon Dobrišek[2(✉)]

[1] Alpineon Research and Development, Alpineon d.o.o., Ulica Iga Grudna 15, 1000 Ljubljana, Slovenia
{ziga.golob,jerneja.gros}@alpineon.si
http://www.alpineon.si
[2] Faculty of Electrical Engineering, University of Ljubljana, Tržaška 25, 1000 Ljubljana, Slovenia
{vitomir.struc,france.mihelic,simon.dobrisek}@fe.uni-lj.si
http://luks.fe.uni-lj.si

Abstract. Minimal deterministic finite-state transducers (MDFSTs) are powerful models that can be used to represent pronunciation dictionaries in a compact form. Intuitively, we would assume that by increasing the size of the dictionary, the size of the MDFSTs would increase as well. However, as we show in the paper, this intuition does not hold for highly inflected languages. With such languages the size of the MDFSTs begins to decrease once the number of words in the represented dictionary reaches a certain threshold. Motivated by this observation, we have developed a new type of FST, called a finite-state super transducer (FSST), and show experimentally that the FSST is capable of representing pronunciation dictionaries with fewer states and transitions than MDFSTs. Furthermore, we show that (unlike MDFSTs) our FSSTs can also accept words that are not part of the represented dictionary. The phonetic transcriptions of these out-of-dictionary words may not always be correct, but the observed error rates are comparable to the error rates of the traditional methods for grapheme-to-phoneme conversion.

Keywords: Pronunciation dictionaries · Automatic grapheme-to-phoneme conversion · Finite-state super transducers · Out-of-dictionary words

1 Introduction

The grapheme-to-phoneme conversion process typically comprises two stages. During the first stage, a pronunciation dictionary look-up is performed for the given word. However, due to the evolving nature of the spoken language many words may not be represented in the dictionary. For these out-of-dictionary words, phonetic transcriptions need to be determined during the

© Springer International Publishing Switzerland 2016
P. Sojka et al. (Eds.): TSD 2016, LNAI 9924, pp. 375–382, 2016.
DOI: 10.1007/978-3-319-45510-5_43

second stage using automatic grapheme-to-phoneme methods, such as rule-based techniques [8,9] or statistical and machine-learning methods [2,10].

Therefore, for highly inflected languages the combination of pronunciation dictionary look-up and machine-learning methods does not provide the most efficient solutions. In systems with limited memory resources, e.g., in text-to-speech engines for embedded systems or multilingual text-to-speech engines, the use of large pronunciation dictionaries is often not feasible. To address these limitations, compact dictionary representations are needed. Finite-state transducers (FSTs) [5,11] are one of the most powerful models for compact dictionary representation. In this paper we build on these FST-based methods and present a new type of FSTs, called finite-state super transducers (FSSTs). FSSTs allow for a smaller representation of all the entries of a pronunciation dictionary than the minimal deterministic FSTs (MDFSTs) and are also capable of performing a grapheme-to-phoneme conversion of the words not present in the original dictionary.

In this paper we present the results from experiments performed on pronunciation dictionaries for three different languages belonging to three different language groups. We use the Slovenian SI-PRON pronunciation dictionary comprising 1,239,410 lexical entries [6], the Carnegie Mellon Pronouncing Dictionary (CMU-US) comprising 133,720 lexical entries [7] and the Italian pronunciation dictionary FST-IT taken from the Festival TTS toolkit comprising 402,962 lexical entries [3].

2   FST Representations of Pronunciation Dictionaries of Different Sizes

An FST can be built in such a way that it accepts all the words from a given dictionary, and for each accepted word it outputs a corresponding phonetic transcription. To provide fast dictionary look-ups and a small size, the FST can be converted to a MDFST using efficient and well-understood determinization and minimization algorithms [12]. The resulting MDFST has the smallest number of states and transitions among all the equivalent FSTs [11].

Figure 1 (i) depicts an MDFST for the simple example of a pronunciation dictionary containing nine two-letter words that represent all possible pairs of the alphabet $A = a, b, c$. For the sake of simplicity, the corresponding phonetic transcriptions are assumed to be identical to the words. As can be seen from Fig. 1, the nine lexical entries can be represented with three states and six transitions. However, when the entry $cc{:}cc$ is removed from the dictionary, a more complex MDFST with more states and transitions is required, as shown in Fig. 1 (ii). From the presented examples we can deduce that in some cases pronunciation dictionaries comprising more words can be represented with smaller and simpler MDFSTs.

In order to test the dependence of the MDFST size on the size of the represented pronunciation dictionaries, eleven sub-dictionaries with different sizes were built for each of the three pronunciation dictionaries mentioned in the

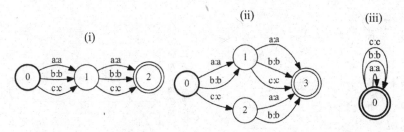

Fig. 1. The MDFSTs representing simple example dictionaries with two-letter lexical entries. The first MDFST (i) represents a dictionary with all nine possible two-letter lexical entries. The second MDFST (ii) represents a dictionary with only eight two-letter lexical entries (without the lexical entry *cc:cc*). The third FST (iii) is a finite-state super transducer (FSST) of the previous two MDFSTs (i) and (ii) that is discussed later in the paper.

Introduction section. The sub-dictionary lexical entries were randomly chosen from the original dictionaries. A MDFST was then built for each sub-dictionary. Figure 2 shows the number of states for all the resulting MDFSTs in relation to the number of states of the MDFST representing the entire dictionary. We could expect that the MDFST with the largest number of states would correspond to the dictionary covering 100 % of the lexical entries. It is interesting to observe that this is not the case for all three languages (Fig. 2). For the Slovenian and Italian languages, which are both highly inflected, the MDFST size is up to 150 % larger for smaller dictionaries, reaching the highest number of states for dictionaries that comprise roughly 60 % of the lexical entries of the original dictionary. On the other hand, the size of the MDFST for the English language is almost linearly dependent on the size of the pronunciation dictionary. It seems that the inflected word forms have a very strong impact on this phenomenon. A possible explanation could be that for many languages the inflection rules follow a similar pattern, with rare exceptions, and in this way inflected forms may often be represented with a similar FST structure for many different lemmas. If an inflected form is missing for a specific lemma, the remaining forms must be represented with a different FST structure, which may result in a larger final size of the MDFST.

To further test this hypothesis, we repeated the experiment for the Slovenian pronunciation dictionary SI-PRON. This time the sub-dictionaries were built in such a way that only the lemmas were chosen randomly, and afterwards all the corresponding inflected forms from the original dictionary were added. The results are presented in Fig. 3. In this case the size of the MDFST for Slovenian dictionaries increases monotonically. Thus, it can be concluded that the missing inflected forms greatly increase the complexity of the MDFST representing the dictionary.

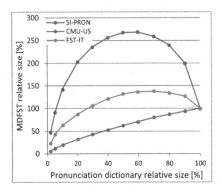

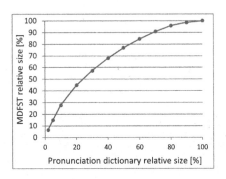

Fig. 2. Number of MDFST states representing dictionaries for three different languages. For each language, 12 sub-dictionaries with different sizes were used, whereby their lexical entries were randomly chosen from the original dictionaries.

Fig. 3. The number of MDFST states representing SI-PRON sub-dictionaries with different sizes, where initially only lemmas were chosen randomly from the original dictionary, and later all the corresponding inflected forms were added.

3 Finite-State Super Transducers

From the observation in the previous section, we can conclude that it is beneficial for the size of the MDFST representing a dictionary to include all the possible inflected forms for a given lemma, even if the inflected forms are very rare or not needed in a particular application. Another question that emerges from the observation is, could the size of the MDFST representing a pronunciation dictionary be even smaller, provided it also accepts other words that are not part of the original pronunciation dictionary? We will demonstrate that this can be achieved by defining specific rules for merging FST states. We denote the resulting FST as a finite-state super transducer (FSST), as an analogy to super sets. It should be noted that when new words are represented by such an extended FST, information about which words or lexical entries belong to the original dictionary and which have been added as the result of such an extension is lost.

3.1 Construction of FSSTs

The main idea behind the construction of FSSTs is to find relevant new words or strings that would allow an additional merging of states. Instead of searching for such words, which is a difficult task, the problem can be solved by searching for relevant states that can be merged. For example, in the FST presented in Fig. 1 (ii), intuitively, states 1 and 2 could be merged. All the input and output transitions of both states are part of a new merged state, only the repetitions of identical output transitions should be left out. The resulting FST is an FSST, presented in Fig. 1 (i), which accepts the additional word *cc*. Furthermore, states 0 and 3 could also be merged, resulting in an FSST with only one state, which is the initial as well as the final state with three output transitions, as in Fig. 1 (iii). The third FSST compactly represents the same dictionary as the other two MDFSTs, while it also accepts an infinite number of additional out-of-dictionary words.

The question that arises is which states of a given MDFST are the best candidates to be merged further. Not all the states can be merged since this would result in a non-deterministic FST, causing the slow and ambiguous translation of input words. In order to ensure that the final FSST is a deterministic one, several rules for merging states were determined. Two states that satisfy these rules are denoted as mergeable states. The rules for merging states are:

- Mergeable states do not have output transitions with the same input symbols and different output symbols.
- Mergeable states do not have output transitions with the same input and output symbols and different target states.
- If one of the two mergeable states is final, both states do not have output transitions with input symbols that are empty strings or ε.

The above rules are stricter than necessary to preserve the determinism of the final FSST, and are very simple to verify as well. The aim of this paper is to demonstrate the potential of FSSTs and not to pursue building the smallest possible FSST.

To build an FSST from an MDFST we developed an algorithm that searches for mergeable states by checking all the possible combinations of two states of the input MDFST. Every time a pair of mergeable states is found, they are merged into a single state. Since merging influences other states, which have already been verified against the mentioned rules, several iterations are normally needed until no new mergeable states are found.

In our experiments, we noticed that the final size of an FSST depends on the order of the state merging. The implementation size or the memory footprint of an FST mostly depends on the number of transitions and less on the number of states [4]. Our experiments have shown that the smallest final number of transitions is obtained if only the states with the highest number of identical transitions are merged in the first iterations, as the repetitions of all the identical transitions can be immediately removed from the transducer. To reduce the number of additional words that are accepted by the resulting FSST, mergeable states whose merging did not decrease the number of transitions are not merged.

3.2   Experimental Results

For our experiments two MDFSTs were built for every available pronunciation dictionary using the open-source toolkit OpenFST [1]. For the second type of MDFSTs, denoted as MDFST2, the output strings of the transitions were constrained to the length 1, in contrast to the first type denoted as MDFST1, which does not have this constraint. The second type of MDFST normally has more states and transitions than the first type. However, it boasts a simpler implementation structure, which results in a smaller implementation size.

Table 1. A comparison of the number of states and transitions between MDFSTs and FSSTs obtained with the two types of FSTs.

(1)

	MDFST1		FSST	
Dictionary	States	Trans.	States	Trans.
SI-PRON	70,681	241,879	57,190	212,765
CMU-US	76,035	184,232	60,477	164,235
FST-IT	57,323	172,210	44,563	148,657

(2)

MDFST2		FSST	
States	Trans.	States	Trans.
226,363	534,061	172,833	428,114
185,053	307,027	153,746	269,677
157,379	318,805	124,717	260,447

The FSSTs were then built from all the MDFSTs using the algorithm presented in Sect. 3.1. The results are presented in Table 1. It is clear from the presented results that the number of states and transitions is significantly smaller for the FSSTs, where the reduction of states ranges from 17 % to 24 % and the reduction of transitions from 11 % to 20 % when compared to the corresponding MDFSTs.

4   Out-of-dictionary Words

A system for grapheme-to-phoneme conversion usually consists of two parts. The first part checks whether an input word is contained in the pronunciation dictionary. If this is not the case, then the second part determines a phonetic transcription of the input word using suitable statistical or machine-learning methods. On the other hand, if an FSST is used to represent a pronunciation dictionary and the input word is accepted by the FSST, then it is not possible to determine whether the output phonetic transcription is correct, as FSSTs can also accept words that are not part of the original pronunciation dictionary, and these out-of-dictionary words may have incorrect phonetic transcriptions.

In order to evaluate this error the SI-PRON pronunciation dictionary was divided into a training and a test set. The training set contained about 90 % of the lexical entries from the original dictionary, which were chosen randomly. The remaining entries represented the test set. An MDFST2 and an FSST were then built from the training set. In this case, the reduction in the number of FSST states was 39.5 % and the reduction in the number of FSST transitions was 40.2 % in comparison to the corresponding MDFST2. These results show that the reduction is much higher when not all the inflected forms are contained in the dictionary.

After an FSST has been built from the training set, the words from the test set, which represent out-of-dictionary words, were used as the input. The accepted and rejected words were then enumerated, counted and checked for the correct resulting phonetic transcriptions at the output. The results of the experiment are shown in Fig. 4. Interestingly, only 8.2 % of the out-of-dictionary words were not accepted by the FSST and for only 5.8 % of the accepted words was the output phonetic transcription incorrect. For the Slovenian language, state-of-the-art statistical and machine-learning methods correctly determine the phonetic transcription for up to 83.5 % of the words [13].

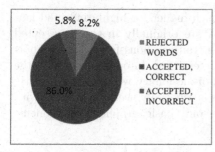
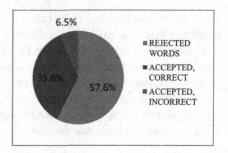

Fig. 4. Results for the out-of-dictionary test words used as the input to the FSST.

Fig. 5. Results for the out-of-dictionary test words when only lemmas were randomly chosen

Nevertheless, it should be noted that since the test entries were chosen randomly from the original dictionary, the inflected forms that belong to the same lemma could be present in the training as well as in the test dictionary. In this way, both dictionaries were partly similar, despite the fact that the word forms were different. Therefore, we repeated the experiment by first choosing the lemmas randomly and then all the corresponding inflected forms were added to the training dictionary. Figure 5 shows the results. In this experiment just 42.3 % of the words were accepted by the FSST. Among the accepted words, the phonetic transcription was still correct for 85 % of the words, which is similar to the accuracy of state-of-the-art methods for grapheme-to-phoneme transcriptions, which in any case need to be used to process the rejected words.

5   Discussion and Conclusions

The obtained results suggest that by using the proposed final-state super transducers (FSSTs) their size in terms of the number of states and transitions can be reduced by up to 40 %. The highest reduction rate was observed for the highly inflected Slovenian language, when some inflected forms were missing from the original dictionary. As can be seen from Fig. 2, the missing inflected forms can greatly increase the size of an MDFST. When the missing inflected forms are not added to the dictionary, an FSST can always be built from an MDFST, and in this way most of the missing inflected forms are automatically represented by an FSST. The proportion of the accepted out-of-dictionary words with correct phonetic transcriptions is surprisingly high for FSSTs. For rejected words the phonetic transcription can always be determined using other approaches. This is especially important when there is a high probability that the input word will be rejected by an FSST. This tends to happen when a word belongs to a new lemma for which the inflected forms are not represented in the dictionary.

FSSTs can be used for a very compact grapheme-to-allophone transcription. The reduction of their representation size compared to MDFSTs can be up to

20 %. Furthermore, this reduction can be much higher for highly inflected languages, when not all of the inflected forms are originally in the represented dictionary. In addition, by using FSSTs, the transcriptions for a large proportion of out-of-dictionary words can also be determined. In our experiments, the phonetic transcriptions of the accepted out-of-dictionary words were correctly determined for up to 86 % of the words. This is especially useful in systems with limited memory resources where no other, or only basic, grapheme-to-phoneme transcription methods can be used.

References

1. Allauzen, C., Riley, M.D., Schalkwyk, J., Skut, W., Mohri, M.: OpenFst: a general and efficient weighted finite-state transducer library. In: Holub, J., Žd'árek, J. (eds.) CIAA 2007. LNCS, vol. 4783, pp. 11–23. Springer, Heidelberg (2007). http://www.openfst.org
2. Bisani, M., Ney, H.: Joint-sequence models for grapheme-to-phoneme conversion. Speech Commun. **50**(5), 434–451 (2008)
3. Black, A., Taylor, P., Caley, R.: The festival speech synthesis system: system documentation (2.4.0). Technical report, Human Communication Research Centre, December 2014
4. Golob, Ž.: Reducing redundancy of finite-state transducers in automatic speech synthesis for embedded systems. Ph.D. thesis, University of Ljubljana, Faculty of Electrical Engineering, Tržaska 25, SI-1000 Ljubljana, Slovenia (2014)
5. Golob, Ž., Žganec Gros, J., Žganec, M., Vesnicer, B., Dobrišek, S.: FST-based pronunciation lexicon compression for speech engines. Int. J. Adv. Rob. Syst. **9**(211), 1–9 (2012)
6. Žganec Gros, J., Cvetko-Orešnik, V., Jakopin, P.: SI-PRON pronunciation lexicon: a new language resource for Slovenian. Informatica (Slovenia) **30**(4), 447–452 (2006)
7. The Carnegie Mellon Speech Group: The Carnegie Mellon University Pronouncing Dictionary (Version 0.7b) [Electronic database]. Carnegie Mellon University, Pittsburgh (1995). http://svn.code.sf.net/p/cmusphinx/code/trunk/cmudict
8. Hahn, S., Vozila, P., Bisani, M.: Comparison of grapheme-to-phoneme methods on large pronunciation dictionaries and LVCSR tasks. In: Interspeech, Portland, OR, USA, pp. 2538–2541, September 2012
9. Jiampojamarn, S., Kondrak, G.: Letter-phoneme alignment: an exploration. In: Proceedings of the 48th Annual Meeting of the Association for Computational Linguistics, ACL 2010, pp. 780–788. Association for Computational Linguistics, Stroudsburg, PA, USA (2010)
10. Lehnen, P., Allauzen, A., Lavergne, T., Yvon, F., Hahn, S., Ney, H.: Structure learning in hidden conditional random fields for grapheme-to-phoneme conversion. In: Interspeech, Lyon, France, pp. 2326–2330, August 2013
11. Mohri, M.: Finite-state transducers in language and speech processing. Comput. Linguist. **23**(2), 269–311 (1997)
12. Mohri, M.: Minimization algorithms for sequential transducers. Theoret. Comput. Sci. **234**(1–2), 177–201 (2000)
13. Šef, T., Skrjanc, M., Gams, M.: Automatic lexical stress assignment of unknown words for highly inflected slovenian language. In: Sojka, P., Kopeček, I., Pala, K. (eds.) TSD 2002. LNCS (LNAI), vol. 2448, p. 165. Springer, Heidelberg (2002). http://dx.doi.org/10.1007/3-540-46154-X_23

Embedded Learning Segmentation Approach for Arabic Speech Recognition

Hamza Frihia$^{(\boxtimes)}$ and Halima Bahi

Labged Laboratory, Universite de Badji Mokhtar Annaba, 23000 Annaba, Algeria
frihia@labged.net, halima.bahi@univ-annaba.dz

Abstract. Building an Automatic Speech Recognition (ASR) system requires a well segmented and labeled speech corpus (often transcription is made by an expert). These resources are not always available for languages such as Arabic. This paper presents a system for automatic Arabic speech segmentation for speech recognition purpose. State-of-the-art models in ASR systems are the Hidden Markov Models (HMM), so that for the segmentation, we expect the use of embedded learning approach where an alignment between speech segments and HMMs is done iteratively to refine the segmentation. This approach needs the use of transcribed and labelled data, for this purpose, we built a dedicated corpus. Finally, the obtained results are close to those described in the literature and could be improved by handling more Arabic speech specificities.

Keywords: Automatic Speech Segmentation · Speech recognition · Hidden Markov Models · Embedded learning

1 Introduction

The speech segmentation task is to divide the input speech signal into a set of basic units and identify the beginning and end of these units (words, syllables or phonemes). The segmentation of the speech signal is involved for several applications, such as: speech recognition, speech synthesis, speaker tracking or for keyword spotting, etc. In this paper, we are interested with the speech segmentation for speech recognition. In this field, the data preparation is a very important and sensitive stage. Among tasks of this phase, we underline the segmentation and labelling one. It consists of defining boundaries of different patterns from the speech stream and labelling them by providing a phonetic transcription.

Earliest speech recognizers were dedicated to isolated word recognition, thus modeling of speech was based on word models [1]. Later, advances in automatic speech recognition (ASR) field lead to the modeling of smaller units such as syllables, phones, diphones, etc. [2]. Such units allow the building of continuous speech recognizers. However, models of small units need segmentation of speech into the chosen units. This task is more difficult than the segmentation of the speech stream into words due to the co-articulation effect of small units (Fig. 1).

© Springer International Publishing Switzerland 2016
P. Sojka et al. (Eds.): TSD 2016, LNAI 9924, pp. 383–390, 2016.
DOI: 10.1007/978-3-319-45510-5_44

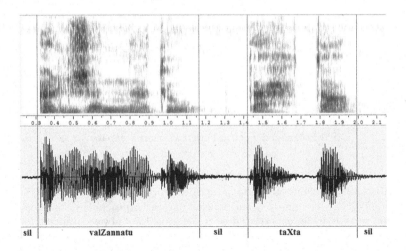

Fig. 1. Example of the co-articulation effect for the word boundaries

The segmentation algorithms estimate the phonemes' boundaries. This is usually done according to two trends: the first one, called implicit segmentation, takes advantages of the signal discontinuities without linguistic knowledge (in particular speech orthographic or phonetic transcription), while the second, called explicit segmentation, takes advantage of language particularities. Algorithms from the first approach, segment the speech signal without any prior knowledge of linguistic information (internal properties). This segmentation is independent of speaker, text and language. Due to the absence of external information, the first phase of the segmentation is based entirely on the analyzing acoustic characteristics present in the signal. These segmentation algorithms are based on signal processing characteristics in the time domain such as: Short Term Energy or Zero Crossing Rate [3] and in the frequency domain such as: Spectral centroid and spectral flux [4] and in these two areas, thresholding methods are used to detect the beginning and the end of each unit. Another possible algorithm relying to this approach was introduced by [5], in this approach there is no linguistic knowledge about the given speech data even the number of the phonemes. Authors suggest the use of clustering methods to determine this number. Recently [6] published new research on spectral clustering methods for acoustic segment modeling where they use expect the use vector quantization techniques. Phonemic segmentation also called explicit segmentation methods operate the signal so that the segments obtained are defined by the phonetic transcription. In the earliest segmentation systems the force-alignment between the speech segment and its model was done using the Dynamic Time Warping (DTW) algorithms, but since the arise of the Hidden Markov Models as the state-of-the-art models in speech recognition; the force-alignment is done using related algorithms (such as Viterbi one) [7]. In the present work, we expect the achievement of an automatic segmentation and labeling of Arabic speech data based on the embedded training approach. It consists on iteratively training a speech recognizer, then using the models to make a new set of training labels via forced alignment, then training a new recognizer from these labels, then re-aligning again, and carrying on until the resulting system stops getting better. To carry on this study, we designed and we built a dedicated dataset of Arabic speech records (ArabPhone); indeed, even if Arabic speech corpora are available, the speech stream is neither segmented nor labelled. The remainder of the paper is organized as follows: in the next section, we introduce the embedded learning with HMMs, then, we describe our development protocol where we combine twice the use of HMMs, first HMMs are used to segment data then they are re-trained to be used in recognition stage; we particularly use a dedicated dataset (ArabPhone) were phonemes occur in several positions of the word to allow segmentation to be more accurate. In Sect. 3, we present our experiments to assess performances of embedded learning segmentation in the context of Arabic speech recognition. Finally, a conclusion ended the paper.

2   Embedded Learning

In the classical learning approaches, HMMs are trained with fixed boundaries according to manual segmentation. While with the embedded learning, boundaries are aligned iteratively by the forced alignment. The idea is that the relabeling is made in each iteration, and the classifier in the next iteration will have more coherent sets of labelled data. The Embedded training uses the same procedure as Baum Welch classical learning, but instead to make the learning of each model individually, all models are learned in parallel. The a priori initializing of the HMMs of phonemes by embedded re-estimation can be performed in two ways. The first is to use a dataset segmented manually to initialize the HMM of each phoneme individually. When used in this manner, the algorithm uses the labeled speech to extract all the segments corresponding to each phoneme HMM to realize learning models. The second way uses the "Flat Start" technique of initially segmenting each learning signal evenly. Then, the phoneme models are aligned with current realizations of this phoneme.

2.1   Proposition

Given a set of signals, features extraction stage is performed which produces a collection of acoustic vectors. Besides them, we consider a dictionary which contains the transcription of phonemes and files containing the transcription into phonemes of each input signal. Acoustic vectors, dictionary and transcribed files are submitted to the embedded learning algorithm. Results of this stage are files which contain labeling of each phoneme and the beginning and the end of each portion of the signal. The speech recognizer including the segmentation stage was implemented using the Hidden Markov Model Toolkit (HTK) [8] functionalities and particularly, the HERest function which implements the embedded learning (estimation and re-estimation) of the HMMs parameters. Practically, the speech segmentation is done as follows: given the known phonetic transcription of speech stream and its MFCC coefficients, we built an initial model HMM

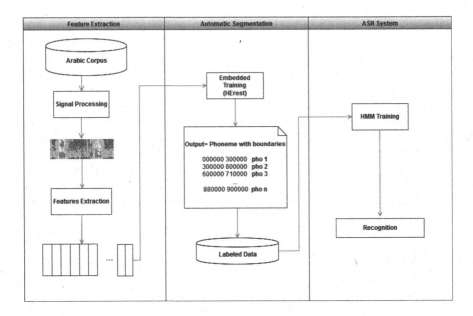

Fig. 2. The recognition system architecture

for each phoneme using the HcompV function of the HTK tool [8], this function did not need any previous labeling of files. Then, we feed the HERest function with the phonemes' sequences and we begin an embedded training session. Once the segmentation stage ended, a second stage consists on the learning. Initialization of HMM models is done with the segmented and labeled data using the HInit function. Then the training stage begins with the HRest function (Fig. 2). Later, the HVite function which performs the recognition provides us with the phonemes' sequence from the input signal.

2.2   ArabPhone: A Dedicated Corpus

Even if some Arabic speech corpora are available, particularly from broadcast news, benefits of embedded learning are limited as well as in training and test stages since phonetic or orthographic transcription are not available and corpora were not designed to include phonemes in different positions of words as expected by our proposition. The unavailability of Arabic speech corpus, that shows highlights of the embedded learning, leads us to build "ArabPhone". (Fig. 3), gives an example of included sentences for three phonemes.

Voice recordings are made in different rooms for some variability. In the first version of the corpus, the audio dataset includes 28 sentences uttered in Modern Standard Arabic (MSA) by 28 (Arabic L1) speakers (20 males and 8 females) aged between 20 and 40 years. The chosen sentences contain all the Arabic phonemes which occur in several positions (begin, middle and end of words) (Fig. 4).

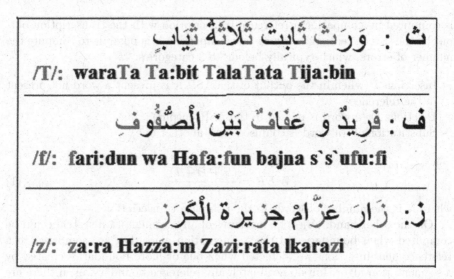

Fig. 3. Example of sentences from ArabPhone corpus

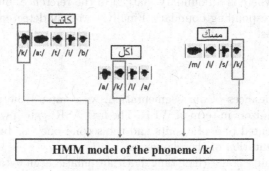

Fig. 4. Learning of the phoneme /k/ from several occurrences

3 Experiments

In order to evaluate the performances of the embedded learning phonetic segmentation, we make some experiments.

3.1 Evaluation Measures

To assess the segmentation performances, a collection of scores may be computed, these measures are issued from two classes [9]: purpose oriented measures and objectives measures. When the segmentation is performed for a specific application, the measures used to evaluate the final system will be applied to evaluate segmentation, in our case we will use the WER (Word Error Rate), a common measure to assess performances of an ASR system. Such a performance is computed by comparing a reference transcription with the transcription output by the speech recognizer. From this comparison it is possible to compute the number of errors, which typically belong to 3 categories:

- Insertions I (when in the output of the ASR it is present a word not present in the reference)
- Deletions D (a word is missed in the ASR output)
- Substitutions S (a word is confused with another one)

$$WER = \frac{(S + D + I)}{N} \qquad (1)$$

where N is the number of words in the reference transcription.

On the other hand, objective measures of any segmentation system could be computed when boundaries of the segmentation algorithm are compared to a reference one. Insertions are detected when one or more boundaries created by a segmentation algorithm do not match any reference boundary, or, if there are several generated boundaries instead of only one reference boundary. Deletions are noted when there is a boundary marked in the reference, but the algorithm produces no corresponding boundary. Finally, correctly detected boundaries are considered as hits.

3.2   Results

To assess performances of our segmentation, we compare the performances of two speech recognizers in term of WER. The first ASR system was built upon to automatic segmented phonetic units and the second one was built on the basis of manual segmentation of the phonemes.

For the learning stage (including both automatic segmentation and learning acoustic models), we use five (5) occurrences for each of the 28 sentences (corpus 1). The test dataset includes 17 occurrences of the sentences (corpus 2) (Table 1).

Table 1. WER of the automatic and manual segmentation

	Automatic segmentation	Manuel segmentation
Corpus 1	0.0278	0.0126
Corpus 2	0.1106	0.0852

These results show a word error rate of about 4 % for training corpus, which is higher of about 1 % from the value issued from the manual segmentation; this score remains interesting. For the test corpus the WER is about 11 % for training corpus, which is higher of about 3 % from the value issued from the manual segmentation. This score is low and should be improved. However, this may be done by providing the training dataset with additional records. For the

Table 2. Correct alignment rates (in %)

	Time tolerance (ms)			
	10 ms	20 ms	30 ms	40 ms
Hit rate	44.83 %	65.52 %	89.65 %	93.10 %

second class of measures, we compute the percentage of correct boundaries with time tolerance ranging from 10 to 40 ms (Table 2).

Although, the dataset is bigger than that used in [10], the obtained values for the correct boundary percentage are in the same range (better). These values are also in the same range as those cited in [7].

4   Conclusion

In this paper, we have presented the two followed trends in speech segmentation both of them provide promising results. In the context of the Arabic speech recognition, we investigate the embedded learning method to segment the data and to build the phoneme acoustic models. For this purpose, we built a dedicated speech waveform corpus: "ArabPhone"?. During the evaluation stage, we compare results issued from speech recognizer whose phonemes' acoustic models were built upon a manual segmentation with those of a speech recognizer whose acoustic models were built upon automatic segmentation. The obtained results in term of correct boundaries are in the range of those available in the literature, however, the WER score should be improved by adding a language model and by the increase of the training dataset.

References

1. Bahi, H., Sellami, M.: Combination of vector quantization and hidden Markov models for Arabic speech recognition. In: Proceedings ACS/IEEE International Conference on Computer Systems and Applications, Beirut, Liban, pp. 96–100 (2001)
2. Bahi, H., Sellami, M.: Neural expert model applied to phonemes recognition. In: Perner, P., Imiya, A. (eds.) MLDM 2005. LNCS (LNAI), vol. 3587, pp. 507–515. Springer, Heidelberg (2005)
3. Sangeetha, J., Jothilakshmi, S.: Robust automatic continuous speech segmentation for indian languages to improve speech to speech translation. Int. J. Comput. Appl. **53**, 13–16 (2012)
4. Khawaja, M.A., Haider, N.G.: Segmentation of Sindhi speech using formants. In: IEEE International Conference on Signal Processing and Communications (ICSPC 2007), Dubai, United Arab Emirates, pp. 24–27 (2017)
5. Sharma, M., Mammone, R.J.: Blind speech segmentation: automatic segmentation of speech without linguistic knowledge. In: ICSLP (1996)
6. Wang, H., Lee, T.: Acoustic segment modeling with spectral clustering methods. IEEE/ACM Trans. Audio Speech Lang. Process. (TASLP), 264–277 (2015)

7. Brognaux, S., Drugman, T.: HMM-based speech segmentation: improvements of fully automatic approaches. IEEE/ACM Trans. Audio Speech Lang., 5–15 (2016)
8. Young, S., et al.: The HTK Book. Cambridge University Engineering Department, Cambridge (2002)
9. Galka, J., Ziolko, B.: Study of performance evaluation methods for non-uniform speech segmentation. Int. J. Circ. Syst. Sig. Process. (2007)
10. Nofal, M., Abdel-Raheem, E., Henawy, H.E., Kader, N.S.A.: Arabic automatic segmentation system and its application for Arabic speech recognition system. In: IEEE 46th Midwest Symposium on Circuits and Systems, vol. 2, pp. 697–700 (2003)

KALDI Recipes for the Czech Speech Recognition Under Various Conditions

Petr Mizera[(✉)], Jiří Fiala, Aleš Brich, and Petr Pollak

Faculty of Electrical Engineering, Czech Technical University in Prague,
K13131, Technicka 2, 166 27 Praha 6, Czech Republic
{mizerpet,fialaji8,brichale,pollak}@fel.cvut.cz
http://noel.feld.cvut.cz/speechlab

Abstract. The paper presents the implementation of Czech ASR system under various conditions using KALDI speech recognition toolkit in two standard state-of-the-art architectures (GMM-HMM and DNN-HMM). We present the recipes for the building of LVCSR using Speech-Dat, SPEECON, CZKCC, and NCCCz corpora with the new update of feature extraction tool CtuCopy which supports currently KALDI format. All presented recipes same as CtuCopy tool are publicly available under the Apache license v2.0. Finally, an extension of KALDI toolkit which supports the running of described LVCSR recipes on MetaCentrum computing facilities (Czech National Grid Infrastructure operated by CESNET) is described. In the experimental part the baseline performance of both GMM-HMM and DNN-HMM LVCSR systems applied on given Czech corpora is presented. These results also demonstrate the behaviour of designed LVCSR under various acoustic conditions same as various speaking styles.

Keywords: Automatic speech recognition · KALDI · Large Vocabulary Continuous Speech Recognition · GMM-HMM · DNN-HMM · CtuCopy

1 Introduction

Automatic speech recognition (ASR) technology has gradually been becoming more prevalent for human-machine communication in daily lives. The ASR technology is the core part of applications such as voice assistants, voice search, dictation systems, various voice control devices, etc. Current ASR systems are mostly based on two architectures (GMM-HMM and DNN-HMM) [3,9] and they can be built using various open-source ASR toolkits such as HTK [23], Sphinx [12], RWTH [21], BAVIECA [1], or KALDI [19] which are available under various licence conditions. The performance and suitability of these toolkits for several voice applications as Large Vocabulary Continuous Speech Recognition (LVCSR), ASR webservice, spoken dialogue system, or phonetic segmentation were compared in many works, e.g. [10,13,16,19].

It was shown in previous works that the KALDI speech recognition toolkit usually outperforms other ASR toolkits, so it is widely used by speech research

© Springer International Publishing Switzerland 2016
P. Sojka et al. (Eds.): TSD 2016, LNAI 9924, pp. 391–399, 2016.
DOI: 10.1007/978-3-319-45510-5_45

community over the world. Consequently it is under intensive development and it supports currently many up-to-date techniques for acoustic modelling (conventional diagonal GMMs, sGMMs, SAT using fMLLR [7], MMI, MPE, or modern hybrid and TANDEM based DNN AM approaches) same as the real-time decoding based on weighted finite-state transducers (WFST) [8,19,22]. The very important and advantageous feature of KALDI toolkit is the availability of so called KALDI recipes which demonstrate how to work with KALDI executable tools using standard bash scripts and how to built a complete ASR systems on the basis of many available speech databases.

This paper presents the new KALDI recipes for the building of Czech LVCSR systems on the basis of speech corpora collected under various acoustic and speaking style conditions. Within these recipes, the recognition of spontaneous and informal speech or a speech with higher level of background noise, serve also as our examples of ASR systems intended for common conditions. It represents challenging tasks, because the accuracy of spontaneous or noisy speech recognition is still rather low in comparison to the accuracy of standard LVCSR [11,14,15]. Finally, the upgrade of CtuCopy [6] feature extraction tool supporting now KALDI format is also described same as the extension of KALDI which enables to run created recipes on MetaCentrum NGI computing facilities.

The paper is organized as follows: Sect. 2 describes new KALDI recipes for Czech ASR, new version of CtuCopy and CTU_Speech_Lab project at Meta-Centrum; the setup of our experiments is described in Sect. 3, the results are discussed in Sect. 4, and the paper is concluded with the summary of the most important contributions of our work.

2   Czech ASR Using KALDI Toolkit

As mentioned above, the KALDI is a modern speech recognition toolkit which allows to build a state-of-the-art ASR system using both standardly used architectures, i.e. GMM-HMM and DNN-HMM. The non-restrictive Apache licence v2.0 and the available KALDI recipes allow to build ASR systems on freely available corpora or corpora provided by the ELRA or LDC. It motivates strongly the speech researchers (mainly junior ones) to use KALDI components for a building of own speech recognition systems. KALDI is written in C++, it requires several external libraries such as OpenFst for WFST support, linear algebra libraries as ATLAS, BLAS and LAPACK. KALDI supports both UNIX and Window systems and it also supports a parallelization to a cluster computation using Sun Grid Engine software.

2.1   KALDI Recipes

Current distribution of KALDI contains the recipes for the creation of complete LVCSR using more than 40 corpora supporting various languages: English (wsj, timit, swbd, rm, etc.), Danish (sprakbanken), Spanish (fisher_callhome), Egyptian (callhome_egyptian), Arabic (gale_arabic), Mandarin Chinese (gale_mandarin, hkust, thchs30), Swahili (swahili), Japanese (csj), Persian (farsdat), Czech (vys-tadial_cz [11]), or several multilingual corpora (GlobalPhone and Babel). Some KALDI recipes were also created as ASR baseline systems for various evaluation tasks such as ASPIRE, CHIME3, REVERB challenges.

Analogous to currently available KALDI recipes we have created several new ones supporting the design of Czech LVCSR using four Czech speech corpora which are available under various license conditions to the speech research community. Our *s5*-recipes are inspired by the stable WSJ (Wall Street Journal) and RM (Resource Management) recipes and they follow the proposed format of scripts. New recipes support the next four corpora which are briefly described in the following paragraphs.

Czech SpeechDat contain the read utterances from 1000 speakers collected via fixed telephone network and it is available via ELRA [18]. Signals were recorded at 8 kHz and quantized using 8bit a-law format.

Czech SPEECON database was designed for a creation of ASR acoustic models and Czech adults part consists of 590 speakers covering various environments such as the office, entertainment, public place, or car. The 16 kHz speech were through 4 different channels (0 - headset microphone, 1 - close distance, 2 - medium distance, 3 - far distance) and saved in raw 16bit PCM format. It also is available via ELRA [17].

Czech car speech (CZKCC) is a private database of Czech speech from 1000 speaker recorded in car environment[1]. It contains speech recorded in 2 channels under various driving conditions using three different microphone setups. In our recipe we use only the subpart containing the speech recorded by headset microphone only.

NCCCz consists of 30 h of spontaneous conversations of 30 males and 30 females speakers[2]. 20 speakers acted as confederates who asked two friends to participate in recordings of natural conversations. The amount of collected data is huge, since every group of three speakers was recorded for approximately 90 min. The detailed description of the corpus can be found in [4] and it is freely available after a signature of a license agreement.

The created recipes are located in exp/ directory and contains the top-level run.sh script, corpus specific sub-directory such as local/, conf/ and steps/, utils/ shared directory across all recipes. The main run.sh script is always specific for particular corpora and contains particular steps for data preparation[3] (creating of train/test/dev lists, text/lexicon normalization, etc.), training/decoding HCLG graphs creation, feature extraction, AM training, decoding and evaluation.

---

[1] The corpus was collected for TEMIC Speech Dialogue Systems GmbH in Ulm at Czech Technical University in Prague in co-operation with Brno University of Technology and University of West Bohemia in Plzen.

[2] The corpus was collected with focus on understanding of very informal speaking style in the collaborative research realized at CTU in Prague and Radboud University Nijmegen.

[3] More information can be found in official KALDI documentation http://kaldi.sourceforge.net/data_prep.html.

We have prepared mainly the corpus specific scripts which are located in local/ directory according to KALDI conventions. These scripts prepare the train, test and dev data directories with KALDI specific format of the required files. They also enable to select various subsets for available environments, channels or particular utterances and convert the corpus specific pronunciation lexica to required KALDI format, etc. The final train, dev and test sets created for particular corpora by mentioned data preparation scripts are summarized in Table 1. For NCCCz, test_13_17 is a subpart of test set containing the utterances with the length of 13–17 words which are more suitable for experiments with LVCSR.

Table 1. Created data sets for available corpora

Corpora	Set	Speakers	Sentences	Words	Hours
SPEECON	train	81f + 79m	43396	107571	38.2
	test	13f + 8m	620	5418	1
	dev	9f + 11m	589	5179	1
CZKCC	train	129f + 115m	10379	76517	16.9
	test	16f + 14m	581	6273	1.1
	dev	14f + 13m	499	5349	1
NCCCz	train	20f + 20m	18192	171997	16.1
	test	10f + 10m	8885	94372	8.1
	test_13_17	10f + 10m	863	12812	1.1
SPEECHDAT_CS	train	426f + 426m	43139	306787	73.4
	test	55f + 45m	943	8542	2.1
	dev	45f + 55m	951	8566	2.1

2.2 CtuCopy Tool

CtuCopy is our private tool for feature extraction which supports standard features such as MFCC, PLP, DCT-TRAP and allows to use a frequency-domain noise reduction [5]. This is the main contribution in comparison to feature extraction tools available in KALDI (i.e. compute-mfcc-feats and compute-plp-feats). Current version of CtuCopy supports also various cepstral normalizations [2] and it is now compatible with KALDI. The simple bash wrapper script for the CtuCopy-based feature extraction was created and located in the *local* directory of created recipes. Both the *CtuCopy* and the wrapper script *make_ctucopy.sh* are publicly available under APACHE 2.0 licence.

2.3 KALDI on Metacentrum

MetaCentrum is the National Grid Infrastructure operated by CESNET which offers computing and storage facilities, many types of development software such as MATLAB, OpenFst, Cuda, etc. Especially it is very suitable as a facility supporting the research which consumes very high computational costs, as it is our case of experiments with DNN-based speech recognition.

We have created CTU Speech Lab project at MetaCentrum where the collection of various types ASR tools and corpora is available. To run the prepared recipes at Metacentrum NGI, it was necessary to modify the standard KALDI tool *queue.pl* to support the Metacentrum PBS planning system. The distributed *pbs.pl* script also couldn't be used directly due to the parameters and parsing incompatibilities. Finally, we prepared the new *mqueue.pl* tool which enables to run the Kaldi at Metacentrum the standard way. Created KALDI recipes same as the KALDI toolkit are available there and they can be used by members who have the licences to access the available corpora and who join this project.

3   Experimental Part and Baseline Results

The experimental part describes mainly the baseline results for the Czech ASR systems built on the basis of available databases containing the speech data of variable quality (clean vs. noisy, wide-band vs. telephone-band) or speaking style (read vs. spontaneous speech).

3.1   LVCSR Setup

All experiments were carried out on the basis of LVCSR with the following rather standard setups for both GMM-HMM and DNN-HMM architectures.

Front-End Processing. Mel-frequency cepstral coefficients (MFCC) with slightly different setups for 16 kHz (wide-band) and 8 kHz (telephone-band) speech were used as speech features, i.e. pre-emphasis coefficient of 0.97, short-time frame length of 25 ms with 10 ms frame shift, 30 bands of mel-frequency filter bank in the range 100–7940 Hz, 13 cepstral coefficients including $c[0]$, cepstral mean normalization over the speaker, and standard delta and delta-delta coefficients for wide-band speech; the setup was the same for telephone-band speech excepting the usage of 23 band mel-frequency filter bank covering the range 125–3800 Hz. These short-time features were stacked with the context of 3 frames to both sides and obtained high dimension vector was then reduced and decorrelated by LDA+MLLT transforms. In the end, we work with the feature vector of the size 40.

Acoustic Modelling. The first AM of a *conventional GMM-HMM system* was based on the top 45 phones (*mono*) expanded to the context-dependent cross-word triphones (*tri1* - trained on MFCC delta-delta features) and built using the above mentioned LDA+MLLT features (*tri2*). It was followed by feature-space maximum likelihood linear regression (fMLLR) per each speaker (speaker adaptive training - *tri3*). The further system was based on UBM in the combination with SGMM, and finally, the best GMM-HMM system used bMMI discriminative retraining of AM.

The second AM based on *DNN-HMM hybrid approach* was built on fMLLR features (exactly MFCC-LDA-MLLT-fMLLR with CMN) where fMLLR was estimated by LDA+MLLT+SAT GMM-HMM system. The DNN topology consisted of input layer with 440 units (the context of 5 frames with 40 dimensional fMLLR features with cepstral mean and variance normalization) followed by 6 hidden layers with 2048 neurons per layer and the sigmoid activation function. The process of building of DNN-HMM system started with the initialization of hidden layers of used network by Restricted Boltzmann Machines (RBMs) and then the output layer was added. The frame cross-entropy training and sMBR sequence-discriminative training were then performed to finalize the process of DNN training [8].

Language Modelling and Lexicon. The language models of baseline systems were built separately for particular corpora using available transcriptions from the train subsets (only the subpart containing the sentences is used for this purpose). Created LMs are based on bigrams with Witten-Bell smoothing technique using SRILM toolkit. Supporting scripts *createLm.sh* are also located in *local* directory of particular recipes.

Another LM for the LVCSR task created from Czech National Corpus and marked as CNC340k was used to present more realistic results. It is a trigram-LM containing 340k of unigrams and more information about this CNC340k LM can be found in [20].

3.2 Results and Discussion

The results of realized experiments represent the base-line results which should be achieved when available recipes are used. They also summarize the behaviour of our LVCSR for Czech under various acoustic conditions and speaking styles.

Baseline Results for Particular Databases. The first results in Table 2 summarize baseline WER which should be achieved when the systems were built using particular database only, i.e. both acoustic and language models were trained from train set of available database. We can observe the results between 10–44 % WER depending on the system setup. The best result was achieved for Speech-Dat setup and we can observe the increase of WER for SPEECON, more noisy car speech from CZKCC, moreover, a serious increase of WER for spontaneous speech from NCCCz. Concerning DNN-HMM architecture, WER was reduced for all analysed acoustic conditions and the next decrease of WER was done by sMBR discriminative technique.

Results for LVCSR Using CNC Language Model. Above described results were achieved in the optimal and non-realistic setup from the point of view language modelling because LMs were created on the basis of available transcriptions, moreover, with possible appearance of some sentences both in the training and testing sets for SpeechDat, SPEECON, and CZKCC databases.

The following Table 3 summarizes the results obtained for LVCSR using 340k-word language model created on the basis Czech National Corpus (CNC340k LM). We can see the increase of WERs in comparison to results in Table 2. Slightly higher WERs was obtained also due to the fact that recognized utterances in test set contained phonetically rich sentences with slightly enhanced appearance of words with rare phones which may be missed by a general language model.

Table 2. Baseline results for particular databases

		GMM-HMM						DNN-HMM	
	data set	mono	tri1	tri2	tri3	sgmm	bmmi	dnn	sMBR
SPEECON	test	24.36	17.52	16.90	16.86	15.78	15.43	15.00	13.73
	dev	26.68	19.18	17.90	17.31	16.39	16.30	15.96	14.95
CZKCC	test	39.58	32.10	31.89	31.64	30.02	29.87	28.01	27.13
	dev	29.80	24.08	24.18	24.87	23.41	23.38	22.11	21.22
NCCCz	test	76.35	59.02	57.89	52.35	48.87	46.95	46.49	43.64
	test_13_17	76.86	58.73	57.78	51.71	48.40	46.84	46.65	43.45
SPEECHDAT_CS	test	22.46	14.33	14.36	14.78	14.05	13.97	13.04	10.86
	dev	20.06	13.96	13.96	14.08	13.44	13.32	13.28	11.15

Table 3. Baseline results for particular databases

		GMM-HMM						DNN-HMM	
	data set	mono	tri1	tri2	tri3	sgmm	bmmi	dnn	sMBR
SPEECON	test	51.32	31.37	27.48	23.54	19.59	19.06	18.99	17.46
CZKCC	test	29.20	15.90	14.79	11.57	9.56	9.43	9.39	8.49
NCCCz	test_13_17	88.09	69.32	66.29	59.92	57.63	55.63	51.15	48.79
SPEECHDAT_CS	test	53.34	29.47	27.26	22.24	19.57	18.86	16.77	13.71

4   Conclusions

The extension of the KALDI speech recognition toolkit by new recipes which enable to work with Czech corpora SpeechDat, SPEECON, CZKCC, and NCCCz has been proposed in this paper. The baseline results for LVCSR systems built on the basis of both GMM-HMM and DNN-HMM for above mentioned databases were presented and they gave an information about the performance of Czech LVCSR under various acoustic conditions and speaking styles. We have described also CTU_Speech_Lab project at Metacentrum which enables to run designed recipes at Czech National Grid Infrastructure. Finally, the updated version of CtuCopy tool compatible with KALDI feature format was introduced. All of described recipes and tools are publicly available and they can be used by other researches in the speech community. Concerning new KALDI recipes, we expect their integration into official version of KALDI on GitHub server.

Acknowledgments. The research described in this paper was supported by internal CTU grant SGS14 /191/OHK3/3T/13 "Advanced Algorithms of Digital Signal Processing and their Applications". Access to computing and storage facilities owned by parties and projects contributing to the National Grid Infrastructure MetaCentrum provided under the programme "Projects of Projects of Large Research, Development, and Innovations Infrastructures" (CESNET LM2015042), is greatly appreciated.

References

1. Bolaños, D.: The BAVIECA open-source speech recognition toolkit. In: 2012 IEEE Spoken Language Technology Workshop (SLT), pp. 354–359, December 2012
2. Borsky, M., Mizera, P., Pollak, P.: Noise and channel normalized cepstral features for far-speech recognition. In: Železný, M., Habernal, I., Ronzhin, A. (eds.) SPECOM 2013. LNCS, vol. 8113, pp. 241–248. Springer, Heidelberg (2013)
3. Dahl, G., Yu, D., Deng, L., Acero, A.: Context-dependent pre-trained deep neural networks for large-vocabulary speech recognition. IEEE Trans. Audio Speech Lang. Process. **20**(1), 30–42 (2012)
4. Ernestus, M., Kockova-Amortova, L., Pollak, P.: The Nijmegen corpus of casual Czech. In: Proceedings of the LREC 2014: 9th International Conference on Language Resources and Evaluation, Reykjavik, Iceland, pp. 365–370 (2014)
5. Fousek, P., Pollak, P.: Efficient and reliable measurement and simulation of noisy speech background. In: Proceedings of the EUROSPEECH 2003, 8-th European Conference on Speech Communication and Technology, Geneve, Switzerland (2003)
6. Fousek, P., Mizera, P., Pollak, P.: CtuCopy feature extraction tool. http://noel.feld.cvut.cz/speechlab/
7. Gales, M.J.F., Woodland, P.C.: Mean and variance adaptation within the MLLR framework. Comput. Speech Lang. **10**, 249–264 (1996)
8. Ghoshal, A., Povey, D.: Sequence-discriminative training of deep neural networks. In: Proceedings of INTERSPEECH (2013)
9. Hinton, G., Deng, L., Yu, D., Dahl, G., Mohamed, A., Jaitly, N., Senior, A., Vanhoucke, V., Nguyen, P., Sainath, T., Kingsbury, B.: Deep neural networks for acoustic modeling in speech recognition: the shared views of four research groups. IEEE Signal Process. Mag. **29**(6), 82–97 (2012)
10. Klejch, O., Plátek, O., Žilka, L., Jurcícek, F.: CloudASR: platform and service. In: Král, P., et al. (eds.) TSD 2015. LNCS, vol. 9302, pp. 334–341. Springer, Heidelberg (2015). doi:10.1007/978-3-319-24033-6_38
11. Korvas, M., Platek, O., Duvsek, O., Zilka, L., Jurcicek, F.: Free English and Czech telephone speech corpus shared under the CC-BY-SA 3.0 license. In: Proceedings of the LREC 2014: 9th International Conference on Language Resources and Evaluation, Reykjavik, Iceland (2014)
12. Lamere, P., Kwok, P., Gouvea, E., Raj, B., Singh, R., Walker, W., Warmuth, M., Wolf, P.: The CMU SPHINX-4 speech recognition system. In: IEEE International Conference on Acoustics, Speech and Signal Processing, ICASSP 2003, Hong Kong, China (2003)
13. Morbini, F., Audhkhasi, K., Sagae, K., Artstein, R., Can, D., Georgiou, P., Narayanan, S., Leuski, A., Traum, D.: Which ASR should I choose for my dialogue system? In: SIGDIAL, Reykjavik, Iceland (2013)
14. Nouza, J., Blavka, K., Bohac, M., Červa, P., Malek, J.: System for producing subtitles to internet audio-visual documents. In: 2015 38th International Conference on Telecommunications and Signal Processing (TSP), pp. 1–5, July 2015

15. Nouza, J., Ždansky, J., Červa, P.: System for automatic collection, annotation and indexing of Czech broadcast speech with full-text search. In: Proceedings of 15th IEEE MELECON Conference, pp. 202–205, La Valleta, Malta (2010)
16. Patc, Z., Mizera, P., Pollak, P.: Phonetic segmentation using KALDI and reduced pronunciation detection in causal Czech speech. In: Král, P., et al. (eds.) TSD 2015. LNCS, vol. 9302, pp. 433–441. Springer, Heidelberg (2015). doi:10.1007/978-3-319-24033-6_49
17. Pollak, P., Černocký, J.: Czech SPEECON adult database. Technical report, April 2004
18. Pollák, P., Boudy, J., Choukri, K., Heuvel, H.V.D., Vicsi, K., Virag, A., Siemund, R., Majewski, W., Staroniewicz, P., Tropf, H., Kochanina, J., Ostroukhov, E., Rusko, M., Trnka, M.: SpeechDat(E)- Eastern European telephone speech databases. In: Proceedings of the XLDB 2000, Workshop on Very Large Telephone Speech Databases (2000)
19. Povey, D., Ghoshal, A., Boulianne, G., Burget, L., Glembek, O., Goel, N., Hannemann, M., Motlicek, P., Qian, Y., Schwarz, P., Silovsky, J., Stemmer, G., Vesely, K.: The KALDI speech recognition toolkit. In: IEEE 2011 Workshop on Automatic Speech Recognition and Understanding. IEEE Signal Processing Society, December 2011
20. Procházka, V., Pollak, P., Ždansky, J., Nouza, J.: Performance of Czech speech recognition with language models created from public resources. Radioengineering 20, 1002–1008 (2011)
21. Rybach, D., Hahn, S., Lehnen, P., Nolden, D., Sundermeyer, M., Tüske, Z., Wiesler, S., Schlüter, R., Ney, H.: Rasr-the RWTH Aachen university open source speech recognition toolkit
22. Veselý, K., Karafiát, M., Grezl, F.: Convolutive bottleneck network features for LVCSR. In: 2011 IEEE Workshop on Automatic Speech Recognition and Understanding (2011)
23. Young, S., et al.: The HTK Book, Version 3.4.1. Cambridge (2009)

Glottal Flow Patterns Analyses for Parkinson's Disease Detection: Acoustic and Nonlinear Approaches

Elkyn Alexander Belalcázar-Bolaños[1(✉)], Juan Rafael Orozco-Arroyave[1,2],
Jesús Francisco Vargas-Bonilla[1], Tino Haderlein[2], and Elmar Nöth[2]

[1] Faculty of Engineering, Universidad de Antioquia UdeA, Medellín, Colombia
`elkyn82@gmail.com`
[2] Pattern Recognition Lab,
Friedrich-Alexander-Universität Erlangen-Nürnberg (FAU), Erlangen, Germany

Abstract. In this paper we propose a methodology for the automatic detection of Parkinson's Disease (PD) by using several glottal flow measures including different time-frequency (TF) parameters and nonlinear behavior of the vocal folds. Additionally, the nonlinear behavior of the vocal tract is characterized using the residual wave. The proposed approach allows modeling phonation (glottal flow) and articulation (residual wave) properties of speech separately, which opens the possibility to address symptoms related to dysphonia and dysarthria in PD, independently. Speech recordings of the five Spanish vowels uttered by a total of 100 speakers (50 with PD and 50 Healthy Controls) are considered. The results indicate that the proposed approach allows the automatic discrimination of PD patients and healthy controls with accuracies of up to 78 % when using the TF-based measures.

Keywords: Dysarthria · Nonlinear behavior · Glottal flow · Parkinson's Disease · Dysphonia · Time-frequency

1 Introduction

Parkinson's Disease (PD) is the most common neurodegenerative disorder in patients older than 65, it affects about 1.5 million of people in the United States of America, and the cost of their treatment will rise up to 50 million dollars in 2040 [14]. The speech symptoms of people with PD (PPD) include problems in respiration, phonation, articulation, and prosody [4]. Usually research related to PD is focused on measuring and identifying patterns in speech, using computational intelligence and pattern recognition techniques [2,16], and it is showing a feasible detection. Such techniques also allow to model phonation, articulation, and prosody phenomena [11]. However, the phonation and articulation processes are clearly defined. When they are analyzed from speech recordings, the information from both processes are combined, making it difficult to conclude whether the results come from phonatory or articulatory impairments. This is the

© Springer International Publishing Switzerland 2016
P. Sojka et al. (Eds.): TSD 2016, LNAI 9924, pp. 400–407, 2016.
DOI: 10.1007/978-3-319-45510-5_46

case in studies analyzing the nonlinear behavior directly from the speech signal [7,13], where the authors did not conclude which impairment causes the non-linear behavior in PD detection. In order to analyze only the phonation process by means of glottal closure patterns from the speech signal, it is necessary to apply techniques that separate the information contributed by the articulators and the glottis.

Abnormalities in the phonation process have been observed in the glottal closure pattern of PPD through laryngeal videoscopic examination [10]. It has revealed that the irregular glottal closure pattern is the most frequent symptom in PD speech, leading to a perceptual impression of breathy voice. Vocal fold bowing and slowed vibration are also observed [9]. These changes are caused by impairments in the movements of various muscles, tissues, and organs, which are involved in the voice production process [5], showing a highly nonlinear behavior. Laryngeal videoscopy is expensive and time consuming, thus the analysis of glottal patterns from speech signals is a good alternative to perform similar screenings. The glottal signal can be extracted from speech by means of Glottal Inverse Filtering (GIF) techniques [17]. GIF allows to estimate the glottal volume velocity waveform, i.e. the glottal flow from a speech signal, in which the effects of the vocal tract and lip radiation are cancelled; once the glottal flow is estimated, it is possible to reconstruct a residual wave by subtracting the glottal spectral components from the speech signal, and then phonation and articulation phenomena are considered separately.

This paper is focused on automatic detection of PD, considering glottal and residual flows with acoustic and nonlinear approaches. The aim of the work is to address two comparisons. First, the PD detection task by acoustic measures from the glottal flow and the nonlinear behavior produced by the glottal and residual flows are compared. Second, only nonlinear behavior is considered in order to analyze if the phonation process (glottal flow) or the articulation process (residual wave) provide more nonlinear information.

The rest of the paper is organized as follows. Section 2 describes the methodology, Sect. 3 provides details of the experimental framework, Sect. 3.3 contains the results, and Sect. 4 comprises the conclusions derived from this work.

2   Methodology

Figure 1 displays the general stages of the proposed methodology.

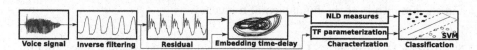

Fig. 1. General methodology

The recordings are considered in time frames using Hamming windows with different lengths, one for time-frequency (TF) and the other for nonlinear dynamics (NLD) measures. The length and time shift will be described when each feature set is introduced. The Parkinson's vs. Healthy decision is made using a Support Vector Machine (SVM). More details of the methodology are provided in the following subsections.

2.1 Glottal Inverse Filtering and Residual Wave Estimation

The *Iterative and/or Adaptive Inverse Filtering (IAIF)* [1] is used to estimate the glottal flow. This method estimates the contribution of the glottal excitation on the speech spectrum with a low-order linear prediction (LP) model that is computed with a two-stage procedure. The vocal tract is then estimated using either conventional LP or discrete all-pole modeling (DAP). This method is based on an iterative refinement of both the vocal tract and the glottal components. As a result of the IAIF process, the glottal waveform $g(n)$ is obtained from the speech signal $s(n)$. Additionally, the residual waveform $r(n)$ can be estimated from $s(n)$ by subtracting the glottal log-spectral components.

2.2 Characterization

Time-Frequency (TF) Glottal Flow Parameterization. Glottal flows are preprocessed using windows with 200 ms length with an overlap of 50 %, and each parameter is calculated for every glottal closure instant (GCI). The GCI is located using residual excitation and a mean-based signal algorithm [3]. Typically, time-domain features are estimated regarding the critical time instants, such as glottal opening and the glottal closure phases in the glottal flow pulse. From these critical instants, five time-domain features are obtained: **Open Quotient** (OQ), which is the ratio of the duration of the opening phase and the duration of the glottal cycle. **Closing Quotient** (CQ), defined as the ratio of closing phase duration and the glottal cycle duration. **Speed Quotient** (SQ), expressed as the ratio of opening and closing phase duration. **Amplitude Quotient** (AQ), which is defined as the ratio of the maximum of the glottal flow and the minimum of its derivative. Finally, **normalized AQ** with respect to the glottal period (NAQ). Figure 2 illustrates the described features extracted from the glottal flow. Besides, considering the spectrum of the glottal flow, some features are introduced: **H1H2**, defined as $H1 - H2$, where $H1$ and $H2$ are the first two harmonics of the glottal flow signal, and the **Harmonic Richness Factor** (HRF) is calculated as the ratio of the sum of the harmonics amplitude and the amplitude of the fundamental frequency.

Nonlinear Dynamics (NLD) Measures. Glottal and residual flows are preprocessed by means of a short-time analysis using windows of 55 ms length with an overlap of 50 %, where the glottal inverse filtering process has been applied previously. Before estimation of the nonlinear features, an embedding attractor has to be reconstructed from each flow. The state-space reconstruction is based

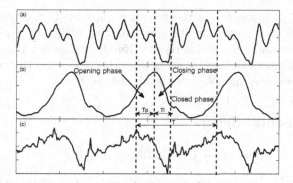

Fig. 2. Features extracted from the glottal flow. (a) Three periods of the voice signal, (b) the glottal flow, (c) derivative of the glottal flow signal.

on the time-delay embedding theorem [5]. A set of eight NLD measures is calculated after the embedding process: **Correlation dimension** (D_2) measures the space dimensionality occupied by the points in the reconstructed attractor. It is implemented according to the Takens estimator method [5]. **Largest Lyapunov exponent** (LLE) is estimated as the average divergence rate of neighboring trajectories in the attractor, according to the Rosenstein method [5]. **Lempel-Ziv complexity** (LZC) is used for complexity estimation in time series, its implementation consists of finding the number of different "patterns" present in a reconstructed binary string sequence. **Hurst exponent** (H) estimates the long-term dependencies in a time series, defined as the relation between the variation rank (R) of the signal and its standard deviation S, $\frac{R}{S} = cT^H$, where c is a scaling constant, and T is the duration of the segment, the estimation following the rank scaling method [5]. Moreover, entropy measures based on the uncertainty of a random variable are considered. By taking into account that in practical terms the Kolmogorov-Sinai entropy can not be computed, different estimation methods are used. One of them is the **approximate entropy** (A_E), which is designed for measuring the average conditional information generated by diverging points on a trajectory in the state space [6]. The main drawback of A_E is its dependence on the signal length due to the self-comparison of points in the attractor. In order to overcome this problem, the **sample entropy** (E_S) is proposed. The only difference lies in the non-comparison of embedding vectors with themselves. Another modification of A_E is the **approximate entropy with Gaussian kernel** E_AGK. It exploits the fact that a Gaussian kernel function can be used to give greater weight to nearby points by replacing the Heaviside function. Finally, the same procedure of changing the distance measure can be applied to define the **sample entropy with Gaussian kernel** E_SGK.

2.3   Classification

An SVM classifier is trained using a radial basis Gaussian kernel with bandwidth σ. To achieve a more robust classifier, the number of support vectors is also optimized with respect to the accuracy obtained in the training process avoiding over-fitting, increasing the generalization ability of the classifier and exhibiting better and more stable results [15]. This classifier is considered here due to its validated success in similar studies that addressed the problem of the automatic detection of pathological speech signals [8].

3   Experimental Setup

3.1   Corpus of Speakers – PC-GITA

This database [12] contains speech recordings of 50 patients with PD and 50 healthy controls (HC) sampled at 44.1 kHz with 16-bit resolution. The speakers in this database are balanced by gender and age between the two subgroups. All of the patients were diagnosed by neurologist experts; the mean values of their evaluation according to the UPDRS-III and Hoehn and Yahr scales are 38.2 and 2.3, respectively. None of people in the HC group has a history of symptoms related to PD or any other kind of movement disorder. The recordings consist of sustained phonation of the five Spanish vowels: /a/, /e/, /i/, /o/, and /u/. Every person repeated the five vowels three times, thus in total the database is composed of 150 recordings per vowel on each class, e.g., PD or HC, respectively.

3.2   Experiment

TF glottal flow parameters and nonlinear behavior from the glottal and residuals waves are considered. First, each one of the seven TF parameters described above is calculated for every glottal closure instant for every person. Second, eight NLD measures, also described above, are obtained from every frame of each one of the signals. Finally, by considering the fact that every measure has a dynamic representation, four functionals are estimated on each parameter per recording: mean value (m), standard deviation (std), kurtosis (k), and skewness (sk).

The classification is performed using a soft margin support vector machine (SVM) with margin parameter C and a Gaussian kernel with parameter γ. The parameters of the SVM are optimized using steps of powers of ten through a gridsearch with $10^{-1} < C < 10^4$ and $1 < \gamma < 10^3$, and the accuracy on the test data as a selection criterion. Note that the optimization criterion could lead to slightly optimistic accuracy estimates, but as there are only two parameters to be optimized, the bias effect should be minimal. The SVM is tested following a 10-fold cross-validation strategy. The folds were formed randomly but ensuring speaker independence and balance in age and gender per fold.

3.3   Results

The results are presented in terms of accuracy, sensitivity, and specificity. The area under the ROC curve (AUC) is also presented in order to give compact information regarding the general performance of the system. Table 1 shows the

Table 1. Performance considering TF and NLD measures

	Vowel	Accuracy (%)	Sensitivity (%)	Specificity (%)	AUC
Glot. TF	/a/	76 ± 8	73 ± 19	79 ± 14	0.81
Glot. NLD		73 ± 8	76 ± 10	70 ± 19	0.75
Res. NLD		77 ± 7	74 ± 12	81 ± 12	0.75
Glot. TF	/e/	77 ± 9	85 ± 14	69 ± 14	0.81
Glot. NLD		75 ± 10	78 ± 14	73 ± 21	0.76
Res. NLD		74 ± 8	73 ± 13	75 ± 21	0.72
Glot. TF	/i/	72 ± 11	73 ± 15	71 ± 28	0.76
Glot. NLD		72 ± 10	71 ± 19	73 ± 20	0.71
Res. NLD		72 ± 6	73 ± 16	71 ± 20	0.74
Glot. TF	/o/	75 ± 8	67 ± 13	84 ± 16	0.78
Glot. NLD		73 ± 7	74 ± 10	71 ± 17	0.71
Res. NLD		71 ± 12	67 ± 15	75 ± 20	0.70
Glot. TF	/u/	75 ± 8	69 ± 17	82 ± 8	0.77
Glot. NLD		73 ± 5	75 ± 12	71 ± 16	0.72
Res. NLD		75 ± 7	70 ± 17	81 ± 24	0.76
Glot. TF	Union	78 ± 10	78 ± 15	77 ± 19	0.81
Glot. NLD		75 ± 9	79 ± 16	71 ± 22	0.79
Res. NLD		75 ± 9	69 ± 12	81 ± 22	0.77

Glot. TF: Glottal Time-Frequency, Glot. NLD: Glottal Nonlinear Dynamics, Res. NLD: Residual Nonlinear Dynamics

results obtained when each Spanish vowel is modeled using the TF parameters over the glottal flow (Glot. TF), and using nonlinear features extracted from the glottal and residual flows (Glot. NLD and Res. NLD, respectively). Additionally, the last 3 rows of the table contain the results when each measure was obtained considering the union of the five Spanish vowels.

Note that TF parameters achieve the best performance in most of the cases; only with the vowel /a/, the best performance is presented by NLD features from the residual flow with 76 %. Although TF parameters present the best accuracy values, it can be noted that the sensitivity and specificity values are not strongly similar; thus, the ability to detect a person with PD or a HC is not the same. But when the vowels are considered jointly, the accuracy value achieves the best performance, and both sensitivity and specificity are sufficiently similar to detect pathological speakers or healthy persons.

Furthermore, when the AUC is considered, it can be noted that TF parameters achieved the best performance in all the vowels and with their union. Figure 3 shows the best ROC curves obtained when the vowels are considered jointly.

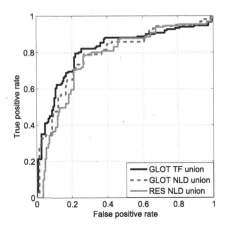

Fig. 3. ROC curves obtained when the vowels are merged

4   Conclusions

Problems in PPD related to vocal bowing and incomplete close of vocal folds are analyzed by means of the automatic separation of source, i.e., glottal flow, from the speech signal. Besides, when the source was estimated, it could obtain the residual wave, which gives information related to the articulation process. It could also be able to give some clues about the nonlinear behavior in the vocal tract, possibly due to the turbulent flow. One of the aims of the work was to determine which set of features offers more discriminatory capability to detect PD, either the TF parameters from the glottal flow, or the NLD measures estimated from the glottal and residual flows. In this sense, by means of accuracy and also with AUC measures of performance, the TF parameters are the best set in this task. It could be due to the ability of representing the glottal phases, describing in detail the phonatory process which is strongly involved in the speech impairments in PPD.

The second aim of this work was to analyze whether the nonlinearity in speech signals comes from the phonation or articulation process. The results show a similar behavior when NLD measures of glottal and residual flows are compared, thus it seems like phonation is not the only phenomenon in speech that is providing nonlinearities; there should be a nonlinear effect in the articulation process when a turbulent flow passes through the vocal tract. This work is our first approach to PD detection using nonlinear behavior of the glottal flow. For future work more nonlinear features will be considered to improve the accuracy and robustness of the models.

Acknowledgments. This work was financed by COLCIENCIAS, project Nº 11155-6933858.

References

1. Alku, P., Svec, J., Vilkman, E., Sram, F.: Glottal wave analysis with pitch synchronous iterative adaptive inverse filtering. Speech Commun. 11, 109–118 (1992)
2. Bocklet, T., Nöth, E., Stemmer, G., Ruzickova, H., Rusz, J.: Detection of persons with Parkinson's disease by acoustic, vocal, and prosodic analysis. In: Proceedings of IEEE ASRU, pp. 478–483 (2011)
3. Drugman, T., Thomas, M., Gudnason, J., Naylor, P., Dutoit, T.: Detection of glottal closure instants from speech signals: a quantitative review. IEEE Trans. Audio Speech Lang. Process. 20(3), 994–1006 (2012)
4. Ho, A., Jansek, R., Marigliani, C., Bradshaw, J., Gates, S.: Speech impairment in a large sample of patients with Parkinson's disease. Behav. Neurol. 11, 131–137 (1998)
5. Kantz, H., Schreiber, T.: Nonlinear Time Series Analysis, 2nd edn. Cambridge University Press, Cambridge (2004)
6. Kaspar, F., Schuster, H.: Easily calculable measure for complexity of spatiotemporal patterns. Phys. Rev. A Gen. Phys. 36(2), 842–848 (1987)
7. Little, M., McSharry, P., Hunter, E., Spielman, J., Ramig, L.: Suitability of dysphonia measurements for telemonitoring of Parkinson's disease. IEEE Trans. Biomed. Eng. 56(4), 1015–1022 (2009)
8. Maier, A., Haderlein, T., Eysholdt, U., Rosanowski, F., Batliner, A., Schuster, M., Nöth, E.: PEAKS - a system for the automatic evaluation of voice and speech disorders. Speech Commun. 51(5), 425–437 (2009)
9. Merati, A., Heman-Ackah, Y., Abaza, M., Altman, K., Sulica, L., Belamowicz, S.: Common movement disorders affecting the larynx: a report from the neurolaryngology committee of the AAO-HNS. Otolaryngol. Head Neck Surg. 133, 654–665 (2005)
10. Midi, I., Dogan, M., Koseoglu, M., Can, G., Sehitoglu, M., Gunal, D.: Voice abnormalities and their relation with motor dysfunction in Parkinson's disease. Acta Neurol. Scand. 117(1), 26–34 (2008)
11. Orozco-Arroyave, J., Arias-Londoño, J., Vargas-Bonilla, J., Daqrouq, K., Skodda, S., Rusz, J., Hönig, F., Nöth, E.: Automatic detection of Parkinson's disease in continuous speech spoken in three different languages. J. Acoust. Soc. Am. 139(1), 481–500 (2016)
12. Orozco-Arroyave, J., Arias-Londoño, J., Vargas-Bonilla, J., Gonzáles-Rátiva, M., Nöth, E.: New Spanish speech corpus database for the analysis of people suffering from Parkinson's disease. In: Proceedings of the 9th LREC, pp. 342–347 (2014)
13. Orozco-Arroyave, J.R., Arias-Londoño, J.D., Vargas-Bonilla, J.F., Nöth, E.: Analysis of speech from people with Parkinson's disease through nonlinear dynamics. In: Drugman, T., Dutoit, T. (eds.) NOLISP 2013. LNCS, vol. 7911, pp. 112–119. Springer, Heidelberg (2013)
14. Ramig, L., Fox, C., Sapir, S.: Speech treatment for Parkinson's disease. Expert Rev. Neurother. 8(2), 297–309 (2008)
15. Schölkopf, B., Smola, A.: Learning with Kernels. The MIT Press, Cambridge (2002)
16. Tsanas, A., Little, M., McSharry, P., Spielman, J., Ramig, L.: Novel speech signal processing algorithms for high-accuracy classification of Parkinson's disease. IEEE Trans. Biomed. Eng. 59(5), 1264–1271 (2012)
17. Walker, J., Murphy, P.: A review of glottal waveform analysis. In: Stylianou, Y., Faundez-Zanuy, M., Esposito, A. (eds.) COST 277. LNCS, vol. 4391, pp. 1–21. Springer, Heidelberg (2007)

Correction of Prosodic Phrases in Large Speech Corpora

Zdeněk Hanzlíček[(✉)]

NTIS – New Technology for the Information Society, Faculty of Applied Sciences,
University of West Bohemia, Univerzitní 22, 306 14 Plze, Czech Republic
zhanzlic@ntis.zcu.cz
http://www.ntis.zcu.cz/en

Abstract. Nowadays, in many speech processing tasks, such as speech
recognition and synthesis, really large speech corpora are utilized. These
speech corpora usually contain several hours of speech or even more.
To achieve possibly best results, an appropriate annotation of the
recorded utterances is often necessary. This paper is focused on problems
related to the prosodic annotation of the Czech speech corpora. In the
Czech language, the utterances are supposed to be split by pauses into
so-called prosodic clauses containing one or more prosodic phrases. The
types of particular phrases are linked to their last prosodic words corresponding to various functionally involved prosodemes. The clause/phrase
structure is substantially determined by the sentence composition. However, in real speech data, different prosodeme type or even phrase/clause
borders can be present. This paper deals with 2 basic problems: the correction of the improper prosodeme/phrase type and the detection of new
phrase borders. For both tasks, we proposed new procedures utilizing
hidden Markov models. Experiments were performed on 4 large speech
corpora recorded by professional speakers for the purpose of speech synthesis. These experiments were limited to the declarative sentences. The
results were successfully verified by listening tests.

Keywords: Speech corpora · Prosody · Annotation

1 Introduction

Nowadays, in many speech processing tasks, such as speech recognition and synthesis, really large speech corpora are utilized. These speech corpora usually
contain several hours of speech or even more. To achieve possibly best results,
an appropriate annotation of the recorded utterances is often necessary. In connection with using the large speech corpora, the automatic phonetic and prosodic

This research was supported by the Czech Science Foundation (GA CR), project
No. GA16-04420S. Access to computing and storage facilities owned by parties and
projects contributing to the National Grid Infrastructure MetaCentrum, provided
under the programme CESNET LM2015042, is greatly appreciated.

© Springer International Publishing Switzerland 2016
P. Sojka et al. (Eds.): TSD 2016, LNAI 9924, pp. 408–417, 2016.
DOI: 10.1007/978-3-319-45510-5_47

annotation of speech [7] became an important task. This paper deals with 2 basic problems: the correction of the improper prosodeme/phrase type and the detection of new phrase borders.

1.1 Prosody Model

For our purposes, we used the formal prosody model proposed by Romportl [5]. On the basis of this model, an utterance is divided into prosodic clauses separated by short pauses. Each prosodic clause includes one or more prosodic phrases containing certain continuous intonation schemes. Furthermore, phrases are composed of prosodic words. The communication function the speaker intends the phrase to have (the type of the phrase) is supposed to be linked with the last prosodic word in the phrase. For this purposes, so called prosodemes are defined. The last prosodic word is linked with a functionally involved prosodeme, other words with null prosodemes. For the Czech language, the following basic classes of functionally involved prosodemes were defined [5]:

P1 – terminating satisfactorily (in declarative sentences)
P2 – terminating unsatisfactorily (in questions)
P3 – non-terminating (in non-terminal phrases of compound sentences)

Since this research is limited to the declarative sentences and neutral speech (i.e. without emphasis, expressions etc.), prosodemes P0, P1.1 and P3.1 were applied. According to the theoretical assumption, all the compound sentences consist of several phrases, where the last one is terminated with prosodeme P1.1 and the other phrases end with P3.1; see a simple example in Fig. 1.

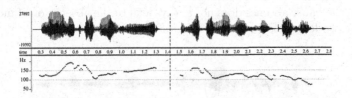

Fig. 1. Declarative compound sentence *"Málokdo věří, že by mohl zvítězit."* (*"Few believe that he could win."*). This prosodeme combination corresponds to the prosody model: the first phrase ends with P3.1 prosodeme and the last one with P1.1.

Particular prosodemes are linked with specific speech features: P1.1 is characteristic with a pitch decrease within its last syllable and a pitch increase is typical for P3.1. Beside the pitch shape (which is the most relevant), spectral features, duration and energy can be important for particular prosodemes. Naturally, particular types of phrases do not vary solely within their last prosodic words. Some specific prosodic differences can be present throughout the whole utterances. However, those differencies are often rather content-related (e.g. emphasis on some key words) and a more complex prosody model would be required. The utilized prosody model seems to be adequate for the phrase classification [1].

1.2   Problems in Real Speech

In real speech data, a different prosodeme than expected could be present. A typical example is a compound sentence split into several independent sentences. Within the compound sentence, all phrases (except the last one) should be terminated with the prosodeme P3.1. However, when the link between particular phrases is weak, the utterance can be split into independent sentences which are naturally terminated by the prosodeme P1.1. This is illustrated in Fig. 2.

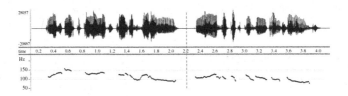

Fig. 2. Declarative compound sentence *"My jsme ekonomické oddělení, ne detektivní kancelář."* (*"We are the economic department, not a detective agency."*). The first phrase is terminated by an evident prosodeme P1.1.

In the Czech text, particular phrases are supposed to be separated by punctuation marks, usually commas[1]. Corresponding segments of speech are supposed to be prosodic phrases ended by functionally involved prosodemes. However, this theoretical assumption is not always fulfilled: Pauses can appear inside text phrases, especially when they are long. Or contrarily, more text phrases can be uttered together without indication of any functionally involved prosodeme. Moreover, the pause absence does not always lead to the absence of a functionally involved prosodeme; please compare Figs. 3 and 4.

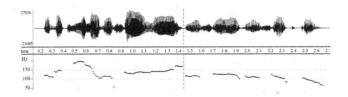

Fig. 3. Declarative compound sentence *"Aby cíle dosáhl, musí mít výsledky."* (*"To achieve the goal, the results are necessary."*). Though there is no pause, the prosodeme P3.1 terminating the first part is obvious.

Badly annotated speech corpora can be a source of various troubles. In speech synthesis (specifically, in unit selection method), prosodeme labels are important

---

[1] This is in contrast with English, where using commas has more complex rules. However, some copulative conjunctions in Czech are also used without a comma, e.g. *"a"*, *"nebo"*, *"ani"*, etc. (*"and"*, *"or"*, *"nor"*, respectively).

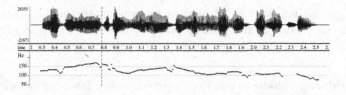

Fig. 4. Declarative compound sentence *"Nevím, kdo jiný by jim mohl pomoci."* (*"I don't know who else could help them."*). The punctuation in text has no evident impact on prosody realization; neither pause nor functional prosodeme are present.

attributes for selecting the optimal sequence of speech units for building resulting speech [6]. Using units from an inappropriate prosodeme or mixing units from various prosodemes can cause a degradation of the overall speech quality.

2   Proposed Approach

To model the prosodic properties of speech we employed a similar HMM framework as it is specific for the HMM-based speech synthesis [8]. Speech was described by a sequence of parameter vectors containing 40 mel cepstral coefficients obtained by STRAIGHT analysis method [2] and the pitch extracted by using the PRAAT software[2]. The speech parameter vectors were modelled by a set of multi-stream context dependent HMMs by using the HTS toolkit[3].

In the HMM-based speech synthesis framework, the phonetic, prosodic and linguistic context is taken into account, i.e. a speech unit is given as a phone with its phonetic, prosodic and linguistic context information. In this manner, the language prosody is modelled implicitly – in various contexts different units/models can be used. In our experiments, each unit is represented by a string

$$a_\ell\text{-}a_c\text{+}a_r@P:p_f\text{-}p_b@S:s_{f1}\,|\,s_{f2}\text{-}s_{b1}\,|\,s_{b2}@W:w_f\text{-}w_b \sim p_x$$

where all subscripted bold letters are contextual factors defined as

a_ℓ, a_c, a_r ... left context, current phoneme and right context
p_f, p_b ... forward and backward position of phone in prosodic word
s_{f1}, s_{b1} ... forward and backward position of syllable in prosodic word
s_{f2}, s_{b2} ... forward and backward position of syllable in phrase
w_f, w_b ... forward and backward position of prosodic word in phrase
p_x ... prosodeme type

2.1   Training Stage

For our experiments, we used 4 large speech corpora recorded for the purposes of speech synthesis. At the beginning, all utterances were segmented to phrases

---

[2] Praat: doing phonetics by computer, www.praat.org.
[3] HMM-based Speech Synthesis System (HTS), http://hts.sp.nitech.ac.jp.

only by detected pauses, i.e. all phrases correspond to clauses. This manner of phonetic annotation is also used in our unit selection TTS system [3], since functionally involved prosodemes are ensured at the end of all phrases.

Model Training. Model parameters were estimated from the speech data by using maximum likelihood criterion. 3-state left-to-right MSD-HSMM with single Gaussian output distributions were used. For a more robust model parameter estimation, the context clustering based on the MDL criterion was performed. In this stage, the default prosodic annotation of particular phrases is used.

Prosodeme Correction. This procedure is a modified version of a more general method introduced in [1]. First, each individual phrase terminated by the prosodeme P3.1 is transcribed by using the prosodeme P1.1, i.e. both transcriptions differ only in the prosodeme contextual factors within the last prosodic word; it is analogous to the example in Table 1, but simpler. Then corresponding speech features are forced-aligned with both transcriptions and the transcription with the best value of alignment score is selected for the given phrase.

When a corrected transcription of all utterances is available, the whole process can be run iteratively. The procedure works on the assumption that most utterances correspond to the theoretical prosody model with some rare exceptions. Then, the trained HMMs are correct and can be used to reveal those exceptions. Problems occur in the case of less consistent prosody: the inconsistencies can cumulate, a part of models can be badly trained and some performed corrections are wrong. To cope with that, an additional step is performed at the

Table 1. An example of splitting utterances by the punctuation: *"Řekl, že přijde."* (*"He said that he will come."*, its phonetic transcription: RekL Ze pQijde). Changes are underlined.

Phones	Default sentence (one phrase)	Sentence split into 2 phrases
R	$-R+e@P:1_4@S:0|0_2|5@W:1_2 ~ 0	$-R+e@P:1_4@S:0|0_2|2@W:1_1 ~ 31
e	R-e+k@P:2_3@S:1|1_2|5@W:1_2 ~ 0	R-e+k@P:2_3@S:1|1_2|2@W:1_1 ~ 31
k	e-k+L@P:3_2@S:1|1_1|4@W:1_2 ~ 0	e-k+L@P:3_2@S:1|1_1|1@W:1_1 ~ 31
L	k-L+Z@P:4_1@S:2|2_1|4@W:1_2 ~ 0	k-L+Z@P:4_1@S:2|2_1|1@W:1_1 ~ 31
Z	L-Z+e@P:1_8@S:0|2_3|3@W:2_1~11	L-Z+e@P:1_8@S:0|0_3|3@W:1_1~11
e	Z-e+p@P:2_7@S:1|3_3|3@W:2_1~11	Z-e+p@P:2_7@S:1|1_3|3@W:1_1~11
p	e-p+Q@P:3_6@S:1|3_2|2@W:2_1~11	e-p+Q@P:3_6@S:1|1_2|2@W:1_1~11
Q	p-Q+i@P:4_5@S:1|3_2|2@W:2_1~11	p-Q+i@P:4_5@S:1|1_2|2@W:1_1~11
i	Q-i+j@P:5_4@S:2|4_2|2@W:2_1~11	Q-i+j@P:5_4@S:2|2_2|2@W:1_1~11
j	i-j+d@P:6_3@S:2|4_1|1@W:2_1~11	i-j+d@P:6_3@S:2|2_1|1@W:1_1~11
d	j-d+e@P:7_2@S:2|4_1|1@W:2_1~11	j-d+e@P:7_2@S:2|2_1|1@W:1_1~11
e	d-e+$@P:8_1@S:3|5_1|1@W:2_1~11	d-e+$@P:8_1@S:3|3_1|1@W:1_1~11

.end of each iteration: the prosodeme correction procedure is performed by using HMMs from another speaker. Only corrections performed in both cases are kept, other changes are annulled, therefore this step is referred to as the *annulling step*.

Splitting Phrases by Punctuation. First, all phrases containing punctuation marks are further split into phrases terminated by P3.1 (excluding the last one, naturally). A simple example is presented in Table 1. When more commas are present in the phrase, all possible split combinations are taken into account. Again, the corresponding speech features are forced-aligned with all transcriptions and the transcription with the best alignment score is selected.

3 Evaluation and Results

For our experiments, 4 large speech corpora recorded for the purposes of speech synthesis [4] were used: 2 male voices (denoted as AJ and JS) and 2 female voices (denoted as KI and MR). Each corpus contained about 10,000 declarative sentences. The detailed description of experimental data is present in Table 2.

Although all corpora are almost equal, some statistics are very different. This indicates various speaking styles of particular speakers. For example, the number of commas inside phrases corresponds how often speakers join text segments separated by a comma into one phrase. By contrast, the number of phrases without any end punctuation tells how often speakers make pauses inside continuous text segments. To illustrate the prosody variability, we performed one iteration of the correction procedure without the annulling step. The higher number of changes is, the lower the consistency is supposed to be – see Table 3.

The iterative correction procedure with annulling step was tested only on voices AJ and MR. The annulling step was performed by using models from JS and KI, since these voices seem to be more consistent and their models are expected to be more robust. Results are presented in Table 4. Splitting phrases by punctuation was performed for all speakers, results are presented in Table 5. Since this splitting procedure is presented as fully new, we did not perform iterations, nor the annulling step in our experiments.

Table 2. Description of experimental data. Please note that the total number of phrases is given as phrases ended by a comma + ended by a dot + without any end punctuation.

Speaker		AJ	JS	KI	MR
Utterances		9,996	9,846	9.896	9,878
Commas	Total	11,400	10,851	10,841	11,249
	Inside phrases	1,998	1,001	8,503	5,013
Phrases	Total	22,971	20,097	13,166	18,236
	Ended by comma	9,381	9,847	2,332	6,217
	Without end punctuation	3,594	404	938	2,141

Table 3. The initial number of prosodemes and the number of P3.1 → P1.1 changes.

Speaker	AJ	JS	KI	MR
# P1.1 prosodemes	9,996	9,846	9,896	9,878
# P3.1 prosodemes	12,974	10,250	3,269	8,358
# changes	114	60	21	452

Table 4. Changing prosodemes P3.1 → P1.1: the initial number of prosodemes and the number of changes in particular iterations. Please remember that the corrections are always performed on the default corpora (the correction procedure is not cumulative).

Speaker	# P1.1	# P3.1	# changes		
			iter.1	iter.2	iter.3
AJ	9,996	12,974	49	56	59
MR	9,878	8,358	223	257	273

Table 5. Splitting utterances by the punctuation. The number of changes affects is equal for both number of phrases and prosodemes since each splitting produces a new phrase ended with the P3.1 prosodeme.

Speaker		AJ	JS	KI	MR
default	# phrases	9,996	9,846	9,896	9,878
	# P3.1 prosodemes	12,974	10,250	3,269	8,358
# changes	annulled	154	47	412	813
	performed	245	116	524	714

3.1 Listening Tests

The suitability of the performed corrections was verified by one overall listening test. It contained 120 utterances with one selected prosodic word. Listeners picked one of 5 choices: definitely P1.1, probably P1.1, definitely P3.1, probably P3.1, null prosodeme. Sentences were selected to be short and simple like the examples in Figs. 1, 2, 3 and 4. Five participants took part in this test, all of them were speech processing experts capable to distinguish various prosodeme types.

The test contained 40 utterances (20 for both AJ and MR) for the evaluation of the prosodeme changing procedure: 2 × 10 utterances with P3.1 to P1.1 corrections and 2 × 10 utterances with corrections discarded in the annulling step. The remaining 80 utterances (20 for each speaker) were intent for the evaluation of the splitting procedure: 4 × 10 utterances that were additionally split by a comma (split utterances) and 4 × 10 utterances that contain a comma, but the splitting was not performed (non-split utterances).

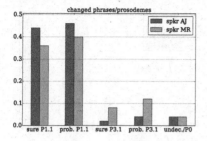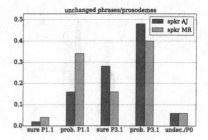

Fig. 5. Results of listening test on changing prosodemes P3.1 → P1.1.

Changed Prosodemes. The distribution of listeners' choices is present in Fig. 5 and Table 6. The most relevant entries are the percentages of changed prosodemes that were marked as P3.1: 90 % and 76 % for AJ and MR, respectively. The other 6 % and 20 % were marked as P1.1 and the remaining 4 % (for both speakers) were indecisive cases. Since only 3 iterations of correction procedure were performed and it wasn't the final state, further improvement could be expected.

As explained in Sect. 2, the purpose of annulling step is to increase the robustness within several initial iterations of the correction procedure. All changes can be still applied in the latter stage without the annulling step. Anyway, the more annulled cases really does not match the desired prosodeme, the more beneficial this step is. In our case, this rate is about 82 % and 62 % (all non-P1.1 cases).

Table 6. Results of listening test on changing prosodemes P3.1 → P1.1: percentage of particular listeners' choices. The agreement between human listeners and the proposed correction procedure is expressed mainly by the bold values.

phrases	speaker	prosodeme P1.1			prosodeme P3.1			P0
		sure	probably	total	sure	probably	total	
changed	AJ	44.0	46.0	**90.0**	2.0	4.0	6.0	4.0
	MR	36.0	40.0	**76.0**	8.0	12.0	20.0	4.0
	all	40.0	43.0	**83.0**	5.0	8.0	13.0	4.0
unchanged	AJ	2.0	16.0	18.0	28.0	48.0	76.0	6.0
	MR	4.0	34.0	38.0	16.0	40.0	56.0	6.0
	all	3.0	25.0	28.0	22.0	44.0	66.0	6.0

Split Phrases. Results of listening test are presented in Fig. 6 and Table 7. A high consistency between listeners and the proposed procedure is evident: prosodemes in split utterances were annotated as definitely or probably P3.1 in about 88 % cases for all speakers (ranged between 84 % for KI and 91 %

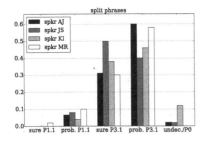
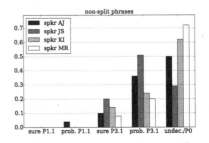

Fig. 6. Results of listening test: splitting utterances into phrases by punctuation.

Table 7. Results of listening test on splitting phrases by punctuation: percentage of particular listeners' choices. The agreement between human listeners and the proposed splitting procedure is expressed mainly by the bold values.

Phrases	Speaker	Prosodeme P1.1			Prosodeme P3.1			P0
		Sure	Probably	Total	Sure	Probably	Total	
Split	AJ	0.0	7.0	7.0	31.0	60.0	**91.0**	2.0
	JS	0.0	8.0	8.0	50.0	40.0	**90.0**	2.0
	KI	0.0	4.0	4.0	38.0	46.0	**84.0**	12.0
	MR	2.0	10.0	12.0	30.0	58.0	**88.0**	0.0
	all	0.5	7.3	7.8	37.3	51.0	**88.3**	4.0
Non-split	AJ	0.0	4.0	4.0	10.0	36.0	46.0	50.0
	JS	0.0	0.0	0.0	21.0	51.0	72.0	28.0
	KI	0.0	0.0	0.0	14.0	24.0	38.0	62.0
	MR	0.0	0.0	0.0	8.0	20.0	28.0	72.0
	all	0.0	1.0	1.0	13.5	32.8	46.3	53.3

for AJ). Surprisingly, an appreciable amount of P1.1 prosodemes appeared in listeners' selections. Actually, it is in accordance with the experiment on changing prosodemes and some P1.1s could be expected here, too.

The actual benefit of the splitting procedure should be also apparent by a comparison of results for the split and non-split utterances. Above all, significantly less P3.1s and more null prosodemes should be present in non-split sentences. This is true; nevertheless, the number of P3.1s in non-split utterances is higher than expected, especially 72 % for JS. The reason could be the influence of the sentence structure on the listeners' decision. Evidently, it depends on the actual speaker, too.

4   Conclusion

This paper presented 2 procedures for the correction of the type and borders of prosodic phrases in large speech corpora. Experiments were performed on

4 corpora. The results have been verified in a listening test. The agreement between the listeners and the proposed procedures was about 83 % for changing the prosodeme type and 88 % for splitting utterances into phrases by the punctuation.

In our future work, both proposed procedures should be joint together into one iterative correction process. The robustness could be improved by employing speaker-independent models and their adaptation. Other types of phrases (e.g. various types of questions) will be included, too. A big challenge is the automatic prosody annotation of speech data, especially of non-professional speakers whose prosody could be problematic due to its bad consistency.

References

1. Hanzlíček, Z.: Classification of prosodic phrases by using HMMs. In: Král, P., et al. (eds.) TSD 2015. LNCS, vol. 9302, pp. 497–505. Springer, Heidelberg (2015). doi:10. 1007/978-3-319-24033-6_56
2. Kawahara, H., Masuda-Katsuse, I., de Cheveigne, A.: Restructuring speech representations using a pitch-adaptive time-frequency smoothing and an instantaneous-frequency-based F0 extraction: possible role of a repetitive structure in sounds. Speech Commun. **27**, 187–207 (1999)
3. Matoušek, J., Tihelka, D., Romportl, J.: Current state of Czech text-to-speech system ARTIC. In: Sojka, P., Kopeček, I., Pala, K. (eds.) TSD 2006. LNCS (LNAI), vol. 4188, pp. 439–446. Springer, Heidelberg (2006)
4. Matoušek, J., Tihelka, D., Romportl, J.: Building of a speech corpus optimised for unit selection TTS synthesis. In: Proceedings of LREC 2008 (2008)
5. Romportl, J., Matoušek, J., Tihelka, D.: Advanced prosody modelling. In: Sojka, P., Kopeček, I., Pala, K. (eds.) TSD 2004. LNCS (LNAI), vol. 3206, pp. 441–447. Springer, Heidelberg (2004)
6. Tihelka, D., Matoušek, J.: Unit selection and its relation to symbolic prosody: a new approach. In: Proceedings of Interspeech 2006, pp. 2042–2045 (2006)
7. Wightman, C., Ostendorf, M.: Automatic labeling of prosodic patterns. IEEE Trans. Speech Audio Process. **2**, 469–481 (1994)
8. Zen, H., Tokuda, K., Black, A.W.: Statistical parametric speech synthesis. Speech Commun. **51**(11), 1039–1064 (2009)

Relevant Documents Selection for Blind Relevance Feedback in Speech Information Retrieval

Lucie Skorkovská[(✉)]

New Technologies for the Information Society and Department of Cybernetics,
Faculty of Applied Sciences, University of West Bohemia,
Univerzitní 8, 306 14 Plzeň, Czech Republic
lskorkov@ntis.zcu.cz

Abstract. The experiments presented in this paper were aimed at the selection of documents to be used in the blind or pseudo relevance feedback in spoken document retrieval. The previous experiments with the automatic selection of the relevant documents for the blind relevance feedback method have shown the possibilities of the dynamical selection of the relevant documents for each query depending on the content of the retrieved documents instead of just blindly defining the number of the relevant documents to be used in advance. The score normalization techniques commonly used in the speaker identification task are used for the dynamical selection of the relevant documents. In the previous experiments, the language modeling information retrieval method was used. In the experiments presented in this paper, we have derived the score normalization technique also for the vector space information retrieval method. The results of our experiments show, that these normalization techniques are not method-dependent and can be successfully used in several information retrieval system settings.

Keywords: Query expansion · Blind relevance feedback · Spoken document retrieval · Score normalization

1 Introduction

In the past years, the focus of the information retrieval (IR) field slowly shifts from the text IR to the speech IR. With the large audio-video databases available on-line it is only natural, that researchers from many fields like history, arts or culture request comfortable and easy access to the documents contained in them. Going through every audio document by listening to it is not possible. The most frequent approach to handling this problem is the use of automatic speech recognition (ASR) to transcribe the speech data into the text data and then use the classic IR methods to search in them. Experiments with the speech retrieval collections containing conversational speech [1] suggest that these IR methods alone are not sufficient enough for successful retrieval. The biggest issue here is that the query words are often not found in the documents from the collection.

© Springer International Publishing Switzerland 2016
P. Sojka et al. (Eds.): TSD 2016, LNAI 9924, pp. 418–425, 2016.
DOI: 10.1007/978-3-319-45510-5_48

One cause of this problem is the high word error rate of the ASR system causing the query words to be misrecognized. The second cause is that the query words was actually not spoken and thus are not contained in the documents. To deal with this issue the query expansion techniques are often used. One of the query expansion methods often used in the IR field is the relevance feedback method. The idea behind this method is that the relevant documents retrieved in the first run of the search are used to enrich the user query for the second run. In most cases, the retrieval system does not have the feedback from the user and thus it does not know which documents are relevant. The blind relevance feedback method can be used, where the system "blindly" selects some documents, which it considers to be relevant and uses them for the enrichment of the user query. Mostly the N best scoring documents are selected, and the number N is chosen beforehand based on expert knowledge or previous experiments.

In this paper, the thorough experiments aimed at the better automatic selection of the relevant documents for the blind relevance feedback method are presented. Our idea is to apply the score normalization techniques used in the speaker identification task [2,3] to dynamically select the relevant documents for each query depending on the content of the retrieved documents instead of just experimentally defining the number of the relevant documents to be used for the blind relevance feedback in advance. In the previous experiments [4,5], the normalization methods were used in the language modeling information retrieval system. In the experiments presented in this paper, we have derived the score normalization techniques also for the vector space information retrieval method in order to find out if these normalization techniques are not method-dependent and can be successfully used in several information retrieval system settings.

2 Blind Relevance Feedback

Query expansion techniques based on the blind relevance feedback (BRF) method has been shown to improve the results of the IR [6]. The idea behind the BRF is that amongst the top retrieved documents most of them are relevant to the query and the information contained in them can be used to enhance the query for acquiring better retrieval results. First, the initial retrieval run is performed, documents are ranked according to some similarity or likelihood function. Then the top N documents are selected as relevant and the top k terms (according to some term importance weight L_t, for example *TF-IDF*) from them is extracted and used to enhance the query. The second retrieval run is then performed.

In the standard approach to the BRF, the number of documents and terms is defined experimentally in advance the same for all queries. In our experiments, we would like to find the number of relevant documents for each query automatically.

3 Previous Experiments

For the previous experiments [4,5], the language modeling (LM) approach [7] was used - the query likelihood model with a linear interpolation of the unigram language model of the document M_d with the model of the collection M_c. The idea of this method is to create a language model from each document d and then for each query q to rank the documents according to the probability $P(d|q)$:

$$P(d|q) \propto \prod_{t \in q} (\lambda P(t|M_d) + (1 - \lambda)P(t|M_c)), \tag{1}$$

where t is a term in a query and λ is the interpolation parameter ($\lambda = 0.1$ [8]). The importance weight L_t [7] was selected, R is the set of relevant documents:

$$L_t = \sum_{d \in R} \log \frac{P(t|M_d)}{P(t|M_c)}. \tag{2}$$

4   Vector Space Model

Vector space model (VSM) [9] is one of the most known and still most used models for IR. In the VSM the document d_j and query q are represented as vectors containing the importance weights $w_{i,j}$ of each of its terms:

$$d_j = (w_{1,j}, w_{2,j}, ..., w_{n,j}) \qquad q = (w_{1,q}, w_{2,q}, ..., w_{n,q})$$

For the $w_{i,j}$ we have used the TF-IDF weighting scheme, where tf_{t_i,d_j} is the term frequency and inverse document frequency is computed:

$$w_{i,j} = tf_{t_i,d_j} \cdot idf_{t_i}, \qquad idf_{t_i} = \log \frac{N}{n_i}, \tag{3}$$

where N is the number of documents and n_i is the number of documents containing t_i. The similarity of d_j and q is then computed using the cosine similarity:

$$sim_{d_j,q} = \frac{d_j \cdot q}{\|d_j\| \, \|q\|} = \frac{\sum_{i=1}^{t} w_{i,j} w_{i,q}}{\sqrt{\sum_{i=1}^{t} w_{i,j}^2} \sqrt{\sum_{i=1}^{t} w_{i,j}^2}} \tag{4}$$

The documents with the highest similarity are considered to be the most relevant.

4.1   Blind Relevance Feedback in VSM

For each document its similarity $sim_{d_j,q}$ is computed and the documents are sorted accordingly. The first N documents are considered to be relevant. For the selection of terms we have used the TF-IDF weight defined in (3).

5   Score Normalization for Relevant Documents Selection

The score normalization methods were used in the open-set text-independent speaker identification (OSTI-SI) problem. The OSTI-SI can be described as a twofold problem: First, the speaker model best matching the utterance has to be found and secondly, it has to be decided if the utterance has really been produced by this best-matching model. The relevant documents selection in IR can be described in the same way: First, we need to retrieve the documents which have the best scores for the query and second, we have to choose only the relevant documents. The only difference is that we try to find more than one relevant document. The normalization methods from OSTI-SI can be used in the same way, but have to be applied to all documents scores.

5.1 Score Normalization Methods

In the previous work [5], the score normalization methods were derived for the language modeling IR system described in Sect. 3 since its principle is the most similar to the OSTI-SI. In the following, the derivation process will be summarized and then it will be shown how the normalization methods are used in the vector space model system. After the initial run of the retrieval system, we have the ranked list of documents with their likelihoods $p(d|q)$. Similarly as in the OSTI-SI [2], we can define the decision formula:

$$p(d_R|q) > p(d_I|q) \rightarrow q \in d_R \quad \text{else} \quad q \in d_I, \tag{5}$$

where $p(d_R|q)$ is the score given by the relevant document model d_R and $p(d_I|q)$ is the score given by the irrelevant document model d_I. By the application of the Bayes' theorem, formula (5) can be rewritten as:

$$\frac{p(q|d_R)}{p(q|d_I)} > \frac{P(d_I)}{P(d_R)} \rightarrow q \in d_R \quad \text{else} \quad q \in d_I, \tag{6}$$

where $l(q) = \frac{p(q|d_R)}{p(q|d_I)}$ is the normalized likelihood score and $\theta = \frac{P(d_I)}{P(d_R)}$ is a threshold that has to be determined. Setting this threshold θ a priori is a difficult task, since we do not know the prior probabilities $P(d_I)$ and $P(d_R)$. Similarly as in the OSTI-SI task the document set can be open - a query belonging to a document not contained in our set can easily occur. A frequently used form to represent the normalization process [2] can be modified for the IR task:

$$L(q) = \log p(q|d_R) - \log p(q|d_I), \tag{7}$$

where $p(q|d_R)$ is the score given by the relevant document and $p(q|d_I)$ is the score given by the irrelevant document. The normalization score $\log p(q|d_I)$ is not known, there are several possibilities how to approximate it:

World Model Normalization (WMN). The unknown model d_I can be approximated by the collection model M_c (language model from all documents in the retrieval collection). This technique was inspired by the World Model normalization [10]. The normalization score of a model d_I is defined as:

$$\log p(q|d_I) = \log p(q|M_c). \tag{8}$$

Unconstrained Cohort Normalization (UCN). For every document model, a set (cohort) of N similar models $C = \{d_1, ..., d_N\}$ is chosen [11]. These models are the most competitive models, i.e. models which yield the next N highest likelihood scores. The normalization score is given by:

$$\log p(q|d_I) = \log p(q|d_{UCN}) = \frac{1}{N} \sum_{n=1}^{N} \log p(q|d_n). \tag{9}$$

Standardizing a Score Distribution. Another solution called Test normalization (T-norm) stated in [11] is to transform a score distribution into a standard form. The formula (7) now has the form:

$$L(q) = (\log p(q|d_R) - \mu(q))/\sigma(q), \tag{10}$$

where $\mu(q)$ and $\sigma(q)$ are the mean and standard deviation of the whole document likelihood distribution.

5.2   Score Normalization in VSM

In the previous text, the derivation of the score normalization methods for IR has been shown. Now we have to alter them for the use in the vector space model for IR. We will start with the normalization formula (7). The likelihood $p(d|q)$ can be replaced with the similarity $sim_{d_j,q}$, but since the likelihoods are in logarithms of probabilities the formula has to be changed to the form:

$$l(q) = sim_{d_R,q}/sim_{d_I,q}. \tag{11}$$

Then the actual score normalization methods can also be rewritten. We have done our experiments with the UCN and the T-norm methods since they are easily transformed for the use in VSM system. The WMN on the other hand, requires replacing the "world" model defined with the collection model M_c with some equivalent in the vector space.

The UCN method can be rewritten as:

$$sim_{d_I,q} = \frac{1}{N} \sum_{n=1}^{N} sim_{d_n,q}, \tag{12}$$

and the T-norm method will now have the form:

$$l(q) = (sim_{d_R,q} - \mu(q))/\sigma(q). \tag{13}$$

Threshold Selection. Even when we have the scores normalized, we still have to set the threshold for verifying the relevance of each document in the list. Selecting a threshold defining the boundary between the relevant and the irrelevant documents in a list of normalized scores is more robust because the normalization removes the influence of the various query characteristics. Since in the former experiments the threshold was successfully defined as a percentage of the normalized score of the best scoring document, the threshold θ will be similarly defined as the ratio k of the best normalized score.

6   Experiments

Since the previous experiments with the use of the score normalization methods for the selection of relevant documents for the BRF in language modeling (LM) IR system have shown very good results (summarized in Table 1), the experiments with the vector space system were performed. We have done thorough experiments with the setting of the standard blind relevance feedback method - the selection of the number of documents and the number of terms. We have found the best parameters settings and selected it for our baseline. Then detailed experiments with the score normalization methods were performed. The results of all the experiments are presented in Sect. 6.2.

Table 1. LM IR results (mGAP score) for no blind relevance feedback, with standard BRF and BRF with score normalization. 30 terms were used to enhance each query.

Method	No BRF	Standard BRF	BRF-WMN	BRF-UCN	BRF-T-norm
Params	-	# of doc. = 20	k = 0.5	k = 0.25, C = 85	k = 0.55
TDN set	0.0392	0.0513	**0.0568**	**0.0570**	0.0564

6.1   Information Retrieval Collection

The experiments were performed on the spoken document retrieval collection used in the Czech task of the Cross-Language Speech Retrieval track organized in the CLEF 2007 evaluation campaign [1]. The collection contains automatically transcribed spontaneous interviews (segmented into 22 581 "documents") and two sets of TREC-like topics, consisting of 3 fields - <title> (T), <desc> (D) and <narr> (N). The training topic set was used for our experiments and the queries were created from all terms from the fields T, D and N. Stop words were omitted, all the terms were also lemmatized [8]. The mean Generalized Average Precision (mGAP) measure that was used in the CLEF 2007 Czech task was used as an evaluation measure. The measure is described in detail in [12].

6.2   Results

Number of Terms. We have done experiments with the number of terms to select with all the described methods in this paper. The number of terms was selected from 5 to 45 terms, with 5 term interval (5, 10, 15...). In the previous experiments [5], all methods have shown best results when selecting around 20–30 terms. The premise that more terms are always better has shown not to be true. The experiments with the VSM model also shows that best results are achieved with a moderate number of terms selected (according to the weight (3)). For the normalization methods, it was again around 20–30 terms. For the standard BRF, it was around 10–25 terms when selecting from a higher count of documents (70–100) and more terms (30–40) when selecting from a lower count of documents - around 30. We think it is because when selecting a smaller amount of documents, the terms in them are all closely related to one subject, so they are more relevant.

But when selecting from more documents, the documents on the lower positions can be topically little shifted from the original query topic, so the number of truly relevant terms is smaller.

Standard BRF. We have experimented with the number of documents to select equal to 5, 10, 20, 30, 40, 50, 100, 200, 300, 500. Best results were achieved with a higher amount of documents, around 70–100. When selecting more documents the mGAP score dropped down, and also when selecting fewer documents.

Score Normalization. In score normalization methods, the number of documents to select for the BRF is dependent on the threshold θ defined as the ratio k of the best normalized score. The final number of documents selected this way is different for each query in the set. The experiments with the different ratio setting (from 0.1 to 0.95 with 0.05 distance) were done for all the score normalization methods presented.

In the **UCN method** apart from the ratio k also the size C of the cohort has to be set. Experiments with C from 5 to 500 with distance 10 were performed. The ratio k and the cohort size C depends on each other directly, because the normalization score in (12) is bigger (an average from the higher likelihoods) for a smaller cohort size. The best setting was $k = 0.95$ with $C = 295$ (also for other close sizes of C).

The **T-norm method** was also experimented with, the best results were obtained with $k = 0.35$. The setting is not very sensitive, the results for $k = 0.35$ or $k = 0.45$ was almost the same. It also stands for the UCN method.

The final comparison can be seen in Table 2. For all methods, 25 terms were selected. As can be seen from the table the BRF methods achieved a better score than without BRF. All the score normalization methods achieved a better mGAP score than the standard BRF, the best score achieved the UCN method.

Table 2. VSM IR results (mGAP score) for no blind relevance feedback, with standard BRF and BRF with score normalization. 25 terms were used to enhance each query.

Query set/method	No BRF	Standard BRF	BRF - UCN	BRF - T-norm
Parameters	-	# of doc. = 100	k = 0.95, C = 295	k = 0.35
Train TDN	0.0456	0.0560	**0.0602**	**0.0597**

7   Conclusions

The experiments showed in this paper, supports our claim that score normalization methods are very useful in the blind relevance feedback in IR. We have previously done experiments with the language modeling information retrieval system and now the experiments with the vector space model system. In both cases, the results were better with the use of the score normalization methods for the selection of documents for the BRF. Although the results of the VSM system has not shown to be significantly better with the score normalization, in the language modeling experiments the results were significantly better (with the Wilcoxon Matched-Pairs Signed-Ranks Test). For the future experiments, we have to derive also the World model normalization for the VSM system. It has to be found out if the "world" model will be some equivalent of the average of all vectors, or vector of all texts in all documents in the collection.

Acknowledgments. The work was supported by the Ministry of Education, Youth and Sports of the Czech Republic project No. LM2015071 and by the grant of the University of West Bohemia, project No. SGS-2016-039.

References

1. Ircing, P., Pecina, P., Oard, D.W., Wang, J., White, R.W., Hoidekr, J.: Information retrieval test collection for searching spontaneous Czech speech. In: Matoušek, V., Mautner, P. (eds.) TSD 2007. LNCS (LNAI), vol. 4629, pp. 439–446. Springer, Heidelberg (2007)
2. Sivakumaran, P., Fortuna, J., Ariyaeeinia, M.A.: Score normalisation applied to open-set, text-independent speaker identification. In: Proceedings of Eurospeech, Geneva, pp. 2669–2672 (2003)
3. Zajíc, Z., Machlica, L., Padrta, A., Vaněk, J., Radová, V.: An expert system in speaker verification task. In: Proceedings of Interspeech, vol. 9, pp. 355–358. International Speech Communication Association, Brisbane (2008)
4. Skorkovská, L.: First experiments with relevant documents selection for blind relevance feedback in spoken document retrieval. In: Ronzhin, A., Potapova, R., Delic, V. (eds.) SPECOM 2014. LNCS, vol. 8773, pp. 235–242. Springer, Heidelberg (2014)
5. Skorkovská, L.: Score normalization methods for relevant documents selection for blind relevance feedback in speech information retrieval. In: Král, P., Matoušek, V. (eds.) TSD 2015. LNCS, vol. 9302, pp. 316–324. Springer, Heidelberg (2015)
6. Ircing, P., Psutka, J.V., Vavruška, J.: What can and cannot be found in Czech spontaneous speech using document-oriented IR methods — UWB at CLEF 2007 CL-SR track. In: Peters, C., Jijkoun, V., Mandl, T., Müller, H., Oard, D.W., Peñas, A., Petras, V., Santos, D. (eds.) CLEF 2007. LNCS, vol. 5152, pp. 712–718. Springer, Heidelberg (2008)
7. Ponte, J.M., Croft, W.B.: A language modeling approach to information retrieval. In: Proceedings of SIGIR 1998, pp. 275–281. ACM, New York (1998)
8. Kanis, J., Skorkovská, L.: Comparison of different lemmatization approaches through the means of information retrieval performance. In: Sojka, P., Horák, A., Kopeček, I., Pala, K. (eds.) TSD 2010. LNCS, vol. 6231, pp. 93–100. Springer, Heidelberg (2010)
9. Salton, G., Wong, A., Yang, C.S.: A vector space model for automatic indexing. Commun. ACM **18**(11), 613–620 (1975)
10. Reynolds, D.A., Quatieri, T.F., Dunn, R.B.: Speaker verification using adapted Gaussian mixture models. Digit. Sig. Process. **10**, 19–41 (2000)
11. Auckenthaler, R., Carey, M., Lloyd-Thomas, H.: Score normalization for text-independent speaker verification systems. Digit. Signal Process. **10**(1–3), 42–54 (2000)
12. Liu, B., Oard, D.W.: One-sided measures for evaluating ranked retrieval effectiveness with spontaneous conversational speech. In: Proceedings of ACM SIGIR 2006, SIGIR 2006, pp. 673–674. ACM, New York (2006)

Investigation of Bottle-Neck Features for Emotion Recognition

Anna Popková[1(✉)], Filip Povolný[2], Pavel Matějka[1,2], Ondřej Glembek[1], František Grézl[1], and Jan "Honza" Černocký[1]

[1] Speech@FIT Group, Brno University of Technology, Brno, Czech Republic
xpopko00@stud.fit.vutbr.cz
[2] Phonexia s.r.o., Brno, Czech Republic

Abstract. This paper describes several systems for emotion recognition developed for the AV+EC 2015 Emotion Recognition Challenge. A complete system, making use of all three modalities (audio, video, and physiological data), was submitted to the evaluation. The focus of our work was, however, on the so called *Bottle-Neck* features used to complement the audio features. For the recognition of arousal, we improved the results of the delivered audio features and combined them favorably with the Bottle-Neck features. For valence, the best results were obtained with video, but a two-output Bottle-Neck structure is not far behind, which is especially appealing for applications where only audio is available.

Keywords: Emotion recognition · Bottle-Neck features · Context · Fusion

1 Introduction

The Speech@FIT group at Brno University of Technology and Phonexia are active and have been successful in multiple aspects of speech data mining. Recently, mainly with the EC-sponsored projects BISON[1] and MixedEmotions[2]

This work has been funded by the European Union's Horizon 2020 programme under grant agreement No. 644632 MixedEmotions and No. 645523 BISON, and by Technology Agency of the Czech Republic project No. TA04011311 "MINT". It was also supported by the Intelligence Advanced Research Projects Activity (IARPA) via Department of Defense US Army Research Laboratory contract number W911NF-12-C-0013. The U.S. Government is authorized to reproduce and distribute reprints for Governmental purposes notwithstanding any copyright annotation thereon. Disclaimer: The views and conclusions contained herein are those of the authors and should not be interpreted as necessarily representing the official policies or endorsements, either expressed or implied, of IARPA, DoD/ARL, or the U.S. Government. Thanks to Fabien Ringeval for scoring several other systems after the deadline of AVEC 2015 which allowed us to make proper analysis for this paper.

[1] http://bison-project.eu/.
[2] http://mixedemotions-project.eu/.

© Springer International Publishing Switzerland 2016
P. Sojka et al. (Eds.): TSD 2016, LNAI 9924, pp. 426–434, 2016.
DOI: 10.1007/978-3-319-45510-5_49

and with an interest both in academia and industry, emotion recognition has become increasingly important.

This paper presents our systems based on the material provided by Audio–Visual+Emotion Recognition Challenge (AV+EC 2015)[3] [1]. AVEC is an annual challenge held since 2011. Its main purpose is emotion recognition from multi-modal data—audio, video, and newly also physiological data. Emotion is understood here as a two-point continuous values on 2D plane according to arousal-valence model [11].

The data comes with three sets of features for audio, video and physiological signals. While the latter two were used as-is (the work concentrated on their post-processing, regressor training and fusion), in audio, we have complemented the provided material by Bottle-Neck (BN) features generated from a narrow hidden layer of a neural network trained toward phonetic targets. BN features were designed for automatic speech recognition [3] and have been included into most top-performing ASR systems including their multi-lingual variants [3]. Recently, BN features (and more general feature extraction schemes based on DNNs) were found very effective in other areas of speech processing, such as language recognition [4,5] and speaker identification [6,10]. Due to their ability to suppress nuisance variability in the speech data, we consider them a promising candidate also for emotion recognition, especially for the AV+EC challenge where very limited amount of labeled data (only 27 speakers) is available.

The rest of the paper provides a description of experiments leading to our submission for the AV+EC challenge and concentrates on BN features used for the audio modality.

2   Provided Material

2.1   Data

The data-set comes from the RECOLA multimodal database [2]. It contains spontaneous interactions in French. Participants were recorded in dyads during a video conference while resolving a collaborative task (winter survival task). Data was collected from 46 participants, but due to consent issues, only 5.5 hours of fully multimodal recordings from 27 participants are usable. The database is gender balanced and the mother tongues of speakers are French, Italian and German. The first 5 min of each recording were annotated by 6 French-speaking emotion annotators in the continuous arousal–valence space, leading to 135 min of data with the emotion ground truth. These recordings are divided into training, development and test sets, where annotations are provided only for training and development ones. The database is freely available[4] and full details are provided in [1,2].

---

[3] http://sspnet.eu/avec2015/.
[4] https://diuf.unifr.ch/diva/recola/.

2.2 AV+EC Features

Five sets of features were provided by the organizers (please refer to the challenge summary paper [1] for full description and references):

- **Audio** 102-dimensional feature set is extended Geneva Minimalistic Acoustic Parameter Set (eGeMAPS). These features are generated from short fixed length segments (3 s) shifted by 40 ms.
- **Video** features include two types of facial descriptors: appearance and geometry based. The former were extracted by Gabor Binary Patterns from Three Orthogonal Planes (LGBP-TOP) leading to total vector size of 84, the latter are facial landmarks leading to vector size of 316. Again, overlapping 3 s segments with 40 ms shift were used. The problem with video features was that for parts of the data, the face was not recognized and no information was provided. For certain recordings, the amounts of unrecognized frames were up to 40 %.
- **Physiological** sets include Electrocardiogram (ECG, 54 parameters) derived features, based on heart rate, its measure of variability, and derived parameters and statistics, and Electrodermal activity (EDA, 60 parameters) including skin conductance response (SCR), skin conductance level (SCL), as well as a number of derived parameters.

2.3 Evaluation and Baselines

The results were evaluated using the concordance correlation coefficient (CCC) to measure the correlation between the prediction and the reference. CCC combines the Pearson correlation coefficient of two time series ρ with mean square error:

$$CCC = \frac{2\rho\sigma_x\sigma_y}{\sigma_x^2\sigma_y^2 + (\mu_x - \mu_x)^2}. \tag{1}$$

CCC produces values from -1 to 1. The value of 1 means that the two variables are identical, -1 means that they are opposite, and 0 means that they are totally uncorrelated.

The organizers experimented with several emotion recognition schemes and provided the best obtained values in [1]. These serve as baselines for our work and are mentioned in the tables.

3   Bottle-Neck Features

We used Stacked Bottle-Neck (SBN) features as our additional feature set. The architecture for this kind of feature extraction consists of two NNs trained towards phonetic targets. The output of the first network is stacked in time, defining context-dependent input features for the second NN, hence the term Stacked Bottleneck Features [4].

The NN input features are filter-bank energies concatenated with fundamental frequency (F0) features produced by four different estimators: BUT F0 detector produces 2 coefficients (F0 and probability of voicing), Snack F0 gives a single F0, and Kaldi F0 estimator outputs 3 coefficients (Normalized F0 across sliding window, probability of voicing and F0 delta). Fundamental frequency variation (FFV) estimator [7] produces a 7-dimensional vector. Therefore, the whole feature vector has $24 + 2 + 1 + 3 + 7 = 37$ coefficients [8].

The conversation-side based mean subtraction is applied on the whole feature vector. 11 frames of log filter bank outputs and fundamental frequency features are stacked together. Hamming window followed by DCT consisting of 0th to 5th base are applied on the time trajectory of each parameter resulting in $(24+13) \times 6 = 222$ coefficients on the first-stage NN input [5].

The first-stage NN has four hidden layers with 1500 units each except the BN layer. BN layer's size is 80 neurons and it is the third hidden layer. Its outputs are stacked over 21 frames and downsampled (every 5th is taken) before they enter the second-stage NN, which has the same structure as the first-stage NN. The outputs from 80 neurons in BN layer are the final BN features for the recognition system [8].

For training the neural networks, the IARPA Babel Program data[5] were used. 11 languages were used to train the multilingual SBN feature extractor [3]: Cantonese, Pashto, Turkish, Tágalog, Vietnamese, Assamese, Bengali, Haitian, Lao, Tamil, Zulu. Details about the characteristics of the languages can be found in [9]. The training speech was force-aligned using our BABEL ASR system [8].

4   Systems and Experiments

The general scheme of our system is shown in Fig. 1. The following subsections deal with individual building blocks and results of experiments therewith in detail. The regressor producing arousal and/or valence values is linear, except for indicated cases, where a neural network is used.

First, a number of single systems was built: for each feature set (5 supplied ones + bottlenecks) and each dimension of emotion (arousal vs. valence), making up 12 systems in total. Each of them was investigated for optimum preprocessing, regressor training, and post-processing. All systems were trained on the training set and evaluated on the development set.

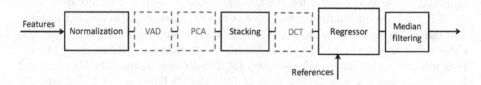

Fig. 1. Emotion recognition system scheme

---

[5] Collected by Appen, http://www.appenbutlerhill.com.

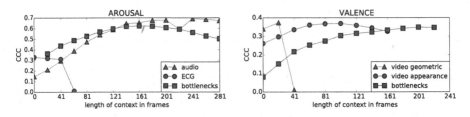

Fig. 2. Dependency of CCC on the number of stacked frames.

4.1   Pre-processing

The data-set from RECOLA includes 27 recordings from 27 different subjects. To prepare the data for the following regression, the whole set was globally mean- and variance-normalized.

We tried to use Voice Activity Detection (VAD) on the training data: frames with detected silence were dropped, and the system was trained only on the remaining speech frames. However, in video, there is always an indication of emotion even in case of silence, therefore, we tested on whole test recordings without silence removal. It is also necessary to note that the recordings are dialogues and the result of the emotion recognition of the observed person could be disturbed by speech of the second person, whose emotions we do not want to recognize. For these two reasons, the VAD does not help to improve our results, and was not used in our final systems.

Principal Component Analysis (PCA) was tested for dimensionality reduction of the feature sets, and had good results in experiments with supplied audio features for arousal. Reduction from the baseline dimensionality of 102 to 13 dimensions performed the best. On contrary, no or only very little reduction helps for both video features, which are used mainly for valence.

In our experiments, we train regression models for valence and arousal values for each frame (every 40 ms). In many other classification and recognition tasks, we have seen the need of adding larger temporal context to make a good prediction. The results with changing context size are shown in Fig. 2. The context of 141 stacked frames (70 to the left and 70 to the right of the current frame + current frame) was found optimal for arousal recognition from audio. Shorter context is necessary for valence while it is recognizing from video.

A further dimensionality reduction can follow the stacking of context frames. A standard technique is to project the temporal trajectories of features to the discrete cosine transform (DCT) bases. We observed, that for systems using bottleneck features, it is beneficial to perform DCT reduction to the first 7 coefficients for arousal, and the first 30 coefficients for valence. For recognition of valence from video appearance features, the first 3 DCT coefficients were found optimal on down-sampled feature trajectories (only every second frame was retained). For all DCT projections, Hamming windowing of the trajectories was applied first.

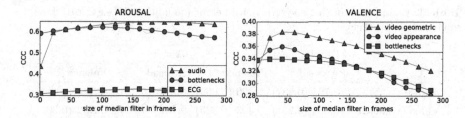

Fig. 3. Influence of the length of median filter applied after the regression.

4.2 Regression Model and Its Training

Linear regression is used on all single systems for arousal and all single systems for valence except for processing video geometric features, where neural network with one hidden layer is used (topology: 948–474–3). The regression can be trained in different ways, as six different labels from 6 different annotators plus another label as the gold standard (normalized and averaged as described in [1]) are available. After experimenting, we empirically found that for training arousal recognizer, data from annotators one and three give the best performance. Those annotations seem diverse and with bigger variation. For valence, we choose mostly annotators one, two and three, whose annotations seem more consistent. Our single systems produce estimates of two values for arousal, and three values for valence (trained to match the best annotators). A weighted average of these is then taken to produce one single value for arousal and one for valence. The weights were determined experimentally.

4.3 Post-processing

The outputs from our initial regression models seemed very noisy with respect to the reference. Median filter was used for smoothing and its optimum length was investigated, see Fig. 3. For arousal, a longer filter (over approx. 7 s) than for valence (over approx. 2 s) is needed.

4.4 Results and Comments

The final results of all investigated systems are summarized in Table 1 along with the baselines in brackets. It is evident that in most cases, our results are outperforming the baseline. They confirm our expectations—the recognition of arousal is better from audio, while for valence, video features perform better. Linear regression was used in all cases except for system processing video geometric features. We trained a neural network with one hidden layer with topology 945–474–3 in this case.

The major improvements are listed below. Using long context—in particular systems as long as 6–7 s, applying the median filter on the output and training on the data from the particular annotators instead of training on the Gold Standard.

Table 1. Comparison of single systems of different modalities, AV+EC 2015 baseline results are in brackets.

CCC	Development		Test	
	Arousal	Valence	Arousal	Valence
Audio	0.704 (0.287)	0.190 (0.069)	**0.595** (0.228)	0.160 (0.068)
Video geometric	0.054 (0.231)	0.403 (0.325)	0.151 (0.162)	**0.302** (0.292)
Video appearance	0.126 (0.103)	0.346 (0.273)	0.110 (0.114)	**0.334** (0.234)
ECG	0.305 (0.275)	0.231 (0.183)	-* (0.192)	-* (0.139)
EDA	0.117 (0.078)	0.235 (0.204)	0.118 (0.079)	0.226 (0.195)
Bottlenecks	0.625	0.344	**0.525**	0.176

*Numeric error prevented us from finishing this evaluation, unfortunately we have no longer access to the references of test data to re-evaluate them.

5   Bottle-Neck System Investigation

Inspired by the positive results of BN features on emotion recognition from audio, we created another system using only bottleneck features for simultaneous recognition of both arousal and valence (multi-task). In this system, long—up to 7 s (181 frames)—but downsampled context is used (only every 4th frame was taken), then DCT is applied and 30 first bases are retained. This system is also based on a simple linear regression. As a post-processing, median filter over 183 frames (7 s) for arousal and over 145 frames (6 s) for valence is used. The results in Table 2 indicate, that multi-task training is more efficient than having two single-task systems especially for the arousal prediction.

Table 2. Comparison of single- and multi-task systems based on Bottle-Neck features.

CCC	Development		Test	
	Arousal	Valence	Arousal	Valence
Single-task*	0.625	0.344	0.525	0.176
Single-task**	0.390	0.343	0.296	0.174
Multi-task	0.699	0.376	0.596	0.293

*Parameters tuned for each modality,
**Identical parameters as in multitask.

6   Fusion

Tuning and optimization of all individual systems was performed, as described in Sect. 4. Because of different sets of features and also different dimensions of emotion, we ended up with systems different in the pre-processing, training and post-processing. For fusion, we chose the two best single systems on arousal and valence separately, the real outputs (not weighted averages) from those best single systems were inputs for the fusion:

Table 3. Parameters of the best single systems used for final fusion.

	Arousal		Valence	
	Audio	Bottlenecks	Video geometric	Video appearance
Annotators	1 + 3	1 + 3	1 + 2 + 3	1 + 2 + 3
PCA	From 102 to 13	Not reduced (80)	Not reduced (316)	From 84 to 70
Length of context	141 (5.6 s)	161 (6.4 s)	3 (0.1 s)	30 (1.2 s)
DCT coefficients	20	7	No reduction	3
Downsampling	-	-	-	2

Table 4. Fusion system, AV+EC 2015 baseline results are in brackets

CCC	Arousal	Valence
Development	0.772 (0.476)	0.518 (0.461)
Test	0.660 (0.444)	0.504 (0.382)

- For *arousal*, we fused systems using audio features and Bottle-Necks. The fusion was a linear regression.
- For *valence*, both video system outputs were used. The fusion was done on the score level with a neural network with one hidden layer (topology: 486–243–1).

The system parameters selected for the final fusion are listed in Tables 3 and 4 contains the final fusion results. In all cases, comparison to baselines is favorable.

7 Conclusions

While the whole AV+EC evaluation was of interest for us, our focus was on the audio, as this is the main modality we are working with—most of our work is done in cooperation with contact centers that have no access to video or physiological data. Arousal is well recognizable from audio feature sets provided with AV+EC 2015 baselines, and we have improved the results by working on the context, regressor training and post-processing. The AV+EC features also combine favorably with the newly introduced Bottle-Neck features. For valence, the best evaluation results were obtained with video features, but the two-output Bottle-Neck structure is not far behind. We have also confirmed that simple regressors (linear or NN) can be used for the emotion prediction task.

References

1. Ringeval, F., Schuller, B., Valstar, M., Jaiswal, S., Marchi, E., Lalanne, D., Cowie, R., Pantic, M.: Av+ec 2015: the first affect recognition challenge bridging across audio, video, and physiological data. In: Proceedings of AVEC 2015, Satellite Workshop of ACM-Multimedia 2015, Brisbane, Australia, October 2015

2. Ringevaland, F., Sonderegger, A., Sauer, J., Lalanne, D.: Introducing the RECOLA multimodal corpus of remote collaborative and affective interactions. In: Proceedings of Face and Gestures, Workshop on Emotion Representation, Analysis and Synthesis in Continuous Time and Space (EmoSPACE) (2013)
3. Grézl, F., Egorova, E., Karafiát, M.: Further investigation into multilingual training and adaptation of stacked bottle-neck neural network structure. In: Proceedings of Spoken Language Technology Workshop, pp. 48–53 (2014)
4. Matějka, P., Zhang, L., Ng, T., Mallidi, H.S., Glembek, O., Ma, J., Zhang, B.: Neural network bottleneck features for language identification. In: Proceedings of Odyssey 2014, pp. 299–304 (2014)
5. Fér, R., Matějka, P., Grézl, F., Plchot, O., Černocký, J.: Multilingual bottleneck features for language recognition. In: Proceedings of Interspeech 2015, pp. 389–393 (2015)
6. Cumani, S., Laface, P., Kulsoom, F.: Speaker recognition by means of acoustic and phonetically informed GMMs. In: Proceedings of Interspeech 2015 (2015)
7. Heldner, M., Laskowski, K., Edlund, J.: The fundamental frequency variation spectrum. In: Proceedings of FONETIK (2008)
8. Karafiát, M., Veselý, K., Szoke, I., Burget, L., Grézl, F., Hannemann, M., Černocký, J.: BUT ASR system for BABEL surprise evaluation. In: 2014 IEEE Spoken Language Technology Workshop (SLT), NV, USA, December 2014
9. Harper, M.: The BABEL program and low resource speech technology. In: ASRU 2013, December 2013
10. Garcia-Romero, D., McCree, A.: Insights into deep neural networks for speaker recognition. In: Proceedings of Interspeech 2015 (2015)
11. Gunes, H., Schuller, B.: Categorical and dimensional affect analysis in continuous input: current trends and future directions. Image Vis. Comput. Affect Anal. Continuous Input **31**(2), 120–136 (2013)

Classification of Speaker Intoxication Using a Bidirectional Recurrent Neural Network

Kim Berninger[✉], Jannis Hoppe, and Benjamin Milde

Language Technology Group Computer Science Departement,
Technische Universität Darmstadt, Darmstadt, Germany
{kim_werner.berninger,jannis_manuel.hoppe}@stud.tu-darmstadt.de,
milde@lt.informatik.tu-darmstadt.de

Abstract. With the increasing popularity of deep learning approaches in the field of speech recognition and classification many of such problems are encountering a paradigm shift from classic approaches, such as hidden Markov models, to *recurrent neural networks* (RNN). In this paper we are going to examine that transition for the ALC corpus which had been used in the Interspeech 2011 Speaker State Challenge. *Filter bank* (FBANK) features are used alongside two types of bidirectional RNNs, each using *gated recurrent units* (GRU). Those models are used to classify the intoxication state of people just by recordings of their voices and outperform humans with state-of-the-art results.

Keywords: Deep learning · Speech classification · Bidirectional recurrent neural network · Filter bank · GRU · BRNN

1 Introduction

Although *neural networks* have already been used for decades in speech specific research [1], they have recently pushed the state of the art in many tasks due to their re-inception through *deep learning* [2].

In this paper, we re-evaluate the speech classification task of speaker intoxication detection with the help of deep neural networks. We use the *ALC corpus* [3] from the Bavarian Archive for Speech Signals[1]. It contains audio samples of German speakers who have been recorded being as well intoxicated as sober in one session each.

A classifier being able to predict speaker intoxication by speech alone would have the potential to enable quick sobriety checks if it worked with high enough accuracy. Such a system could be used in speech-controlled cars in order to prevent driving under the influence or in speech communication systems to assess the accountability of the opposite caller.

Our approach uses a *bidirectional recurrent neural network* (BRNN) [4] since speech has the form of a sequential signal with complex dependencies between the

---

[1] https://www.phonetik.uni-muenchen.de/Bas/BasALCeng.html.

© Springer International Publishing Switzerland 2016
P. Sojka et al. (Eds.): TSD 2016, LNAI 9924, pp. 435–442, 2016.
DOI: 10.1007/978-3-319-45510-5_50

different time steps in both directions. Such a network can learn discriminative features directly from the data, so we only use minimal feature engineering by providing a basic spectrogram representation of the audio signal. Additionally we propose an alternative layout for the BRNN and show that it performs better on that specific task.

In the following section, we discuss related work. In Sect. 3 we provide a short introduction to the used features and models. Afterwards in Sect. 4 we are going to have a look at our implementation and the applied preparations on the ALC corpus. Finally, the results of the experiments will be examined in Sect. 5, followed by a short outlook into possible future modifications in Sect. 6.

2   Related Work

The ALC corpus has already been subject of the *Interspeech Speaker State Challenge 2011* [5] where the participants' task had been to develop a classifier which was able to determine, given the audio recordings of various people, whether those have been intoxicated or sober at the time of recording.

A baseline which uses *Support Vector Machines* (SVMs) along with sequential minimal optimization has been provided by Schuller, Steidl, Batliner, Schiel and Krajewski, achieving an unweighted accuracy score of 65.9 percent. The winning contestants achieved an unweighted average recall (UAR) of 70.54 and an accuracy of 70.47 percent using speaker normalized hierarchical features and GMM supervectors [6].

Finally, Baumeister and Schiel also tested the human performance on the challenge data which averaged on a surprisingly low score of 63.1 percent [7]. That result showed that speech alone is not quite sufficient for humans to tell whether a person is sober or not and that humans heavily rely on the visual impressions and other perceptual clues (e.g. olfactory) of their opposite to make a better guess.

Usage of deep bidirectional GRU layers can be found in the work of Amodei et al. [8] where, for example, up to seven bidirectional GRU layers are used and even combined with up to three convolutional layers which altogether improved the performance of the network. An approach for speech classification solely using *Convolutional Neural Networks* (CNN) instead of recurrent ones can be seen in the work of Milde and Biemann [9].

Trigeorgis et al. used combinations of CNNs and RNNs [10] to skip the hand-crafting of features altogether and successfully trained a speech classifier on data in raw wave form.

3   Background

3.1   Recurrent Neural Networks

Similar to conventional neural networks, RNNs consist of artificial neurons which take an input vector $\mathbf{x}$ and yield an output $\mathbf{o}$ as the inner product of $\mathbf{x}$ and a weight matrix $\mathbf{U}$ to which an inner activation function has been applied in the hidden state h.

However, in contrast to the conventional neurons, the recurrent ones receive their input as a time-dependent sequence $\mathbf{x}_{1:T}$. The input $\mathbf{x}_t$ at each time step t gets combined with the result of the hidden step h_{t-1} for time step $t-1$, which has been transformed using some weight $\mathbf{W}$, before being used as the next input for the inner activation function as it can be seen in Fig. 1. Finally the outer activation function gets applied to the output of each, or alternatively only the last, hidden state after it has been transformed using another weight $\mathbf{V}$.

The time dependent behavior can alternatively be viewed using an unfolded representation of the network as in Fig. 2.

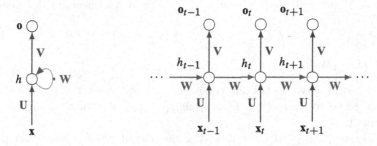

Fig. 1. A simple RNN.   **Fig. 2.** The unfolded view of the network in Fig. 1.

3.2 Gated Recurrent Unit Networks

GRU Networks [11] are a variation of the *long short term memory* (LSTM) networks [12]. Similarly to those, the mechanisms used in a GRU help to avoid the so-called *vanishing gradient problem* [13], however an LSTM unit uses more parameters. Therefore the former proved to be more suitable for our experiments since the resulting network demanded less space memory for that reason, and was able to fit into graphics memory.

3.3 Bidirectional RNNs

Conventional RNN structures propagate information only forwards in time. However, for some applications, such as human language, there can be dependencies in both time directions. In this context, bidirectional RNNs [4] can be helpful by having separate layers processing the two different directions and feeding each others output into the same output layer as it is depicted in Fig. 3.

In addition to that we also used a different approach to BRNNs shown in Fig. 4. Here, the output of the forward layer is not directly propagated towards the output layer, instead it serves as the input of the backward layer.

In order to distinguish between those two forms, we will call the former *merged BRNNs* and the latter *sequential BRNNs* from now on.

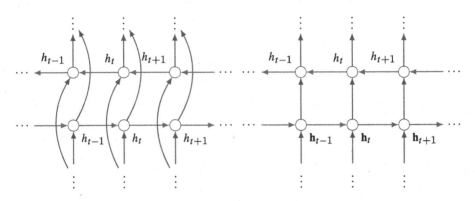

Fig. 3. A ("merged") bidirectional RNN.

Fig. 4. A sequential bidirectional RNN.

3.4 Dropout

Deep neural networks usually come with a large amount of parameters, leaving them prone to overfitting. One straightforward way of avoiding that is using Dropout [14].

Those are applied by multiplying the output of each node that propagates towards a dropout layer by some random noise with each batch of training data. Usually that noise is drawn from a Bernoulli distribution and therefore either leaves the output unchanged or deactivates it altogether. The amount of deactivated nodes depends on the dropout rate p which can be seen as the distribution's parameter in the binary case.

The positive effect of better generalization capabilities can be explained in two ways. First, it essentially produces an ensemble of neural networks and therefore producing an equally averaged result over those. Second, it can also be understood as forcing the neural network to learn more robust features since it can not rely on any node to be activated.

Gaussian dropout is a modification to the normal dropout scheme. Here, the noise is drawn from a Gaussian distribution $\mathcal{N}\left(1, \sqrt{\frac{p}{1-p}}\right)$. Those two types of dropout shall be compared side by side in our experiments.

4 Experimental Setup

4.1 Alcohol Language Corpus

ALC was recorded from 2007 until 2009 and contains speech from 162 German speaking people, 78 female and 84 male. Of those, a gender balanced selection of 154 was chosen to be examined in the Interspeech Speaker State Challenge which is the same dataset we are using for this paper. Each subject was free to choose the blood alcohol concentration under which the first recording would take place, the second recording usually occurred three weeks later. For each person, several recordings have been gathered which differed in type and content as, for example, there are dialogues, monologues or readings with content types ranging from picture descriptions to number reading.

We use the training, test and development-test sets which have originally been provided in the challenge.

4.2 Data Preparation

Since the audio files provided with the ALC corpus come in the WAVE RIFF format with a sampling rate of 44100 Hz and 16 bit per sample, we first compute a feature representation for those. In contrast to the original challenge, where the features have been generated using the openSMILE feature extractor [15], we use 40-dimensional FBANK features provided by the CMU Sphinx speech recognition toolkit[2]. Those are computed with standard parameters.

After extracting the features, the data has the form

$$\mathcal{D} = \left\{ \left(\mathbf{x}_{1:T^{[1]}}^{[1]}, y^{[1]} \right), \ldots, \left(\mathbf{x}_{1:T^{[N]}}^{[N]}, y^{[N]} \right) \right\}, \tag{1}$$

with features $\mathbf{x}_t^{[i]} \in \mathbb{R}^{40}$ for each $t \in \{1, \ldots, T^{[i]}\}$, labels $y^{[i]} \in \{0,1\}$ and with $T^{[i]}$ being the length of sequence $\mathbf{x}_{1:T^{[i]}}^{[i]}$, for each $i \in \{1, \ldots, N\}$.

From those we generate the following tensor, which we pad to the maximum sequence length $T_{\max} = \max_i \left(T^{[i]} \right)$

$$\mathbf{X} = (x_{i,t,j})_{i \in \{1,\ldots,N\}, t \in \{1,\ldots,T_{\max}\}, j \in \{1,\ldots,40\}}, \tag{2}$$

$$\text{with} \quad x_{i,t,j} = \begin{cases} \mathbf{x}_{t,j}^{[i]} & t \leq T^{[i]}, \\ 0 & \text{otherwise.} \end{cases} \tag{3}$$

Since the resulting tensor is quite sparse, we standardize it by subtracting the mean of each spatial dimension from the according vector elements of all time steps and dividing the results by the respective standard deviation as in the work of Laurent et al. [16] where it is also used in order to improve the learning performance.

Most importantly, we avoid having zero-entries in our tensor which would result in vanishing signals throughout the time dimension and, hence, prevent the network from being trained at all.

4.3 Network Layout

The RNNs are built in Python using the Keras[3] framework. We use two GRU layers with 128 hidden states each. One of those uses a forward orientation, the other one propagates its internal outputs backward in time. Using one-directional RNNs did not show good results in early experiments.

---

[2] http://cmusphinx.sourceforge.net.
[3] http://keras.io.

For the merged BRNN those are both connected to the input and combine their results to a dropout layer with $p = 0.4$. That combination is realized using addition of the 128-dimensional output vectors.

For the sequential BRNN the first GRU layer produces another time-dependent sequence which is then fed into the backward GRU layer. After each of those follows one layer of dropout with $p = 0.4$.

For both types of networks we finally use one to three fully connected (dense) layers with, again, 128 hidden states each, eventually followed by another single-state layer which produces the output by applying the *sigmoid* function which is also used as the inner activation of the GRU nodes. The remaining activation functions between each two nodes are defined as the rectifier or relu function.

As optimizer we use *Adam* [17]. In earlier stages, *RMSprop* [18] has also been tried, but it clearly proved to perform worse on that learning task and also oscillated a lot, making it useless for the early stopping described in Sect. 4.4.

The loss which is minimized in the training stage is the binary cross-entropy.

4.4 Training

Training is performed on the training set containing 5400 recordings with about two thirds representing the sober label as well as the first 3060 samples of the development-test which has a similar sober to intoxicated ratio. Thus, the ZeroR baseline of both sets already lies above the original baseline provided for the challenge. For that reason, like Bone et al. [6], we use UAR for model validation.

The model is then trained for a maximum of 50 epochs and evaluated on the remaining 900 samples of the development-test set after each iteration. Training is stopped when the UAR of the validation set does not increase for 10 epochs. A model checkpoint is used to keep track of the weights which have been used to reach the highest UAR on the validation set up to the last epoch.

Finally we obtain a measure of general performance by using the highest scoring model to predict the labels on the testing set. For that, the weights saved by the checkpoint are reloaded beforehand.

5 Results

The performance of the networks is depicted in Table 1. It can be seen that the alternative BRNN layout achieves much better results than the conventional one.

Also, the Gaussian dropout leads to slightly higher performance than the binary one. In most cases the network with two fully connected layers after the recurrent ones achieves the highest measures.

Eventually we used the sequential BRNN with two dense layers and Gaussian dropout after being trained on the sets specified in Sect. 4.4 to predict the labels of the testing set and reached a UAR of 71.03 and an accuracy of 71.30 percent. The resulting confusion matrix can be seen in Table 2.

Table 1. The accuracy and the average recall on the validation set predictions after training using the different models in percent.

BRNN type	Merged						Sequential					
Dropout	Bernoullian			Gaussian			Bernoullian			Gaussian		
Dense layers	1	2	3	1	2	3	1	2	3	1	2	3
Accuracy	69.1	69.4	67.7	71.3	72.8	67.4	67.1	64.8	67.8	**75.9**	70.8	73.9
UAR	65.0	67.9	66.4	66.8	66.6	66.5	66.3	67.8	66.1	67.8	**69.2**	67.9

Table 2. The confusion matrix of the test set predictions. The rows show the true labels, the columns the predicted ones.

True/pred.	Sober	Intoxicated
Sober	1204	416
Intoxicated	445	935

6 Future Work

To conclude, we can say that modern neural network approaches can perform well on a speech classification task, without investing that much time into feature engineering. To go even further we could try to avoid the creation of features altogether by applying the end-to-end approach of Trigeorgis et al. [10] to the intoxication classification problem.

With only two GRU layers, it is easily imaginable that a deeper network, similar to the one of Amodei et al. [8], could provide better results.

Also it became clear, that there exists a variant of BRNNs which has, to our knowledge, not been researched thoroughly yet. It would be interesting to see whether the sequential BRNN could be able to achieve better results than the merged one in different tasks as well.

Acknowledgments. We want to thank Chris Biemann from the Language Technology Group at TU Darmstadt for encouraging us to submit this paper. Furthermore we thank the team at the HHLR for providing us with the computational resources on the Lichtenberg computing cluster. This work was also supported by QSL funds to TU Darmstadt.

References

1. Bourlard, H., Morgan, N.: Connectionist speech recognition: a hybrid approach. Technical report, Kluwer Academic Publishers (1994)
2. Hinton, G.: Learning multiple layers of representation. Trends Cogn. Sci. **11**(10), 428–434 (2007)
3. Schiel, F., Heinrich, C., Barfüsser, S.: Alcohol language corpus: the first public corpus of alcoholized German speech. Lang. Resour. Eval. **46**(3), 503–521 (2012)

4. Schuster, M., Paliwal, K.K.: Bidirectional recurrent neural networks. IEEE Trans. Sign. Proces. **45**(11), 2673–2681 (1997)
5. Schuller, B., Steidl, S., Batliner, A., Schiel, F., Krajewski, J.: The interspeech 2011 speaker state challenge. In: Proceedings of the Interspeech, pp. 3201–3204 (2011)
6. Bone, D., Black, M., Li, M., Metallinou, A., Lee, S., Narayanan, S.S.: Intoxicated speech detection by fusion of speaker normalized hierarchical features and GMM supervectors. In: Proceedings of the Interspeech, pp. 3217–3220 (2011)
7. Baumeister, B., Schiel, F.: Human perception of alcoholic intoxication in speech. In: Proceedings of the Interspeech, pp. 1419–1423 (2013)
8. Amodei, D., Anubhai, R., Battenberg, E., Case, C., Casper, J., Catanzaro, B., Chen, J., Chrzanowski, M., Coates, A., Diamos, G., et al.: Deep speech 2: end-to-end speech recognition in English and mandarin. arXiv preprint arXiv:1512.02595 (2015)
9. Milde, B., Biemann, C.: Using representation learning and out-of-domain data for a paralinguistic speech task. In: Proceedings of the Interspeech, pp. 904–908 (2015)
10. Trigeorgis, G., Ringeval, F., Brueckner, R., Marchi, E., Nicolaou, M.A., Schuller, B., Zafeiriou, S.: Adieu features? End-to-end speech emotion recognition using a deep convolutional recurrent network. In: Proceedings ICASSP (2016)
11. Cho, K., van Merrienboer, B., Gulcehre, C., Bougares, F., Schwenk, H., Bengio, Y.: Learning phrase representations using RNN encoder-decoder for statistical machine translation. In: Proceedings EMNLP, pp. 1724–1734 (2014)
12. Hochreiter, S., Schmidhuber, J.: Long short-term memory. Neural Comput. **9**(8), 1735–1780 (1997)
13. Hochreiter, S.: Untersuchungen zu dynamischen neuronalen Netzen. Diplomarbeit, Institut fur Informatik, Technische Universität, München (1991)
14. Srivastava, N., Hinton, G., Krizhevsky, A., Sutskever, I., Salakhutdinov, R.: Dropout: a simple way to prevent neural networks from overfitting. J. Mach. Learn. Res. **15**(1), 1929–1958 (2014)
15. Eyben, F., Weninger, F., Gross, F., Schuller, B.: Recent developments in openSMILE, the Munich open-source multimedia feature extractor. In: Proceedings ACM International Conference on Multimedia, pp. 835–838 (2013)
16. Laurent, C., Pereyra, G., Brakel, P., Zhang, Y., Bengio, Y.: Batch normalized recurrent neural networks. arXiv preprint arXiv:1510.01378 (2015)
17. Kingma, D., Ba, J.: Adam: a method for stochastic optimization. In: International Conference on Learning Representation (2015)
18. Hinton, G.: Lecture 6e: RMSprop: divide the gradient by a running average of its recent magnitude. University of Toronto via Coursera: Neural Networks for Machine Learning, October 2012

Training Maxout Neural Networks for Speech Recognition Tasks

Aleksey Prudnikov[1,2] and Maxim Korenevsky[1,2(✉)]

[1] ITMO University, Saint Petersburg, Russia
{prudnikov,korenevsky}@speechpro.com
[2] Speech Technology Center, Saint Petersburg, Russia

Abstract. The topic of the paper is the training of deep neural networks which use tunable piecewise-linear activation functions called "maxout" for speech recognition tasks. Maxout networks are compared to the conventional fully-connected DNNs in case of training with both cross-entropy and sequence discriminative (sMBR) criteria. Experiments are carried out on the CHiME Challenge 2015 corpus of multi-microphone noisy dictation speech and the Switchboard corpus of conversational telephone speech. The clear advantage of maxout networks over DNNs is demonstrated when using the cross-entropy criterion on both corpora. It is also argued that maxout networks are prone to overfitting during sequence training but in some cases it can be successfully overcome with the use of the KL-divergence based regularization.

Keywords: Dropout · Regularization · Rectified linear units · Maxout · Cross-entropy · Sequence training · CHiME · Switchboard · KL-divergence

1 Introduction

Over the past decade, automatic speech recognition (ASR) systems showed a significant increase of recognition accuracy. This is mainly related to the widespread introduction of deep neural networks (DNN) [3] into the ASR systems.

Until 2006 methods of DNNs training were not widespread. The conventional error back-propagation method [20] was unable to learn parameters of the layers far from the output one. Much attention to the deep learning was attracted after the series of papers by Hinton and his colleagues [7,12], where the first breakthrough results in deep networks training were obtained[1]. The greedy layerwise pretraining method [4,12] became an impulse for tempestuous development of DNN training. The pretraining results in DNN weights initialization that facilitates the subsequent finetuning and improves its quality.

The first large success of DNNs application to speech recognition tasks is dated by 2011 when DNNs were used to compute posteriors of the decision

---

[1] Similar ideas of deep networks training were proposed much earlier (see for example review in [25]) but for some reasons they were not paid proper attention.

© Springer International Publishing Switzerland 2016
P. Sojka et al. (Eds.): TSD 2016, LNAI 9924, pp. 443–451, 2016.
DOI: 10.1007/978-3-319-45510-5_51

tree tied states of context-dependent triphones (CD-DNN-HMM) [26]. The success of feedforward networks with fully connected layers resulted in application to ASR other well-known neural network architectures such as convolutional neural networks (CNN) widely used in computer vision [1,22] and recurrent neural networks (RNN) [10,11]. Lately researchers have started to integrate together different kinds of networks [21] in order to combine the merits of different architectures. Nevertheless, the fully connected feedforward deep neural networks (hereinafter just DNNs for brevity) still remain the workhorses of the large majority of ASR systems. And research in this field has a significant impact on the development of other types of neural networks.

The introduction of piecewise-linear ReLU (rectified linear units) activation functions [15] made it possible to simplify and improve the optimization process during DNN training significantly. It was shown [31], [8] that DNNs with ReLU activations may be successfully learned without layerwise pretraining. However, the fact that ReLU and its analogues are linear almost everywhere may also result in overfitting and instability of the training process. This implies the necessity of using effective regularization techniques. One of such techniques is *dropout* proposed in [13]. Dropout can be treated as an approximate way to learn the exponentially large ensemble of different neural nets with subsequent averaging [27].

To improve efficiency of the dropout regularization a new sort of tunable piecewise-linear activation function called *maxout* was proposed in [9]. Using maxout neurons makes the approximation of ensemble averaging during dropout more accurate and demonstrates the impressive results on several popular benchmark tasks in image recognition. In a short time the deep maxout networks (DMN) were applied to speech recognition tasks [14,28]. It was also shown [19,24] that dropout for DMNs can be very effective in an *annealed* mode, i.e. if the dropout rate gradually decreases epoch-by-epoch.

This work continues the investigation of using DMNs for speech recognition tasks. We apply DMNs to two significantly different tasks and demonstrate their clear superiority over conventional DNNs under cross-entropy (CE) training. We also discuss the difficulties of sequence discriminative training of maxout networks and propose to use the KL-divergence based regularization to avoid overfitting and improve accuracy.

The rest of the paper is organized as follows. Section 2 is devoted to the more detailed description of dropout regularization, piecewise-linear activation functions, maxout and annealed dropout. Section 3 describes experiments and discusses their result. Section 4 concludes the paper.

2   Dropout and Maxout

2.1   Dropout

Dropout proposed in [13] is a regularization technique for deep feedforward network training which is effective in particular for training network with a large amount of parameters on a limited size dataset. In the original form of dropout it was proposed to randomly "turn off" half of neurons per every training example[2]. And the remained neurons try to learn from the current example. As a result, different neurons tend to learn different features of the input data so the dropout reduces the co-adaptation of neurons.

The dropout may be treated as an approximate way to train and average an exponentially large ensemble of different networks with a half number of neurons. If the layers were linear this averaging would be exact. When the neurons with piecewise-linear activation functions (like $ReLU(x) = \max(0, x)$) are used the input space is divided into multiple regions where data is linearly transformed by the network. From this point of view using of dropout with piecewise-linear activation functions performs the more exact ensemble averaging than with activation functions of nonzero curvature (such as sigmoid). The effectiveness of DNNs with ReLU neurons trained with dropout was demonstrated [27] on many tasks from different domains.

2.2 Maxout

Maxout [9] is a piecewise-linear activation function which was proposed to improve neural network training with dropout. The idea of maxout is to divide all neurons of a linear layer into the groups of size k and pass only the maximum value of activations from each group to the next layer:

$$v_i^l = \max_{j \in [1,k]} z_{ij}^l, \quad \text{where} \quad z_{ij}^l = W_{ij...}^l \mathbf{v}^{l-1} + b_{ij}. \tag{1}$$

(The weights tensor W and the bias matrix b are trainable parameters.) So, the "maxout neuron" is in fact a group of linear neurons with a single output equal to the maximal output of the neurons group. This makes maxout activation to be piecewise linear. Thus, for maxout neural networks the same conclusions hold about the more exact ensemble averaging under dropout as for rectifier networks do. The advantage of the maxout activation function over ReLU is the possibility to adjust its form by means of parameters tuning (although it comes at the cost of k-fold increasing of the parameters number). Maxout networks (both fully-connected and convolutional) demonstrated impressive results on several benchmark tasks from the computer vision domain.

2.3 Annealing Dropout

The first application of Deep Maxout Networks (DMN) to speech recognition task seems to be done in [14] and in [5], where it was shown that training of DMNs can be effective even without using dropout. Similar results were obtained in [32] were several generalization of maxout were proposed. Moreover, the results of DMNs training with using dropout in [6] are even worse than without it.

Nevertheless, the training of DMNs can be successfully combined with dropout regularization as it was shown in [19,24]. There it was proposed to

---

[2] The fraction of the neurons turned off is called dropout rate and can differ from 0.5.

use not the conventional dropout where the dropout rate is constant during the entire training but the *annealed dropout* (AD) which consists of the gradually decreasing dropout rate according to the linear schedule. The combination of maxout and AD provided a significant WER reduction on such benchmark datasets as Aurora4, Open Voice Search and Hub5'2000. However, the theoretical justification of this technique success is still to be made, therefore the investigation of its applicability to a wide range of tasks is of importance.

3   Experiments and Results

3.1   CHiME 2015

The CHiME Challenge 2015 [2] database consists of multi-microphone recordings made in real-life acoustic conditions from the distance of about 0.5 meters. The size of the training dataset is about 25 h, a quarter of which consists of real recordings and the remaining part of artificially simulated. The STC + ITMO ASR system [17] developed for CHiME uses MFCC-features normalized with CMVN and transformed with LDA-MLLT and fMLLR for DNN acoustic model training. The input vector of DNN is extended by the i-vector [23] which is extracted from the utterance for the additional adaptation to the speaker and environment. We compared the cross-entropy training of the conventional sigmoidal DNNs and DMNs with AD regularization[3]. The recognition results are presented in Table 1[4].

Table 1. WER% comparison of the cross-entropy models for the CHiME2015

Model	Hidden layer size	Number of parameters, mln	Dev		Eval	
			Real	Simu	Real	Simu
DNN	1024	7.7	8.2	9.9	14.8	14.7
	1536	14.8	8.0	9.8	14.4	14.1
	2048	25.9	8.0	9.7	14.4	14.1
DMN	1024	13.4	7.5	9.0	13.2	13.0

The following conclusions can be made from the table. With the equal number of layers' output neurons the combination of maxout and AD provides the significant WER reduction on cross-entropy training. Increasing the number of DNN parameters up to the amount comparable to that of DMN improves results

---

[3] Hereinafter the maxout group size $k = 2$ is used unless otherwise is stated. The increase of the group size makes both training and recognition slower and makes training less stable but does not provide significant improvements in our experiments.

[4] We do not compare Maxout + AD to ReLU + AD because, in our experience, AD training of ReLU networks does not provide significant WER reduction (it is also observed in [19, Sect. 4.8]) whilst carefully tuned sigmoidal DNNs with L_2 weight decay often outperform ReLU DNNs with dropout and other types of regularization.

slightly but further increasing does not improve results anymore. We also compared the dynamics of DNNs and DMNs training with SGD. We found that 4x larger minibatches can be used for DMN training compared to DNN training without loss of accuracy. Therefore, although DMN requires more epochs for cross-entropy to saturate, it takes significantly less time to converge.

3.2 Switchboard

For the experiments on the Switchboard corpus the acoustic DNN model was trained on about 262 hours of speech. Baseline cross-entropy DNN based on MFCC features transformed with LDA-MLLT and fMLLR was trained with using Kaldi recipe s5c [16]. For the sake of comparison two more kinds of models were trained as well. The first variant uses i-vectors like in the CHiME2015 system. When the model was trained the low-rank factorization (based on SVD) of the last hidden layer was performed and the model was fine-tuned to provide bottleneck features, hereinafter SDBNs (Speaker-Dependent BottleNeck). The detailed description of the procedure can be found in [18]. The second variant of the acoustic model was trained on SDBN features of dimension 80 spliced over 31 frames (instead of 11 frames in baseline DNN) with the 5 frame hop.

For each model type the DMN analogue with the same layer sizes was trained using AD. The results of testing all these models on two subsets of the Hub5'2000 corpus, namely SWB and SWB + CH, are presented in Table 2. These results show that combination of maxout and AD provides the relative WER reduction of 5–9 % for all kinds of features.

Table 2. WER% comparison of the Switchboard-trained cross-entropy models on the Hub5'2000

Features	DNN		DMN	
	SWB	SWB + CH	SWB	SWB + CH
fMLLR	14.6	20.4	13.6	19.3
fMLLR + iVec	14.1	19.8	13.1	18.5
SDBN	13.6	19.5	12.5	17.8

3.3 Using Sequence Training

It is a common practice to use sequence discriminative training (ST) to finetune cross-entropy trained models. Usually this provides a significant WER reduction. We also applied ST based on the state-level Minimum Bayes Risk (sMBR) criterion to our previously trained cross-entropy models to assess its effectiveness for DMNs. In our first attempts we did not use regularization.

Experiments on CHiME2015 showed that while ST improves the DMNs accuracy over cross-entropy models, it results in no meaningful difference from the conventional DNNs in terms of system performance (see Table 3). Moreover, the Switchboard experiments showed no improvement by using ST for DMNs compared to cross-entropy networks. Since we observed that the cross-validation value of the sMBR criterion either increases or remains constant epoch-by-epoch we concluded that the model is overfit. So to improve ST performance an effective regularization is required.

We tried to use the L1 and L2 penalty as well as dropout and F-smoothing [29] to make ST work on Switchboard, however none of these approaches succeeded in overfitting reduction. The solution we found is to use the KLD-regularization [30] which changes the training criteria by adding a penalty proportional to the KL divergence between current and reference CE model output distributions.

The results from Table 3 show that ST with KLD-regularization is effective for the Switchboard task and provides a significant accuracy gain, although less than that of ST on conventional DNNs. However, the gain from KLD was negligible for the CHiME task so the more effective regularization technique for ST of maxout networks is still to be found.

Table 3. Effect of ST on CHiME2015 and Hub5'2000, WER% (the Switchboard sMBR DMNs are trained with KLD-regularization)

Model	Criterion	CHiME2015 %		Hub5'2000 %	
		Dev real	Eval real	SWB	SWB + CH
DNN	CE	8.0	14.4	13.6	19.5
	sMBR	6.7	12.5	12.1	17.8
DMN	CE	7.5	13.2	12.5	17.8
	sMBR	6.8	12.4	11.6	17.0

4   Conclusions

We considered the use of maxout activation functions for training deep neural networks as acoustic models for ASR. Using two English speech corpora namely CHiME Challenge 2015 dataset and Switchboard we demonstrated that in case of training with the cross-entropy criterion Deep Maxout Networks (DMN) are superior to conventional fully-connected feedforward sigmoidal DNNs. For the same layer sizes the number of DMN parameters is larger than that of DNN but the increase in DNN layer sizes is unable to provide the comparable accuracy gain. This confirms the effectiveness of maxout networks which was demonstrated earlier on different speech datasets. We also found that sequence discriminative training of maxout networks is prone to overfitting which can be reduced with the use of KLD-regularization. However, this approach does not work equally well for different tasks so finding the effective regularization technique for sequence training of maxout networks is an issue of future work.

Acknowledgments. This work was financially supported by the Ministry of Education and Science of the Russian Federation, Contract 14.575.21.0033 (ID RFMEFI575 14X0033)

References

1. Abdel-Hamid, O.,Mohamed, A.R., Jiang, H., Penn, G.: Applying convolutional neural networks concepts to hybrid NN-HMM model for speech recognition. In: 2012 IEEE International Conference on Acoustics, Speech and Signal Processing (ICASSP), pp. 4277–4280. IEEE (2012)
2. Barker, J., Marxer, R., Vincent, E., Watanabe, S.: The third 'chime' speech separation and recognition challenge: dataset, task and baselines. In: Proceedings of 2015 IEEE Automatic Speech Recognition and Understanding Workshop (ASRU 2015), pp. 504–511 (2015)
3. Bengio, Y.: Deep learning architectures for AI. Found. Trends Mach. Learn. **2**(1), 1–127 (2009)
4. Bengio, Y., Lamblin, P., Popovici, D., Larochelle, H.: Greedy layer-wise training of deep networks. In: Schölkopf, B., Platt, J.C., Hoffman, T. (eds.) Advances in Neural Information Processing Systems, vol. 19, pp. 153–160. MIT Press, Cambridge (2007)
5. Cai, M., Shi, Y., Liu, J.: Deep maxout neural networks for speech recognition. In: 2013 IEEE Workshop on Automatic Speech Recognition and Understanding (ASRU), pp. 291–296. IEEE (2013)
6. de-la-Calle-Silos, F., Gallardo-Antolín, A., Peláez-Moreno, C.: Deep maxout networks applied to noise-robust speech recognition. IberSPEECH 2014. LNCS, vol. 8854, pp. 109–118. Springer, Heidelberg (2014)
7. Carreira-Perpinan, M.A., Hinton, G.: On contrastive divergence learning. In: AISTATS, vol. 10, pp. 33–40. Citeseer (2005)
8. Dahl, G.E., Sainath, T.N., Hinton, G.E.: Improving deep neural networks for LVCSR using rectified linear units and dropout. In: 2013 IEEE International Conference on Acoustics, Speech and Signal Processing (ICASSP), pp. 8609–8613. IEEE (2013)
9. Goodfellow, I.J., Warde-Farley, D., Mirza, M., Courville, A., Bengio, Y.: Maxout networks. arXiv preprint arXiv:1302.4389 (2013)
10. Graves, A., Mohamed, A.R., Hinton, G.: Speech recognition with deep recurrent neural networks. In: 2013 IEEE International Conference on Acoustics, Speech and Signal Processing (ICASSP), pp. 6645–6649. IEEE (2013)
11. Graves, A., Jaitly, N.: Towards end-to-end speech recognition with recurrent neural networks. In: Proceedings of the 31st International Conference on Machine Learning (ICML-14), pp. 1764–1772 (2014)
12. Hinton, G.E., Osindero, S., Teh, Y.W.: A fast learning algorithm for deep belief nets. Neural Comput. **18**(7), 1527–1554 (2006)
13. Hinton, G.E., Srivastava, N., Krizhevsky, A., Sutskever, I., Salakhutdinov, R.R.: Improving neural networks by preventing co-adaptation of feature detectors. arXiv preprint arXiv:1207.0580 (2012)

14. Miao, Y., Metze, F., Rawat, S.: Deep maxout networks for low-resource speech recognition. In: 2013 IEEE Workshop on Automatic Speech Recognition and Understanding (ASRU), pp. 398–403. IEEE (2013)
15. Nair, V., Hinton, G.E.: Rectified linear units improve restricted boltzmann machines. In: Proceedings of the 27th International Conference on Machine Learning (ICML-10), pp. 807–814 (2010)
16. Povey, D., Ghoshal, A., Boulianne, G., Burget, L., Glembek, O., Goel, N., Hannemann, M., Motlicek, P., Qian, Y., Schwarz, P., Silovsky, J., Stemmer, G., Vesely, K.: The kaldi speech recognition toolkit. In: Proceedings of 2011 IEEE Workshop on Automatic Speech Recognition and Understanding (ASRU 2011) (2011)
17. Prudnikov, A., Korenevsky, M., Aleinik, S.: Adaptive beamforming and adaptive training of DNN acoustic models for enhanced multichannel noisy speech recognition. In: Proceedings of 2015 IEEE Automatic Speech Recognition and Understanding Workshop (ASRU 2015), pp. 401–408 (2015)
18. Prudnikov, A., Medennikov, I., Mendelev, V., Korenevsky, M., Khokhlov, Y.: Improving acoustic models for Russian spontaneous speech recognition. In: Ronzhin, A., Potapova, R., Fakotakis, N. (eds.) SPECOM 2015. LNCS, vol. 9319, pp. 234–242. Springer, Heidelberg (2015)
19. Rennie, S.J., Goel, V., Thomas, S.: Annealed dropout training of deep networks. In: 2014 IEEE Spoken Language Technology Workshop (SLT), pp. 159–164. IEEE (2014)
20. Rumelhart, D., Hinton, G., Williams, R.: Learning internal representations by error propagation. Parallel Distrib. Process. **1**, 318–362 (1986)
21. Sainath, T., Rao, K., et al.: Acoustic modelling with CD-CTC-SMBR LSTM RNNS. In: 2015 IEEE Workshop on Automatic Speech Recognition and Understanding (ASRU), pp. 604–609. IEEE (2015)
22. Sainath, T.N., Mohamed, A.R., Kingsbury, B., Ramabhadran, B.: Deep convolutional neural networks for LVCSR. In: 2013 IEEE International Conference on Acoustics, Speech and Signal Processing (ICASSP), pp. 8614–8618. IEEE (2013)
23. Saon, G., Soltau, H., Nahamoo, D., Picheny, M.: Speaker adaptation of neural network acoustic models using i-vectors. In: Proceedings of 2013 IEEE Workshop on Automatic Speech Recognition and Understanding (ASRU 2013), pp. 55–59 (2013)
24. Saon, G., Kuo, H.K.J., Rennie, S., Picheny, M.: The IBM 2015 English conversational telephone speech recognition system. arXiv preprint arXiv:1505.05899 (2015)
25. Schmidhuber, J.: Deep learning in neural networks: an overview. Neural Netw. **61**, 85–117 (2015)
26. Seide, F., Li, G., Chen, X., Yu, D.: Feature engineering in context-dependent deep neural networks for conversational speech transcription. In: 2011 IEEE Workshop on Automatic Speech Recognition and Understanding (ASRU), pp. 24–29. IEEE (2011)
27. Srivastava, N., Hinton, G., Krizhevsky, A., Sutskever, I., Salakhutdinov, R.: Dropout: a simple way to prevent neural networks from overfitting. J. Mach. Learn. Res. **15**(1), 1929–1958 (2014)
28. Swietojanski, P., Li, J., Huang, J.T.: Investigation of maxout networks for speech recognition. In: 2014 IEEE International Conference on Acoustics, Speech and Signal Processing (ICASSP), pp. 7699–7703. IEEE (2014)
29. Yu, D., Deng, L.: Automatic Speech Recognition. A Deep Learning Approach. Springer, London (2015)

30. Yu, D., Yao, K., Su, H., Li, G., Seide, F.: KL-divergence regularized deep neural network adaptation for improved large vocabulary speech recognition. In: 2013 IEEE International Conference on Acoustics, Speech and Signal Processing (ICASSP) (2013)

31. Zeiler, M.D., Ranzato, M., Monga, R., Mao, M., Yang, K., Le, Q.V., Nguyen, P., Senior, A., Vanhoucke, V., Dean, J., et al.: On rectified linear units for speech processing. In: 2013 IEEE International Conference on Acoustics, Speech and Signal Processing (ICASSP), pp. 3517–3521. IEEE (2013)

32. Zhang, X., Trmal, J., Povey, D., Khudanpur, S.: Improving deep neural network acoustic models using generalized maxout networks. In: 2014 IEEE International Conference on Acoustics, Speech and Signal Processing (ICASSP), pp. 215–219. IEEE (2014)

An Efficient Method for Vocabulary Addition to WFST Graphs

Anna Bulusheva[1(✉)], Alexander Zatvornitskiy[1,2], and Maxim Korenevsky[1]

[1] STC-Innovations Ltd., St. Petersburg, Russia
{bulusheva,zatvornitskiy,korenevsky}@speechpro.com
[2] Speech Technology Center Ltd., St. Petersburg, Russia

Abstract. A successful application of automatic speech recognition often requires the ability to recognize words that are not known during system building phase. Modern speech recognition decoders employ large WFST graphs so to add new words one needs to recompile the graph. · The compilation process requires a lot of memory and consumes a lot of time. In this paper a method to add new words into a speech recognition graph is presented. The method requires significantly less memory, takes less time than a full recompilation and doesn't affect recognition accuracy or speed.

Keywords: ASR · WFST · Decoding architecture · Dynamic vocabulary

1 Introduction

Weighted finite state-transducers theory has become the most widely used framework for creating decoding graphs in automated speech recognition field. WFST-based representation allows a single set of tools to be applied to all sources of the information that are needed for decoding [1]. A graph for WFST-based decoder is a cascade of transducers that combine information from lexicon, language model and the mapping of context-independent phonemes to context-dependent identifiers of acoustic models [2,3]. This cascade is compiled to large static or dynamic graphs, which are usually optimized for fast access during the recognition phase [4]. However, this approach has several disadvantages. For instance, it is difficult to add new information to a compiled WFST graph, whereas a full recompilation of the graph with new information added requires significant time and memory resources. Therefore, in recognition tasks that require an ability to supplement the vocabulary fast one faces certain problems. The problem for dynamic graphs has already been solved in several articles [5–8] but recognition with a dynamic graph operates slower than with a static one since on-the-fly composition takes time. Consequently, not all ASR applications can use a dynamic graph.

It is much more difficult to add new words to a static graph because of context-dependent phonemes between words [1]. The simplest approach is to use biphones at the word boundaries instead of cross-word triphones, but it leads

P. Sojka et al. (Eds.): TSD 2016, LNAI 9924, pp. 452–458, 2016.
DOI: 10.1007/978-3-319-45510-5_52

to the degradation of recognition accuracy. The article [9] suggests a technique of adding new words which can maintain cross-word context dependencies. It operates similarly to the method in earlier work [8], but also concerns static graphs. In this paper we propose a new method of word addition that works for both static and dynamic decoding graphs. It is more effective in terms of memory consumption compared to [9], because it does not rely on additional sources of information (like so called "affix graphs" in [9]) and therefore may lead to a possible occurrence of non-determinized areas in graph (in our experiments we have shown that it does not affect neither the speed nor the accuracy of recognition).

Experiments have shown that our method requires significantly less memory and operates much faster than the method with complete recompilation of the graph, while the recognition WER and RTF do not increase.

The rest of the paper is organized as follows. Section 2 presents the details of the method, Sect. 3 shows the results of our experiments and Sect. 4 summarizes the results.

2   Method Description

Speech recognition graph is represented as a weighted finite-state transducer (WFST) [10] that can be constructed as an optimized cascade (1)

$$N = (H * mindet(C * L)) \circ min(G), \tag{1}$$

where G is a language model (an automaton over words), L is a pronunciation dictionary or lexicon (a transducer mapping phones to words), C is the context-dependency specification (a transducer mapping context-dependent to context-independent phones) and H transforms sequences of senones (tied triphone states) to triphones. "*" represents the composition operator. "∘" represents the composition operator with a look-ahead filter.

The new method consists of two steps: the first step is designing the graph in a certain way so that later it would be easy to add new words into. The second step is adding the new information. The first step can be performed at the developer's side, and the second at the client's side. In order to make the analysis less complicated, we describe our method for a CLG cascade. The application of the method for HCLG cascades is similar to CLG case.

2.1   Construction of a Basic Graph

At the design stage we have two requirements for the basic graph:

- There must be a way to determine the exact spot in the graph where to insert new words;
- The cross-word context-dependencies should be maintained after the addition procedure.

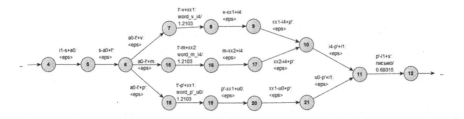

Fig. 1. Fragment of the source graph with inserted dummy words.

In order to meet the requirements above, the language model in G transducer, must be supplemented with unigram "dummy" words $word\_ph_b\_ph_e$ (where ph_b and ph_e are phonemes) for each possible pair of ph_b and ph_e. The transcription of these words must be ph_b xx_i ph_e, where xx_i is an auxiliary symbol ("dummy" phoneme; index i is explained later). Figure 1 shows a fragment of the graph with inserted dummy words.

The main idea of the algorithm is replacing the unigrams $word\_ph_b\_ph_e$ later with new words with ph_b and ph_e phonemes at the begin and at the end, respectively. At the same time context-dependencies are preserved at the new word boundaries, because the "dummy" words and the newly added words begin and end with the same phonemes.

It is vital for the arcs of "dummy" words not to merge with each other during the graph assembling. To avoid it, xx_i should be different for words with matching ph_b or ph_e. For example, for our speech recognition system [1] with 52 phonemes we have added 2704 dummy words $word\_ph_b\_ph_e$ and 52 xx_i symbols into the language model and lexicon. It leads to increasing the duration of the cascade construction and memory consumption during the process, but it does not matter since this graph is compiled in advance at the developer's side.

2.2   Construction of the New Words Graph

In order to add new words we initially divide them into groups of words with the same first and last phonemes in the transcription. Then an individual CL graph is constructed for each of these groups. This graphs will be inserted to the basic graph instead of dummy words. Several little CL graphs (one graph for each pair ph_b, ph_e) or one large graph can be constructed, but in the case of one graph it is important to ensure that arcs from subgraphs for different ph_b, ph_e are not merged during the optimization. The merge can be avoided by appending the special symbol $\#_j$ at the beginning and the end of each new word's transcription. $\#_j$ are auxiliary symbols, usually used in lexicon at the end of each transcription to make L transducer determinizable if the graph has identical transcriptions. For our purposes j must be different for words with different ph_b, ph_e. Figure 2 shows a new words graph with transcriptions starting with the phoneme "v" and ending with the phoneme "i4".

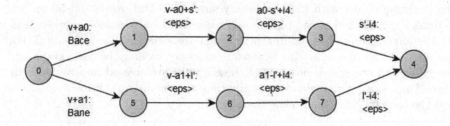

Fig. 2. A new words graph with transcriptions starting with the phoneme "v" and ending with the phoneme "i4".

2.3 Inserting the New Words Graphs to the Basic Graph

Now "dummy" words in the basic graph should be replaced with our graphs consisting of new words. Since instead of $word\_ph_b\_ph_e$ we insert words also starting with the phoneme ph_b and ending with the phoneme ph_e, correct contextual dependencies are preserved in adjacent words within the graph. The first and the last triphones of a dummy word in the basic graph match the first and the last triphones in the graph of new words. The phoneme xx_i in the triphone should be replaced with the relevant phoneme of the new word, while the arcs with triphones $ph_b - xx_i + ph_e$ should be replaced with inner arcs of the new words graph.

2.4 Generalization to Dynamic Graphs

The main difference between dynamic and static graphs is that for dynamic graphs the composition of model-level or state-level lexicon transducer (CL or HCL) with language model (G) is performed during the recognition, on-the-fly. Therefore we need to add new words to both CL and G graphs. The construction of the new words graph and the addition it to the CL is similar to the above-described method for CLG. To deal with G transducer, we add dummy word "new_word" to the language model and then compile G in the usual way. At the stage of the new word addition the arcs with "new_word" are replaced with arcs with new words.

3 Experiments

To evaluate our method, we used the Russian ASR system described in [1]. One hour of news records (5,000 words) was used as the test data. In order to increase OOV rate 1,000 words found in the test data were deleted from the recognition dictionary. Experiments were carried out for both static and dynamic graphs. To test the static graph, we used the language model with 5 million n-grams. The size of the lexicon was 244,000 words. For the dynamic graph, we took the language model with the vocabulary size of 377,000 words and 50 million n-grams. The method of complete graph reconstruction was used as a baseline.

Resources required for adding new words are shown in Tables 1 and 2. For our method, the time of adding new words does not include the basic graph and the HC graph creation time because these graphs are created in advance, only once. Thus, this value includes only the time to generate graph for new words, and the time to insert it into the basic graph.

Table 1. Resources required to add new words to a static graph

Number of added words	Baseline		Proposed method	
	Memory	Time	Memory	Time
300	12.2 GB	570 s	10.1 GB	180 s
500	12.2 GB	570 s	10.1 GB	180 s
1000	12.2 GB	570 s	10.1 GB	180 s

Table 2. Resources required to add new words to a dynamic graph

Number of added words	Baseline		Proposed method	
	Memory	Time	Memory	Time
300	13 GB	1050 s	5.5 GB	125 s
500	13 GB	1050 s	5.5 GB	125 s
1000	13 GB	1050 s	5.5 GB	125 s

As shown in Tables 1 and 2, speed has increased 8 times for the dynamic graph, and the memory consumption has decreased more than 2 times. In the experiment with the static graph the time of creation and memory usage have also decreased.

Changes in WER and RTF during the recognition are shown in Tables 3 and 4. Values are very close to the baseline.

Table 3. Changes in WER and RTF during the recognition, static graph

Number of added words	Baseline		Proposed method	
	WER %	RTF	WER %	RTF
300	36.2	0.39	36.3	0.42
500	33.8	0.38	33.9	0.42
1000	28.3	0.4	28.3	0.43

Table 4. Changes in WER and RTF during the recognition, dynamic graph

Number of added words	Baseline		Proposed method	
	WER %	RTF	WER %	RTF
300	34.9	0.76	34.9	0.81
500	32.7	0.78	32.7	0.83
1000	27.5	0.79	27.6	0.83

4   Conclusion

In this article we have introduced a method allowing to add words to a WFST-graph quickly and without its total reconstruction. The method can be applied to both static and dynamic graphs. Our experiments have shown significant reduction of time and RAM spent on adding new words. At the same time, there were practically no changes in word error rate and recognition speed.

Acknowledgements. The work was financially supported by the Ministry of Education and Science of the Russian Federation. Contract 14.579.21.0121, ID RFMEFI57915X0121.

References

1. Khomitsevich, O., Mendelev, V., Tomashenko, N., Rybin, S., Medennikov, I., Kudubayeva, S.: A bilingual Kazakh-Russian system for automatic speech recognition and synthesis. In: Ronzhin, A., Potapova, R., Fakotakis, N. (eds.) SPECOM 2015. LNCS, vol. 9319, pp. 25–33. Springer, Heidelberg (2015)
2. Novak, J., Minematsu, N., Hirose, K.: Open source WFST tools for LVCSR cascade development. In: Proceedings of the 9th International Workshop on Finite State Methods and Natural Language Processing, Blois, France, pp. 65–73 (2011)
3. Allauzen, C., Mohri, M., Riley, M., Roark, B.: A generalized construction of integrated speech recognition transducers. In: Proceedings of the ICASSP, pp. 761–764 (2004)
4. Saon, G., Povey, D., Zweig, G.: Anatomy of an extremely fast LVCSR decoder. In: Proceedings of the Interspeech, Lisbon, Portugal (2005)
5. Dixon, P.R., Hori, C., Kashioka, H.: A specialized WFST approach for class models and dynamic vocabulary. In: INTERSPEECH, pp. 1075–1078 (2012)
6. Novak, J.R., Minematsu, N., Hirose, K.: Dynamic grammars with lookahead composition for WFST-based speech recognition. In: INTERSPEECH, pp. 1079–1082 (2012)
7. Aleksic, P., Allauzen, C., Elson, D., Kracun, A., Casado, D.M., Moreno, P.: Improved recognition of contact names in voice commands. In: Proceedings of the ICASSP (2015)
8. Schalkwyk, J., Hetherington, I.L., Story, E.: Speech recognition with dynamic grammars using finite-state transducers. In: Proceedings of the Eurospeech, pp. 1969–1972 (2003)

9. Allauzen, C., Riley, M.: Rapid vocabulary addition to context-dependent decoder-graphs. In: INTERSPEECH, pp. 2112–2116 (2015)
10. Mohri, M., Pereira, F., Riley, M.: Speech recognition with weighted finite-state transducers. In: Benesty, J., Sondhi, M., Huang, Y. (eds.) Handbook of Speech Processing, pp. 559–584. Springer, Heidelberg (2008)

Dialogue

Influence of Reverberation on Automatic Evaluation of Intelligibility with Prosodic Features

Tino Haderlein[1]([✉]), Michael Döllinger[2], Anne Schützenberger[2],
and Elmar Nöth[1]

[1] Lehrstuhl für Informatik 5 (Mustererkennung),
Friedrich-Alexander-Universität Erlangen-Nürnberg (FAU),
Martensstraße 3, 91058 Erlangen, Germany
Tino.Haderlein@fau.de
http://www5.cs.fau.de
[2] Phoniatrische und pädaudiologische Abteilung in der HNO-Klinik,
Klinikum der Universität Erlangen-Nürnberg,
Bohlenplatz 21, 91054 Erlangen, Germany

Abstract. Objective analysis of intelligibility by a speech recognizer and prosodic features was performed for close-talking recordings before. This study examined whether this is also possible for reverberated speech. In order to ensure that only the room acoustics are different, artificial reverberation was used. 82 patients after partial laryngectomy read a standardized text, 5 experienced raters assessed intelligibility perceptually on a 5-point scale. The best feature subset, determined by Support Vector Regression, consists of the word correctness of a speech recognizer, the average duration of silent pauses, the standard deviation of the F_0 on the entire sample, the standard deviation of jitter, and the ratio of the durations of the voiced sections and the entire recording. A human-machine correlation of $r = 0.80$ was achieved for the close-talking recordings and $r = 0.72$ for the worst case of the examined signal qualities. By adding three more features, also $r = 0.80$ was reached for the reverberated scenario.

Keywords: Intelligibility · Automatic assessment · Prosody · SVR · Reverberation

1 Introduction

Automatic evaluation of voice and speech impairment allows to obtain objective assessment of the current state and temporal changes of distorted communication by voice [1]. In this study, the topic is the evaluation of intelligibility after partial removal of the larynx. Earlier work has shown that this can be performed using an automatic speech recognition (ASR) system, supported by a prosodic analysis module [2]. As a reference evaluation, the perceptual assessment by a speech therapist is the standard.

© Springer International Publishing Switzerland 2016
P. Sojka et al. (Eds.): TSD 2016, LNAI 9924, pp. 461–469, 2016.
DOI: 10.1007/978-3-319-45510-5_53

In order to get the best possible acoustic quality of the recordings to be analyzed, usually a headset or another close-talking microphone is used. However, this recording situation might have a negative influence on the patient. The patient might feel watched or controlled when he or she is aware that other people could get access to the recording. For patients after head or neck surgery, wearing a headset can also be painful. If the microphone is somewhere else in the room, both effects are attenuated, but the samples will be affected by reverberation then. It has been shown that, with according training data for the ASR system, speech recognition does also work in reverberated environment, even if the properties in the recording environment, such as the room impulse response, are not exactly known [3]. It has, however, not been examined whether a close-talking ASR system, supported by prosodic analysis, can be also used for intelligibility assessment in reverberated environment, i.e. for larger distances between mouth and microphone, and whether the same prosodic features are optimal for different distances. This will be addressed in this paper. The reference evaluation for all scenarios, however, is the human evaluation of a close-talking speech sample, because in a therapy session the patient and the therapist sit close to each other.

This paper is organized as follows: Sect. 2 introduces the speech data used for the experiments, Sect. 3 describes the artificial reverberation of these data. The processing of the features from the speech recognizer (Sect. 4) and the prosody module (Sect. 5) by Support Vector Regression follows in Sect. 6. The results will be discussed in Sect. 7.

2   Test Data and Subjective Evaluation

Two sets of speech data have been used for this study. The samples of the first set, further denoted as set *t82*, were used to determine prosodic measures of reverberated recordings that can describe perceptual intelligibility results. The second set – here called set *t85* – was used to test whether the findings on set *t82* also hold for other data with the same type of voice disorder. Set *t82* contains recordings of 82 persons (68 men and 14 women) after partial laryngectomy due to laryngeal cancer. Their average age was 62.3 with a standard deviation of 8.8 years; the youngest speaker was 41, the oldest one was 86 years old. Set *t85* was recorded from 85 patients (75 men, 10 women) suffering from cancer in different regions of the larynx. 65 of them had already undergone partial laryngectomy, 20 speakers were still awaiting surgery. The average age of the speakers was 60.7 years with a standard deviation of 9.7 years. The youngest and the oldest person were 34 and 83 years old, respectively.

Informed consent had been obtained by all participants prior to the examination. The study respected the principles of the World Medical Association (WMA) Declaration of Helsinki on ethical principles for medical research involving human subjects and has been approved by the ethics committee of our university. All persons read the German version of the tale "The North Wind and the Sun" [4], which is widely used in medical speech evaluation in German-speaking and other countries. It consists of 71 disjoint words and 108 words in total (172 syllables). The patients were recorded by a close-talking microphone (Logitech Premium Stereo Headset 980369-0914) with 16 kHz sampling frequency and 16 bit linear amplitude resolution.

Five experienced voice professionals (ear-nose-throat doctors, speech therapists) evaluated the intelligibility of each recording. The samples were played to the experts once via loudspeakers in a quiet seminar room without disturbing noise or echoes. Rating was performed on a five-point Likert scale. For computation of average scores for each patient, the grades were converted to integer values (1 = 'very high', 2 = 'rather high', 3 = 'medium', 4 = 'rather low', 5 = 'very low'). The sets *t82* and *t85* were evaluated by two different rater groups. Three of the therapists were part of both groups; however, set *t85* had already been evaluated about one year before set *t82*. The human raters evaluated only the original close-talking recordings and not the artificially reverberated samples that will be introduced in Sect. 3.

3   Artificial Reverberation of Speech Samples

In order to obtain reverberated speech with a large variety of acoustic quality, speech data must be collected in many rooms with different impulse responses. This leads to the problem that not always the same speakers are available, and if so, they will not be able to reproduce a given text exactly as in the other recording sessions. Hence, the recordings will not only be different with respect to the room impulse response, but also with respect to speaking style and maybe vocabulary. Reverberating close-talking speech artificially with previously measured room impulse responses can avoid this problem and also drastically reduce the effort of data acquisition.

For this study, the required impulse responses were obtained in a room where the reverberation time could be changed from $T_{60} = 250$ ms to $T_{60} = 400$ ms by removing sound absorbing carpets and curtains. 12 impulse responses were measured for loudspeaker positions on three semi-circles in front of the microphone

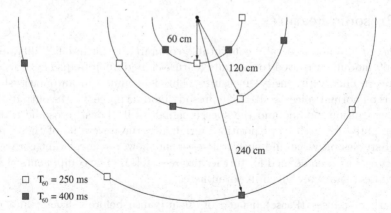

Fig. 1. Measuring room impulse responses for artificial reverberation (microphone position: •) at 12 speaker positions in a room with variable reverberation times T_{60}

at distances 60 cm, 120 cm, and 240 cm (see Fig. 1 and Table 2). The recording angle was counted clockwise from 0° to 165°. Six impulse responses each were measured with $T_{60} = 250$ ms and $T_{60} = 400$ ms. The available close-talking speech data (Sect. 2) were reverberated with each of them so that 12 reverberated versions of the original samples were available.

4   The Speech Recognition System

The speech recognition system used for the experiments has been described in [5]. It is based on semi-continuous Hidden Markov Models (HMM) and was trained with close-talking speech only. For each 16 ms frame, a 24-dimensional feature vector is computed. It contains short-time energy, 11 Mel-frequency cepstral coefficients, and the first-order derivatives of these 12 static features. The recognition vocabulary of the recognizer was changed to the 71 words of the standard text. The word accuracy and the word correctness were used as basic automatic measures for intelligibility since they had been successful for other voice and speech pathologies [6,7]. They are computed from the comparison between the recognized word sequence and the reference text consisting of the $n_{all} = 108$ words of the read text. With the number of words that were wrongly substituted (n_{sub}), deleted (n_{del}) and inserted (n_{ins}) by the recognizer, the word accuracy in percent is given as

$$\text{WA} = [1 - (n_{sub} + n_{del} + n_{ins})/n_{all}] \cdot 100$$

while the word correctness omits the wrongly inserted words:

$$\text{WR} = [1 - (n_{sub} + n_{del})/n_{all}] \cdot 100$$

Only a unigram language model was used so that the results mainly depend on the acoustic models. A higher-order model would correct too many recognition errors and thus make WA and WR useless as measures for intelligibility.

5   Prosodic Features

In order to find automatically computable counterparts for intelligibility, also a 'prosody module' was used to compute features based upon frequency, duration, and speech energy (intensity) measures. This is common in automatic speech analysis on normal voices [8–10]. The prosody module processes the output of the word recognition module and the speech signal itself. 'Local' prosodic features are computed for each word position. Originally, there were 95 of them. After several studies on voice and speech assessment, however, a relevant core set of 33 features has been defined for further processing [11]. The components of their abbreviated names are given in parentheses:

– Length of pauses (Pause): length of silent pause before (–before) and after (–after), and filled pause before (Fill-before) and after (Fill-after) the respective word

- Energy features (En): regression coefficient (RegCoeff) and the mean square error (MseReg) of the energy curve with respect to the regression curve; mean (Mean) and maximum energy (Max) with its position on the time axis (MaxPos); absolute (Abs) and normalized (Norm) energy values
- Duration features (Dur): absolute (Abs) and normalized (Norm) duration
- F_0 features (F0): regression coefficient (RegCoeff) and mean square error (MseReg) of the F_0 curve with respect to its regression curve; mean (Mean), maximum (Max), minimum (Min), voice onset (On), and offset (Off) values as well as the position of Max (MaxPos), Min (MinPos), On (OnPos), and Off (OffPos) on the time axis; all F_0 values are normalized.

The last part of the feature name denotes the context size, i.e. the interval of words on which the features are computed (see Table 1). They can be computed on the current word (W) or in the interval that contains the second and first word before the current word and the pause between them (WPW). A full description of the features used is beyond the scope of this paper; details and further references are given in [5, 12].

Besides the 33 local features per word, 16 'global' features were computed for intervals of 15 words length each. They were derived from jitter, shimmer, and the number of detected voiced and unvoiced sections in the speech signal [12]. They covered the means and standard deviations of jitter and shimmer, the number, length, and maximum length of voiced and unvoiced sections, the ratio of the numbers of voiced and unvoiced sections, the ratio of the length of the voiced sections to the length of the signal, and the same for unvoiced sections. The standard deviation of the F_0 was measured in two ways: it was computed for all voiced sections only and also over all sections of the speech recordings. In the latter case, each unvoiced frame contributed a value of 0. Hence, it incorporated also information about the percentage of frames where no regular voice signal was detected. Since all patients read the same text, this was supposed to indicate the degree of pathology.

The human listeners gave ratings for the entire text. In order to receive also one single value for each feature that could be compared to the human ratings, the average of each prosodic feature over the entire recording served as final feature value.

Table 1. Local prosodic features; the context size denotes the interval of words on which the features are computed (W: one word, WPW: word-pause-word interval)

Features	Context size	
	WPW	W
Pause: before, Fill-before, after, Fill-after		•
En: RegCoeff, MseReg, Abs, Norm, Mean	•	•
En: Max, MaxPos		•
Dur: Abs, Norm	•	•
F0: RegCoeff, MseReg	•	•
F0: Mean, Max, MaxPos, Min, MinPos, Off, OffPos, On, OnPos		•

6   Support Vector Regression (SVR)

In order to determine the best subset of WA, WR, and prosodic features to model the human intelligibility rating, Support Vector Regression (SVR, [13]) was used. The underlying SVM used a linear kernel. The complexity constant C for the SVR was set to 1. Each training example for the regression consisted of a set of features from the close-talking *t82* data set (the inputs) and a human intelligibility score (the target output). The sequential minimal optimization algorithm (SMO, [13]) of the Weka toolbox [14] was applied in a 10-fold cross-validation manner. For the attribute selection, the Greedy Stepwise algorithm was applied. The standard settings were not changed. In contrast to [2], all input features were not normalized but standardized (mean value: $\mu = 0$, standard deviation: $\sigma = 1$) for the analysis.

7   Results and Discussion

Former experiments on data set *t82* had revealed optimal features for close-talking analysis [2]. The human-machine correlation was $r = 0.79$ when prosodic features and the word correctness (WR) were combined. Using the WA instead gave worse results. The best feature subset consisted of WR, the average duration of the silent pauses before a word (Pause–before), the standard deviation of the fundamental frequency on the entire sample (StandDevF0/Sig), the standard deviation of jitter (StandDevJitter), and the ratio of the durations of the voiced sections and the entire recording (RelDur+Voiced/Sig). This set revealed also the best results on the same data in this study ($r = 0.80$; see Table 2), although the setup of the regression has been slightly changed.

In earlier experiments on the analysis of chronic hoarseness, also the absolute energy measured in a word-pause-word interval (EnAbsWPW) was among the candidates for the best feature set [15]. When this feature was added in this study, however, no significant rise of human-machine correlations could be obtained, except for impulse response h421045 ($r = 0.78$ instead of 0.77) and h423075 ($r = 0.74$ instead of 0.72). All other correlations were equal to those of the smaller set or even lower. For this reason, EnAbsWPW has been removed from the feature set again.

As expected, the human-machine correlation got lower in a higher reverberation time and with rising microphone distance, but still the lowest measured correlation was as high as 0.72 for 240 cm microphone distance and $T_{60} = 400$ ms where it had been 0.80 for the close-talking case. The angle at which the speaker spoke towards the microphone did not show consistent influence. The standard situation would be 90°, i.e. right in front of the microphone. All available pairs of the same T_{60} and microphone distance showed a relative difference in the angle of 90°, but very often not the one, which was closer to the absolute 90° position, was better with respect to human-machine correlation. Additionally, the influence of the angle was usually only $\Delta r = 0.02$.

The weights of the single features in the regression formulae for the *t82* data (see Table 2) do not show a unique behavior that would make it easy to

Table 2. Feature weights (columns 5 to 9) and human machine correlation r for artificially reverberated and close-talking *t82* data (last line: *t85*); reverberation time T_{60}, microphone distance ('dist.') and recording angle for the impulse responses are given on the left side

Impulse Response	T_{60} (ms)	dist. (cm)	angle (°)	Pause–before	StandDev Jitter	RelDur+ Voiced/Sig	WR	StandDev F0/Sig	r
h411000	250	60	0	0.207	0.074	−0.542	−0.419	0.524	0.78
h411090	250	60	90	0.253	0.151	−0.676	−0.356	0.521	0.79
h412060	250	120	60	0.313	0.136	−0.699	−0.377	0.468	0.78
h412150	250	120	150	0.256	−0.067	−0.593	−0.323	0.560	0.80
h413030	250	240	30	0.260	−0.234	−0.436	−0.376	0.573	0.77
h413120	250	240	120	0.238	−0.194	−0.451	−0.304	0.511	0.75
h421045	400	60	45	0.286	0.179	−0.700	−0.382	0.471	0.77
h421135	400	60	135	0.245	0.225	−0.782	−0.362	0.544	0.80
h422015	400	120	15	0.232	−0.265	−0.414	−0.154	0.524	0.76
h422105	400	120	105	0.322	−0.174	−0.471	−0.267	0.549	0.74
h423075	400	240	75	0.387	−0.222	−0.328	−0.116	0.472	0.72
h423165	400	240	165	0.292	−0.325	−0.364	−0.237	0.570	0.74
– (close-talk)	—	3–5	90	0.191	0.223	−0.881	−0.412	0.511	0.80
– (close-t., t85)	*—*	*3–5*	*90*	*0.485*	*−0.013*	*−0.551*	*−0.313*	*0.554*	*0.73*

relate them to T_{60} or microphone distance. The weight for the pauses between words (Pause–before) is relatively stable at about 0.2 to 0.4 among the simulated recording situations. The standard deviation of jitter (StandDevJitter) tends to get higher absolute values for more reverberant environments. The ratio of the durations of the voiced segments and the whole recording (RelDur+Voiced/Sig) loses influence with rising reverberation. So does also the word correctness (WR). The reason is obviously the mismatch between the acoustic properties of training and test environment of the recognizer. For the F_0 and jitter features, also the worse acoustic quality and hence the unreliable detection of F_0 is the most probable reason. However, the weight for StandDevF0/Sig is at about the same level in all experiments, and the StandDevF0, which is not related to the overall duration, but only to the voiced segments, does not occur in the best feature sets at all.

In general, the feature set, which was best for the close-talking case, can also basically be considered suitable for reverberated environment. In another experiment, a feature set has been determined, which is best for the acoustic scenario most deviant from the close-talking case (impulse response h423075; $T_{60} = 400$ ms, microphone distance 240 cm, angle 165°; see Table 3). Here, for the *t82* data the best feature set is a superset of the one for the close-talking

Table 3. Feature weights and human-machine correlation r for the set optimal for reverberated (impulse response h423165) *t82* data, tested on reverberated and close-talking *t82* and *t85* data

Impulse Response	Pause– before	Mean Jitter	StandDev Jitter	#+Voiced	RelNum +/–Voiced	RelDur+ Voiced/Sig	StandDev F0/Sig	r
h423165 *(t82)*	0.290	0.895	−0.970	−0.422	−0.206	−0.200	0.608	0.80
h423165 *(t85)*	0.504	0.640	−0.553	−0.121	−0.349	−0.292	0.452	0.75
– (close-t., *t82*)	0.432	0.027	0.147	−0.072	0.056	−0.847	0.508	0.73
– *(close-t., t85)*	0.377	0.658	−0.659	−0.215	−0.288	−0.403	0.493	0.74

case. It contains additionally the mean of jitter, the number of voiced sections in the recording (#+Voiced), and the ratio of the numbers of voiced and unvoiced sections (RelNum+/–Voiced). This lifts the human-machine correlation from 0.74 to 0.80. For the close-talking recordings, however, the correlation drops from 0.80 to 0.73 with this set, and the additional features show very low regression weights. The results on the *t85* data set (Tables 2 and 3) confirm the suitability of the selected features for intelligibility assessment both in good and bad acoustic conditions.

The inter-rater correlation between a single rater's intelligibility scores and the average of the 4 other raters was $r = 0.84$ for the *t82* recordings (for details, see [2]) and $r = 0.81$ for the *t85* data [16]. These are the reference values that an automatic system should reach to be regarded as reliable as an average human rater. The current results are almost at this level. Due to slight differences in correlations and regression weights for different features that occurred in this study, the experiments have to be continued with larger data sets in order to reassure the relevance of the selected features and to add other features where applicable. Nevertheless, the conclusion of this study is that automatic evaluation of intelligibility can be done on reverberated speech samples as reliable as for close-talking samples.

Acknowledgments. We would like to thank Dr. Wolfgang Herbordt for his kind support with the software and data for artificial reverberation. Dr. Döllinger's contribution was supported by Deutsche Krebshilfe grant no. 111332.

References

1. Baghai-Ravary, L., Beet, S.: Automatic Speech Signal Analysis for Clinical Diagnosis and Assessment of Speech Disorders. Springer, New York (2013)
2. Haderlein, T., Nöth, E., Batliner, A., Eysholdt, U., Rosanowski, F.: Automatic intelligibility assessment of pathologic speech over the telephone. Logopedics Phoniatrics Vocology **36**, 175–181 (2011)
3. Couvreur, L., Couvreur, C., Ris, C.: A corpus-based approach for robust ASR in reverberant environments. In: Proceedings of ICSLP, Beijing, vol. 1, pp. 397–400 (2000)

4. International Phonetic Association (IPA): Handbook of the International Phonetic Association. Cambridge University Press, Cambridge (1999)
5. Haderlein, T., Moers, C., Möbius, B., Rosanowski, F., Nöth, E.: Intelligibility rating with automatic speech recognition, prosodic, and cepstral evaluation. In: Habernal, I., Matoušek, V. (eds.) TSD 2011. LNCS, vol. 6836, pp. 195–202. Springer, Heidelberg (2011)
6. Haderlein, T.: Automatic Evaluation of Tracheoesophageal Substitute Voices. Studien zur Mustererkennung, vol. 25. Logos Verlag, Berlin (2007)
7. Maier, A.: Speech of Children with Cleft Lip and Palate: Automatic Assessment. Studien zur Mustererkennung, vol. 29. Logos Verlag, Berlin (2009)
8. Nöth, E., Batliner, A., Kießling, A., Kompe, R., Niemann, H.: Verbmobil: the use of prosody in the linguistic components of a speech understanding system. IEEE Trans. Speech Audio Process. **8**, 519–532 (2000)
9. Rosenberg, A.: Automatic detection and classification of prosodic events. Ph.D. thesis, Columbia University, New York (2009)
10. Origlia, A., Alfano, I.: Prosomarker: a prosodic analysis tool based on optimal pitch stylization and automatic syllabification. In Calzolari, N., et al. (eds.) In: Proceedings of the 8th International Conference on Language Resources and Evaluation (LREC 2012), pp. 997–1002 (2012)
11. Haderlein, T., Schwemmle, C., Döllinger, M., Matoušek, V., Ptok, M., Nöth, E.: Automatic evaluation of voice quality using text-based laryngograph measurements and prosodic analysis. Comput. Math. Methods Med. **2015**, 11 p. Published 2 June 2015 (2015)
12. Batliner, A., Buckow, J., Niemann, H., Nöth, E., Warnke, V.: The prosody module. In: Wahlster, W. (ed.) Verbmobil: Foundations of Speech-to-Speech Translation, pp. 106–121. Springer, Berlin (2000)
13. Smola, A.J., Schölkopf, B.: A tutorial on support vector regression. Stat. Comput. **14**, 199–222 (2004)
14. Witten, I.H., Frank, E.: Data Mining: Practical Machine Learning Tools and Techniques, 2nd edn. Morgan Kaufmann, San Francisco (2005)
15. Haderlein, T., Döllinger, M., Matoušek, V., Nöth, E.: Objective voice and speech analysis of persons with chronic hoarseness by prosodic analysis of speech samples. Logopedics Phoniatrics Vocology **41**, 106–116 (2016)
16. Bocklet, T., Haderlein, T., Hönig, F., Rosanowski, F., Nöth, E.: Evaluation and assessment of speech intelligibility on pathologic voices based upon acoustic speaker models. In: 3rd Advanced Voice Function Assessment International Workshop (AVFA2009), pp. 89–92. Universidad Politécnica de Madrid, Madrid (2009)

Automatic Scoring of a Sentence Repetition Task from Voice Recordings

Meysam Asgari$^{(\boxtimes)}$, Allison Sliter, and Jan Van Santen

Center for Spoken Language Understanding, Oregon Health and Science University,
Portland, OR, USA
{asgari,sliter,vansantj}@ohsu.edu

Abstract. In this paper, we propose an automatic scoring approach for assessing the language deficit in a sentence repetition task used to evaluate children with language disorders. From ASR-transcribed sentences, we extract sentence similarity measures, including WER and Levenshtein distance, and use them as the input features in a regression model to predict the reference scores manually rated by experts. Our experimental analysis on subject-level scores of 46 children, 33 diagnosed with autism spectrum disorders (ASD), and 13 with specific language impairment (SLI) show that proposed approach is successful in prediction of scores with averaged product-moment correlations of 0.84 between observed and predicted ratings across test folds.

Keywords: Automatic language assessment · Autism spectrum disorders · Language impairment

1 Introduction

Language disorders (LD) in childhood are associated with social stress and increased difficulties with peer relations in adolescence [3], as well as with risk for poor social adaptation and psychiatric disorders in adulthood [2]. Children with LD have problems with communication that negatively impacting long-term cognitive, academic, and psychosocial development [1]. There is a significant need for language assessment for early detection, diagnosis, screening, and progress tracking of language difficulties. However, assessment involves face-to-face sessions with a professional, which may not always be available or affordable. Clearly, there is a need for computer-based systems for automated speech-based language assessment.

Researchers have investigated some forms of automated speech assessment. Automated speech assessment has been conducted for assessment of "speech intelligibility" for children post-cleft palate repair surgery and adults post laryngectomy [4]. Some effort has been made in automatic assessment of disordered speech by looking at intermediate phonological models and specifying automatic speech recognition (ASR) models by main pathology for improved computed and expert rated intelligibility [5]. Automated prosody assessment has also been

© Springer International Publishing Switzerland 2016
P. Sojka et al. (Eds.): TSD 2016, LNAI 9924, pp. 470–477, 2016.
DOI: 10.1007/978-3-319-45510-5_54

conducted in studies of children with autism spectrum disorder (ASD) [12]. This paper showed that different forms of prosodic stress could be detected automatically.

Starting with describing the childrens' speech corpus, in Sect. 2, we formulate an approach on language assessment of recalling sentences by creating automatic scoring methods that optimally predict gold standard human ratings. We use automatic speech recognition (ASR) to transcribe the spoken responses of children and then we predict item scores via a regression algorithm from ASR output. In our corpus, items are scored on a 0–3 scale. Subject-level scores are then predicted from standard summary statistics derived from estimated sentence-level item scores, as described in Sect. 3. The machine learning experiments and the results are reported and discussed in Sect. 4.

2   Corpus

The corpus was generated by the audio recordings of the "Recalling Sentences" portion of the CELF-4 with a time-aligned transcription of the study subjects' response. The study group includes 46 children, ages 6 to 9 years (mean age of 7 years 3 months), 18 of which were diagnosed as having an autism spectrum disorder (ASD) as well as a language deficit, 15 with ASD but no language deficit, and 13 with specific language impairment (SLI) but no ASD. The study group included 36 males and 10 females. ASD can present with and without language impairment. To tease apart these differences, it is often necessary to perform a comprehensive clinical language evaluation. The Clinical Evaluation of Language Fundamentals edition 4 (CELF-4) does precisely this. However, it's made up of 18 subtests, all of which are currently scored on paper, by speech language pathologist [6].

2.1   Task

Each member of the study group was evaluated using the Clinical Evaluation of Language Fundamentals edition 4 (CELF-4). [9] CELF-4 is an individually administered test, designed by The Psychological Corporation and used by speech language pathologists to determine if a student (ages 5–21 years) has a language disorder or delay. The specific task that was recorded and provided the speech data for this experiment is the *Recalling Sentences* task where an examiner recites with increasingly long and syntactically complex sentences (32 sentences in English) (the *prompt*) that the child is asked to repeat verbatim. This task contributes to a core language score, an expressive language score, and language structure score. Scores vary from zero to three, with a single error (a word omission, a repetition, a word addition, transposition or substitution) results in a single point reduction. Two to three errors result in a score of 1. Four or more errors earn a zero. We note that a word transposition is scored as one error unless it changes the meaning, such as a subject-object reversal, in which case two errors are counted. As in other tests, items increase in difficulty, and to minimize frustration a stopping rule is used after a certain number of errors. The examination is discontinued if the child receives five consecutive scores of zero. Scores were generated by a certified expert and independently verified for reliability.

2.2   Manual Scoring

The CELF examinations were conducted by a speech language pathologist and audio recorded. Those recordings were then time-align transcribed by linguists. Neologisms were phonetically transcribed. Regular developmentally appropriate phoneme replacement was disregarded. Using the transcript's time alignment, the portions of the audio where the child was responding to the prompt were selected and paired with their transcription. Table 1 shows the distribution of manually rated scores across all 1083 sentences in the corpus.

Table 1. The frequency of scores across all recordings

Scores	0	1	2	3
# of sentences	380	175	158	370

3   Method

3.1   Automatic Scoring of a Sentence Repetition Task

Our proposed method consists of two main components, an ASR system and a machine learning based scoring algorithm as described in the following. First, we employ an ASR system to automatically transcribe the repeated sentences. Typically, an ASR system is used to produce the single, highest-likelihood transcription of the child's spoken response. However, it is often the case that the most accurate transcript is not the one receiving the highest likelihood. Therefore, instead of relying on the highest-likelihood transcription, we generate, via ASR, the 10 highest-likelihood transcriptions of each response, and compute the Levenshtein distance of each transcription to the stimulus sentence. For example, the Levenshtein distance of the response *the boy ate cookie* to the stimulus sentence *the boy dropped the cookie* is 2, because there is one substitution (*dropped* to *ate*) and one deletion (*the*). This gives us the transcription with the lowest WER, often known as the *oracle*.

Next, we encode the transcription with the lowest word error rate (WER), *oracle*, using the following features: numbers of insertions, deletions, and substitutions; WER; and Levenshtein distance itself. The results in a five dimensional per-sentence features vector. For a given subject, we then compute subject-level features by applying standard summary statistics including mean, median, standard deviation, and entropy over the aforementioned per-sentence feature vectors derived from the first N (12 in our experiments) items presented to that subject. We also capture interaction between per-sentence features by computing the covariance matrix (upper triangular elements) of features. This generates a global feature vector of fixed dimension for each subject. Finally, we use these features to predict the total score (here defined as the average item score for the first N items), using support vector regression (SVR) models.

3.2 Automatic Speech Recognition

Learning acoustic models in ASR systems requires a fairly large amount of training data, which is mostly beyond the scope of data collection for specialized populations. We tackle this issue by adding a large children's speech database to our small corpus for learning acoustic models. For automatic transcription of the recordings, we built a context-dependent HMM-GMM system with 39-dimensional MFCC features with delta and delta-delta coefficients, using the state-of-the-art Kaldi speech recognition toolkit [8]. We used the OGI Kids Speech Corpus, consisting of 27 h of spontaneous speech from 1100 children, from kindergarten through grade 10 [10] for training acoustic models. After cepstral mean and variance normalization and LDA, we employed model space adaptation using maximum likelihood linear regression (MLLR). Also, speaker adaptive training (SAT) of the acoustic models was performed by both vocal tract length normalization (VTLN) and feature-space adaptation using feature space MLLR (fMLLR). A trigram language model was built on OGI Kids Speech Corpus using the SRILM toolkit [11]. The WER on a 2-h held-out test corpus was approximately 26 %, and adding another path of recognition to decode with the oracle reduced the WER to 14.27 %.

4 Experiments

As mentioned above, the corpus included children ages 6–9 years, 36 males and 10 females. We evaluated the performance of our proposed method on two tasks on our corpus: (1) a four-class classification task to classify sentence-level ratings and (2) a regression task to predict subject-level scores. The manual scoring describes how precisely the prompt has been repeated by participants and provides a reference for the automatic scoring. From the transcriptions generated by described ASR system, we extracted the number of insertions, deletions, substitutions, WER, and Levenshtein distance from a pair of prompt and its associated automated transcription constructing a feature vector of five dimension per each recording. For comparison, we re-performed the same analysis over manual transcripts and extracted the same set of features from them. We also excluded sentences for which the raters took semantic changes into account for scoring the recordings.

4.1 Multi-class Classification

We evaluated the performance of the proposed method on learning a 4-way classification model to predict the sentence-level scores employing support vector machine (SVM) classifier using radial basis function (RBF) and linear kernels implemented in scikit-learn toolkit [7]. Manually rated scores vary from zero to three according to the number of errors (a word omission, a repetition, a word addition, transposition or substitution) seen in the repeated sentence. Class labels of 3, 2, 1, and 0 correspond to zero, one, two or three, and four or more errors, respectively. In order to estimate the optimal set of model parameters, we used a five-fold cross validation scheme, setting all model parameter using four of the five sets as training set, and using the fifth ones only at the testing time. Parameters of the optimal SVM model were estimated on the training set separately for each fold, via grid search and cross-validation. As shown in Table 1, the class distributions are not balanced and thus the *classification accuracy* may not well describe the performance of the classifier. In order to address this drawback, we adopt unweighted average recall (UAR) as the performance criteria that normalizes the effect of skewed classes. Table 2 reports the performance of different classifiers measured in terms of UAR for classifying sentence ratings into four classes. From the results, it is clear that our proposed method applied on all three forms of transcription significantly outperforms the chance model in terms of UAR. Also, results suggest that SVM model with linear kernel is more suitable for this task than the non-linear RBF kernel. Furthermore, the comparable UAR between *ASR-Oracle* and *Manual* suggests that ASR system can be employed effectively used for this task.

Table 2. Unweighted average recall

Model	Chance	ASR	ASR-Oracle	Manual
SVM-linear	0.24	0.55	0.59	0.61
SVM-RBF	0.24	0.54	0.57	0.58

We also report the detailed performance of our 4-way classification system (*Linear SVM*) in terms of *precision, recall*, and *F1-score* in Tables 3, 4, and 5 for predicting the sentence-level true scores using ASR, ASR-Oracle, and manual transcriptions, respectively.

Table 3. ASR

Class label	Precision	Recall	F1-score
0	0.77	0.85	0.80
1	0.36	0.26	0.30
2	0.29	0.17	0.20
3	0.70	0.87	0.77

Table 4. ASR-Oracle

Class label	Precision	Recall	F1-score
0	0.81	0.85	0.83
1	0.32	0.35	0.38
2	0.29	0.18	0.21
3	0.76	0.88	0.81

Table 5. Manual transcription

Class label	Precision	Recall	F1-score
0	0.84	0.91	0.87
1	0.51	0.39	0.43
2	0.34	0.30	0.30
3	0.77	0.86	0.80

4.2 Regression

Subject-Level Scores. According to the distribution of manual ratings plotted in Fig. 1, sentence difficulty varies across sentences with less variation in small sentences (1 to 12). In the other words, number of errors seen in the repeated sentences is directly proportional to the length of the *prompt*. In order to assess subject-level performance in the repetition task, we took the first 12 sentences presented to the subject and averaged across their sentence-level ratings. This gave us a unique per-subject metric describing overall performance of the subject. Through a regression model, we aim to predict this score for every subject in our corpus.

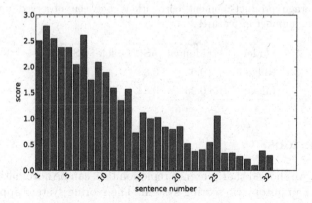

Fig. 1. Distribution of manually rated scores as a function of number of words in the sentence

Features. Per-subject features are computed across all $N = 12$ sentences by applying standard summary statistics including mean, median, standard deviation, and entropy to the 5-dimensional per-sentence feature vectors as described earlier. We also capture interaction between per-sentence features by computing the covariance matrix (upper triangular elements) of features. This results in a global 35 dimensional feature vector for each subject.

Learning Strategies. We investigated two forms of regularization, L2-norm in ridge regression, and hinge loss function in support vector regression. These two learning strategies were evaluated on our data set using five cross-validation scheme with the scikit-learn toolkit [7]. The averaged mean absolute error (MAE) between predicted and gold standard scores across five test folds employing two learning strategies are presented in Table 6. We also show in Table 7, the performance of learning strategies in predicting the subject-level scores in terms of averaged product-moment correlations between observed and predicted ratings across test folds. From the results, it is observed that *Linear SVR* outperforms ridge regression. Also, slightly higher MAE using *ASR-Oracle* compare to *Manual* transcription suggests that proposed automatic method using an ASR system can be effectively employed for assessing the sentence repetition tasks.

Table 6. Mean Absolute Error (MAE) between observed and predicted ratings. The K is the number of features.

Model	Manual	ASR-Oracle	ASR	K
Ridge	0.10	0.11	0.21	35
Linear SVR	0.035	0.10	0.13	35

Table 7. Average product-moment correlations between observed and predicted ratings. K is the number of features.

Model	Manual	ASR-Oracle	ASR	K
Ridge	0.93	0.83	0.73	35
Linear SVR	0.96	0.84	0.75	35

5   Conclusions

In this paper, we showed that sentence repletion task can be automatically scored and described an automatic scoring system. The scoring system applies ASR to the spoken responses, optionally computes estimated item scores via machine learning (ML) from ASR output, and computes estimated total scores by applying ML either to estimated item scores or directly to ASR output. Confining the analysis to those 41 children (12 with ALN, 18 with ALI, and 11 with SLI) who had been given at least the first $N = 12$ items, and computing each child's average of their scores on these items, we found that the Mean Absolute Error of observed and estimated total scores was 0.10 (observed score range of 0.83–3.0); the product moment correlation was 0.84.

Acknowledgements. We thank Katina Papadakis for manually transcribing the corpus for this project. This research was supported by NIH award 1R01DC013996-01A1. Any opinions, findings, conclusions or recommendations expressed in this publication are those of the authors and do not reflect the views of the funding agencies.

References

1. Beitchman, J.H.: Language, Learning, and Behavior Disorders: Developmental, Biological, and Clinical Perspectives. Cambridge University Press, Cambridge (1996)
2. Clegg, J., Hollis, C., Mawhood, L., Rutter, M.: Developmental language disorders-a follow-up in later adult life. Cognitive, language and psychosocial outcomes. J. Child Psychol. Psychiatry **46**(2), 128–149 (2005)
3. Conti-Ramsden, G., Mok, P.L., Pickles, A., Durkin, K.: Adolescents with a history of specific language impairment (SLI): strengths and difficulties in social, emotional and behavioral functioning. Res. Dev. Disabil. **34**(11), 4161–4169 (2013)
4. Maier, A., Haderlein, T., Eysholdt, U., Rosanowski, F., Batliner, A., Schuster, M., Nöth, E.: Peaks-a system for the automatic evaluation of voice and speech disorders. Speech Commun. **51**(5), 425–437 (2009)
5. Middag, C., Martens, J.P., Van Nuffelen, G., De Bodt, M.: Automated intelligibility assessment of pathological speech using phonological features. EURASIP J. Adv. Sig. Process. **2009**(1), 1–9 (2009)
6. Paslawski, T.: The clinical evaluation of language fundamentals (CELF-4): a review. Can. J. Sch. Psychol. **20**(1/2), 129 (2005)
7. Pedregosa, F., Varoquaux, G., Gramfort, A., Michel, V., Thirion, B., Grisel, O., Blondel, M., Prettenhofer, P., Weiss, R., Dubourg, V., Vanderplas, J., Passos, A., Cournapeau, D., Brucher, M., Perrot, M., Duchesnay, E.: Scikit-learn: machine learning in Python. J. Mach. Learn. Res. **12**, 2825–2830 (2011)
8. Povey, D., Ghoshal, A., Boulianne, G., Burget, L., Glembek, O., Goel, N., Hannemann, M., Motlicek, P., Qian, Y., Schwarz, P., Silovsky, J., Stemmer, G., Vesely, K.: The kaldi speech recognition toolkit. In: IEEE 2011 Workshop on Automatic Speech Recognition and Understanding. IEEE Signal Processing Society, December 2011. iEEE Catalog No.: CFP11SRW-USB
9. Semel, E.M., Wiig, E.H., Secord, W.: CELF 4: Clinical Evaluation of Language Fundamentals. Psychological Corporation, Pearson (2006)
10. Shobaki, K., Hosom, J.P., Cole, R.: The OGI kids' speech corpus and recognizers. In: Proceedings of ICSLP, pp. 564–567 (2000)
11. Stolcke, A.: SRILM - an extensible language modeling toolkit, pp. 901–904 (2002)
12. Van Santen, J.P., Prud'hommeaux, E.T., Black, L.M.: Automated assessment of prosody production. Speech Commun. **51**(11), 1082–1097 (2009)

Platon: Dialog Management and Rapid Prototyping for Multilingual Multi-user Dialog Systems

Martin Gropp, Anna Schmidt, Thomas Kleinbauer$^{(\boxtimes)}$, and Dietrich Klakow

Spoken Language Systems Group, Saarland University, Saarbrücken, Germany
{martin.gropp,anna.schmidt,thomas.kleinbauer,
dietrich.klakow}@lsv.uni-saarland.de

Abstract. We introduce Platon, a domain-specific language for authoring dialog systems based on Groovy, a dynamic programming language for the Java Virtual Machine (JVM). It is a fully-featured tool for dialog management that is also particularly suitable for, but not limited to, rapid prototyping making it possible to create a basic multilingual dialog system with minimal overhead and then gradually extend it to a complete system. It supports multilinguality, multiple users in a single session, and has built-in support for interacting with objects in the dialog environment. It is possible to integrate external components for natural language understanding and generation, while Platon can itself be integrated even in non-JVM projects or run in a stand-alone debugging tool for testing. In this paper we describe important elements of the language and present two scenarios Platon has been used in.

Keywords: Dialog framework · Dialog systems · Dialog management · Multilingual · Multi-user · Rapid prototyping

1 Introduction

Platon is a domain-specific language for authoring dialog systems. It was designed with rapid prototyping in mind and has integrated facilities to interact with discourse and world objects. Platon is able to support multi-linguality and can handle multi-user scenarios. Its modular architecture allows building dialog systems that can either be used as a component in a more complex application, or function as a stand-alone system with built-in support for speech recognition and text synthesis.

Platon is designed for both technical and non-technical users. The language is designed for readability and maintainability, yet offers advanced users flexible extension options. Being based on Groovy, a dynamic language for the Java Virtual Machine (JVM), Platon dialogs can draw on the full set of features of an established powerful programming language, integrate existing Java classes, and utilize the vast amount of existing libraries for the JVM. This ability allows, for example, to incorporate existing parsers, or to connect to external databases or services.

© Springer International Publishing Switzerland 2016
P. Sojka et al. (Eds.): TSD 2016, LNAI 9924, pp. 478–485, 2016.
DOI: 10.1007/978-3-319-45510-5_55

This article introduces the Platon language and exemplifies typical use cases.[1] The main contributions of Platon to the landscape of already existing solutions to dialog management are:

- Accessibility: Easy to use script format for implementing dialog behavior
- Extensibility: Backed by a dynamic programming language (Groovy)
- Integration: Modular interaction with third-party components
- Flexibility: Not tied in with a specific dialog management model
- Speed: Rapid prototyping through compact and expressive syntax
- Availability: OS-independent open source package ready for download

2   Related Work

Platon is agnostic to the underlying dialog management *model*, hence we concentrate in this section on relevant alternative *systems*. (A general introduction to different models of dialog management is given by [7].) A number of approaches for implementing dialog managers have been proposed in the past, which can be classified according to the complexity of interactions they allow.

Declarative interpreted languages, such as e.g. AIML[2] or VoiceXML[3], can be appealing due to their ease of use, allowing non-expert users to model simple dialog behavior. When based on widely accepted standards, such as XML, the availability of sophisticated editing tools can facilitate speedy development. However, this often comes at the price of limited functionality. AIML, for instance, interprets user input with a simple pattern matching language, provides only rudimentary management of the dialog history through its <that> and <topic> tags, and is restricted to purely user-initiative dialog behavior. Similarly, VoiceXML has strengths in form-filling applications but has been criticized for its inflexible handling of dialog initiative and shallow dialog model [3,8].

NPCEditor [6] is a tool that allows to model question/answer pairs and uses information retrieval techniques to process user input that matches only partially. Some more elaborate tools, such as e.g. RavenClaw [1], aim for higher flexibility and task independence. To this end, RavenClaw employs a two-tiered architecture that separates domain and dialog management aspects, and has been successfully used in a number of applications. However, the system expects dialogs to be modeled as a hierarchical plan which may be too rigid a constraint for dialogs that are not purely task-driven.

In general, the more opinionated a framework or tool is with respect to expressing possible interactions, the more it runs the risk of hampering what was not anticipated by the framework's developers. This is one criticism that motivated the development of the DIPPER [2] architecture as an alternative to TrindiKit [5]. Both systems are inherently based on the Information State

---

[1] A complete documentation is available as a separate technical report [4] with details about the language definition and the implementation of Platon.

[2] http://www.alicebot.org/aiml.html.

[3] http://www.w3.org/TR/voicexml21.

Update approach [9] which, however, restricts their use to situations where this model is applicable.

In comparison, Platon unites the strengths of the above systems while avoiding some of their shortcomings. Basic keyword-based dialog behavior can be easily implemented, similar to simple scripting languages, but based on very flexible pattern matching capabilities. In addition, Platon also contains a sophisticated task model based on agents to allow modeling more complex dialogs. Moreover, being tightly coupled with Groovy/Java, Platon makes it especially easy to extend the off-the-shelf functionality, e.g. to integrate external resources. These points will be highlighted in more detail in the following sections.

3 Elementary Features

3.1 Platon for Rapid Prototyping

Platon can be used to realize a basic chatbot-like dialog system as a stand-alone application, as a first impression of a planned bigger system, or for early integration with other components.

A developer can directly define and test reactions to a limited number of textual inputs using simple matching rules, such as the one in Fig. 1. These rules can, as in this example, use regular expressions, but Platon also supports more complex input analyses. A number of predefined functions can be used in the reactions to an input match, for example for language output, for waiting for user responses, or for interacting with the outside world (see below).

```
input(~/\bhello\b/) {
  tell all, "Hello World!";
}
```

Fig. 1. Monolingual *Hello World* script.

3.2 Multilingual Multi-user Dialog Systems

Especially for prototypes and dialog systems without dedicated NLU and NLG components, Platon scripts can (optionally) provide internationalization support. Figure 2 shows a complete "Hello World" example for English and German: both input matching and reactions are realized bilingually.

Platon supports dialog situations with more than one user. To keep the complexity as low as possible and to make it easy to use the correct resources for users with different languages, a new dialog engine instance is created for each user at first. Interaction

```
input(
  en: ~/\bhello\b/,
  de: ~/\bhallo\b/
) {
  tell all, [
    en: "Hello World!",
    de: "Hallo Welt!"
  ];
}
```

Fig. 2. Multilingual *Hello World* script

between these instances is achieved either by using shared variables or by sending and receiving arbitrary messages. Reacting to such messages works much like receiving text input from users.

4   Complex Platon Systems

Although Platon works well for rapid prototyping, it was built with more complex scenarios in mind. In a more complex system, it is intended to take the role of the dialog manager that interacts with external NLU and NLG components.

4.1   Dialog Management: Task Decomposition and Agents

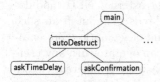

Fig. 3. Part of a task tree

Dialog management in Platon is based on a concept of hierarchical task decomposition similar to that of RavenClaw [1], breaking a complex scenario up into manageable parts. For example, the tree in Fig. 3 is a small excerpt of the task tree for a computer game set on a space station. Here, the station has an auto-destruct system that can be activated by the player. For this, a time delay needs to be specified and the player has to confirm that he/she is really sure about giving the command.

In a dialog script these sub-tasks appear as *agents*, each of which contains a set of rules for input handling and for other reactions, and may keep a local state. Figure 4 shows the script part of the example. The `autoDestruct` agent has two variables, `delay` and `confirmed`, that represent the state of the agent. The following `enter` block is executed when the agent is activated and every time a sub-agent completes (technically: whenever the agent becomes the top element on the stack; see below). In the example, `enter` checks which information is still needed and activates other agents accordingly, or, if all necessary information is available, exits.

```
agent autoDestruct {
  def delay = null;
  def confirmed = false;

  enter {
    if (delay == null) {
      askTimeDelay();
    } else if (!confirmed) {
      askConfirmation();
    } else {
      exit();
    }
  }
  ...
  agent askTimeDelay { ... }
  agent askConfirmation { ... }
}
```

Fig. 4. Outline of the `autoDestruct` agent

```
input(TimeDelay) {
  input ->
  delay = input;
  askConfirmation();
}

agent askTimeDelay {
  input(String) {
    input ->
    delay = new TimeDelay(input);
    exit();
  }
}
...
```

Fig. 5. Input statements from the `autoDestruct` agent

Although Platon can provide basic language understanding tasks as described above, a more complex dialog system typically integrates a separate NLU module that can provide a comprehensive analysis of the user input. Platon's JVM foundation makes the integration of many existing parsers, taggers, dialog act classifiers, etc. straight-forward. Moreover, if necessary, the dialog manager can provide access to certain context information, e.g. about the active agents, the dialog history, or entities in the environment, which can, for example, be used for the context-aware disambiguation of the input.

Platon is able to integrate such a broad range of external NLU modules because it does not impose any restrictions on the kinds of input from such modules. In particular, it does not expect a specific kind of semantic representation, dialog act scheme, domain ontology, etc. Platon can operate with any user-defined input type. For instance, an application can use a set of different classes as in the example of Fig. 5 where the NLU module uses the class TimeDelay for utterances specifying time delays, or opt for a different representation, such as simple strings, if that is considered more suitable.

4.2   Processing Input

Active agents are organized in a stack. When an agent calls another agent, e.g. askTimeDelay() in Fig. 4, the new agent is pushed on the stack, and stays there until it exits[4]. In the example of Fig. 6, the agent autoDestruct has called askTimeDelay, which has consequently been put at the top of the stack. Every time the dialog manager has to determine the system's reaction to an event (e.g. user input), it starts with the agent at the top of the stack and then proceeds downwards until an agent accepts the event. Optionally, an agent can delegate events to another (possibly inactive) agent, either on a case-by-case basis, or as a regular part of its own event processing procedure. This feature makes it easy to integrate agents for common tasks without adding complexity to the general stack-based processing scheme. We are currently working on a standard library of common agents (e.g. for repetitions or confirmations).

Examining the agent stack in Fig. 6, we see that autoDestruct (from Fig. 4 on the preceding page) has already called askTimeDelay, which is now on top of the stack. Its only input statement accepts String objects, but not objects of type TimeDelay. These are matched in the second agent, autoDestruct. This means that objects of type TimeDelay will be handled even if the askTimeDelay agent is not active: as long as autoDestruct is somewhere on the stack, TimeDelay objects can be interpreted as the delay for the self-destruct sequence.

Assuming a user input of "*set the time delay to five minutes*" this string would first be passed to the NLU which recognizes it as a time delay specification and

---

[4] Since all agents on the stack are active and can manipulate the stack, the call semantics are actually more complex than for example with regular functions. By default, agent changes are handled as if the agent executing the operations were on top of the stack, removing other agents covering the caller. This leads to the behavior expected for a regular function call. If required, this "stack cutting" mechanism can be disabled for each call. See [4] for details.

returns a `TimeDelay` object storing the duration. The `input` statement in the `askTimeDelay` agent only matches objects of type `String`, hence we proceed down the stack and find the next agent, `autoDestruct`. Its first `input` rule accepts the `TimeDelay` object and calls the next agent, which is pushed on top of the stack replacing `askTimeDelay`.

4.3   Situated Interaction

Platon was built to interact with objects in the dialog environment, to affect this "world" using voice input, and to react to changes. Platon systems can connect to an external server to exchange information about world objects, either using a direct Java-compatible interface or via an RPC protocol based on Apache Thrift[5]. Such a world object server must implement one function to allow the manipulation of object attributes, plus an additional two if atomic transactions are required. On the other side of the interface, Platon implements functions to receive notifications about added,

Fig. 6. Three active agents on a stack

deleted, and modified objects from the world server, which are transparently cached, and supports transactions as well. From the perspective of a dialog designer, this complexity is completely invisible. Platon provides the statements `objectAdded`, `objectDeleted`, and `objectModified` to react to changes in the world state, which support complex selectors to decide whether or not a given change in an object is relevant.

5   Sample Applications

Platon was originally developed in the context of an interactive multi-user game focusing on collaboration between players speaking different languages. The dialog system plays a central role in this game, acting as the on-board computer controlling a space station in an emergency situation. The players cannot communicate with each other directly. Instead, they interact with the game *environment* using their voice, and external changes to the environment may be communicated via voice output in addition to the graphical user interface and sound effects. Consequently, in addition to being the interface to the space station, the dialog system becomes a mediator between the players when they have to collaborate in order to achieve common goals. This kind of setup requires a flexible dialog system framework which supports (a) multiple users (b) speaking different languages and (c) which is able to interface well with the game world as well as (d) with the other software components. Platon's design meets all

---

[5] https://thrift.apache.org.

S: Alert! Oxygen levels are decreasing rapidly in Module 0. Activate oxygen pump to prevent suffocation of crew members.

P: Then activate this pump?

S: Error 201. Can't turn on oxygen pump. Energy missing. Deactivate systems to have at least 20% energy.

P: Okay, which systems can we switch off?

S: Systems currently online: ...

Fig. 7. Excerpt of a dialog between system (S) and player (P) from the beginning of the adventure game.

of these requirements. Its rapid prototyping capabilities proved to be a crucial feature for integrating the individual parts of the game as early as possible, including external ASR and TTS and world server components. Once the early prototype stages had been established, Platon allowed a seamless progression to a more feature-rich dialog system. Figure 7 shows an example dialog from this game scenario.

To demonstrate Platon's suitability for other domains, we built a second dialog system for a home automation scenario. Here, we control a virtual apartment with a number of devices including lights, heating, door locks, etc. The user can query and manipulate the status of each of these devices. This system does not rely on an external NLU. Instead, the necessary functions for basic reference resolution and keyword spotting were implemented directly in Groovy. Platon's built-in object interaction support proved especially useful here, allowing us to easily react to opening doors or finished washing machines, etc. With custom classes and methods for the world objects it was possible to perform most environment manipulations in a single line of code.

6   Integration and Tools

Platon comes with command line and graphical tools to run and test dialog scripts. Both support input and output of written text, the GUI also has built-in support for speech synthesis[6] and speech recognition[7] and can automatically test a dialog system with prefabricated bulk input.

To run a Platon dialog system outside this tool, a host application needs to manage sessions and take care of handling input and output, as illustrated in Fig. 8. The figure also includes the optional interfaces for natural language understanding and for interacting with

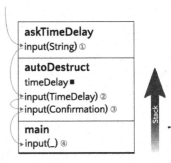

Fig. 8. Platon interfaces (gray: optional)

---

[6] MaryTTS: http://mary.dfki.de/.

[7] Sphinx: http://cmusphinx.sourceforge.net/.

world objects, as described in Subsects. 4.2 and 4.3. In addition to the direct Java-compatible interfaces, Platon provides additional Apache Thrift RPC interfaces to maximize the compatibility with non-JVM applications. When it is ready, a Platon application can be deployed as a single *jar* file including all dialog scripts.

7   Conclusions

We described Platon, a domain-specific language for dialog systems. Its focus ranges from rapid prototyping to the realization of fully-fledged dialog systems. Sophisticated input processing is implemented through a hierarchical task decomposition model based on agents for individual sub-tasks. Platon is agnostic toward the choice of underlying dialog management model as well as to the (semantic or dialog act) representation of system inputs and outputs. As it is based on Groovy, dialog scripts have ready access to third-party software written for the Java Virtual Machine. With two example systems, we further demonstrated how a Platon-based dialog system can interact with an application environment.

Platon is available under the Apache License on https://github.com/uds-lsv/.

Acknowledgments. The research presented in this paper has been funded by the Eureka project number E!7152. https://www.lsv.uni-saarland.de/index.php?id=71

References

1. Bohus, D., Rudnicky, A.I.: The RavenClaw dialog management framework: architecture and systems. Comput. Speech Lang. **23**(3), 332–361 (2009)
2. Bos, J., Klein, E., Lemon, O., Oka, T.: DIPPER: description and formalisation of an information-state update dialogue system architecture. In: Proceedings of the 4th SIGdial Workshop on Discourse and Dialogue, pp. 115–124 (2003)
3. Fabbrizio, G.D., Lewis, C.: Florence: a dialogue manager framework for spoken dialogue systems. In: Proceedings of Interspeech 2004, Jeju Island, Korea, pp. 3065–3068 (2004)
4. Gropp, M.: Platon. Technical report LSV TR 2015–002 (2015). http://www.lsv.uni-saarland.de
5. Larsson, S., Traum, D.: Information state and dialogue management in the trindi dialogue move engine toolkit. Natural Lang. Eng. **5**(3–4), 323–340 (2000)
6. Leuski, A., Traum, D.: NPCEditor: creating virtual human dialogue using information retrieval techniques. AI Mag. **32**(2), 42–56 (2011)
7. McTear, M.F.: Spoken dialogue technology: enabling the conversational user interface. ACM Comput. Surv. **34**(1), 90–169 (2002)
8. Nyberg, E., Mitamura, T., Hataoka, N.: Dialogxml: extending voicexml for dynamic dialog management. In: Proceedings of the Second International Conference on Human Language Technology Research, pp. 298–302 (2002)
9. Traum, D.R., Larsson, S.: The information state approach to dialogue management. In: van Kuppevelt, J., Smith, R.W. (eds.) Current and New Directions in Discourse and Dialogue. TSLT, vol. 22, pp. 325–353. Springer, Netherlands (2003)

How to Add Word Classes to the Kaldi Speech Recognition Toolkit

Axel Horndasch[(✉)], Caroline Kaufhold, and Elmar Nöth

Lehrstuhl für Informatik 5 (Mustererkennung),
Friedrich-Alexander-Universität Erlangen-Nürnberg (FAU),
Martensstraße 3, 91058 Erlangen, Germany
{axel.horndasch,caroline.kaufhold,elmar.noeth}@fau.de
http://www5.cs.fau.de/

Abstract. The paper explains and illustrates how the concept of word classes can be added to the widely used open-source speech recognition toolkit Kaldi. The suggested extensions to existing Kaldi recipes are limited to the word-level grammar (G) and the pronunciation lexicon (L) models. The implementation to modify the weighted finite state transducers employed in Kaldi makes use of the OpenFST library. In experiments on small and mid-sized corpora with vocabulary sizes of 1.5 K and 5.5 K respectively a slight improvement of the word error rate is observed when the approach is tested with (hand-crafted) word classes. Furthermore it is shown that the introduction of sub-word unit models for open word classes can help to robustly detect and classify out-of-vocabulary words without impairing word recognition accuracy.

Keywords: Word classes · Kaldi speech recognition toolkit · OOV detection and classification

1 Introduction

It is a well-known fact that class-based n-gram language models help to cope with the problem of sparse data in language modeling and can reduce test set perplexity as well as the word error rate (WER) in automatic speech recognition [1–3]. But word classes can also be used to support the detection of out-of-vocabulary (OOV) words. For example in [4] multiple Part-of-Speech (POS) and automatically derived word classes are introduced to better model the contextual relationship between OOVs and the neighboring words. The authors of [5,6] focus on semantically motivated word classes which are combined with generic or more generalized word models. In both cases the idea is to focus on open word classes like person or location names for which OOVs are very common.

In so called hierarchical OOV models a sub-word unit (SWU) language model for OOV detection is embedded into a word-based language model; the SWU-based language model can also be inserted into a word class. This approach has been suggested for example in [7] and just recently in [8]. With the solution presented in this paper it is possible to implement a similar strategy.

© Springer International Publishing Switzerland 2016
P. Sojka et al. (Eds.): TSD 2016, LNAI 9924, pp. 486–494, 2016.
DOI: 10.1007/978-3-319-45510-5_56

The speech recognition toolkit we based our work on is the widely used open-source software suite Kaldi [9]. It provides libraries and run-time programs for state-of-the algorithms under an open license as well as complete recipes for building speech recognition systems. All training and decoding algorithms in Kaldi make use of Weighted Finite State Transducers (WFSTs), the fundamentals of which are described in [10]. By modifying the standard Kaldi transducers using the OpenFST library tools [11] we were able to integrate word classes into the decoder, a feature that was missing in Kaldi so far.

Because the topic[1] has been discussed by Kaldi users now and again, we thought it would be worthwhile to write a paper in which we share the experiences we made when we extended the Kaldi recipes to include word classes.

The rest of the paper is structured in the following way: In Sect. 2 we introduce word classes for automatic speech recognition in a bit more detail. How word classes can be modeled in Kaldi is described in Sect. 3. Our experiments and the data sets we used are presented in Sect. 4 and the paper ends with a conclusion and an outlook.

2 Word Classes for Automatic Speech Recognition

2.1 Mathematical Formalism

In the context of this research we only look at non-overlapping word classes which can be defined as a mapping $C : W \to C$ which determines a sequence of word classes $\mathbf{c}$ given a sequence of words $\mathbf{w}$ (definition taken from [12]):

$$\mathbf{w} = w_1 \ldots w_n \rightsquigarrow C(w_1) \ldots C(w_n) = c_1 \ldots c_n = \mathbf{c} \tag{1}$$

When using word classes the probabilities of the language model for word sequences $\mathbf{w}$ have to be adjusted. In the case of bigram modeling the formula

$$P(\mathbf{w}) = P(w_1) \prod_{i=2}^{m} P(w_i|w_{i-1}) \tag{2}$$

needs to be rewritten as

$$P(\mathbf{w}) = P(w_1|c_1)P(c_1) \prod_{i=2}^{m} P(w_i|c_i)P(c_i|c_{i-1}) \tag{3}$$

Following the maximum likelihood principle, the class-related probability of a word $P(w_i|c_i)$ can simply be estimated by counting the number of occurrences of the word divided by the number of all words in the word class (uniform modeling). The (conditional) class probabilities $P(c_i|\ldots)$ can be determined in the same way normal n-gram probabilities are computed. The only difference is that the word sequences used for training must be converted to word class sequences.

---

[1] See for example https://sourceforge.net/p/kaldi/discussion/1355348/thread/c7c5e4f6/.

The class mappings used for the data sets in this paper (see Sect. 4) were created manually, but there are many clustering algorithms to automatically find optimal word (equivalence) classes based on measures like (least) mutual information etc. (see again [1–3]).

2.2 Open vs. Closed Word Classes

Specially designed word classes can be used to limit the number of possible user utterances that need to be considered in the speech recognition module of a spoken dialog system. If it can be assumed that users are cooperative, this will improve recognition accuracy. Imagine for example the task of having to recognize time information like "10:40 am" in an utterance. One way to configure the language model (in essence a grammar) could be, to introduce (closed) word classes which model hours and minutes:

> ...   10   40   am   ...
> ... NUMBER_1_TO_12 NUMBER_0_TO_59 AM_OR_PM ...

In case some (valid) numbers were not seen in the data which was collected to train the speech recognizer, they can be easily added to the language model by putting them into the appropriate word class as an entry. Adding all possible entries for a closed word class will prevent errors caused by the out-of-vocabulary problem. There are of course many more examples for (maybe more intuitive) closed words classes e.g. WEEKDAY, MONTH etc.

For word classes with a virtually unlimited vocabulary it is impossible to rule out the OOV problem. Such open word classes come into play if a task makes it necessary to recognize named entities like person or location names for example. Nevertheless, word classes are helpful in this case too because specialized word models (generic HMMs, sub-word unit language models) can be embedded in them to capture unknown words. This approach is also called hierarchical OOV modeling and has been the subject of quite a number of publications (e.g. [5–8]). In the following section, we show how this can be implemented as part of a Kaldi recipe.

3 Modeling Word Classes in Kaldi

The Kaldi Speech Recognition Toolkit [9] uses Weighted Finite State Transducers (WFSTs) to bridge the gap between word chains and feature vectors. Four different levels of transducers are used to do that: a word-level grammar or language model G to model the probabilities of word chains, a pronunciation lexicon L which provides the transition from letter to phone sequences, a context-dependency transducer C which maps context-independent to context-dependent phones and an HMM transducer H to map context-dependent phones to so called *transition IDs* (please refer to [9] for more details). Our approach to introduce word classes in Kaldi only affects the G and L transducers in the decoding graphs.

3.1   Modifying the Kaldi Transducers

The procedure we came up with to integrate a word class into the existing Kaldi transducers is as follows:

1. Create a language model transducer G_{CAT} with class entries mapped to non-terminal string identifiers (in Fig. 1 that identifier is CITYNAME)
2. Create sub-language models for each word class including the following steps
   - Add transitions with a class-specific disambiguation symbol as input and the empty word (<eps>) as output symbol before and after the actual sub-language model (in Fig. 1 the disambiguation symbol is #CITYNAME)
   - Make sure there is no path through the sub-language model which generates no output (i.e. just empty words)
   - Convert the sub-language model to a WFST
3. Insert the sub-language model WFSTs for the according non-terminal string identifiers in G_{CAT} using fstreplace
4. Remove all transitions with <eps>:<eps> labels (fstrmepsilon) and minimize the resulting graph (fstminimize) to get G
5. Add self-loops for all class-specific disambiguation symbols to the lexicon transducer L so the word classes "survive" the composition of G and L
6. Compose G and L and carry on in the usual way

It is important to note that the non-terminal string identifier (CITYNAME in Fig. 1) and the class-specific disambiguation symbol (#CITYNAME in Fig. 1), which is introduced to keep the decoding graph determinizable, must not be confused. As the name indicates the non-terminal symbol is not present in the final decoding graph any more. The new disambiguation symbol however is essential during the decoding process.

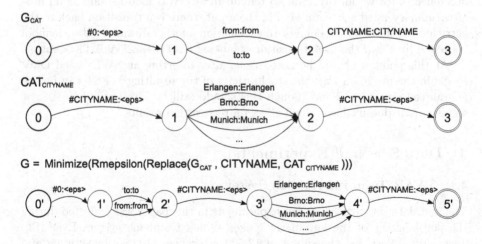

Fig. 1. The root grammar G_{CAT}, the sub-language model $CAT_{CITYNAME}$ and the resulting graph G after the replacement, <eps> removal and minimization steps

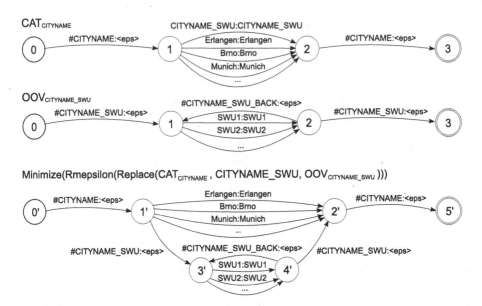

Fig. 2. The sub-language model $CAT_{CITYNAME}$ (which also serves as the root in this case), the OOV model $OOV_{CITYNAME\_SWU}$ and the resulting graph after the replacement, <eps> removal and minimization steps

3.2   Adding a Sub-Word Unit Based OOV Model

The addition of a sub-word unit (SWU) based OOV model to a word class as visualized in Fig. 2 is very similar to adding a word class to the parent language model as described in Sect. 3.1. However, there is one small but important difference: since we ideally want to remain in an SWU loop as long as an out-of-vocabulary word is encountered in the input, there is a transition back to the entry loop of the OOV model. For this transition another disambiguation symbol is needed (in Fig. 2 the back transition is labeled #CITYNAME_SWU_BACK:<eps>).

At this point it should be noted that, after inserting an SWU-based OOV model in the proposed way, the stochasticity of the resulting WFST can not be guaranteed any more. Even though this should still be investigated further, it seems as if it doesn't have a bad effect on recognition results.

4   Data Sets and Experiments

4.1   EVAR Train Information System

The first data set we used for our experiments in this paper was collected during the development of the automatic spoken dialog telephone system Evar [13]. Just as in [14] we use the subset of 12,500 utterances (10,000 for training and 2,500 for testing) which are the recordings of users talking to the live system over a phone line. The Evar corpus used in [6] contains more utterances, but the additional data consists of read data which was initially used to train the speech recognition module and which is quite different from the spontaneous user inquiries.

Table 1 shows that the vocabulary of the Evar corpus is limited to 1,603 words or 1,221 syllables. It also indicates the low number of words per utterance on average (less than 3.5). Nevertheless the data is suitable for our experiments because it contains the open word class CITYNAME.

Table 1. Statistics regarding words and syllables in the training and test set of the Evar corpus

Set	Utterances	Words		Syllables	
		Types	Tokens	Types	Tokens
Training	10,000	1,423	34,934	1,118	55,874
Test	2,500	712	8,286	669	13,154
All	12,500	1,603	43,220	1,221	69,028

The importance of the word class CITYNAME for the Evar corpus can be seen in Table 2: more that an eighth of all words in the test set are city names; 78 city names in the test set are out-of-vocabulary, which is almost one third of all OOV tokens. Overall the OOV rate of 2.98 % is not particularly high, but this is not surprising given the functionality of the system, which was limited to providing the schedule of express trains within Germany. As expected though, the OOV rate for the open word class CITYNAME[2] is much higher (7.05 %). The OOV rate for syllables (1.19 %) is much lower than for words. Syllables are also the type of

Table 2. OOV statistics regarding words, syllables, phones as well as entries of the category CITYNAME in the test set of the Evar corpus.

SWU/word class	Types		Tokens		
	#OOVs	#all	#OOVs	#all	%
Words	180	712	247	8,286	2.98
Syllables	103	669	156	13,160	1.19
Phones	0	43	0	37,610	0.00
CITYNAME	41	115	78	1,106	7.05

---

[2] To save resources it was decided during the design phase of Evar that the system should only be able to provide information about express trains (so called IC/ICE trains). As a consequence only city names with an express train station were included in the vocabulary of the recognizer. While this may seem intuitive at first, it lead to a large number of OOVs because even cooperative users were not always sure in which cities there were express train stops.

sub-word unit which was chosen for the OOV model for the experiments in this paper. Single phones, for which there is no OOV problem any more, have the drawback that the recognition accuracy goes down and if used in a hierarchical OOV model, they often produce very short inaccurate OOV hypotheses.

To test our approach we used three different speech recognizers and compared the resulting word error rates (WER); for our hierarchical OOV model we also looked at recall (percentage of OOVs found), precision (percentage of correct OOV hypotheses) and FAR (false alarm rate, the number of false OOV hypotheses divided by the number of in-vocabulary words):

1. *Baseline*: the standard s5 recipe of the Kaldi toolkit [9] with a tri-gram language model
2. *Category-based*: like baseline but with the word class CITYNAME added to the decoding graph following the procedure described in Sect. 3.1
3. *Category-based + OOV*: like category-based but with a syllable-based OOV model inserted into the word class CITYNAME

The results of our tests on the Evar corpus are summarized in Table 3. It can be seen that both category-based approaches – for Evar the only word class we used was CITYNAME – improve the word error rate compared to the baseline recognizer. If an OOV model is included, the WER goes down even further. The reason for this is that, if no OOV model is present, the speech recognizer often tries to cover the out-of-vocabulary word with short in-vocabulary words. For example the German town "Heilbronn", which was not in the training data and thus out-of-vocabulary, was recognized as two words "Halle Bonn" by the conventional word recognizer. The recognizer which featured an OOV model for the word class CITYNAME produced the result "halp#Un_ORTSNAME.OOV" consisting of the two phonetic syllables "halp" and "Un".

4.2   SmartWeb Handheld Corpus

The second data set, which we used to test our approach, is the *SmartWeb Handheld Corpus*. It was collected during the SmartWeb project, which was carried out from 2004 to 2007 by a consortium of academic and industrial partners [15].

Table 3. Word error rates and OOV detection results (recall, precision, false alarm rate) for all OOVs and for out-of-vocabulary city names on the test set of the Evar corpus.

Model	WER	OOV results			OOV results (CITYNAME)		
		Recall	Precision	FAR	Recall	Precision	FAR
Baseline	14.7	0.0	0.0	0.00	0.0	0.0	0.0
Category-based	14.5	0.0	0.0	0.00	0.0	0.0	0.0
Category-based + OOV	14.1	21.0	75.0	0.02	58.0	63.0	3.0

The recordings of the handheld corpus were carried out using cell phones and the signals underwent *a complex chain of speech transmissions including Bluetooth and UMTS* [16]. The resulting data was given one of three labels: *normal, bad* or *unusable*. For the experiments in this paper only the *normal* recordings were used.

The SmartWeb handheld corpus is more interesting because SmartWeb was designed to be an open-domain question-answering system. As a consequence there are many more open word classes than in Evar and the vocabulary is much larger as well (5,787 word types); most out-of-vocabulary words are encountered for the class CELEBRITY (59 word tokens). Overall the OOV rate for SmartWeb on the word level is 4,86 % (432 out of 8887 test tokens), on the syllable level however it's only 0,60 % (104 out of 17,225 test tokens).

In Table 4 the reduction of the word error rate for both category-based recognizers can be observed again. Also, as for Evar, the very robust OOV detection can be found in the resulting data. The False Alarm Rate (FAR) of 0.00 is not a mistake: only 20 OOV hypotheses were inaccurate, on the other hand there are 8,455 in-vocabulary words. For the experiments on the SmartWeb corpus 14 categories were used for the two enhanced recognizers; the OOV model was again based on the syllables which were extracted from the training set.

Table 4. Word error rates and OOV detection results (recall, precision, false alarm rate) for all OOVs and for out-of-vocabulary names of celebrities on the test set of the SmartWeb corpus.

Model	WER	OOV results			OOV results (CELEBRITY)		
		Recall	Precision	FAR	Recall	Precision	FAR
Baseline	19.7	0.0	0.0	0.00	0.0	0.0	0.0
Category-based	18.8	0.0	0.0	0.00	0.0	0.0	0.0
Category-based + OOV	18.4	23.0	83.0	0.00	36.0	70.0	7.0

5   Conclusion and Outlook

In this paper we have shown how the concept of word classes can be integrated into the Kaldi speech recognition toolkit by modifying the Weighted Finite State Transducers (WFSTs) for the language model G and the pronunciation lexicon L in Kaldi's standard recipes. In experiments on the Evar and SMARTWEB data sets we showed that this approach can help improve the word error rate. Another advantage of the proposed method is the option to add an OOV model based on sub-word units to robustly detect out-of-vocabulary words for open word classes.

While we are certain that our approach will also work for vocabulary sizes beyond 5.5 K words, this still needs to be proven by further experiments. Another point that should be investigated in more depth, even though it didn't impair the recognition performance, is the non-stochasticity of the resulting decoding graph after inserting the OOV model into the sub-language model. Furthermore, to improve the system, it could be explored which other sub-word units apart from syllables are suitable for the hierarchical OOV model. An obvious candidate are data-driven phonetic sub-word units which can be generated automatically based on existing pronunciation lexicons.

References

1. Ney, H., Essen, U.: On smoothing techniques for bigram-based natural language modelling. In: Proceedings of International Conference on Acoustics, Speech and Signal Processing, ICASSP 1991, Toronto, Canada (1991)
2. Brown, P.F., et al.: Class-based N-gram models of natural language. In: Computational Linguistics, vol. 18, pp. 467–479. MIT Press, Cambridge (1992)
3. Kneser, R., Ney, H.: Improved clustering techniques for class-based statistical language modelling. In: Proceedings of EUROSPEECH 1993, Berlin, Germany (1993)
4. Bazzi, I., Glass, J.R.: A multi-class approach for modelling out-of-vocabulary words. In: Proceedings of INTERSPEECH 2002, Denver, Colorado, USA (2002)
5. Schaaf, T.: Detection of OOV words using generalized word models and a semantic class language model. In: Proceedings of INTERSPEECH 2001, Aalborg, Denmark, pp. 2581–2584 (2001)
6. Gallwitz, F.: Integrated stochastic models for spontaneous speech recognition. Ph.D. thesis, Pattern Recognition Lab, Computer Science Department 5, University of Erlangen-Nuremberg, Logos Verlag, Berlin, Germany (2002)
7. Seneff, S., Wang, C., Hetherington, I.L., Chung, G.: A dynamic vocabulary spoken dialogue interface. In: Proceedings of INTERSPEECH 2004, Jeju Island, Korea (2004)
8. Aleksic, P.S., Allauzen, C., Elson, D., Kracun, A., Casado, D.M., Moreno, P.J.: Improved recognition of contact names in voice commands. In: 2015 IEEE International Conference on Acoustics, Speech and Signal Processing, ICASSP 2015, South Brisbane, Queensland, Australia, pp. 5172–5175 (2015)
9. Povey, D., et al.: The Kaldi speech recognition toolkit. In: IEEE 2011 Workshop on Automatic Speech Recognition and Understanding. IEEE Signal Processing Society (2011)
10. Mohri, M., Pereira, F., Riley, M.: Weighted finite-state transducers in speech recognition. Comput. Speech Lang. $16(1)$, 69–88 (2002). Elsevier
11. Allauzen, C., Riley, M.D., Schalkwyk, J., Skut, W., Mohri, M.: OpenFst: a general and efficient weighted finite-state transducer library. In: Holub, J., Žd'árek, J. (eds.) CIAA 2007. LNCS, vol. 4783, pp. 11–23. Springer, Heidelberg (2007)
12. Schukat-Talamazzini, E.G.: Automatische Spracherkennung - Grundlagen, statistische Modelle und effiziente Algorithmen. Vieweg, Braunschweig (1995)
13. Eckert, W., Kuhn, T., Niemann, H., Rieck, S., Scheuer, A., Schukat-Talamazzini, E.G.: A spoken dialogue system for German intercity train timetable inquiries. In: Proceedings of EUROSPEECH 1993, Berlin, Germany, pp. 1871–1874 (1993)
14. Stemmer, G.: Modeling variability in speech recognition. Ph.D. thesis, Pattern Recognition Lab, Computer Science Department 5, University of Erlangen-Nuremberg, Logos Verlag, Berlin, Germany (2005)
15. Wahlster, W.: SmartWeb: Mobile applications of the Semantic Web. GI Jahrestagung 1, 26–27 (2004)
16. Mögele, H., Kaiser, M., Schiel, F.: SmartWeb UMTS speech data collection: the SmartWeb handheld corpus. In: Procedings of LREC 2006, Genova, Italy, pp. 2106–2111 (2006)

Starting a Conversation: Indexical Rhythmical Features Across Age and Gender (A Corpus Study)

Tatiana Sokoreva[✉] and Tatiana Shevchenko

Moscow State Linguistic University, Moscow, Russia
jey-t@yandex.ru, tatashevchenko@mail.ru

Abstract. The study investigates indexical rhythmical features in the first dozen words in 102 American English adult speakers' telephone talks. The main goal is to explore age- and gender-related changes in prosodic characteristics of accented syllables (AS) and non-accented syllables (NAS) which affect speech rhythm in dialogue. The rhythm measures include duration, fundamental frequency and intensity, both mean values and PVI scores in adjacent syllables. The results suggest increasing accentual prominence achieved through growing values of foot and AS mean duration, increasing F0 range values, as well as higher PVI scores for F0 maxima values across three age groups. The accent-based (prototypical "stress-timed") pattern of English proves to be developing with age and varying with gender in AmE spontaneous speech.

Keywords: AmE · Dialogue · Speech rhythm · Age · Gender · Duration · Fundamental frequency · Intensity · Corpus study

1 Introduction

When two people start a telephone conversation their first words may indicate a lot of socially relevant information about their age, gender, social status and where they come from. Thus the interlocutor's voice may either help to identify an old acquaintance or give the first impression about a new one, thanks to its segmental and prosodic characteristics.

Among the personality features in prosody which have been explored so far in spontaneous speech rhythm is the least investigated. The typological division of languages into stress-timed (like English and Dutch) and syllable-timed (like French and Spanish) dates back to the works of Pike and Abercrombie [1,18]. Syllable-timing refers to the situation in which, supposedly, each syllable has approximately the same duration. Conversely, stress-timing refers to the pattern in which each foot reputedly has about the same duration, so that the duration from one stress to the next is approximately the same. Subsequently, mora-timing was added to describe languages such as Japanese in which each mora has roughly the same duration [2,9].

© Springer International Publishing Switzerland 2016
P. Sojka et al. (Eds.): TSD 2016, LNAI 9924, pp. 495–505, 2016.
DOI: 10.1007/978-3-319-45510-5_57

Although the proposed typological division has been a debatable issue as it lacks sufficient acoustic evidence for clear division of languages into two rhythmic groups [4,22], the recently developed rhythm metrics revived the linguists' interest to this typology [5,17,20,26]. Rhythm metrics in current use include Pairwise Variability Index (PVI), an index of mean difference in a given acoustic measure across successive linguistic units, distinctive for normalized tempo [15]. The PVI techniques have mainly been used in language and dialect typology studies [6,15,17,19,24], and then further extended to culturally and ethnically distinctive populations [16]. Most of the research focused on adjacent vowels and consonants, syllables and feet durations, fewer authors introduced fundamental frequency and intensity parameters [14,16,24]. We find it imperative to explore pitch and intensity together with duration of syllables and feet, a whole complex of prosodic features relevant for studying the rhythmic characteristics in the language under analysis.

It was found that even infants respond successfully to the difference in basic types of speech rhythm, they distinguish between stress-based and syllable-based speech [8]. Moreover, English rhythm is known to take longer to acquire than the syllable-based rhythm of the French language, the acquisition study found that English children learn the stress-based pattern by the age of six [25]. However, we don't know how rhythmic prosodic properties develop over the lifespan of an individual. Since prosodic characteristics are reported to have a constant decline with aging, which concerns fundamental frequency and intensity levels, articulation slowing down and voice quality changes [13], we assume that the 'apparent time' techniques of comparing speakers' data from different age groups may be productive in search of rhythmic prosodic properties change [21].

Thus, the main goal of the research is to investigate indexical rhythmical features formed by basic prosodic components in male and female speech across three age groups that manifest themselves in the dialogue onsets.

2   Methodology

2.1   Corpus

Sound material for analysis was taken from the Switchboard Corpus [7], the telephone speech recordings where nobody spoke twice with the same interlocutor and on the same topic.

The pilot study was carried out on carefully selected speech samples of 32 US people equally divided by gender – 2 men and 2 women from each of the 4 regions (North, North Midland, South Midland, South), and representing 2 age groups (16 young, aged 20–39, and 16 middle-aged speakers, aged 40–59).

For the main experiment we downloaded samples of 102 speakers' conversations equally balanced for gender (51 men, 51 women) in three age groups (34 young adult speakers, aged 20–39; 34 middle-aged adults, aged 40–59; 34 old speakers, aged 60–69) representing two main regions in the USA (North – 47 and South – 45 speakers) with ten samples from the Western states.

We assume that even in the dialogue onset the rhythmical characteristics of speech may be indicative of certain age- and gender-related features of the speakers' language therefore the samples were extracted from the beginning of telephone conversations. The items analyzed contained 21 syllables from each speaker, the total is 2,142 syllables (on the average the first dozen words) and did not include hesitations, repetitions, vocalized pauses and the first unstressed syllables in Intonation Phrases.

2.2 Data Segmentation

The prosodic labeling was performed by three trained phoneticians who marked accents. The general agreement on the choice of accent position was 95.6 % among the labelers.

The foot and the syllable are the basic rhythmic units of our research, consequently we focused on the division into feet, comprising accented (AS) and non-accented syllables (NAS). Syllable segmentation was carried out using Praat software v. 6.0.14 2016 (Boersma and Weenik) [3].

Measurements of duration, F0 and intensity were also computed in Praat, with some of Praat's values hand-corrected to remove pitch tracker errors. Low F0 values, typically the result of a creaky voice, were excluded from the analysis. F0 range was calculated using the online converter [23]. Intensity range was computed as the difference between maximum and minimum intensity values over each syllable.

2.3 Measurements

The basic prosodic component of the pilot study was duration – in the course of analysis the mean syllable duration (MSD) and the mean stressed syllable duration (MSSD) of each of the 32 speakers were calculated.

In order to be statistically certain we enlarged the number of informants in the main experiment and added two more prosodic parameters for analysis (fundamental frequency and intensity, apart from duration) as according to certain linguists [14,16,24] they contribute to the results of speech rhythm research. To explore indexical rhythmical features revealed in prosodic characteristics the following measurements were taken in the course of the main experiment.

- Duration of feet, accented (AS) and unaccented syllables (NAS).
- Maximum and minimum F0 values ($F0_{max}$, $F0_{min}$) and $F0_{range}$ in AS and NAS.
- Maximum and minimum Intensity values (Int_{max}, Int_{min}) and Intensity range (Int_{range}) in AS and NAS.

In order to confirm the changes in rhythmic characteristics of speech the PVI formula (1) was used to test the difference in successive feet duration and in successive syllables duration, as well as $F0_{max}$ and Int_{max} variability in successive syllables.

$$nPVI = \frac{100}{m-1} \times \sum_{k=1}^{m-1} \left| \frac{d_k - d_{k+1}}{(d_k + d_{k+1})/2} \right| \tag{1}$$

where m – the number of intervals in an utterance; d_k – duration, $F0_{max}$, Int_{max} of the k_{th} interval.

2.4 Statistical Analysis

In the course of the pilot study the values were averaged and tested by means of one-way analysis of variance (ANOVA) [27, pp. 194–199] the variables being MSD, MSSD, the factors are region and age. The Mann-Whitney U criterion was used for separate comparisons of levels of variables within each of the two factors [27, pp. 188–190].

For statistical analysis in the main experiment the mean values of prosodic parameters and PVI scores were computed and tested by means of two-way analysis of variance [27, pp. 200–202], the factors being age (three levels: young, middle-aged, old) and gender (two levels: male and female). In addition the Spearman correlation coefficient was used to test the relationship between the obtained prosodic characteristics and the speakers' age (date of birth – DOB). The analysis was performed using the statistical programs MINITAB 16.2.2 and PASW Statistics 18.0.0 (2009).

3 Results

3.1 Pilot Study

The pilot study shows that there is no significant difference in either MSD or MSSD values in the speech of Americans from four different regions as well as there is no significant gender-related diversity found in rhythmical characteristics of male and female speech. However, both ANOVA (MSD: $F = 14.111$; $p < .001$; MSSD: $F = 18.196$; $p < .001$) and Mann-Whitney U criterion (MSD: $W = 45422.0$; $p < .001$; MSSD: $W = 14067.0$; $p < .001$) prove the age-related changes in MSD and MSSD revealing the increase in the analyzed parameters' values with age.

3.2 Main Experiment

Given that there is no significant difference in speech rhythm features of American people representing four main regions of the US, we don't study the region factor in the main body of our research, though we take it into account as the number of informants of the main experiment can be almost equally divided into two groups representing the North and the South regions with a number of people coming from the West.

Duration. Figure 1 shows the mean values of AS and NAS durations in the speech of male and female informants across three age groups ($n = 102$). According to the results, AS duration increases with age in the speech of both men and women. This tendency is not consistent across NAS duration values.

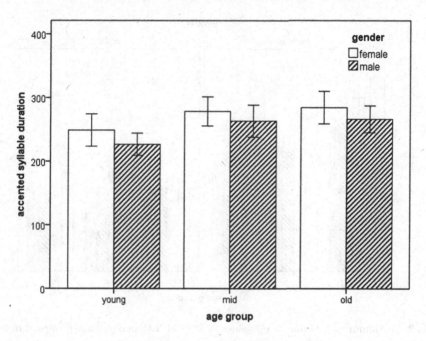

Fig. 1. AS mean duration (ms), ($n = 102$).

The ANOVA results show significant change in mean AS duration increasing with age ($DF = 2$, $F = 7.12$, $p = .001$) and dependent on the speaker's gender ($DF = 1$, $F = 4.36$, $p < .05$). The analysis of mean NAS duration reveals only significant effect of Gender ($DF = 1$, $F = 4.24$, $p < .05$). The correlation analysis discloses a significant negative correlation between the mean AS duration and speakers' dates of birth, with high values of AS duration associated with lower values of their DOB ($rho = -.319$, $n = 102$, $p < .01$).

Although the ANOVA reveals no significant changes in foot duration depending on either age or gender of the speakers, according to correlation analysis there is strong negative correlation between the speakers' DOB and their mean foot values ($rho = -.303$, $n = 102$, $p < .01$).

Fundamental Frequency. The mean values of F0 measures in AS and NAS of men and women across three age groups are represented in Fig. 2. As expected, there is a distinct gender difference in maximum and minimum F0 scores between men and women that is compatible with a well-known fact that men's pitch level is lower than women's one due to physiological differences of the vocal system [11,13]. The general decline with age is observed in $F0_{max}$ and $F0_{min}$ in both types of syllables in women's speech. Contrary to the expectations is the fact that in men's speech middle age is characterized by the highest $F0_{max}$ values.

The two-way ANOVA was conducted with $F0_{max}$ values in AS as the dependent variable and Age and Gender as two factors. The obtained results reveal a significant effect of Age ($DF = 2$, $F = 3.04$, $p = .052$) and Gender ($DF = 1$,

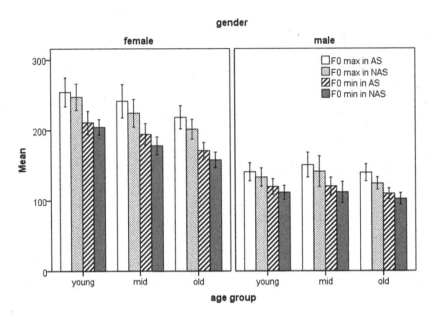

Fig. 2. Maximum and minimum F0 values in AS and NAS across 3 age groups of males and females.

$F = 193.54$, $p < .000$). ANOVA of $F0_{max}$ in NAS as the dependent variable showed the significant effect of both Age ($DF = 2$, $F = 6.54$, $p < 0.05$) and Gender ($DF = 1$, $F = 206.10$, $p < .000$). We find decrease in $F0_{max}$ values both in AS and NAS, thus we observe that the contrast 'AS – NAS values' is sustained.

ANOVA scores for $F0_{min}$ in AS as the dependent variable and Age and Gender as factors registered a significant effect of Age ($DF = 2$, $F = 9.11$, $p < .000$) and Gender ($DF = 1$, $F = 237.83$, $p < .000$) along with significant interaction ($DF = 2$, $F = 3.14$, $p < .05$). Similar results were obtained for $F0_{min}$ in NAS: Age ($DF = 2$, $F = 13.50$, $p < .000$), Gender ($DF = 1$, $F = 264.28$, $p < .000$), interaction ($DF = 2$, $F = 6.43$, $p < .05$).

Another important observation is that $F0_{range}$ values show an increase in every subset of data with age. Figure 3 represents mean $F0_{range}$ values in AS and NAS.

ANOVA results for $F0_{range}$ values in AS as the dependent variable show the significant effect of Age solely ($DF = 2$, $F = 3.92$, $p < .05$). The ANOVA conducted with the $F0_{range}$ values of the NAS as the dependent variable reveals the significant effect of both Age ($DF = 2$, $F = 3.86$, $p < .05$) and Gender ($DF = 1$, $F = 4.06$, $p < .05$).

The relationship between the speakers' DOB and each of the F0 parameters reveals itself in significant correlation between DOB and $F0_{min}$ values both in AS and NAS ($rho = .237$, $n = 102$, $p < .05$ and $rho = .246$, $n = 102$, $p < .05$ correspondingly) and in significant negative correlation between DOB and $F0_{range}$ both in AS and NAS ($rho = -.335$, $n = 102$, $p < .01$ and $rho = -.295$,

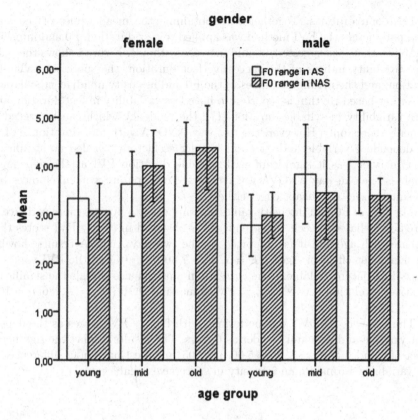

Fig. 3. $F0_{range}$ values in AS and NAS (st).

$n = 102$, $p < .01$ correspondingly). As a result we observe a significant decline in $F0_{min}$ values and an increase in $F0_{range}$ values with age.

Intensity. In the course of intensity analysis maximum and minimum intensity and intensity range values were calculated in AS and NAS of male and female speakers across three age groups.

The two-way ANOVA was performed with first Int_{max} values of AS as the dependent variable and then Int_{max} values of NAS. The results show no significant effect of either factors in either tests. As for Int_{min} values in AS and NAS, according to ANOVA results, merely the interaction scores appear to be significant in both tests: Int_{min} in AS (interaction – $DF = 2$, $F = 4.74$, $p < .05$), Int_{min} in NAS (interaction – $DF = 2$, $F = 3.67$, $p < .05$). Intensity$_{range}$ values in both types of syllables reveal a significant effect of Gender ($DF = 1$, $F = 4.51$, $p < .05$) and interaction ($DF = 2$, $F = 4.31$, $p < .05$) in AS; Age ($DF = 2$, $F = 4.04$, $p < .05$) and interaction ($DF = 2$, $F = 3.63$, $p < .05$) in NAS.

The above-mentioned results are confirmed by correlation analysis as it does not reveal any correlation between the speakers' DOB and each of the intensity parameters.

PVI Measurements. Together with obtaining the mean values of the analyzed parameters the PVI method was applied to test duration, F0 and intensity
variability that, as suggested, should increase with age. Figure 4 shows the pairwise variability indices of the successive feet duration, the successive syllables
duration, and the maximum values of the F0 and intensity in adjacent syllables.

Accent-based rhythm is reported to have less variability in feet duration and
more variability in syllable duration [17], the tendency which can be traced in
women's scores only. However, the two-way ANOVA with foot duration PVI as
the dependent variable and Age and Gender as factors revealed no significant
effect; neither was it significant with syllable duration PVI as the dependent
variable. The one-way ANOVA was conducted on separate women's indices but
it resulted in no significant effect either.

The PVI of F0 and intensity values are supposed to possess more variance in
more accent-based speech. This tendency is observed in $F0_{max}$ PVI scores that
increase with age and it is supported by the two-way ANOVA results showing
the significant effect of Age ($DF = 2$, $F = 7.80$, $p < .01$) on the PVI values of
$F0_{max}$ in adjacent syllables. The correlation analysis also reveals the significant
negative correlation between $F0_{max}$ PVI values and DOB ($rho = -.286$, $n = 102$,
$p < .01$).

The two-way ANOVA was performed with Int_{max} PVI scores as the dependent variable and Age and Gender as factors. The results display the significant
effect of Gender ($DF = 1$, $F = 5.00$, $p < .05$) manifesting gender differences in
the variability of maximum intensity in successive syllables.

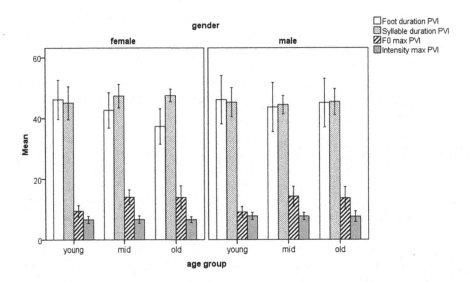

Fig. 4. Mean PVI values for foot and AS syllable duration, maximum F0 and intensity
of men and women across three age groups.

4   Discussion and Conclusion

The pilot study, investigating the rhythmical characteristics in 32 Americans' telephone conversations, shows that there is an increase in the mean stressed syllable duration and consequently in the mean foot duration with age that indicates the slowing down of speech tempo in older people's speech as compared to that of younger people which is consistent with previous findings [12, 13].

Another important observation is that speech rhythm variation does not depend on the territory the speaker resides in. We find that, contrary to the common belief that the Southerners speak slower in the US [12], the articulation rate does not change significantly in relation to the region factor that is compatible with J. Hartman's claim who suggests that the impression of speech slowness in the South is achieved by the lengthening of vowels and the amount of silence and claims that "southern speech is not necessarily slower than that found elsewhere" [10].

The main research results reveal various age and gender peculiarities of American speech rhythm in telephone conversations. As in the pilot study, we observe the increase in mean stressed syllable duration as well as in mean foot duration with age testifying to the tempo deceleration of elders' speech which is consistent with other studies [13]. As for gender distinctions in mean stressed syllable duration we find that the latter is longer in female speech compared to that of male speakers, thus suggesting that female elders speak even slower than male ones.

The analysis showed a significant decrease in fundamental frequency values with age that corresponds to other works on vocal aging [11, 13]. Though both $F0_{max}$ and $F0_{min}$ values drop in each type of syllables due to physiological reasons, the contrast between accented and unaccented syllables' fundamental frequency is sustained, maintaining the prominence that forms the basis of speech rhythm and provides better speech production and perception in general. In addition, we register the increase in fundamental frequency range values with age that help sustain the prominence as well and confirm the age-related improvement of people's speaking skills.

The gender differences of fundamental frequency parameter reveal themselves in significantly higher F0 values in women's speech than those in men's speech that is compatible with numerous research findings attributing this distinction to physiological discrepancy of male and female vocal systems [13].

To support the idea that fundamental frequency is as important in identifying the rhythmic patterns of speech as duration we found that PVI measures of F0 maxima also show an increase with age suggesting growing variability in fundamental frequency values that reach the highest point in the speech of the 'old' group. This fact can be accounted for either by lifespan development of stress-based type of rhythm or by instability of elders' speech [13].

In conclusion we would like to say that even the first phrases of a speaker in a telephone conversation can give you the key information about his/her social characteristics that are manifested by means of certain prosodic and eventually rhythmic features of speech.

References

1. Abercrombie, D.: Elements of General Phonetics. Edinburgh University Press, Edinburgh (1967)
2. Block, B.: Studies in colloquial Japanese IV: phonemics. Language **26**, 86–125 (1950)
3. Boersma, P., Weenink, D.: Praat: doing phonetics by computer [Computer program], Version 6.0.14. http://www.praat.org/. Accessed 11 Feb 2016
4. Dauer, R.M.: Stress-timing and syllable-timing reanalyzed. J. Phon. **11**, 51–62 (1983)
5. Dellwo, V.: Rhythm and speech rate: a variation coefficient for ΔC. In: Karnowski, P., Szigeti, I. (eds.) Language and Language - Processing, pp. 231–241. Peter Lang, Frankfurt am Main (2006)
6. Deterding, D.: Measurements of rhythm: a comparison of Singapore and British English. J. Phon. **29**, 217–230 (2001)
7. Godfrey, J., Holliman, E., McDaniel, J.: SWITCHBOARD: telephone speech corpus for research and development. In: ICASSP 1992 (1992)
8. Grabe, E., Post, B., Watson, I.: The acquisition of rhythmic patterns in English and French. In: Proceedings of the 14th ICPhS, San Francisco, pp. 1201–1204 (1999)
9. Han, M.S.: The feature of duration in Japanese. Onsei no kenkyuu **10**, 65–80 (1962)
10. Hartman, J.: Guide to pronunciation. In: Cassidy, F.J. (ed.) Preface to the Dictionary of American Regional English. Harvard University's Belknap Press, Cambridge (1985)
11. Helfrich, H.: Age markers in speech. In: Scherer, K., Giles, H. (eds.) Social Markers in Speech, pp. 63–93. Cambridge University Press, Cambridge (1979)
12. Jacewicz, E., Fox, R., O'Neil, C., Salmons, J.: Articulation rate across dialect, age and gender. Lang. Var. Change **21**, 233–251 (2009)
13. Linville, S.E.: Vocal Aging. Singular Publishing Group, San Diego (2001)
14. Loukina, A., Kochanski, G., Shih, C., Keane, E., Watson, I.: Rhythm measures with language-independent segmentation. In: Proceedings of Interspeech 2009: Speech and Intelligence, Brighton, UK, pp. 1531–1534 (2009)
15. Low, E., Grabe, E., Nolan, F.: Quantitative characterization of speech rhythm: syllable-timing in Singapore English. Lang. Speech **43**, 377–401 (2000)
16. Nokes, J., Hay, J.: Acoustic correlates of rhythm in New Zealand English: a diachronic study. Lang. Var. Change **24**, 1–31 (2012)
17. Nolan, F., Asu, E.L.: The pairwise variability index and coexisting rhythms in language. Phonetica **66**, 64–77 (2009)
18. Pike, K.L.: The Intonation of American English. University of Michigan Press, Ann Arbor (1945)
19. Prieto, P., Vanrell, M., Astruc, L., Payne, E., Post, B.: Phonotactic and phrasal properties of speech rhythm. Evidence from Catalan, English, and Spanish. Speech Commun. **54**, 681–702 (2012)
20. Ramus, F., Nespor, M., Mehler, J.: Correlates of linguistic rhythm in the speech signal. Cognition **73**, 265–292 (1999)
21. Reubold, U., Harrington, J., Kleber, F.: Vocal aging effects on F0 and the first formant: a longitudinal analysis in adult speakers. Speech Commun. **52**, 638–651 (2010)
22. Roach, P.: On the distinction between 'Stress-timed' and 'Syllable-timed' languages. In: Crystal, D. (ed.) Linguistic Controversies, pp. 73–79. Edward Arnold, London (1982)

23. Semitone conversions. http://users.utu.fi/jyrtuoma/speech/semitone.html
24. Szakay, A.: Rhythm and pitch as markers of ethnicity in New Zealand English. In: Proceedings of 11th Australian International Conference on Speech Science and Technology, New Zealand, pp. 421–426 (2006)
25. Vihman, M.M.: Acquisition of the English Sound System. The Handbook of English Pronunciation, pp. 333–354. Wiley Blackwell, Oxford (2015)
26. White, L., Mattys, S.L.: Calibrating rhythm: first and second language studies. J. Phon. **35**, 501–522 (2007)
27. Woods, A., Fletcher, P., Hughes, A.: Statistics in language studies. Cambridge University Press, Cambridge (2003)

Classification of Utterance Acceptability Based on BLEU Scores for Dialogue-Based CALL Systems

Reiko Kuwa[✉], Xiaoyun Wang, Tsuneo Kato, and Seiichi Yamamoto

Graduate School of Science and Engineering, Doshisha University,
1-3 Tatara Miyakodani, Kyotanabe-shi, Kyoto 610-0394, Japan
{dup0128,euo1001}@mail4.doshisha.ac.jp,
{tsukato,seyamamo}@mail.doshisha.ac.jp

Abstract. We propose a novel classification method of recognized second language learners utterances into three classes of acceptability for dialogue-based computer assisted language learning (CALL) systems. Our method uses a linear classifier trained with three types of bilingual evaluation understudy (BLEU) scores. The three BLEU scores are calculated respectively, referring to three subsets of a learner corpus divided according to the quality of sentences. Our method classifies learner utterances into three classes (*correct*, *acceptable with some modifications* and *out-of-the-scope of assumed erroneous sentences*), since it is suitable for providing effective feedback. Experimental results showed that our proposed classification method could distinguish utterance acceptability with 75.8 % accuracy.

Keywords: Dialogue-based CALL · Classification · Learner corpus · BLEU

1 Introduction

Learning a second language is becoming more important due to globalization. People have more chances than ever to communicate in foreign languages.

Computer assisted language learning (CALL) is gaining interest, as it gives chances to self-learn second languages (L2s) with less constraint of time, space, and cost. While computer assisted pronunciation training (CAPT) has been a typical form of CALL systems, interactive CALL systems, which enable learners to compose utterances by themselves, are getting more attention with the advancement of automatic speech recognition (ASR). Various types of interactive CALL systems have been proposed, such as translation game type [1], dialogue game type [2], and role-playing game type [3].

For these systems, appropriate corrective feedback and accurate evaluation of a learner's ability are necessary. However, correctly detecting erroneous utterances is still a challenge because a learner's speech contains a tremendous variety of lexical, syntactic, and semantic errors, and because ASR of second-language speech is not perfect due to grammatical errors and erroneous pronunciations.

© Springer International Publishing Switzerland 2016
P. Sojka et al. (Eds.): TSD 2016, LNAI 9924, pp. 506–513, 2016.
DOI: 10.1007/978-3-319-45510-5_58

The basic approach to develop dialogue-based CALL systems is designing interactions that match a learner's skills and constructing a task-specific finite state automaton (FSA) grammar which accepts both correct answers and expected errors [4]. However, it is difficult to develop this sort of task-specific FSA grammar since there are still sometimes out-of-the-scope utterances. On the other hand, a versatile system covering a wide range of topics has been studied to motivate all learners by enabling interactions on their interests. Because more out-of-the-scope utterances are made by learners, a versatile system has to first distinguish utterance acceptability in terms of grammaticality and semantics. An n-gram language model (LM) trained with a manipulated corpus to which sentences with typical errors of L2 learners were added was proposed to accurately recognize both correct and erroneous utterances [5]. However, no matter how carefully such an n-gram LM is designed, it would be impossible to handle all the out-of-the-scope utterances well. We previously proposed a method for detecting the out-of-the-scope utterances using an edit distance between the recognized utterance and its reference and the difference in acoustic scores between the task-specific FSA grammar and a general n-gram LM [6]. This method used acoustic and linguistic scores, but did not make the best use of linguistic information.

In this paper, we propose a novel method for classifying learner utterances into three classes (*correct*, *acceptable with some modifications* and *out-of-the-scope of assumed erroneous sentences*) based on a learner corpus. To make the best use of the linguistic output from ASR, we apply "bilingual evaluation understudy (BLEU) scores" [7] as features of a linear classifier for recognized learner utterances that include recognition errors and grammatically incorrect use. BLEU is originally a widely used method for evaluating machine translation, which uses only correct expressions as reference translations. With our method, on the other hand, references of BLEU are not only correct expressions but also inappropriate expressions with some errors and incomprehensible expressions. If one runs a CALL system continuously, a large learner corpus is created. Such a large corpus tends to include erroneous expressions which have similarities since they were generated by learners of the same mother tongue. Therefore, our method applies three types of BLEU scores at the same time as features of a linear classifier to make the best use of this tendency.

This article is structured as follows. We give an overview of our previously proposed dialogue-based CALL system in Sect. 2, and describe our proposed classification method of learner utterances using BLEU scores in Sect. 3. In Sect. 4 we describe the experiments and results, and close with a conclusion and future work in Sect. 5.

2   Overview of the Dialogue-Based CALL System

To improve learner communication competence in an L2, a dialogue-based CALL system prompts learners to construct utterances on their own. We previously proposed a translation-game-type dialogue-based CALL system with which Japanese learners can practice English conversation while enjoying a role-playing game in a shopping situation [6]. The system plays a role of a salesperson and a learner plays a role of a shopper. The system asks a question to a learner. A translated sentence in Japanese of the expected response which a learner should utter in English is also displayed. The purpose of this is to constrain considerable variations of his/her utterances.

After a learner's utterance is recognized with the speech recognizer, the subsequent process branches into the following three patterns. If the learner's utterance is a correct response, the system moves on to the next question. If the learner's utterance is acceptable with some modifications, the system offers, as feedback, one correct answer selected as the closest expression in the corpus. If the learner's utterance is out-of-the-scope of assumed erroneous sentences, the system requests a re-utterance. If the re-utterance is out-of-the-scope of assumed erroneous sentences again, the system gives a model answer and moves on to the next question. Thus, it is crucial for such a dialogue-based CALL system to correctly distinguish acceptability of learner utterances since it affects the effectiveness of feedback.

Even though a translated sentence in Japanese of the expected response is displayed, learners cannot always make a correct utterance. Their utterances often include pronunciation errors and a wide variety of lexical and grammatically incorrect use. Therefore, we classify such utterances into three classes: *correct*, *acceptable with some modifications* and *out-of-the-scope of assumed erroneous sentences* to give learners appropriate feedback. The most important utterance class is *acceptable with some modifications*, since the system should tell a learner of this class how to amend his/her utterance by giving informative feedback. The system must also be able to let a speaker of the *correct* class know that his/her utterance is good enough without the need of any modification. Otherwise, the reliability of the system would decrease, thus, lowering a learner's confidence. From the perspective of system efficiency, it is also important to accurately distinguish utterances of *out-of-the-scope of assumed erroneous sentences* class because the system is supposed to offer an example answer without any corrective instructions to a learner of this class.

3   Utterance Classification Based on Multiple BLEU Scores

To classify learner utterances accurately, our proposed method uses three BLEU scores as features of the classifier. Originally, BLEU evaluates the quality of translated sentences by counting the number of n-grams that match those in reference translations. BLEU is defined by the following formulation;

$$BLEU = BP * \exp(\sum_{n=1}^{N} \frac{1}{N} \log P_n), \qquad (1)$$

where BP denotes the brevity penalty and N denotes the maximum length of word series n of the n-grams; N generally equals to 4. P_n is the modified n-gram

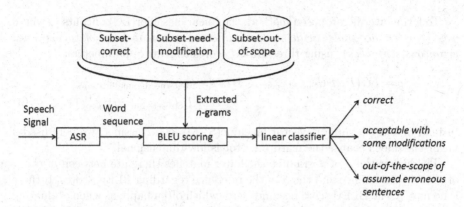

Fig. 1. Schematic classification flow of recognized learner utterances

precision and it is calculated as follows;

$$P_n = \frac{\sum_{n-gram \in U} \min(Max\_Ref\_Count(n-gram),\ Count(n-gram))}{\sum_{n-gram \in U} Count(n-gram)} \quad (2)$$

where U denotes the recognized learner utterance, $Count(n-gram)$ denotes the frequency of an $n-gram$ in U, and $Max\_Ref\_Count(n-gram)$ denotes the maximum number of times an $n-gram$ occurs in any single reference translation. A BLEU score ranges from 0 (low quality) to 1 (high quality). Our method uses BLEU scores as features of a linear classifier to classify recognition results contaminated with recognition errors and lexical and grammatically incorrect use into the three classes. Figure 1 shows the classification flow.

We made an English learner corpus as a reference material, which consists of utterances by Japanese-speaking students and one bilingual English/Japanese speaker. The quality of each sentence was evaluated on a scale of 1 to 5; 5 (Perfect), 4 (Good), 3 (Non-native), 2 (Disfluent), 1 (Incomprehensible), by the bilingual English/Japanese speaker. The English learner corpus is divided into three subsets: Subset-correct consisting of the bilingual speaker's sentences and students' sentences graded as 5, Subset-need-modification consisting of students' sentences graded as 3 or 4, and Subset-out-of-scope consisting of students' sentences graded as 1 or 2.

The reason why our method applies not only the BLEU score from the correct references but also additional BLEU scores from the references of Subset-need-modification and Subset-out-of-scope is as follows. There is a tendency that speakers who have the same mother tongue make similar types of lexical or grammatical mistakes when they speak an L2 affected by the characteristics of their mother tongue. For example, the English spoken by Japanese speakers often includes absence or mistaken use of articles, mistaken use of passive, and so on. Therefore, we assumed that there would be some correlation due to this tendency between the quality of the learner utterance and BLEU scores calculated on Subset-need-modification and Subset-out-of-scope.

To formulate the proposed method, the linear classifier(LC) outputs y, where $y \in \{correct, acceptable\ with\ some\ modifications, out\text{-}of\text{-}the\text{-}scope\ of\ assumed\ erroneous\ sentences\}$, using three BLEU scores as feature parameters;

$$y = LC(BLEU_{\text{Subset}-\text{correct}}, BLEU_{\text{Subset}-\text{need}-\text{modification}},$$
$$BLEU_{\text{Subset}-\text{out}-\text{of}-\text{scope}}) \tag{3}$$

Instead of using a support vector machine(SVM) as a classifier, we here used a linear classifier because the feature vector is low-dimensional.

The higher the n of n-gram is, the fewer matches there are between n-grams of a learner utterance and those of the reference regarding BLEU scores. If there is no match, the BLEU score becomes zero, which often happens when evaluating at sentence level. Instead of standard BLEU, our method uses modified BLEU ("BLEU+1") [8], which was designed to compensate for this phenomenon by smoothing, where P_n is defined as follows;

$$P_n = \frac{1 + \sum_{n-gram \in U} \min(Max\_Ref\_Count(n - gram),\ Count(n - gram))}{1 + \sum_{n-gram \in U} Count(n - gram)} \tag{4}$$

We conducted experiments to verify the accuracy of our classification method.

4   Experiments

4.1   Experimental Setup

We trained a linear classifier for our method with all three types of BLEU scores calculated using Subset-correct, Subset-need-modification, and Subset-out-of-scope respectively. As a reference material, we prepared an English learner corpus which consisted of 2,803 utterances of 41 Japanese-speaking students and one bilingual English/Japanese speaker. Each Japanese student orally produced 66 utterances in English. The bilingual English/Japanese speaker made multiple translations per response, considering that a single Japanese sentence often has more than one possible translation in English. The English learner corpus is divided into three subsets according to the quality of sentences: Subset-correct consisting of 619 sentences, Subset-need-modification consisting of 1,285 sentences, and Subset-out-of-scope consisting of 933 sentences.

The evaluation data set consisted of 924 utterances by 14 university students (7 males and 7 females) who were recruited for the test set, each of whom made 66 utterances responding the target questions. The BLEU scores of each utterance were computed based on its corresponding reference sentences of the target question. Based on the three types of BLEU scores, the system then classified each utterance into *correct* or *acceptable with some modifications* or *out-of-the-scope of assumed erroneous sentences*. Classification was conducted with 10-fold cross validation. We evaluated the developed linear classifier based on classification accuracy. For comparison, we also conducted an additional experiment in which only the BLEU score calculated on Subset-correct was used as a classification feature and each utterance was classified based on two thresholds; 0.4 and 0.9.

As an ASR engine, we used the hidden Markov model (HMM) toolkit (HTK). A tri-phone HMM acoustic model was trained with an L2 English speech database collected from 201 Japanese students (100 males and 101 females) [9]. The pronunciation lexicon consisted of about 35,000 vocabulary words. We developed a bi-gram language model trained with 4,472 English transcriptions of utterances made by 65 Japanese students and an English native. It covered various expressions from correct to unacceptable with grammatical and/or lexical errors.

4.2 Experimental Results

Figure 2 shows the classification accuracies with our proposed method using the three BLEU scores and our baseline method, set for comparison, using a single BLEU score calculated on Subset-correct. The word accuracy of the ASR engine was 91.7 % and the sentence accuracy was 73.7 %. Tables 1 and 2 list the classification results from our classification method and our baseline method, respectively.

Figure 2 shows that the overall accuracy of the proposed method was 9.9 point higher than that of the baseline method. It also shows that our proposed method exhibited 17.1 point higher classification accuracy of *acceptable with some modifications* than the baseline. We believe the most important category for improving dialogue-based CALL systems is the *acceptable with some modification* class since informative corrective feedback is the most necessary for that class. These results suggest that using three types of BLEU scores as classification features contribute to the improvement of the classification accuracy of the *acceptable with some modification* class and that of overall classification accuracy.

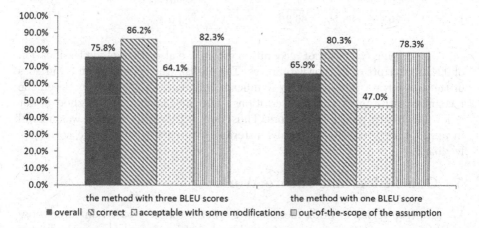

Fig. 2. Accuracy of our proposed method and the baseline with one BLEU score

Table 1. Confusion matrix of linear classifier with three BLEU scores

		Classification results		
		Correct	Acceptable with some modifications	Out-of-the-scope of assumed erroneous sentences
Observed	Correct	188	14	16
	Acceptable with some modifications	51	243	85
	Out-of-the-scope of assumed erroneous sentences	11	47	269

Table 2. Confusion matrix of linear classifier with one BLEU score

		Classification results		
		Correct	Acceptable with some modifications	Out-of-the-scope of assumed erroneous sentences
Observed	Correct	175	33	0
	Acceptable with some modifications	36	178	165
	Out-of-the-scope of assumed erroneous sentences	6	65	256

Table 3. Classification accuracies of transcripts

Overall	Correct	Acceptable with some modifications	Out-of-the-scope of assumed erroneous sentences
76.9 %	88.1 %	66.2 %	82.0 %

For verifying the effect of recognition errors, we also conducted classification of 924 transcripts of learner utterances. Table 3 lists the classification accuracies of the transcripts. There was no significant difference in accuracy between the recognized utterances and transcriptions. This means that our method is not much affected by recognition errors. Thus, our classification method was helpful in appropriately classifying learner utterances to effectively provide corrective feedback.

5   Conclusion and Future Work

We proposed a classification method of learner utterances based on three types of BLEU scores for improving dialogue-based CALL systems. Our method takes advantage of the similarities in grammatical errors made by learners who speak the same mother tongue. That is, to classify utterance acceptability, our method uses, as feature parameters of a linear classifier, three types of BLEU scores computed referring to three subsets of a corpus consisting of a wide variety of expressions. For verifying the proposed method, we examined the classification accuracy of learner utterances of our previously proposed dialogue-based CALL system developed for Japanese people learning English. We found that our method exhibited high performance with 75.8 % accuracy. Since high classification accuracy is a prerequisite for providing effective feedback, our proposed method will be helpful for improving dialogue-based CALL systems.

In the future, we aim to develop a method that can detect inappropriate parts of utterances classified as *acceptable with some modifications* by referring to the error patterns extracted from a learner corpus, and denote how inappropriate they are by providing informative feedback.

References

1. Seneff, S., Wang, C.: Web-based dialogue and translation games for spoken language learning. In: SLaTE, pp. 9–16 (2007)
2. Wik, P., Hjalmarsson, A.: Embodied conversational agents in computer assisted language learning. Speech Commun. **51**(10), 1024–1037 (2009)
3. Kyusong, L., Kweon, S.O., Sungjin, L., Hyungjong, N., Lee, G.G.: Postech immersive english study (POMY): dialog-based language learning game. IEICE TRANS. Inf. Syst. **97**(7), 1830–1841 (2014)
4. Raux, A., Eskenazi, M.: Using task-oriented spoken dialogue systems for language learning: potential, practical applications and challenges. In: InSTIL/ICALL Symposium 2004 (2004)
5. Anzai, T., Ito, A.: Recognition of utterances with grammatical mistakes based on optimization of language model towards interactive CALL systems. In: Signal and Information Processing Association Annual Summit and Conference (APSIPA ASC), Asia-Pacific, pp. 1–4. IEEE (2012)
6. Nagai, Y., Senzai, T., Yamamoto, S., Nishida, M.: Sentence classification with grammatical errors and those out of scope of grammar assumption for dialogue-based CALL systems. In: Sojka, P., Horák, A., Kopeček, I., Pala, K. (eds.) TSD 2012. LNCS, vol. 7499, pp. 616–623. Springer, Heidelberg (2012)
7. Papineni, K., Roukos, S., Ward, T., Zhu, W.J.: BLEU: a method for automatic evaluation of machine translation. In: Proceedings of the 40th Annual Meeting on Association for Computational Linguistics, pp. 311–318. Association for Computational Linguistics (2002)
8. Lin, C.Y., Och, F.J.: Automatic evaluation of machine translation quality using longest common subsequence and skip-bigram statistics. In: Proceedings of the 42nd Annual Meeting on Association for Computational Linguistics, p. 605. Association for Computational Linguistics (2004)
9. Minematsu, N., Tomiyama, Y., Yoshimoto, K., Shimizu, K., Nakagawa, S., Dantsuji, M., Makino, S.: Development of English speech database read by Japanese to support CALL research. In: Proceedings ICA, vol. 1, pp. 557–560 (2004)

A Unified Parser for Developing Indian Language Text to Speech Synthesizers

Arun Baby[✉], Nishanthi N.L., Anju Leela Thomas, and Hema A. Murthy

Computer Science and Engineering, IIT Madras, Chennai, India
{arunbaby,hema}@cse.iitm.ac.in, nlnishanthi@gmail.com,
anjuthomas95@gmail.com

Abstract. This paper describes the design of a language independent parser for text-to-speech synthesis in Indian languages. Indian languages come from 5–6 different language families of the world. Most Indian languages have their own scripts. This makes parsing for text to speech systems for Indian languages a difficult task. In spite of the number of different families which leads to divergence, there is a convergence owing to borrowings across language families. Most importantly Indian languages are more or less phonetic and can be considered to consist broadly of about 35–38 consonants and 15–18 vowels. In this paper, an attempt is made to unify the languages based on this broad list of phones. A common label set is defined to represent the various phones in Indian languages. A uniform parser is designed across all the languages capitalising on the syllable structure of Indian languages. The proposed parser converts UTF-8 text to common label set, applies letter-to-sound rules and generates the corresponding phoneme sequences. The parser is tested against the custom-built parsers for multiple Indian languages. The TTS results show that the accuracy of the phoneme sequences generated by the proposed parser is more accurate than that of language specific parsers.

Keywords: Indian languages · Text-to-speech synthesis (TTS) · Letter to sound (LTS) · Syllable · Common label set · Parser

1 Introduction

India is the second most populous nation with over 1.27 billion people. It has about 22 major languages, written in 13 different scripts, with over 1600 languages/dialects. Further, it is only about 65 percent of this population that is literate, that too primarily in the vernacular. Less than 5 % is English literate thus marginalizing most of the Indian society. Speech interfaces, especially in the vernacular, are enablers in such an environment.

The objective of this paper is to build technology that will enable the building of TTS systems in any Indian language quickly. Hidden Markov model (HMM), a statistical parametric based approach which is found effective in synthesizing speech is employed here [12]. The objective of text to speech (TTS) synthesis

© Springer International Publishing Switzerland 2016
P. Sojka et al. (Eds.): TSD 2016, LNAI 9924, pp. 514–521, 2016.
DOI: 10.1007/978-3-319-45510-5_59

system is to convert an arbitrary input text to its corresponding speech output. Text processing and speech generation are the two major components of a TTS system. The text processing component converts graphemes into a sequence of phonemes while the latter proffers the produced sequence of phonemes to the synthesizer to generate the speech waveform. Determining the appropriate sequence of sounds is very crucial for natural and intelligible speech.

Syllable structures across Indian languages can be similar or vary. Traditional approaches in converting text to speech for a given language make use of language specific parsers. Such approaches use specific rules of a given language and build parsers that are highly customised. This makes the task of creating individual parsers for new languages difficult.

Parsers that work for more than one language focuses on structurally related languages such as English and French or English and German [1]. Bilingual parsers built in Indian multilingual context are also available [9]. This paper introduces a unified parser that can handle Indian languages which are free-word-order and are also morphologically rich. The main challenges are finding the rules for different languages and incorporating the context-sensitive rules. A unified parser is designed that systematically identifies the invariant properties of different Indian languages. The UTF-8 text is converted to a sequence of labels. A common label is first defined for all the languages. Rules that are peculiar to a language are treated as exceptions.

Text to speech synthesis systems built using the common parser show that the parsers are as good as if not better than custom built parsers. Lex and Yacc [4] stands in good stead to build rule-based language parsers as these employ rule-based method for token matching. This paper tries to capture the similarities and resolve the differences in rules across multiple Indian languages so that lexical rules can handle occurrences of all native sentences and pass it to a synthesizer.

Section 2 describes the characteristics of Indian languages. Section 3 gives a brief overview of Letter To Sound (LTS) rules and discusses the design of the common label set. Section 4 discusses the design of the parser. Section 5 details the exceptions in parsing Indian language text. Section 6 discusses experiments and results. Section 7 concludes the work and provides future scope of the work.

2   Characteristics of Indian Languages

Most of the Indian languages can be broadly classified into two language families:

- Indo-Aryan languages
- Dravidian languages

The former is the largest and is spoken mostly in North India while the latter is predominant in South. These classes of languages share some common features and the geographical proximity of the regions where these languages have been spoken, have resulted in significant borrowings [5]. Indian languages are characterized by character set, which is termed as aksharas [7]. These aksharas are the fundamental linguistic units of the writing system in Indian languages [3]. According to the properties of aksharas, syllable boundaries can be marked at vowels at regular intervals for a given sequence of phones. This finding is typically followed in building TTS systems for Indian languages [2].

Indian languages are syllable-timed and a large number of syllables are common across Indian languages [7]. Approximating to the nearest syllable is possible even if the syllable as such is not available [8]. Accounting for the acoustic phonetic properties of different languages, this paper primarily focuses on the generation of phonemes sequences of the form C*VC*, where C is a consonant and V is a vowel.

3   Acoustics and Phonetics

Despite having different scripts there exist a relationship between orthography and sound that is common in the Indian languages. Exploiting this fact, Letter To Sound (LTS) rules and the common label set are introduced.

3.1   LTS

LTS rules are a set of hand-crafted rules that define the relationship between graphemes and phonemes for each language. Most Indian languages consist of about 50 sounds - 15–18 vowels and 35–38 consonants. Appropriateness of using the letter to sound rules is language dependent and requires considerable effort. The rules vary with the context as well and hence the LTS is modified with precise data making it more appropriate and flawless.

3.2   Common Label Set

The acoustic similarity among the same set of phones of different languages suggests the possibility of a compact and common set of labels [10,11]. The common label set is defined using the Latin 1 script. The common label set uses a standard set of labels for speech sounds that are commonly used in Indian languages. The notations of labels and rules for mapping are detailed in [10]. The paper make uses of this label set such that the native script is largely recoverable from the transliteration.

4   Parsers

Each language has its own set of grammar and syllabification rules. Non-phonetic languages like English use Classification and Regression Tree (CART) [6] to predict the phonetic transcription. This approach needs a large dictionary of words and its correct phonetic transcription. Since Indian languages are more or less phonetic in nature, building a rule-based parser is possible.

4.1   Structure of Parser and Its Parsing

The primary task of a parser is to segment the text for TTS systems. However, the parser cannot handle the raw text as it is available in news websites, blogs, documents etc. Standardizing the text input by removing the unwanted char-acters like special characters and emoticons is essential. Once the text is stan-dardized, next phase is language identification. The first character of the word is taken and compared with the Unicode range to detect the language. Identifying whether the language belongs to Aryan or Dravidian is also vital owing to vast differences in pronunciation. For parsing the word, the sequence of graphemes is mapped to labels in the common label set. Having mapped the input to com-mon label set, the next step is to obtain appropriate pronunciation for each of these labels. Although Indian languages are more or less phonetic, occasionally, the one-to-one correspondence between the spoken and written form is absent. These exceptions are handled by rules detailed in Sect. 5.

5   Parsing Indian Language Text

The main issue with parsing is the identification of vowel deletion points, syl-lable boundaries and the manner of applying rules. The unified parser uses the following set of rules.

Schwa Deletion Rules: Phonetically, schwa is a short neutral vowel sound /a/ which is associated with a consonant. For Aryan languages, the implicit mid central vowel (schwa), in each consonant of the script, is obligatorily deleted in certain context while uttering. This is known as schwa deletion or Inherent Vowel Suppression (IVS). Identifying which schwas are to be deleted and which are to be retained makes the process of schwa deletion complex. This is obvious for a native speaker, but for machine processing this decision depends on language specific rules. IVS rules are performed on Free Consonants/Semivowels (FCS) in a word. FCS refers to the consonants/semivowels in a word that do not have a vowel sound adjacent to it in written form. Example: In चटचटाहट, the first and third occurrences of ट are FCS whereas the second occurrence is not. Following are the known IVS rules.

1. Characters present in the first position of a word, never undergo IVS. Exam-ple: चदा (c a dxh aa)
2. Characters in final position always undergo IVS. Example: चदाकर (c a dxh aa k a r)
3. No two successive characters undergo IVS. Example: चटचटाहट (c a tx c a tx aa h a tx)
4. No two vowels come together.
5. The remaining FCS in a word that are not processed by rules 1 and 2 is processed in left-to-right order. IVS occurs for an FCS if its successor in the word is (i) not the last character of the word or (ii) a vowel other than

'*a*'. Application of this rule leads to erroneous parsing of a subset of words. Example: Application of the rule yields the following output.

पागलपन (p aa g a l p a n)

ताजमहल (t aa j a m h a l)

पागलपन is parsed correctly whereas ताजमहल is not. This single rule would not be able to handle such contradictorily parsed words.

This paper proposes 2 new rules - AB (1) and AB (2) - to solve such parsing problems.

AB (1) Rule: A free semivowel at the second position of a word starting with a vowel never undergoes IVS whereas a free consonant at the second position of a word starting with a vowel always undergoes IVS.

AB (2) Rule: The focus of this rule is on the substring of the word which is not processed by rules 1, 2 and AB (1). The inherent vowel sounds in this substring are named unmarked schwa. The rule proposes lexicographically ordered processing of FCS in this substring. An FCS is processed only if it is preceded by an unmarked schwa and succeeded by a vowel, vowel sound or unmarked schwa in the transliterated form. In this case, the predecessor (unmarked schwa) is deleted. If the successor is an unmarked schwa, it is marked as non-deletable in further iterations (marked schwa). Application of AB rule parses पागलपन and ताजमहल correctly. The process is illustrated in Tables 1 and 2. In Table 2, a^* represents unmarked schwa and â represents marked schwa.

Table 1. Pass 1 – apply rules 1, 2 and AB (1)

Word	Transliterated String	Rule 1	Rule 2	AB(1)
ताजमहल	taa *ja* ma ha la	NA	taa *ja* ma ha l	NA
पागलपन	paa *ga* la pa na	NA	paa *ga* la pa n	NA
अकबर	a *ka* ba ra	NA	a *ka* ba r	a k ba r
असफल	a *sa* fa la	NA	a *sa* fa l	a sa fa l

Table 2. Pass 2 – apply AB (2) rule

Substring considered (with unmarked schwa)	Iteration 1			Iteration 2			Iteration 3		
	Character	Process?	Output	Character	Process?	Output	Character	Process?	Output
ja* *ma** *ha**	j	No	*ja** *ma** *ha**	m	Yes	j mâ *ha**	h	No	j mâ ha*
ga* *la** *pa**	g	No	*ga** *la** *pa**	p	Yes	ga l pâ	l	No	ga l pa*
ba*	b	No	*ba**	-	-	-	-	-	-
sa*	s	No	*sa**	-	-	-	-	-	-

Geminate Correction Rules: The term geminate in phonology refers to a long or doubled consonant sound, such as the /kk/ in the Hindi word पक्का that contrasts phonemically with its shorter or singleton counterpart पका. Such contrasts occur frequently in Indian languages. There exist other phonetic cues to geminates besides consonantal duration such as pitch and intensity differences. However, this paper will not focus on the phonological behavior of geminates. Focus is to keep the geminates together, that is, they are always grouped as a syllable, as the sound is distinct. Example: पक्का (p a)(k k aa)

Syllable Parsing Rules: Though each sound is mapped to a corresponding label in the common label set, the label set does not handle the implicit /a/ sound associated with each consonant of the script. Hence, a separate rule is written to add the /a/ sound to the labels of all consonants without a vowel modifier associated with it. Thereafter, schwa deletion is performed for Aryan languages alone. For Dravidian languages schwa deletion rules are not applied. The processed input text is split into a set of sub-syllables, both at vowel and halant positions. These sub-syllables are processed in last to first manner, to ensure that all the consonantal units are suffixed by a vowel. Necessary correction (if required) is done subsequently i.e., if the current unit does not possess a vowel sound, it is appended to the previous unit. For example, ताजमहल is syllabified as (t aa j)(m a)(h a l). This rule is significant in particular for chillaksharas in Malayalam that do not possess an inherent vowel. This rule is also considered while grouping geminates as syllables, as the first occurrence of the consonant does not possess an inherent vowel.

Language-Specific Rules: Only 80 %–95 % accuracy is achieved even after applying all the above rules. This is due to the fact that each language has certain specific rules which cannot be generalized. These language-specific rules are applied during parsing to obtain an accuracy of 95 %–100 %. A few examples of such rules for Tamil are shown in Fig. 1.

Grapheme	Phoneme in CLS	Rule	Phoneme in actual pronunciation	Example
க	k	If previous and next grapheme is a vowel	g	ஆகாயம் (aagayam)
ㄴ	tx	If previous grapheme is ண்	dx	வேண்டும் (weenxdxum)
ச	c	If previous grapheme is ஞ	j	பஞ்சம் (panjjam)
ப	p	If previous grapheme is ம	b	குடும்பம் (kudxumbam)
க	k	If previous grapheme is ங	g	திங்கள் (tinggalx)

Fig. 1. Language specific rules for Tamil

Agglutination is the process of combining words that are formed by stringing together morphemes. The combination is carried without changing the morphemes in either spelling or phonetics. Languages that uses the property of agglutination are called agglutinative languages. Unified parser handles even the agglutinative words that are common in Dravidian languages since it employs a rule-based approach.

6   Experiments and Results

Text to speech synthesis systems are built using the language specific parsers and unified parser for 11 Indian languages. Pairwise Comparison (PC) tests are performed by an average of 12 native listeners to evaluate the performance of the unified parser approach. PC tests reveal the effectiveness of the unified parser. As can be seen from Fig. 2, in most cases the unified parser and native parser have the same preference. Occasionally there is preference for the unified parser. This is primarily because a systematic approach to parsing across various languages has been taken. This has resulted in a consistent set of rules.

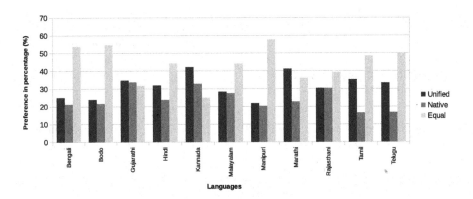

Fig. 2. Pairwise comparison test results

Experimental results show that the unified parser is more robust and accurate than the current systems that require similar supervision. This work is also the first attempt to introduce a unified approach, using the common label set, for parsing Indian languages. Unlike previous systems that require manually-constructed rules, this system requires much less knowledge of the native languages and can be easily scaled to other languages. To build new TTS systems for a new language, the mapping is made to the CLS. Existing language specific rules can be adapted. For example, Hindi and Rajasthani follows mostly the same set of rules.

7 Conclusion

The work presented in this paper is a step towards building an efficient pronunciation generator for Indian languages. Although the objective was primarily to unify various Indian language parsers, it is observed that the unified parser is more robust than custom built parsers. In order to build a parser for a new language, one needs to identify the language family, borrow the standard rule set. Exceptions may be handled incorporating language-specific rules.

References

1. Copestake, A., Flickinger, D.: An open source grammar development environment and broad-coverage English grammar using HPSG. In: Proceedings of LREC 2000, pp. 591–600 (2000)
2. Kishore, S., Kumar, R., Sangal, R.: A data driven synthesis approach for indian languages using syllable as basic unit. In: Proceedings of International Conference on NLP (ICON), pp. 311–316 (2002)
3. Lavanya, P., Kishore, P., Madhavi, G.T.: A simple approach for building transliteration editors for Indian languages. J. Zhejiang Univ. Sci. A 6(11), 1354–1361 (2005)
4. Levine, J.R., Mason, T., Brown, D.: LEX & YACC. O'Reilly Media Inc., Sebastopol (1992)
5. Prakash, A., Reddy, M.R., Nagarajan, T., Murthy, H.A.: An approach to building language-independent text-to-speech synthesis for Indian languages. In: 2014 Twentieth National Conference on Communications (NCC), pp. 1–5. IEEE (2014)
6. Quinlan, J.R.: Induction of decision trees. Mach. Learn. 1(1), 81–106 (1986)
7. Raghavendra, E.V., Desai, S., Yegnanarayana, B., Black, A.W., Prahallad, K.: Global syllable set for building speech synthesis in Indian languages. In: SLT 2008, pp. 49–52. IEEE (2008)
8. Raghavendra, E.V., Yegnanarayana, B., Black, A.W., Prahallad, K.: Building sleek synthesizers for multi-lingual screen reader. In: INTERSPEECH, pp. 1865–1868 (2008)
9. Raina, A.M., Mukerjee, A., Goyal, P., Shukla, P.: A unified computational lexicon for hindi-english code-switching. In: Proceedings International Conference on Natural Language Processing (ICON), Hyderabad, India, December 2004, pp. 19–22 (2004)
10. Ramani, B., Christina, S.L., Rachel, G.A., Solomi, V.S., Nandwana, M.K., Prakash, A., Shanmugam, S.A., Krishnan, R., Kishore, S., Samudravijaya, K., et al.: A common attribute based unified hts framework for speech synthesis in Indian languages. In: 8th ISCA Workshop on Speech Synthesis, pp. 311–316 (2013)
11. Singh, A.K.: A computational phonetic model for Indian language scripts. In: Constraints on Spelling Changes: Fifth International Workshop on Writing Systems (2006)
12. Tokuda, K., Yoshimura, T., Masuko, T., Kobayashi, T., Kitamura, T.: Speech parameter generation algorithms for HMM-based speech synthesis. In: Proceedings of the 2000 IEEE International Conference on Acoustics, Speech, and Signal Processing, 2000, ICASSP 2000, vol. 3, pp. 1315–1318. IEEE (2000)

Influence of Expressive Speech on ASR Performances: Application to Elderly Assistance in Smart Home

Frédéric Aman[1,2(✉)], Véronique Aubergé[1,2], and Michel Vacher[1,2]

[1] LIG, Université Grenoble Alpes, 38000 Grenoble, France
{Frederic.Aman,Veronique.Auberge,Michel.Vacher}@imag.fr
http://www.liglab.fr/
[2] LIG, CNRS, 38000 Grenoble, France

Abstract. Smart homes are discussed as a win-win solution for maintaining the Elderly at home as a better alternative to care homes for dependent elderly people. Such Smart homes are characterized by rich domestic commands devoted to elderly safety and comfort. The vocal command has been identified as an efficient, well accepted, interaction way, it can be directly addressed to the "habitat", or through a robotic interface. In daily use, the challenges of vocal commands recognition are the noisy environment but moreover the reformulation and the expressive change of the strictly authorized commands. This paper focuses (1) to show, on the base of elicited corpus, that expressive speech, in particular distress speech, strongly affects generic state of the art ASR systems (20 to 30 %) (2) how interesting improvement thanks to ASR adaptation can regulate (15 %) this degradation. We conclude on the necessary adaptation of ASR system to expressive speech when they are designed for person's assistance.

Keywords: Expressive speech · Distress call · Ambient assisted living · Home automation

1 Introduction

Smart homes aim at anticipating and responding to the special needs of elderly people living alone at their own home by assisting them in their daily life. A large variety of sensors are used in this framework [8] but one of the best interfaces seems to be the speech interface, that makes possible interaction using natural language and that is well adapted to people with reduced mobility. The purpose of this study is to develop a system able to detect distress calls uttered by elderly people. In a previous study, we reported that elderly voices resulted in degraded performances for generic Automatic Speech Recognition (ASR) systems [2]. The content of test sentences was related to distress but these sentences were read and uttered in a neutral manner without any particular emotion; in this article we are using "neutral" when utterances are not highly perturbated by intentions, attitudes or emotions.

© Springer International Publishing Switzerland 2016
P. Sojka et al. (Eds.): TSD 2016, LNAI 9924, pp. 522–530, 2016.
DOI: 10.1007/978-3-319-45510-5_60

In a real situation and in the framework of home automation, vocal commands can be expressive for 2 reasons:

1. if the user integrates a representation of the home as a communicative entity [5], thus prosody and morphosyntax can implement intentions, attitudes and social affects;
2. the command is motivated by an emotional context (panic, happiness to speak with a relative, etc.).

In a distress situation, a cry for help (i.e., "I fell", "Help me") may be charged in emotion, prosodic modifications influencing voice quality [3,16] so that emotions could be perceived by the interlocutor. Many studies are related to emotion recognition from voice (distress, fear, stress, etc.) [4,11,14,15], in [13], the authors used energy, pitch, jitter and shimmer. Scherer [10] compare effects of different emotions on prosodic parameters with regard to neutral voice.

An immediate and more reachable issue consists in making ASR more robust in presence of emotion instead of recognizing the distress state but few studies were done in this domain excepted the study of Vlasenko et al. [17]. They observed a performance degradation with expressive speech when ASR were trained with neutral voice.

In former studies [1], the different ASR performances between acted expressive voice and spontaneous expressive voice from a corpus recorded using a Wizard of Oz technique was brought to light; in this study, results were compared to those of neutral voice. The aim of our study is to highlight the lack of robustness of ASR systems with expressive voice. For in-depth study, we recorded the *Voix Détresse* corpus which is made of distress sentences uttered in expressive and neutral manner in an elicitation protocol. A reduction of performance decrease was observed thanks to the use of an adapted acoustic model. Expressive voice characterisation is presented at the end.

2 Method

Ideally, this work should have been held on natural dataset. But in the spontaneous corpus of distress, situations (calls to emergency centers, etc.) are: (1) not addressed to a system (human isolation), but to a person devoted to assistance (voice human contact); (2) after a phone call, that is not in the so serious situation that the elderly is not anymore able to give a call. It is why, in this primary study, we decided to be closer to the natural situations but to control the relevant cues that will be the real situations for which there is yet no available data. Thus we proposed a devoted elicitation protocol. The Voix Détresse French corpus was recorded in our laboratory in the following elicitation protocol:

– a list of 20 prototypical utterances was selected from the *AD80* corpus [2], that was validated to be representative of distress commands spontaneously produced by subjects in real distress situations at home;

- to elicitate expressive speech in the specific context of these distress situations, for each utterance is associated a devoted photography showing an elderly in this given distress home situation; the subject had to produce the utterance (as many times as the subject decided) trying to stand in that individual's shoes;
- before to enter the elicitation procedure, in order to get a control sub-corpus, each speaker had to read the 20 distress sentences in a neutral manner, that is without any elicitation support.

The expected emotions for these distress elicitations were mainly negative emotions like fear, anger, sadness. The elderly speakers had some difficulties to enter the elicitation protocol, so that only 5 speakers, all females, had sufficient performances of "quite-natural" distress expressions to be selected for the corpus. It has to be noted that the females are the main users of the Smart homes, because the elderly population over 80 is mainly females, and moreover some French investigations show that the females are more interested by such Smart homes solutions. To complete these low number of elderly subjects, together with collecting a non elderly control sub-set, 20 other subjects, from 23 to 60 years old, were recorded in the same elicitation protocol. *Voix Détresse* is described in Table 1. The corpus is made of 1481 sentences, its length is 24 min 19 s.

One half of the corpus was reserved for model training and the remaining part for testing. Since 25 speakers were recorded and regarding each speaker, approximately 10 neutral sentences were reserved for adapting and 10 for testing, and about 20 expressive sentences for training and 20 for testing (see Table 2).

For our study, ASR was operated using *Sphinx3* [12] based on a context dependant triphone acoustic model. Our HMM model was trained with *BREF120* corpus [6] which is made of 100 hours of texts read by 120 speakers (65 females, 55 males), this generic model was called *BREF120*.

For the decoding, a trigram language model (LM) with a 10 K lexicon was used. It results from the interpolation of a generic LM (with a 10 % weight) and a specialized LM (with a 90 % weight). The generic LM was estimated on about

Table 1. *Voix Détresse* corpus.

Composition	Younger group	Elderly group
All speakers	12F, 8M, 23–60 years old	5F, 67–85 years old
Neutral voice	400 sentences, 5 min 41 s (total)	100 sentences, 1 min 38 s (total)
Expressive voice	782 sentences, 12 min 44 s (total)	199 sentences, 4 min 15 s (total)

Table 2. Number of sentences of the *Training* and *Testing* partitions of the corpus.

Young voice	Training	Test	Elderly voice	Training	Test
Neutral	200	200	Neutral	50	50
Expressive	393	389	Expressive	99	100

1000M of words from Gigaword, a collection of French newspapers[1], this model is unigram and made of 11,018 words. The specialized LM was estimated from the distress and call for help sentences of the *AD80* corpus, it is made of 88 unigrams, 193 bigrams et 223 trigrams. This combination has been shown as leading to the best WER for domain specific application [7]. The interest of such combination is to bias the recognition towards the domain LM but when the speaker deviates from the domain, the general LM makes it possible to correctly recognize the utterances.

3   Experiment

3.1   Decoding with BREF120 Acoustic Model

Decoding was operated on the *Test* part of the *Voix Détresse* corpus. WER results are presented in Fig. 1 for each speaker and as the average for each group in Table 3. A severe performance degradation can be observed: absolute WER difference is 29.95 % for younger speakers and 19.6 % for elderly speakers.

A Welch's t test was operated in order to determine if WER differences between neutral and expressive voices are significant. Test results indicate a significant difference between the corresponding WER ($t = -7.2026$, $df = 21.365$, $p < 0.05$ ($p = 3.834e - 07$) for younger speakers and $t = -2.5165$, $df = 7.874$, $p < 0.05$ ($p = 0.03645$) for elderly speakers).

As shown in Fig. 1, we can observe a WER increase for expressive voice with regard to neutral voice for all speakers, excepted for *Y08F25*, and these variations are more or less important according to the speaker. One explanation of these differences could be due to the difficulty of the participants, who are not professionnal, to "play the game" of distress during the records.

Overall, persons more comfortable to play distress situations are persons with a higher WER for distress sentences. In real situations, it is feared that

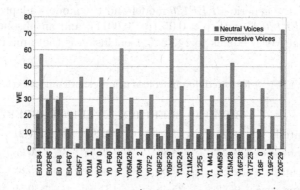

Fig. 1. *WER from Sphinx3 decoding (BREF120 acoustic model) for neutral and expressive voices*

---

[1] http://catalog.ldc.upenn.edu/LDC2006T17.

Table 3. WER (%) from Sphinx3 decoding (BREF120 acoustic model) for the two groups.

Voice	Younger group	Elderly group
Neutral	9.27	18.82
Expressive	39.22	38.42

very expressive sentences will be badly recognized by ASR, while it is essential to act in this situation.

3.2 Decoding with an Adapted Acoustic Model

In order to establish the possibility of improvement, a MLLR adaptation of *BREF120* model to each speaker was operated: (1) *BREF120_N* by using the 10 neutral *Training* sentences uttered by the speaker, (2) *BREF120_E* by using the 20 expressive sentences, and (3) *BREF120_N+E* by using neutral and expressive sentences. For each speaker, 3 different models adapted to his voice were obtained. Decoding was operated as in Sect. 3.1 on the 739 neutral and expressive sentences to the *Test* part of the corpus using the 3 adapted acoustic models. Results are given in Table 4 with regard to the standard *BREF120* model.

ANOVA was operated on the 2 groups *neutral* and *expressive voices* (samples are Gaussian and homogeneity of group variances is verified). There is no significant difference inside the samples of the *neutral voice* group (F(3;96)=1.293; p=0.281) but a significant difference (p-value < 0.05) inside the samples of the *expressive voice* group (F(3;96)=7.828; p=9.96e-05). Results of the Tukey HSD test are given in Table 5 for *neutral voices* and in Table 6 for *expressive voices*. For neutral voices, no significant difference is observed between the WER obtained from the different acoustic models, p > 0.05 for each acoustic model pair. By contrast for expressive voices, a significant WER difference exists between the generic *BREF120* model and the models adapted to expressive voice *BREF120_E*, with p < 0.05 (p = 0.0011), and an absolute difference between average WER equals to 14.78 %. Conclusion is the same for *BREF120*

Table 4. WER (%) for the different voices using generic and adapted models.

Group	BREF120	BREF120_N	BREF120_E	BREF120_N+E
Young neutral	9.27	7.80	11.03	7.21
Young expressive	39.22	28.86	22.45	20.23
Elderly neutral	18.82	17.06	16.48	15.88
Elderly expressive	38.42	35.48	31.64	30.50
Average neutral	11.18	9.65	12.12	8.94
Average expressive	39.06	30.18	24.28	22.28

Table 5. *Neutral voice* group: p-value of Tukey HSD test and WER.

	BREF120	BREF120_N	BREF120_E	BREF120_N+E
BREF120	-	p = 0.8298	p = 0.9529	p = 0.5977
BREF120_N	-	-	p = 0.5174	p = 0.9786
BREF120_E	-	-	-	p = 0.2923
BREF120_N+E	-	-	-	-
WER (%)	11.18	9.65	12.12	8.94

Table 6. *Expressive voice* group: p-value of Tukey HSD test and WER.

	BREF120	BREF120_N	BREF120_E	BREF120_N+E
BREF120	-	p = 0.0976	**p = 0.0011**	**p = 0.0002**
BREF120_N	-	-	p = 0.4120	p = 0.1683
BREF120_E	-	-	-	p = 0.9526
BREF120_N+E	-	-	-	-
WER (%)	39.06	30.18	24.28	22.28

and *BREF120_N+E*, with $p < 0.05$ ($p = 0.0002$), and an absolute difference of 16.78 %. By contrast, there is no significant difference between the other models.

Therefore, with regard to expressive sentences, speaker adaptation from neutral sentences is not sufficient to improve significantly the WER. Models adapted to expressive voice may be used, but adaptation can be done from mixing neutral and expressive sentences because there is no significant difference between models *BREF120_E* and *BREF120_N+E*. In the same way, neutral sentences can be decoded using adapted model *BREF120_N+E* without important performance decrease.

4   Expressive Voice Characterisation

Average values of prosodic parameters, flow, F0, *jitter*, *shimmer* et Harmonic to Noise Ratio (HNR), were measured and compared between neutral and elicited expressive voices. Data were the 739 neutral and expressive sentences of the *Testing* part of *Voix Détresse*.

We observed for expressive voice, with regard to neutral voice and in average, a decrease of the flow, an increase of F0 and HNR, and an increase of *jitter* and *shimmer*. Averages concerning all speakers for each parameter are given in Table 7 with significance tests of the difference between neutral and expressive voices.

In a first step, we considered for our analysis that the distress corpus (i.e., the expressive sentences) of *Voix Détresse* was homogeneous with regard to expressiveness of acted emotion (eventually with variable expressive intensity), and

Table 7. Average prosodic parameters as a function of kind of voice, and Welch's t test p-values.

Parameters	Neutral voice	Expressive voice	Welch's t test (significant difference if p < 0.05)
Flow (Phonemes/s)	13.58	11.71	p < 0.05 (p = 0.0019)
F0 (Hz)	162.74	260.43	p < 0.05 (p = 1.16e-06)
Jitter (%)	3.07	2.30	p < 0.05 (p = 0.00015)
Shimmer (%)	13.63	9.07	p < 0.05 (p = 1.41e-07)
HNR (dB)	12.44	14.76	p < 0.05 (p = 0.00077)

varied possibly only with the speaker (difference of interpretation or idiosyncratic physiologic variation). However, by more precise hearing of the considered utterances, sentence by sentence, we can suppose that the nature of the distress varies according the situation suggested by the picture submitted for elicitation, and that this nature could be homogeneous by situation (i.e., by sentence) without dominant variability of the subject. For example, the sentence "Je ne me sens pas bien" (I don't fell very well) was associated to a picture showing a very demoralized man taking his arm, supported by an other person, this person was very close and had already started to help the man in distress situation. By contrast, the sentence "A moi" (Help me) is associated to a picture showing an isolated drowning person: it is a question of a request for emergency assistance by an isolated person in a critical condition. The two situations are two opposite situations of our corpus: in the first case a helper took care of the person but in the second case a person in a critical situation was calling but there was no identified helper.

That is what we wanted to verify by evaluating prosodic parameters not by speaker but by context, The 489 expressive sentences were spread into 10 contexts. *Jitter*, *shimmer*, F0 and flow were measured, results are given in Table 8. It appears that 2 extreme situations "A moi" (Help me) and "Je ne me sens pas bien" (I don't feel very well) are characterized, throughout all speakers, by extreme values of F0, *jitter* and *shimmer*; this fact could confirm our hypothesis on the variability of the nature of the distress. According to the 2D model of Russel [9] of cognitive psychology, emotions can be represented along 2 axis: valence (positive/negative) and *arousal* (active/passive).

Thus, the context "A moi" could correspond to a very "active" distress, by hearing they are perceived as very stressed. The context "Je ne me sens pas bien" could be more "passive" and is perceived as the less stressed. Measures corroborate these observations: we can observe in Table 8 that the context "A moi" has the lowest *jitter* and the highest F0 with regard to the 9 other contexts, and the context "Je ne me sens pas bien" has the highest *jitter* and the lowest F0.

Table 8. Prosodic parameters for some different French distress sentences.

Distress sentences	Jitter (%)	Shimmer (%)	F0 (Hz)	Intensity (dB)
A moi !	1.028	5.962	317	70.27
Oh la la !	1.367	6.383	224	65.72
J'ai un malaise !	1.411	6.778	232	63.53
Aidez-moi !	1.423	6.801	279	66.46
Du secours s'il vous plaît !	1.550	6.791	276	64.80
Me laissez pas tout seul !	1.550	7.131	276	64.27
Qu'est-ce qu'il m'arrive !	1.654	7.470	239	62.21
Au secours !	1.679	5.414	294	69.79
Je ne peux plus bouger !	1.708	8.181	282	64.72
Je ne me sens pas bien !	1.718	8.372	220	58.08

We can see that some analog expressions are characterized by the same parameter values, as "Du secours s'il vous plaît" and "Ne me laissez pas tout seul" with $jitter$=1.550 % and F0=276 Hz for these 2 contexts.

5 Conclusion

After adaptation, WERs of expressive voices remain higher than WERs of neutral voices with the generic model *BREF120*, the difference is 13.1 %. Moreover, the record of spontaneous expressive voices is very difficult and the *Voix Détresse* corpus is made of acted expressive voice. In all likelihood, performances may be lower in a real case.

Acoustic characteristics of distress are very subtle. It is clear that in the harsh acoustic conditions of a smart home, we have to be extremely vigilant on the physiologic, functional and communicative context of utterances to recognize: a high amount of data must be recorded in order to obtain a robust system. Moreover these data must be finely characterized as regard to their expressiveness. Those conditions are necessary but insufficient ones for the use of speech technologies in real conditions for the assistance of elderly persons.

Acknowledgements. This study was supported by the French funding agencies ANR and CNSA through the project CIRDO - Industrial Research (ANR-2010-TECS-012). The authors would like to thank the persons who agreed to participate in recordings.

References

1. Aman, F., Auberge, V., Vacher, M.: How affects can perturbe the automatic speech recognition of domotic interactions. In: Workshop on Affective Social Speech Signals, pp. 1–5, Grenoble, France, August 2013

2. Aman, F., Vacher, M., Rossato, S., Portet, F.: Analysing the performance of automatic speech recognition for ageing voice: does it correlate with dependency level? In: 4th Workshop on Speech and Language Processing for Assistive Technologies, pp. 9–15, Grenoble, France, August 2013

3. Audibert, N.: Prosodie de la parole expressive: dimensionnalité d'énoncés méthodologiquement contrôlés authentiques et actés. Ph.D. thesis, INPG, Ecole Doctorale "Ingénierie pour la Santé, la Cognition et l'Environnement" (2008)

4. Chastagnol, C.: Reconnaissance automatique des dimensions affectives dans l'interaction orale homme-machine pour des personnes dépendantes. Ph.D. thesis, Université Paris Sud-Paris XI (2013)

5. Clarcke, A.C.: 2001: A Space Odyssey. New American Library, New York (1968)

6. Lamel, L., Gauvain, J., Eskenazi, M.: BREF, a large vocabulary spoken corpus for French. In: Proceedings of EUROSPEECH 1991, vol. 2, pp. 505–508, Geneva, Switzerland (1991)

7. Lecouteux, B., Vacher, M., Portet, F.: Distant speech recognition in a smart home: Comparison of several multisource ASRs in realistic conditions. In: 12th International Conference on Speech Science and Speech Technology (INTERSPEECH 2011), pp. 2273–2276, Florence, Italy, 28–31 August 2011

8. Peetoom, K., Lexis, M., Joore, M., Dirksen, C., De Witte, L.: Literature review on monitoring technologies and their outcomes in independently living elderly people. Disabil. Rehabil. Assist. Technol. 10, 271–294 (2014)

9. Russell, J.A.: A circumplex model of affect. J. Pers. Soc. Psychol. 39(6), 1161–1178 (1980)

10. Scherer, K.R.: Vocal communication of emotion: a review of research paradigms. Speech Commun. 40(1–2), 227–256 (2003). http://www.sciencedirect.com/science/article/pii/S0167639302000845

11. Schuller, B., Batliner, A., Steidl, S., Seppi, D.: Recognising realistic emotions and affect in speech: state of the art and lessons learnt from the first challenge. Speech Commun. 53(9–10), 1062–1087 (2011). http://dx.doi.org/10.1016/j.specom.2011.01.011

12. Seymore, K., Stanley, C., Doh, S., Eskenazi, M., Gouvea, E., Raj, B., Ravishankar, M., Rosenfeld, R., Siegler, M., Stern, R., Thayer, E.: The 1997 CMU Sphinx-3 english broadcast news transcription system. DARPA Broadcast News Transcription and Understanding Workshop (1998)

13. Soury, M., Devillers, L.: Stress detection from audio on multiple window analysis size in a public speaking task. In: Humaine Association Conference on Affective Computing and Intelligent Interaction (ACII), pp. 529–533, September 2013

14. Vaudable, C.: Analyse et reconnaissance des émotions lors de conversations de centres d'appels. Ph.D. thesis, Université Paris Sud-Paris XI (2012)

15. Vidrascu, L.: Analyse et détection des émotions verbales dans les interactions orales. Ph.D. thesis, Université Paris Sud-Paris XI, Discipline: Informatique (2007)

16. Vlasenko, B., Prylipko, D., Philippou-Hübner, D., Wendemuth, A.: Vowels formants analysis allows straightforward detection of high arousal acted and spontaneous emotions. In: Proceedings of INTERSPEECH 2011, pp. 1577–1580 (2011)

17. Vlasenko, B., Prylipko, D., Wendemuth, A.: Towards robust spontaneous speech recognition with emotional speech adapted acoustic models. In: Proceedings of the KI 2012, pp. 103–107 (2012)

From Dialogue Corpora to Dialogue Systems: Generating a Chatbot with Teenager Personality for Preventing Cyber-Pedophilia

Ángel Callejas-Rodríguez[1], Esaú Villatoro-Tello[1(✉)], Ivan Meza[2], and Gabriela Ramírez-de-la-Rosa[1]

[1] Language and Reasoning Research Group, Information Technologies Department, Universidad Autónoma Metropolitana (UAM) Unidad Cuajimalpa, Mexico City, Mexico
acallejas21@gmail.com, {evillatoro,gramirez}@correo.cua.uam.mx
[2] Instituto de Investigaciones en Matematicas Aplicadas y en Sistemas, Universidad Nacional Autonoma de Mexico (UNAM), Mexico City, Mexico
ivanvladimir@turing.iimas.unam.mx

Abstract. A conversational agent, also known as chatbot, is a machine conversational system which interacts with human users via natural language. Traditionally, chatbot technology is built under certain set of "manually" elaborated conversational rules. However, given the availability of large and real examples of humans' interactions in the web, automatically generating these rules is becoming a more feasible option. In this paper we describe an approach for building and training a conversational agent, which holds a teenager personality and it is able to dialogue in Mexican Spanish. By means of this chatter bot we aim at assisting law enforcement officers in the prevention of cyber-pedophilia. Our performed experiments demonstrate that our developed chatbot is able to elaborate comparable lexical and syntactical constructions to those a teenager would produce. As an additional contribution, we compile and release a large dialogue corpus containing real examples of conversations among teenagers.

Keywords: Conversational agent · Dialog systems · Dialogue corpora · Cyber-pedophilia · Natural language processing

1 Introduction

With the continued growth and use of Internet as a tool for communication worldwide, more and more people are enjoying and becoming more dependent on the convenience of its provided services. Particularly, children and teenagers represent a population sector that have been turning into very active users[1].

---

[1] Recent statistics for the American continent are described in a study from the Pew Research Center (http://www.pewinternet.org/2015/04/09/teens-social-media-technology-2015/).

© Springer International Publishing Switzerland 2016
P. Sojka et al. (Eds.): TSD 2016, LNAI 9924, pp. 531–539, 2016.
DOI: 10.1007/978-3-319-45510-5_61

Unfortunately, this phenomenon has opened doors to several types of cyber-crime, such as cyber-pedophilia, which is an increasingly important issue as well as of great social concern.

Traditionally, a term that is used to describe malicious actions with a potential aim of sexual exploitation or emotional connection with a child is referred as "Child Grooming" or "Grooming Attack" [7]. Defined in [6] as: "a communication process by which a perpetrator applies affinity seeking strategies, while simultaneously engaging in sexual desensitization and information acquisition about targeted victims in order to develop relationships that result in need fulfillment" (*e.g.* physical sexual molestation). Sexual predators are catch through police officers or volunteers, whom pose as teenager in chat rooms and provoke sexual predators to approach them[2]. Unfortunately, online sexual predators always outnumber the officers. Therefore, automatic tools that can assist to officers in the process of sexual predators identification (SPI) are highly needed.

With this goal in mind, our proposed method for developing a conversational agent, seeks to provide an automatic chatter bot that is able to behave as a teenager in chats, social networks and similar services. Our chatbot uses Natural Language Processing (NLP) and Pattern Recognition techniques to automatically construct its conversational rules from real dialogue corpora. Thus, our proposed method represents a language independent technique that does not require the formulation of hand-coded knowledge base. For performing our experiments and a proper evaluation we took on the task of collecting a dialogue corpora for the Mexican Spanish from a popular web site among Mexican teenagers. We consider the construction of this corpus as an important contribution to the scientific community doing research on similar areas.

The rest of this document is organized as follows. Section 2 present some related work concerning to the SPI problem and automatic dialogue systems. In Sect. 3 we describe our proposed method for automatically inferring conversational rules from dialogue corpora. Section 4 provides some details regarding the collected dialogue corpus and shows obtained experimental results. Finally, Sect. 5 depicts our conclusions and some future work directions.

2   Related Work

During the last decade several approaches have been proposed for solving the problem of SPI. These can be divided in the following categories: *(i)* identification of predatory chat lines, *(ii)* classification of predatory chat conversations, and *(iii)* identification of the offender and the victim.

On the one hand, there are methods that approach these problems by means of traditional Text Classification (TC) strategies, *i.e.,* they train a machine learning algorithm using a particular suited form of representation of the data. Proposed methods vary from those that use traditional Bag-of-Words (BOW) rep-

---

[2] The American foundation called Perverted Justice (PJ) (http://www.perverted-justice.com/) is one of the most representative organizations that follows the above mentioned approach.

resentations [10], psycho-linguistic features [3], through the inclusion of complex behavioural attributes [9].

On the other hand, some research works have focused on the learning algorithms instead. For instance, in [5] multiple predictions obtained from chained classifiers are used to determine whether or not there is a predator behaviour. Recently, in [4] authors proposed a semi-supervised approach called one-class Support Vector Machine algorithm for anomaly detection of on-line predators.

As can be noticed, most of the previous research work on SPI has been oriented to analyse and classify completed chat conversations by means of text mining strategies. Although these are important developments, it is also important to develop systems that can assist officers and volunteers from the beginning, *i.e.*, from the moment they have to pose as children within chat rooms. Having this type of systems will facilitate the officers task; since being too much time in front of a computer, posing as a victim, might result very upsetting.

In [8] authors proposed a chatter-bot, called "Negobot", which is based on game theory for the detection of pedophile behaviour. Even though Negobot was designed for working in Spanish, its knowledge base comes from the PJ website, meaning it depends on the accuracy of an automatic translation module. This is an important limitation, given that the particular slang used by teenagers on the Internet cannot be processed by an automatic translator. In order to avoid the previous limitations, our proposed method for developing a chatbot considers learning conversational rules from a real teenagers' dialogue corpora. Our performed experiments demonstrate that our chatbot is able to elaborate comparable lexical and syntactic constructions to those a teenager would produce.

3   Automatically Inferring the Conversational Rules

Our main idea was motivated by the work described in [2]. In general, this framework is based on the extraction of a set of rules from a dataset of question-answer pairs $(q : a)$. The rules consists of a *head* word h associated to an answer a. This pairing is built for each question in the dataset $(h : a)$. These rules use the h word as a trigger to the answer a. These pattern are codified into rules in the Artificial Intelligence Markup Language (AIML[3]), which is a common tool among chatbots [8]. Because we are reducing the questions q into a set of heads h it could be that multiple questions reduce to the same h, resulting in a head word h being associated with multiple answers $(h : \{a_1, \ldots, a_k\})$. In these cases, AIML will randomly dispatch one of the associated answers, a functionality associated to AIML that allows to face this type of situations.

In [1], for creating a rule of the form $(h : a)$, authors propose identify a *keyword* (k) within the question q. Hence, once the keyword is found, k will constitute the h word of the pattern. Intuitively, if k is identified correctly, the provided answers by the chatbot would be more coherent. Accordingly, in [1] two main strategies were considered for determining k, namely *first word* and *least*

---

[3] AIMLss was developed by Wallace and the Alicebot community in 1995–2000 [11].

frequent word. The intuitive idea behind the *first word* approach was that the first word within a question could be a good element defining the main topic of the question. The *least frequent word* method considers the word with the lower frequency value as the more informative term. Although their reported results showed a good performance, authors employed domain specific corpora for their experiments, which was formed by formal and well written pairs of $(q : a)$ texts.

Given that our main goal is to have a teenager chatbot, *i.e.*, a chatter bot that writes with orthographic mistakes, typos and that uses slang; we consider that including contextual information is fundamental for the proper construction of pairs $(h : a)$. Therefore, in this work we propose to explore new alternative strategies to chose the head (h) of the pattern:

Most Frequent Word. The use of *least frequent words* model is not as direct when the domain includes informal language. The presence of multiple orthographic errors, several typos and the extended use of slang, most of them with a low frequency value; would mislead the process of patterns construction. Hence, we explore to use the most frequent words for defining h.

Least Frequent Bigram. This strategy uses word 2-grams[4] rather than a single word. We hypothesize that the least frequent sequence of words would capture better the meaning of the question and therefore we will trigger a better answer. Using word n-grams represents a form of including contextual information.

Most Frequent Bigram. Similarly to the *most frequent word* model, this strategy is also motivated by the fact that our goal is to effectively handle informal language, which is typical in chat sessions. Our intuition is that by means of adding context information (*i.e.*, word n-grams), the extracted pattern could trigger a better answer.

An advantage of the proposed strategies is that they can complement each other. For instance a *bigram* strategy can be used together with *word* strategy in a back-up fashion. In this case, if a question given by an user does not triggers any bigram, the system could try to trigger a rule based on a word.

4   Experimental Results

4.1   Data Set

In order to provide our chatbot with a teenager personality, we took on the task of collecting a corpus from which such personality could be inferred. We collected a total of 304399 pairs of *question-answer* examples from a popular social network among Mexican teenagers called ASK[5]. Table 1 shows some basic information from the collected corpus.

The built corpus has been automatically processed in order to protect the identity of the involved users; personal names, pictures, URLs, and sensitive information was removed. In the experiments we used the *filtered* partition[6].

---

[4] An n-gram is a contiguous sequence of n words from a given sequence of text.

[5] http://ask.fm/.

[6] http://ccd.cua.uam.mx/~evillatoro/Resources/Corpus_Ask_MX.tar.gz.

Table 1. Some statistics regarding the collected corpora.

Number of ...	Original data	Filtered data
Users	1300	1300
Avg. interventions (per user)	782	234
Avg. length per intervention (words)	10	7
Vocabulary size (distinct tokens)	711854	186857

4.2 Evaluation Metrics

For evaluation we use three different metrics, namely: *lexical richness* (LR), *syntactic richness* (SR) and *perplexity* (P). The lexical richness is defined as the ratio between the number of distinct lexical units and the total number of lexical elements used. This metric intents to capture the vocabulary diversity used in the answers. We calculate this for three lexical types: unigrams, bigrams and trigrams. Syntactic richness is similar to the previous metric but it is based on Part-Of-Speech (POS) tags. This metric tries to complement the previous metric by paying attention to the sequence the syntactic information. Similarly to the previous metric we perform this metric for: single POS, bigram POS and trigrams of POS. Finally perplexity give us an idea of how predictable is the language we are producing in the answer. Since we are trying to capture informal language used by teenagers we expect a high perplexity and that a good system is not that predictable but random. For all the performed experiments we employed as a validation method a 10-*fold cross validation* strategy.

4.3 Results

In order to evaluate our proposed strategies (see Sect. 3) we performed six different configurations for inferring the conversation rules ($h : a$) for our chatbot: *(i)* **1W-MF**: most frequent word, *(ii)* **1W-LF**: least frequent word (this method corresponds to the one proposed in [1]), *(iii)* **2W-MF**: most frequent word bigram, *(iv)* **2W-LF**: least frequent word bigram, *(v)* **1W-MF-2W-LF**: most frequent word and least frequent word bigram, and, *(vi)* **1W-LF-2W-MF**: least frequent word and most frequent word bigram. As mentioned in previous sections, by means of using word n-grams we aim at capturing contextual information (configuration *(iii)* and *(iv)*), whilst experiments *(v)* and *(vi)* represent the configuration where a bigram strategy is combined with a single word model.

Table 2 summarizes the obtained results and allows to compare the naturalness of our chatbot by comparing human-to-chatbot dialogues versus human-to-human conversations. In general, we prefer those models that allow our chatbot to behave as humans do. Hence, we applied the proposed metrics to evaluate the performance of the human-to-humans dialogues. Obtained results in human-to-humans dialogues are: $LR-1gram = 0.273$, $LR-2gram = 0.407$, $LR-3gram = 0.735$, $SR-1gram = 0.382$, $SR-2gram = 0.134$, $SR-3gram = 0.126$ and $P = 1618$.

Table 2. Obtained results of the different versions of the developed chatbot. Results are reported in terms of the obtained difference against the performance in a human-to-human conversation

Method	Lexical richness (LR)			Syntactical richness (SR)			Perplexity (P)
	1-gram	2-grams	3-grams	1-gram	2-grams	3-grams	
1W-MF	0.078	0.135	0.256	−0.097	−0.022	0.000	772
1W-LF	0.130	0.213	0.412	−0.144	−0.016	0.031	917
2W-MF	**0.019**	0.076	0.151	−0.043	−0.011	**0.001**	632
2W-LF	0.079	0.181	0.344	−0.049	−0.001	0.021	928
1W-MF-2W-LF	0.086	0.157	0.306	−0.037	0.000	0.019	654
1W-LF-2W-MF	0.039	**0.066**	**0.128**	**0.016**	**0.009**	0.011	**213**

Notice that in terms of *LR* humans have richer vocabulary. However, our experiments show that the combination of using the most frequent bigram (2W-MF) and the least frequent word (1W-LF) allow our chatbot to behave very similar to humans. Regarding the *SR*, we can observe that most of the proposed methods obtained a greater value of *SR* than humans, thus the negative differences. Contrary to the *LR*, this is not a desirable situation since having a large value might indicate that the chatbot is creating very confusing even complex sentences. A low performance in this metric is preferable. Accordingly, the method that gets the overall smallest differences is the 1W-LF-2W-MF strategy. Finally, even though a low perplexity value is preferable when evaluating language models, this is achievable on very specialized corpora. Our data set does not represent a domain-specific set of documents but a large corpus of conversations among teenagers, hence we would prefer a model that gets to be surprised as humans do, thus the model that obtains similar perplexity values to the humans-to-humans dialogues is the 1W-LF-2W-MF configuration.

Figure 1 shows additional comparison results. Generally speaking, depicted figures provide an idea of how our chatbot behaves during the evaluation stage. Left graph compares the use frequency of lexical units employed by our chatbot

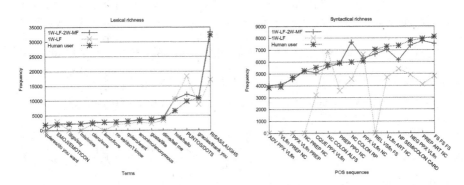

Fig. 1. Most commonly used lexical units and syntactic constructions by two versions of the developed chatbot.

(1W-LF and 1W-LF-2W-MF) against those used by humans. Similarly, right graph shows what are the most common syntactic constructions employed by our chatbot and by humans. As it is possible to observe, the model that behaves the most similar to humans is the 1W-LF-2W-MF method. Finally, Table 3 contains two samples of the conversation produced by two chatbot models (1W-LF and

Table 3. Conversations between a human user (u) and the chatbot (c). We compared versions 1W-LF and 1W-LF-2W-MF. The closest English translation appears between parenthesis in *Italics*. You can download the GitHub project at: https://github.com/Angel2113/Chatbot

Chatbot model: 1W-LF	Chatbot model: 1W-LF-2W-MF
u: hola *(hello)*	u: hola *(hello)*
c: hola *(hello)*	c: hola *(hello)*
u: como te llamas? *(what is your name?)*	u: como te llamas? *(what is your name?)*
c: muy bien :) *(very good :))*	c: No soy famosa *(I'm not famous)*
u: que edad tienes? *(how old are you?)*	u: que edad tienes? *(how old are you?)*
c: D: *(D:)*	c: 17 yirs ol *(17 years old)*
u: de donde eres? *(where are you from?)*	u: de donde eres? *(where are you from?)*
c: aslajshhas? C: *(aslajshhas? C:)*	c: Por que? *(Why?)*
u: que te gusta? *(what do you like?)*	u: que te gusta? *(what do you like?)*
c: D: *(D:)*	c: Ser rechazado un y otra vez por la misma persona xDDD Aun asi seguire XDDY una golpiza :c *(Being rejected over and over again by the same person xDDD still keep up XDDY beating :c)*
u: en donde estudias? *(where do you study?)*	u: en donde estudias? *(where do you study?)*
c: 7:39 en mi casa jaja *(7:39 in my home haha)*	c: En un hospital ah que :33 *(In a hospital ah :33)*
u: dime algo *(tell me something)*	u: dime algo *(tell me something)*
c: Bien supongo... *(Ok I guess...)*	c: Jajaja ok ok 7w7 *(Hahaha ok ok 7w7)*
u: lo que sea *(Anything)*	u: lo que sea *(Anything)*
c: Un detalle muy especial *(A very special detail)*	c: Que se rumora? *(What is rumored?)*
u: me tengo que ir *got to go*	u: me tengo que ir *(got to go)*
c: ...	c: Clase a estas horas? Pues bueno bueno. Nos seguimos contactando c: *(Class at this hour? Well then good. We'll keep in contact c:)*
u: Adios *(Goodbye)*	u: Adios *(Goodbye)*
c:	c: Bye :3 *(Bye :3)*

1W-LF-2W-MF). As can be noticed, the chatbot that uses the later approach produces a more ordinary conversation.

5   Conclusions

In this paper we have described a method for developing a conversational agent that is able to behave as a teenager. Our proposed method is able to infer its conversational rules from a real dialogue corpora. Performed experiments demonstrate that the model that uses the method of the least frequent word (1W-LF) in combination with the most frequent bigram (2W-MF) allows to fulfill our goal. In general, the combination of these two models accurately identify the main topic and allows to handle the informal language (orthographic mistakes, typos and slang) of a given question.

As an additional contribution, we collected and released a dialogue corpora in Mexican Spanish. Such resource contains real examples of conversations among teenagers and represents a valuable resource to the NLP community interested in carrying out future research on this field. Finally, given that our proposal represents a language independent method, as future work we want to test our developed chatbot under a PJ scenario, *i.e.*, evaluate if our chatter bot could assist law enforcement by effectively posing as a victim in chat rooms.

Acknowledgments. This work was partially funded by CONACYT under the Thematic Networks program (Language Technologies Thematic Network projects 260178, 271622). We thank to UAM Cuajimalpa and SNI-CONACyT for their support.

References

1. Abu Shawar, B.A.: A Corpus Based Approach to Generalising a Chatbot System. Ph.D. thesis, School of Computing, University of Leeds (2005)
2. Abu Shawar, B.A., Atwell, E.: Using dialogue corpora to train a chatbot. In: Proceedings of the Corpus Linguistics 2003 Conference, pp. 681–690 (2003)
3. Cano, A.E., Fernandez, M., Alani, H.: Detecting child grooming behaviour patterns on social media. In: Aiello, L.M., McFarland, D. (eds.) SocInfo 2014. LNCS, vol. 8851, pp. 412–427. Springer, Heidelberg (2014). http://dx.doi.org/10.1007/978-3-319-13734-6_30
4. Ebrahimi, M., Suen, C.Y., Ormandjieva, O., Krzyzak, A.: Recognizing predatory chat documents using semi-supervised anomaly detection. Electron. Imag. **2016**(17), 1–9 (2016)
5. Escalante, H.J., Juarez, A., Villatoro, E., Montes, M., Villaseñor, L.: Sexual predator detection in chats with chained classifiers. In: WASSA 2013, p. 46 (2013)
6. Harms, C.M.: Grooming: an operational definition and coding scheme. Sex Offender Law Rep. **8**(1), 1–6 (2007)
7. Kucukyilmaz, T., Cambazoglu, B.B., Aykanat, C., Can, F.: Chat mining: predicting user and message attributes in computer-mediated communication. Inf. Process. Manage. **44**(4), 1448–1466 (2008). http://dx.doi.org/10.1016/j.ipm.2007.12.009

8. Laorden, C., Galán-García, P., Santos, I., Sanz, B., Hidalgo, J.M.G., Bringas, P.G.: Negobot: a conversational agent based on game theory for the detection of paedophile behaviour. In: Herrero, Á., Snášel, V., Abraham, A., Zelinka, I., Baruque, B., Quintián, H., Calvo, J.L., Sedano, J., Corchado, E. (eds.) CISIS 2012-ICEUTE 2012-SOCO 2012. AISC, vol. 189, pp. 261–270. Springer, Heidelberg (2013)
9. Vartapetiance, A., Gillam, L.: "our little secret": pinpointing potential predators. Secur. Inform. **3**(1), 1–19 (2014). http://dx.doi.org/10.1186/s13388-014-0003-7
10. Villatoro-Tello, E., Juárez-González, A., Escalante, H.J., Montes-y Gómez, M., Villaseñor-Pineda, L.: A two-step approach for effective detection of misbehaving users in chats. In: CLEF (Online Working Notes/Labs/Workshop) (2012)
11. Wallace, R.S.: The elements of aiml style. Technical report, Alice A.I. Fdn., Inc (2003)

Automatic Syllabification and Syllable Timing of Automatically Recognized Speech – for Czech

Marek Boháč[(✉)], Lukáš Matějů, Michal Rott, and Radek Šafařík

Institute of Information Technology and Electronics, Technical University of Liberec,
Studentská 2/1402, 461 17 Liberec, Czech Republic
{marek.bohac,lukas.mateju,michal.rott,radek.safarik}@tul.cz
https://www.ite.tul.cz/itee/

Abstract. Our recent work was focused on automatic speech recognition (ASR) of spoken word archive documents [6,7]. One of the important tasks was to structuralize the recognized document (to segment the document and to detect sentence boundaries). Prosodic features play significant role in the spoken document structuralization. In our previous work we bound the prosodic information on the ASR events – words and noises. Many prosodic features (e.g. speech rate, vowel prominence or prolongation of last syllables) require higher time resolution than word-level [1]. For that reason we propose a scheme that is able to automatically syllabify the recognized words and by forced-alignment of its phonetic content provide the syllables (and its phonemes) with time-stamps. We presume that words, non-speech events, syllables and phonemes represent an appropriate hierarchical set of structuralization units for processing various prosodic features.

Keywords: Automatic syllabification · Forced-alignment of phonemes · Automatic speech recognition

1   Introduction

In the last decade automatic speech recognition technologies (ASR) started to face the challenge of processing very complex data. This complexity comes from the structure of processed documents (e.g. discussions of multiple participants) and from the usage of unprepared natural language. Processing of such data relies on many information sources – with great importance of the speech prosody.

Speech prosody can be described by a large set of partial features – for example fundamental frequency of speech, short time signal energy, timbre (higher harmonic structure), non-speech events in the utterance (taking breath, silence) or speech rate. When we want to incorporate multiple prosodic features in one tool we need to bound the computed features to suitable elements of speech transcription. Some of the features can be computed (and observed) in very short time frames – e.g. fundamental frequency, signal energy. Other features provide the information in longer time context (e.g. speech rate expressed as number of

© Springer International Publishing Switzerland 2016
P. Sojka et al. (Eds.): TSD 2016, LNAI 9924, pp. 540–547, 2016.
DOI: 10.1007/978-3-319-45510-5_62

syllables per time unit). The range of time scales used for analysis of prosodic features shows that we need a multi-level resolution of the transcription units to utilize the prosodic information.

In this paper we propose to use three levels of time resolution: (1) ASR system events (words and non-speech), (2) syllables of the recognized words and (3) recognized phonemes. Non-speech events are directly usable as prosodic features (hesitation sound, taking breath), syllables can be used to determine the speech rate [3,11] and the duration of phonemes (vowels) may be used to determine the prominence of the syllable or syllable prolongation [1].

To automatically obtain all the above mentioned levels of transcription segmentation, we proposed the following solution. ASR system [9] is used to recognize the recording, which provides us with time stamps of all recognized events, category of non-speech events and the phonetic form of recognized speech events. Our scheme separately syllabifies the orthographic form of recognized words and obtains time stamps of recognized phonemes via DNN-based forced-alignment. Third tool assigns timed phonemes to orthographic syllables.

In the next section we propose our scheme for automatic syllabification and timing of recognized documents. In Sect. 3 the partial tasks are experimentally evaluated and in Sect. 4 conclusions and discussion take place.

2 Proposed Scheme

The proposed scheme works in three stages. These stages are: (1) orthographic-form word syllabification (described in Subsect. 2.1), (2) forced-alignment of phonetic transcription (described in Subsect 2.2) and (3) alignment between syllables and time-aligned phonemes (described in Subsect. 2.3). Overall structure of the scheme is shown in Fig. 1. *AM*, *LM* and *VOC* mark the acoustic and language model and vocabulary of the ASR system, *audio* stands for the input spoken document and *rules* marks statistical models for syllabification and syllable-to-phonemes alignment modules.

2.1 Syllabification Tool

The syllabification tool works in three layers – all implemented via Weighted Finite State Transducers[1] (WFST). The first layer split the processed word into

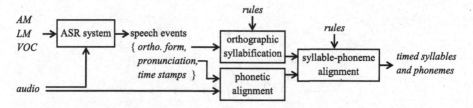

Fig. 1. Overall illustration of the proposed syllabification and timing scheme

---

[1] http://www.openfst.org/twiki/bin/view/FST/WebHome.

word	layer 1	layer 2	layer 3
aerolinkou	aero-li-nko-u	aero-lin-ko-u	aero-lin-kou
agilními	a-gi-lní-mi	a-gil-ní-mi	a-gil-ní-mi

Fig. 2. Example of Czech word syllabification (aerolinkou - by aerline; agilními - agile)

syllable-like units mostly ending with (or containing) a vowel which is typical for Czech. We defined 2,500 units ending with a vowel (less penalized), 7,000 units ending with a consonant (more penalized) and 45 special groups coming from foreign languages. Consonants will be further denoted c, vowels v, border between syllables will be marked by a dash, multiple repetition of symbol by + and asterisk stand for "anything" – vowel or consonant. This first layer does not produce a good syllabification, but produces a starting position with all reasonable units containing a vowel. The second layer implements corrections very similar to voicing assimilation – replace groups $*v-c(c)+v$ with $*vc-(c)+v$ while respecting existence of 30 defined Czech prefixes. Occurrence of single consonant or vowel as a last unit is solved by the third layer – it merges single grapheme with the preceding unit (syllable). The following figure (Fig. 2) shows examples of Czech words syllabification.

2.2   Phoneme-Level Forced Alignment

Automatic timing of phonemes consists of two subsystems. First one is a Deep Neural Network (DNN) recognizing the phonemes in the parametrized recording. Second one is a WFST-based decoder which makes alignment between a sequence of phonemes (recognized by preceding ASR system) and phoneme likelihoods (determined by the DNN). The function of the decoder is similar to the computation of dynamic time warping alignment method.

The aligner is represented by two transducers. The first one models the input signal (DNN output). The second one represents the allowed transitions of the aligner (ASR-recognized phonemes). Figure 3 shows simplified second transducer for a phoneme sequence of 'word'. The transition likelihoods are obtained by forward pass of the DNN. The alignment is done by the composition of transducers, the single best-path represents the phonetic alignment.

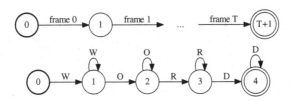

Fig. 3. Illustration of the phoneme alignment implemented via WFST

The DNN is trained using phonetic transcriptions of 270 h of Czech recordings (originally prepared for training of our ASR system with DNN-HMM acoustic models as proposed in [2]). The recordings are phonetically aligned using GMM-HMM aligner – part of the HTK toolbox[2]. Within this paper we use these alignments as a baseline for our system and a start point for creating mono-phone aligners. Considering the training of DNNs is done on level of tied-state triphones, the alignments are thus mapped to monophones.

All four compared aligners (Table 1) are based on DNNs trained with following settings: the networks have 5 hidden layers each consisting of 512 neurons. ReLU activation function and mini-batches of size 1024 are utilized. The training is done within 25 epochs using the learning rate set to 0.08. Log filter banks of size 39 are used for feature extraction. The input vector is a concatenation of 5 previous frames, current frame and 5 following frames. Frame length is 25 ms with 10 ms overlap. The input data are locally normalized by 1 s window. The size of the output layer is 48 meaning that each neuron represents corresponding acoustic inventory item (monophone or nos-speech event). The torch library[3] is employed for training of DNNs.

By employing the DNN-WFST-based aligner technique we created 4 new alignments of the training data labels. The first one is achieved by using the monophone GMM-HMM alignments as labels for DNN. The trained network is than used by the aligner to create alignments labeled as 'mono-1'. The re-aligned training data are then used as labels for another DNN iteration to gather alignments labeled as 'mono-2'.

The third alignments are gained by re-aligning the training data by our best tied-state triphone DNN-HMM ASR model [5]. The alignments are then converted back to monophones and labeled as 'tri-1'. Alignment 'tri-1' was used as labels for training 'tri-2' DNN. The process of obtaining the 'tri' alignments is significantly more time consuming than obtaining the 'mono' ones. Effects of the training label re-alignment are commented in the next chapters.

2.3 Syllable-Phoneme Alignment

The alignment between syllables (a sequence of groups of graphemes) and phonetic form of the word (a sequence of single phonemes) is obtained via a dynamic decoder with pruning (derived from the Minimum Edit Distance algorithm [10]). The pruning demands ability to rate any proposed alignment between syllables and phonemes. We use Phonetic Alphabet for Czech (PAC – proposed in [8]), which mostly marks the phonetic form of grapheme with the grapheme character itself. Therefore we can simply count hits between graphemes and phonemes to obtain a basic score. The score is enhanced by a set of expert-verified rules (pronunciation(s) of syllables) which override the basic scoring. The best-rated alignment usually contains the highest number of rules.

---

[2] http://htk.eng.cam.ac.uk.
[3] http://torch.ch.

3 Conducted Experiments

In this section we present three experiments which show how accurate are the three components of the introduced scheme. Our first experiment evaluates the overall accuracy of the phoneme-level forced alignment and also shows the impact of the DNN training data re-alignment. The second experiment compares our syllabification tool with state-of-the-art solution. The third experiment tests the combined performance of syllabification and syllable-phoneme alignment.

3.1 Evaluation of Phoneme Forced-Alignment

For the evaluation of the phoneme forced-alignment we used 15 h of test recordings (with labeled phonetic content). Thus the re-alignment of the training data (described in Subsect. 2.2) is basically the same process as the forced-alignment of the known test phonemes. As the only kind of error that can be produced by the WFST-aligner is the shift of the border between two subsequent phonemes, the accuracy (number of correctly labeled frames divided by the total number of frames) is a sufficient metric. The baseline model (reference data) may have some of the phoneme borders slightly shifted. We compare such combinations of reference-result data that it is possible to make assessment not only about the accuracy of the aligner but also about the baseline references. The results are shown in Table 1.

Table 1. The degree of similarity between phoneme alignments

Reference	Alignment	Accuracy [%]	Reference	Alignment	Accuracy [%]
Baseline	mono-1	90.35	mono-1	mono-2	95.75
Baseline	mono-2	89.47	tri-1	tri-2	92.98
Baseline	tri-1	92.89	mono-1	tri-1	92.56
Baseline	tri-2	90.11	mono-2	tri-2	95.20

3.2 Evaluation of Syllabification

In this experiment we evaluate the accuracy of our syllabification tool comparing it with Franklin Liang's algorithm [4] implemented by Ned Batchelder[4]. A list of Czech rules is provided with GNU GPL licence, which limits the use of the Liang's algorithm (the last actualization of rules was done on 28 December 2003). The comparison is done amongst our ASR lexicon (containing 577k Czech words and its pronunciations). Before we compare both tools (on our vocabulary mentioned above), we should say that in some cases even annotators are not able to agree on proper syllabification. We should also say that we exclude abbreviations (e.g. SMS, USB) and foreign trademarks (e.g. pocketpc, motogp) from the comparison.

---

[4] http://nedbatchelder.com/code/modules/hyphenate.html.

After comparing 570,000 syllabified words both tools agreed on 264,000 words. Remaining 306,000 syllabification differed, so we analyzed the errors made by both tools.

The most common errors of the Liang's algorithm were placing a syllable boundary inside syllables (inserting extra boundary) and missing rules causing missing boundaries. Examples of the extra boundaries are given in the Table 2. The missing rules can be illustrated with following examples: kriminálech : kri-minálech, buržoazie : buržo-a-zie or pachatel : pacha-tel.

Our tool produced incorrect syllabification when processing foreign words (the vocabulary contains approx. 1.5 % words of German, French and English origin). We are already experimenting with making enhanced set of rules for processing words with foreign origin. The most common error of our tool is a shift of syllable boundary. This shift mostly happens when second processing layer shifts consonants from one syllable to the preceding one. In some cases the shift should not be done and in some other cases one more consonants should be shifted. We believe these errors may be removed by adding more specific rules to the second layer.

The overall comparison of both tools shows that both of them produce approximately same number of errors (and both of them can be enhanced by proper changes in the rule sets).

Table 2. Liang's algorithm: examples of syllabification within a syllable

Error	Word examples	Number of errors
-p-ro-	postprodukce, doprodat	999
-š-ko-	zaškolovat, škoda	318
-š-tě-	zaštěká, štěně	296

3.3 Evaluation of Syllabification and Syllable-to-Phoneme Alignment

It is very important to evaluate the accuracy of the syllabification and following alignment between syllables and phonemes. We plan to semi-manually verify the syllabification (and syllable-phoneme alignment) of the whole ASR system vocabulary. The verification will give us better estimation of the proposed system error rate. The vocabulary contains over 577,000 items (some with multiple pronunciations) so the manual evaluation of the system performance would have to be performed only on a fraction of the vocabulary. As we want to evaluate the overall accuracy of the scheme we designed another experiment utilizing the whole processed vocabulary.

This experiment assumes that correctly syllabified and aligned vocabulary should contain all the information needed for designing grapheme-to-phoneme (G2P) conversion of words. In this experiment we have found 1,500 Czech words not contained in the vocabulary (all chosen from contemporary news). We syllabified these test words and trained a tri-gram model of the syllable pronunciations (from the already processed 577k vocabulary). Then the best pronunciation was found for the syllabified words which we compared with a reference (gained via our well proven G2P tool).

From the 1,500 test words 60 could not be processed because some of the syllables were not contained in the model (some of the test words were of foreign origin or had non-canonical written form inadmissible in the ASR vocabulary). The remaining words were evaluated (1,440 words). The phonetic form of 1,300 words had correctly determined pronunciation (90 % of processable words). In the remaining words (with length of 5 to 12 phonemes) there were typically 1–2 incorrectly determined phonemes. Many of the errors were caused by incorrect voicing assimilation, so we added a simple layer fixing these errors. With such enhancement 1,370 were processed correctly (95 % of the words).

4   Conclusions and Future Work

The scheme we proposed in this paper is able to syllabify words recognized by an ASR system and to provide the timing information on both levels – syllables and phonemes. This way we can prepare structural units in all time-levels needed for annotation of prosodic features.

The accuracy of syllabification tool is evaluated in experiments shown in Subsect. 3.2. The alignment between proposed syllables and word phonemes is evaluated in Subsect. 3.3. However both experiments show that our tools are quite accurate, we want to further exploit the fact that ASR systems recognize only words contained in its vocabulary. We are going to process our ASR vocabulary, to find potentially incorrectly processed words and to replace the introduced tools by (automatically pre-prepared) verified vocabulary. The suspected words are already marked by the experiment shown in Subsect. 3.2 and we believe, that phonetic rules described in Subsect. 3.3 can reveal most of the remaining errors. Using a vocabulary of pre-processed words should also speed-up the system and allow it to correctly process abbreviations, symbols and other special vocabulary items. ·

In Subsect. 3.1 we evaluated the accuracy of the phoneme-level forced alignment. If we analyze the results shown in Table 1, we can formulate some conclusions. The first one says that iterative re-alignment of the training data differs from the baseline alignment (left half of the table). The second one says that correlation between 'mono' and 'tri' iterations increases. Our explanation is that the initial baseline alignment is slightly inaccurate and the iterative re-alignment converges to a more accurate labeling. If our explanation is valid, the timing accuracy (on the level of phonemes) is about 95 %. This correlates with the high accuracy of our ASR system trained from using above mentioned acoustic data annotations.

In our future work we plan to use the proposed syllabification and timing scheme for automatic extraction of prosodic features. We want to find the proper timing-level for using the prosodic features for document segmentation and for finding sentence boundaries. This mainly involves the fundamental frequency of speech and phoneme/syllable prolongation, possibly the speech rate. As we already have access to manually transcribed utterances (including the punctuation marks indicating the sentence boundaries) we should be able to find correlation between particular prosodic features and sentence boundaries.

We want to find the important words in the utterance (words with prominent syllables) to enhance tasks like spoken-word document summarization. Our last planned experiment is estimating speaker-wise distribution of the lengths of concrete phonemes. If there is some general model of phoneme lengths it could be potentially used for computing the syllable prominence.

Acknowledgment. This work was partly supported by the Student's Grant Scheme at the Technical University of Liberec (SGS 2016).

References

1. Bachan, J., Wagner, A., Klessa, K., Demenko, G.: Consistency of prosodic annotation of spontaneous speech for technology needs. In: 7th Language & Technology Conference, pp. 125–129 (2015)
2. Dahl, G.E., Yu, D., Deng, L., Acero, A.: Context-dependent pre-trained deep neural networks for large-vocabulary speech recognition. IEEE Trans. Audio Speech Lang. Process. **20**(1), 30–42 (2012)
3. Huici, H., Kairuz, H.A., Martens, H., Van Nuffelen, G., De Bodt, M.: Speech rate estimation in disordered speech based on spectral landmark detection. Biomed. Signal Process. Control **27**, 1–6 (2016). http://www.sciencedirect.com/science/article/pii/S1746809416000069
4. Liang, F.M.: Word Hy-phen-a-tion by Com-put-er (hyphenation, computer). Ph.D. thesis, Stanford University, Stanford, CA, USA (1983). aAI8329742
5. Mateju, L., Červa, P., Ždánský, J.: Investigation into the use of deep neural networks for LVCSR of Czech. In: ECMSM 2015, pp. 1–4 (2015)
6. Nouza, J., Blavka, K., Boháč, M., Červa, P., Ždánský, J., Silovský, J., Pražák, J.: Voice technology to enable sophisticated access to historical audio archive of the Czech radio. In: Grana, C., Cucchiara, R. (eds.) MM4CH 2011. CCIS, vol. 247, pp. 27–38. Springer, Heidelberg (2012)
7. Nouza, J., et al.: Making Czech historical radio archive accessible and searchable for wide public. J. Multimedia **7**(2), 159–169 (2012). http://ojs.academy publisher.com/index.php/jmm/article/view/jmm0702159169
8. Nouza, J., Psutka, J., Uhlír, J.: Phonetic alphabet for speech recognition of Czech. Radioengineering **6**(4), 16–20 (1997)
9. Seps, L., Málek, J., Červa, P., Nouza, J.: Investigation of deep neural networks for robust recognition of nonlinearly distorted speech. In: INTERSPEECH, pp. 363–367 (2014)
10. Wagner, R.A., Fischer, M.J.: The string-to-string correction problem. J. ACM **21**(1), 168–173 (1974)
11. Yarra, C., Deshmukh, O.D., Ghosh, P.K.: A mode-shape classification technique for robust speech rate estimation and syllable nuclei detection. Speech Commun. **78**, 62–71 (2016). http://www.sciencedirect.com/science/article/pii/S01 6763931600025X

Author Index

Printed in the United States
By Bookmasters